68/–

D0414715

QUANTUM THEORY
OF ANGULAR MOMENTUM

PERSPECTIVES IN PHYSICS

A Series of Reprint Collections

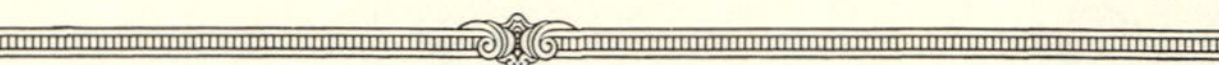

STATISTICAL THEORIES OF SPECTRA: FLUCTUATIONS, Charles E. Porter (Ed.). 1965

QUANTUM THEORY OF ANGULAR MOMENTUM, L. C. Biedenharn and H. C. van Dam (Eds.). 1965

In Preparation

MAGNETOHYDRODYNAMIC STABILITY AND THERMONUCLEAR CONTAINMENT, A. Jeffrey and T. Taniuti (Eds.).

MATHEMATICAL PHYSICS IN ONE DIMENSION: EXACTLY SOLUBLE MODELS OF INTERACTING PARTICLES, E. Lieb and D. Mattis (Eds.).

COULOMB EXCITATION, K. Alder and A. Winther (Eds.).

QUANTUM THEORY OF ANGULAR MOMENTUM

A Collection of Reprints and Original Papers

Edited by

L. C. BIEDENHARN

DEPARTMENT OF PHYSICS
DUKE UNIVERSITY
DURHAM, NORTH CAROLINA

H. VAN DAM

DEPARTMENT OF PHYSICS
UNIVERSITY OF NORTH CAROLINA
CHAPEL HILL, NORTH CAROLINA

1965

ACADEMIC PRESS New York and London

ACADEMIC PRESS INC.
111 Fifth Avenue, New York, New York 10003

United Kingdom Edition published by
ACADEMIC PRESS INC. (LONDON) LTD.
Berkeley Square House, London W.1

LIBRARY OF CONGRESS CATALOG CARD NUMBER: 65-18424

PRINTED IN THE UNITED STATES OF AMERICA

CONTENTS

QUANTUM THEORY
OF ANGULAR MOMENTUM

INTRODUCTION

L. C. Biedenharn and H. van Dam

Angular momentum played an important role in classical mechanics—Kepler's second law bears witness to this—but it was not until the advent of present day quantum mechanics that a deeper insight was gained into the fundamental and essential nature of angular momentum considerations. The quantum theory of angular momentum is now a discipline—familiarly known as "Racah algebra"—indispensable to the working physicist in atomic, molecular, and nuclear structure.

One may trace this change in emphasis to many sources, but probably the single most important is the connection between the conservation of angular momentum and rotational invariance. Angular momentum considerations are thus contained in the wider frame of symmetry and invariance principles so characteristic of modern physical theory.

The present collection of papers traces the development of the quantum theory of angular momentum from its origins in atomic and nuclear spectroscopy. Our intent is not so much to offer the reader angular momentum theory as a fully developed and ready-to-use discipline, as it is to present ideas and concepts in angular momentum theory in the words used by its creators; it is this which we feel can be of genuine value for future developments and generalizations. We would feel inadequate to this task of selection were it not for the singular circumstance that two of the most important and fundamental papers in the field have never been properly published. These are: the famous unpublished manuscript of Wigner (1940; p. 87), and the monograph of Schwinger on angular momentum (1952; p. 229)—each of which inspired further work despite the handicap of limited circulation. These works of Wigner and Schwinger, combined with Racah's classic papers (p. 134 ff.) and that of Bargmann (p. 300), form the core of the present selection and enable us to be fully confident of the value of our collection.

In order to enhance the usefulness of the papers selected, we shall attempt to survey, and within our limitations to assess, the development of the quantum theory of angular momentum, and thereby to fit the selected papers into a suitable frame of reference. In this same spirit, we have incorporated a fairly extensive bibliography designed to complete our survey of the theory.

Brief Survey of the Development of the Quantum Theory of Angular Momentum

The quantization of orbital angular momentum was one of the Bohr postulates in 1913, and indeed Planck's quantum of action is dimensionally an angular momentum, but it would be a mis-emphasis to attribute, initially, too much significance for angular momentum theory to these facts.[1] A clearer view of the nature of quantization, in terms of action-angle variables (for which angular momentum is the prototype) came with the Wilson-Sommerfeld quantization rules in 1915.

With the clarity born of hindsight, one may now discern that the period of the old quantum mechanics was by-and-large a groping to induce from bewildering facts concepts whose true nature is that of the quantum theory of angular momentum. (This can be seen rather naturally from the fact that the stationary states of an atom are characterized within a shell essentially by angular momentum alone.) The fundamental concepts of the spatial quantization of angular momentum (Sommerfeld, 1916)[2] and the selection rules imposed by the conservation of angular momentum in (dipole) radiation (Rubinowicz, 1918)[3] were the next advances; it is to Sommerfeld (1923)[4] that we owe the concept of the total angular momentum J and its projection M as the essential elements to characterize the stationary states. This was also the period (1921–1925), which saw the empirical formulation of the "vector model" which characterized states in terms of semiclassical ideas on angular momentum vectors and their coupling.[5] The familiar Landé formulas for the g-factor and for the spin-orbit coupling—now recognized as one of Racah's coefficients, $W(L1JS; LS)$—stem from this period (1923), as does the initial idea of a vector operator $\mathbf{V} \to \mathbf{V} \cdot \hat{\mathbf{J}}\hat{\mathbf{J}}$. [It is interesting to note, too, that it was Landé who determined the result ${J^2}_{\text{classical}} \to (J(J+1))_{\text{quantum}}$ based on the *empirical* agreement of this form with experimental data.] A primary task was to unravel the Zeeman effect on spectral lines, to determine particularly the selection rules, polarization, and intensities—problems all solved in present day language by the Wigner coefficients. The achievements of this period of the old quantum theory are of the first rank and all too easy to underestimate in the light of current knowledge.

Heisenberg's matrix mechanics (1925) was the step that brought order. The fundamental commutation rules for angular momentum, $[J_i, J_j] = i\hbar\, \varepsilon_{ijk}\, J_k$, were stated almost immediately thereafter by Born, Heisenberg, and Jordan (1926)[6] and simultaneously by Dirac[7] (1926). The paper of Born, Heisenberg,

[1] The wonderfully thorough discussion of the history of the old quantum theory (1900–1926) by E. T. Whittaker ("A History of the Theories of Aether and Electricity," Vol. II. Thomas Nelson, New York, 1953) is recommended here.

[2] A. Sommerfeld, *Ann. Physik* **51**, 1 (1916).

[3] A. Rubinowicz, *Phys. Z.* **19**, 441, 465 (1918).

[4] A. Sommerfeld, *Ann. Physik* **70**, 32 (1923).

[5] See, for instance, the text by E. Back and A. Landé, "Zeemaneffekt und Multiplettstruktur der Spektrallinien," Springer Verlag, Berlin (1924).

[6] Born, Heisenberg, and Jordan, *Z. Physik* **35**, 557 (1926).

[7] P. A. M. Dirac, *Proc. Roy. Soc.* **A111**, 281–305 (1926).

and Jordan is especially noteworthy since it was in this remarkable paper that the complete algebraic technique of raising and lowering operators for angular momentum was first employed in physics.[8] By this algebraic method the integral, and half-integral, values for J and M were established, including the matrices of the angular momentum,

$$\mathbf{J}_{\pm}|J, M> = [(J \mp M)(J \pm M + 1)]^{1/2}|J, M \pm 1>,$$

$$\mathbf{J}_z|J, M> = M|J, M>.$$

This is an important achievement since the whole of angular momentum theory can be based directly on these algebraic results.[9] Moreover in the paper of Born, Heisenberg, and Jordan, the commutation rules were employed to obtain the matrix elements for vector operators, thereby deriving intensity relations and selection rules for atomic transitions.

Digression on Hamilton's "turns" and the Commutation Relations

The commutation relations for angular momentum are the foundation upon which the entire theory is based, and it is worthwhile to examine these relations a bit more critically. The original introduction of these relations, referred to above, is based on the *classical* orbital angular momentum relation $\mathbf{L} = \mathbf{r} \times \mathbf{p}$ and the Heisenberg commutation law. Yet these rules so deduced suffice to include half-integer angular momenta as well—a point which indicates that classical considerations may possibly be deeper than is immediately evident. This is indeed the case, and it is from Hamilton's researches that such ideas first emerge clearly.

Consider first the linear momentum operator, $\mathbf{p}$. The significance of this operator, as emphasized very early by Dirac, is that it is the operator which generates displacements—displacement by an amount $\mathbf{a}$ is given by the operator exp $(i\hbar^{-1}\mathbf{a} \cdot \mathbf{p})$. The commutation law for the components p_i is obtained from the parallelogram law for combining finite displacements; that the p_i all commute is basically the postulate that the displacements obey Euclidean geometry.

In order to obtain a uniform way in which to treat rotations as well as

[8] For mathematics, the technique was fully developed already in the thesis of E. Cartan (Paris, 1894).

[9] The matrices of the angular momentum can be used, for example, for a direct (though cumbersome) determination of the rotation matrices. This is very familiar for spin-$\frac{1}{2}$ case; less familiar is the spin-1 result:

$$\begin{aligned} d^{(1)}(\theta) &\equiv \exp(-i\theta J_y) \\ &= 1 - iJ_y \sin\theta + J_y^2(\cos\theta - 1), \end{aligned}$$

based on $J_i^3 = J_i$ for spin 1. [This procedure has been carried very far recently by Lehrer-Ilamed (see bibliography).]

displacements, Hamilton based both operations on *reflections*,[10] a point of view later elaborated (for physicists) in the text of Kelvin and Tait. Two reflections in parallel planes lead to a finite displacement; two reflections in intersecting planes lead to a rotation.

We have seen that for linear momentum the essential object is the displacement vector. Is there an elementary object for rotations which corresponds to the displacement vector? Such an object is Hamilton's "turn." Consider the rotation now as generated by two reflections; the essential items are the *axis* (defined by the intersection of the planes of the two reflections) and the *angle* ϕ (given by half the dihedral angle of the planes). We may represent this information by a directed arc (length $\phi/2$) of a great circle on a unit sphere—this object is Hamilton's turn.[11] Two turns are considered to be the same if by translation along their common great circle they may be brought into coincidence. Note also that a given rotation may be represented by two distinct turns, as a result of the peculiar fact that there are *two* distinct turns (0 and π) equivalent to *no* rotation (the double covering of the rotation group).

To find the commutation rules one examines the analog of the parallelogram law. Turns are composed by placing them head to tail where their great circles intersect; it is clear that $AB \neq BA$. As the separate turns A and B get smaller, however, the turns AB and BA approach parallel vectors, which, on a sphere, implies that the turns AB and BA intersect as far away as possible ($\pi/2$). A diagram will readily convince one that this leads to the law $\mathbf{J} \times \mathbf{J} = i\mathbf{J}$ for the generators of infinitesimal rotations.

This simple and intuitive "explanation" of the commutation relations for angular momentum is rather more fruitful than it might appear. For example, it shows rather clearly that for a sphere of very large radius the infinitesimal turns approach infinitesimal displacement vectors on the appropriate tangent plane to the sphere. Thus one sees that [taking the point of tangency (and the center) to define the $\hat{z}$-axis] the generators J_x and J_y become in the limit the *displacement* operators p_x and p_y, while J_z still remains J_z. In this way one has derived intuitively the asymptotic relationship between the spherical harmonics (the eigenfunctions of angular momentum appropriate to the sphere) and the Bessel and circular functions (the eigenfunctions appropriate to the tangent plane).

The more usual discussion of rotations, and spinors, (such as in the books

[10] This idea is very fruitful; Coxeter, for example, has shown how the vector diagrams for Lie groups can be expressed in terms of reflections [see H. S. M. Coxeter and W. O. J. Moser, "Generators and Relations for Discrete Groups," Ergebnisse der Mathematik, No. 14, Springer Verlag, Berlin (1957)].

[11] It is interesting to note that a very similar method has been used by Wigner for characterizing Lorentz transformations [E. P. Wigner, *Ann. Math.* **40**, 149, (1939)].

by Weyl and by Wigner) is rather different, and involves the stereographic projection of the sphere onto the complex plane; this leads to the familiar "$\xi - \eta$ calculus."

The intuitive approach, as sketched above, can be made the basis of a more rigorous investigation. Let us note, however, that the full generality of the commutation relations has been investigated most carefully by van der Waerden[12]; the commutation relations are shown to result from the basic requirements of rotational invariance and linearity. Earlier von Neumann and Wigner[13] had derived the differentiability of the angular momentum functions —and the commutation relations—by postulating continuity alone.

* * *

The basis for the quantum theory of angular momentum was established, as we have seen, at the very beginning of quantum mechanics. This particular period is truly amazing and one can hardly read the journals of 1925–1926 without astonishment at the wealth of profound ideas that were developed in such a brief span.[14] Pauli's exclusion principle; the matrix mechanics of Born, Heisenberg, Jordan, and Dirac; and the wave mechanics of Schrödinger —all the necessary basic tools now stood at hand for atomic spectroscopy. We have chosen to begin our collection with Pauli's paper "Zur Quantenmechanik des magnetischen Elektrons." In this paper the meaning of two component wave functions is clarified (Darwin also discussed this),[15] the Pauli spin-$\frac{1}{2}$ matrices are explicitly given and the transformation matrix for "spinor" functions (the word stems from Ehrenfest) developed. Pauli's paper is characteristically elegant, but in point of fact it followed by over a year the paper of Heisenberg and Jordan[16] which in some respects went further (spin-$\frac{1}{2}$ matrices deduced from the general matrices of the angular momentum, addition of angular momentum, and proof of the Landé formula).

As the reader will note, Pauli's paper refers to Wigner's group theoretical investigations. Wigner had earlier discussed the terms arising from three equivalent electrons, to improve the original discussion of Heisenberg, and of Dirac. Three equivalent objects define the symmetric group S_3 (alternatively known as the dihedral group of the triangle) for which there exists not only a

[12] B. L. van der Waerden, *Math. Z.* **36**, 780 (1932); a recent discussion is contained in Section 9, p. 223–228 of van der Waerden's contribution to the Pauli memorial volume [edited by F. Fierz and V. F. Weisskopf, Wiley (Interscience), New York (1960)].

[13] J. von Neumann, *Math. Z.* **30**, 3 (1927); J. von Neumann and E. Wigner, *Z. Phys.* **47**, 203 (1928).

[14] Let us note in passing that a most engaging study of the development of the ideas of the exclusion principle and spin is contained in the article by van der Waerden in the Pauli memorial volume, loc. cit.

[15] C. G. Darwin, *Proc. Roy. Soc.* **A116**, 227 (1927).

[16] W. Heisenberg and P. Jordan, *Z. Phys.* **37**, 263 (1926).

symmetric, and an antisymmetric (irreducible) representation, but a doubly degenerate representation as well. [The triangle group is the still common example in Wigner's books.] However, Wigner did *not* use group theory at all in this first paper; in a second paper (the one referred to by Pauli) he availed himself of the full machinery of the symmetric group (Frobenius, Schur) and thanks von Neumann for calling this to his attention.

The first application of group theory to analyze the significance of *rotational invariance* for atomic spectroscopy was given by Wigner in 1927. Although this paper is limited in scope (since it did not consider spin), it is nonetheless of great importance, for it gave a systematic treatment of representation theory as applied to quantum mechanics. Of particular importance is the fact that the rotation matrices are defined and discussed, including recursion relations (derived from the reduction of the direct product $D^{(j)} \times D^{(1)}$). The Zeeman effect and the Stark effect are both discussed, the intensity relations are derived, as well as the selection rules; Laporte's rule is shown to depend on the reflection character ("parity") of the wave functions. We can discern, in retrospect, the beginning of the "Wigner-Eckart theorem" [compare remark after Wigner's Eq. (16e)], and the beginning of the idea of the "Wigner coefficients" (compare the explicit factorization now given in the paper correcting earlier errors).

We have chosen to include Wigner's paper as the second item in this collection. In many respects it was a landmark for physics; certainly it marked the beginning of a great many applications of group theory to quantum mechanics, for example, papers by Delbrück, Heitler, Hund, London, and Weyl.

Among the several papers on this subject, there is a remarkable paper by Hermann Weyl, "Quantenmechanik und Gruppentheorie," *Z. Physik* **46,** 1–46 (1927). To understand the background of this paper one should realize that Weyl had but recently completed two mathematical accomplishments of the first rank: (1) The representation theory of the semi-simple Lie Groups (1925–1926)[17] and (2) The completeness of the irreducible representations of the compact continuous groups (1927)—the justly famous Peter-Weyl theorem.[18] The 1927 paper on group theory and quantum mechanics was simply too difficult; it was a brilliant tour-de-force. After some critical introductory remarks, Weyl interprets the true meaning of group theory for physics to be the insight that *quantum kinematics forms a group.* (This deep, but Delphic, remark was interpreted only recently by Schwinger).[19] Weyl's paper was followed a year later by an equally difficult but very complete book, "Group Theory and Quantum Mechanics" (1928).

[17] H. Weyl, I: *Math. Z.* **23**, 271–309 (1925). II: *ibid.* **24**, 328–376 (1926). III: *ibid.* **24**, 377–395 (1926). Nachtrag: *ibid.* **24**, 789–791 (1926).

[18] F. Peter and H. Weyl, *Math. Ann.* **97**, 737–755 (1927).

[19] J. Schwinger, *Proc. Natl. Acad. Sci. U.S.* **46**, 570–579 (1960).

With von Neumann, Wigner continued the application of group theory to spectroscopy in a series of three papers entitled "Zur Erklärung einiger Eigenschaften der Spektren aus der Quantenmechanik des Drehelektrons," *Z. Physik*, **47**, 203 (1928); **49**, 73 (1928); **51**, 844 (1928) [reprinted in Volume I of the von Neumann series]. These papers extended Wigner's earlier treatment of spectra to include spin, and furnished a comprehensive and fully worked out taxonomy for the whole of atomic spectroscopy. Noteworthy for angular momentum theory were the explicit rotation matrices for n electrons (equivalent to a complete determination of the representation matrices of the SU_2 group) in paper I, and the explicit introduction of the Wigner coefficients (defined by the reduction of the Kronecker product $D^{(j)} \times D^{(j')}$), in an appendix to paper III. The case of $j = \frac{1}{2}$ and 1 were given explicitly and used (in the text) to derive the Landé g formula.

This was the period when the application of group theory to physics was fast becoming the accepted technique, but a succéssful reaction soon put the subject in disfavor.[20]

von Neumann and Wigner, as well as Weyl, applied group theoretical techniques in the approximation that the space and spin coordinates may be treated separately. The space and spin wave functions so obtained are each representations of the symmetric group and in order for the complete wave function to be antisymmetric, as required by the Pauli principle, the space and spin representations must be dual.

Slater made a major simplification in the theory, applying the Pauli principle right from the beginning in the form of a determinantal wave function. (These determinantal wave functions had an additional significance in the Hartree-Fock-Slater approximation.) The net result of this simplification (aided by Dirac's vector model) was to eliminate completely any necessity for group theoretical techniques, substituting instead explicit algebraic manipulations.

Wigner's book (1931) was a complete and self-contained treatment of atomic spectra by the group theoretical method, and was noteworthy for its detailed treatment of angular momentum theory (including the Wigner-Eckart theorem).[21] Weyl's expanded second edition (of the same year), included field quantization, the Dirac theory, and the Lorentz group as well. Nevertheless group theoretical methods were to be in disfavor for several years to come.[22]

[20] van der Waerden (Reference 12) expresses the mood of that period extremely well; see also the introduction to Condon and Shortley's book for the famous anecdote concerning Dirac and Weyl on this subject.

[21] In this connection see also: C. Eckart, *Rev. Modern Phys.* **2**, 305–380 (1930) and Bauer's book (1933) (cited in bibliography).

[22] In fact purely algebraic methods were now applied (by Casimir and van der Waerden, see bibliography) to the Lie groups themselves. This work, now familiar in terms of the *Casimir operator*, was extended and formally completed by Racah (1950).

Algebraic methods were particularly well suited to the earlier matrix mechanics of Born, Heisenberg, and Jordan and were the basis for the chapter on angular momentum in the text of Born and Jordan (1930); Pauli was the past master of such techniques (his quantization of the nonrelativistic hydrogen atom by such methods in 1926 is magnificent). In 1931, with Güttinger, Pauli applied algebraic methods to angular momentum; in a mathematical appendix Güttinger and Pauli, starting from the commutation relations alone, introduced first the idea of an arbitrary vector operator (defined by the commutation relations) and next the idea of reduced matrix elements (explicit independence of m). By commutator algebra, and by considering the composition of three angular momenta, they derived all the matrix elements for what we now recognize as the complete Racah coefficient, $W(l1sj; l'j')$.

It was on this elegant work of Pauli and Güttinger, and the earlier text of Born and Jordan, that Condon and Shortley (1935) based the angular momentum discussion—their Chapter III—which set the standard for angular momentum techniques in atomic spectra for almost the next two decades.

Wigner continued his group theoretical considerations in angular momentum, focusing attention on categorizing the essential characteristics of the rotation group that are responsible for the simple structure of the vector addition coefficients. This led to the idea of a *simply reducible group*[23] [ambivalent (each class contains its inverse elements) and multiplicity free (the Kronecker product contains no representation more than once)]. These considerations culminated in an extraordinary (unpublished) manuscript ($\sim$1941) wherein the entire structure of the $3n$-j symbols is elaborated in full detail. This manuscript, and the earlier 1941 paper of which it is a continuation, are included in this collection.

In order to indicate something of the scope of this manuscript let us begin with the Wigner-Eckart theorem, which as noted earlier was one of the generalizations contained in Wigner's book (1931). The essential content of the Wigner-Eckart theorem is the assertion that there exists a set of *unit tensor operators* (call them $\mathbf{S}_k^{q,\Delta j}$) *that form a complete orthonormal basis for tensor operators in the quantum theory of angular momentum.* These unit tensor operators are precisely those operators whose matrix elements are the Wigner coefficients. The *reduced matrix element* for a given tensor operator, $\mathbf{T}_k^q$, is simply the projection of T on the Wigner operator, i.e., $\mathbf{T}_k \cdot \mathbf{S}_k$—an *invariant* operator. Customarily one paraphrases this by a remark due to Racah, that the Wigner-Eckart theorem separates the *physical* aspects of the problem (the reduced matrix elements, i.e., the physical invariants) from the purely *geometrical* aspects. [These fundamental ideas are capable of further

[23] E. P. Wigner, *Ann. Math.* **63**, 57 (1941).

generalization to geometrical structures other than that of the rotations in 3-space; an important new feature is that of multiplicity].[24]

A logically equivalent role for the Wigner coefficients is that of a *coupling coefficient* (hence the earlier name "vector coupling coefficients"). The coupling, however, involves two *different* spaces (bra and ket space); one is led in this way to investigate that coupling which produces an invariant. The particular Wigner coefficients which accomplish this coupling play the role of a *metric* relation; these are the "1-j symbols."

One may mix these two aspects of the Wigner coefficients and consider the coupling of two Wigner (unit tensor) operators. This defines a new tensor operator which—by the Wigner-Eckart theorem—is equivalent [within normalization, i.e., a reduced matrix element] to a unit tensor operator. The reduced matrix element $[\mathbf{S}_a \otimes \mathbf{S}_b] \cdot \mathbf{S}_c$ so defined is an invariant operator formed of five Wigner coefficients (two are required to express the coupling). This invariant operator has matrix elements which are just the familiar Racah coefficients.

It is clear that this process of generating invariants may be continued indefinitely; by this pedestrian path we have but paraphrased Wigner's introduction of the *3n-j symbols* that were defined and studied for the first time in that remarkable paper which forms the fourth item in this collection.

To the same period belongs Racah's epoch-making series on the theory of complex spectra. Racah's emphasis is on atomic spectroscopy; his techniques are completely algebraic and in the same tradition as the work of Pauli and Güttinger. The first step is the deduction of the Wigner coefficients by algebraic methods (§2, page 146) and then the definition, by commutators, of the general tensor operator in terms of the matrices of the angular momentum. These lead (§3, p. 149) to the *algebra of tensor operators* (a structure already implicit as we have seen in the Wigner-Eckart theorem). The Racah coefficients are introduced first as the coupling coefficients for evaluating the scalar product of tensor operators (eq. 38, p. 152), and later as transformation coefficients between coupling schemes of three angular momenta (p. 171). (The explicit algebraic formula for *all* the Racah coefficients, the (S_4) symmetry properties, and two of the three independent sum rules were all given in II.) Having forged his tools, Racah then turns to give a detailed account of spectroscopic applications of the formalism. Ideas begun in II are further elaborated in III: the coefficients of fractional parentage, explicit conjugation (phase) relations, and the introduction of the seniority operator. In developing

[24] There is a rapidly growing literature; see for example: W. T. Sharp, Thesis, Princeton (1960), available as AECL 1098 Chalk River, Ont., (1960); A. P. Stone, *Proc. Cambridge Phil. Soc.* **57**, 460–468 (1961); L. C. Biedenharn, *Phys. Letters* **3**, 254 (1963); B. Diu, *Nuovo Cimento* **28**, 466 (1963); J. Ginibre, *J. Math. Phys.* **4**, 720 (1963); M. Moshinsky, *J. Math. Phys.* **4**, 1128 (1963).

these spectroscopic concepts Racah has gone much beyond his starting point of algebraic methods in angular momentum, and in the fourth and final paper of the series (1949, some eight years after I), Racah turns to group theoretical methods to facilitate and comprehend the often formidable analysis.

These advances in angular momentum techniques achieved by Racah proved to be of the most immediate interest, not in atomic spectroscopy, but in the theory of the angular correlation of nuclear radiations, where Racah's techniques were first applied by Gardner (1949), by Lloyd and later by Racah and by Blatt (references may be found in reviews mentioned in the bibliography). It was in such applications that Fano pointed out that the basic element in the algebra of tensor operators is not Racah's coefficient but rather the 9-j symbol (as we now know it) which arises from matrix elements of the general tensor product of two tensor operators.

Angular correlation theory was the stimulus that led to the first tabulations of the Racah coefficients. The search for recursion relations to aid this tabulation led to the third independent Racah relation (p. 221); a similar result was found about the same time (in nuclear structure calculations, however) by Elliott (p. 228). It goes without saying that such a relation was already contained in Wigner's manuscript.

The elegant derivation by Schwinger (p. 229) of the entire quantum theory of angular momentum from the point of view of a second quantized boson assembly of spin-$\frac{1}{2}$ systems also stems from the same period, 1952. The treatment of a general angular momentum as composed from spin-$\frac{1}{2}$ systems was used earlier, as we have seen, by von Neumann and Wigner, and by Weyl, but the consideration of second quantized boson systems adds enormously to the power of the technique.[25]

The applications of the quantum theory of angular momentum, in particular the exploitation of the Racah-Wigner techniques, have by now become so extensive that it is hardly possible for us to attempt even a synoptic survey. Besides the field of angular correlations already noted, one may single out the two fields of atomic and nuclear spectroscopy where these newer methods have had the greatest impact. One of the basic problems if not *the* basic problem in spectroscopy, both atomic and nuclear, is the construction of antisymmetric n-particle wave functions from the (degenerate) states of a given energy shell. Following Slater the standard procedure employed determinantal states to achieve antisymmetry, but the resulting wave functions are not necessarily eigenstates of the angular momentum. One

[25] It seems to have been generally unnoticed that this same idea was proposed quite early (1935) by Jordan, in a very brief paper wherein the general unitary group was discussed in terms of *both* fermion and boson second quantized structures. Jordan was specifically interested in the many body problem, and his was the first application of second quantized techniques to this field [P. Jordan, *Z. Physik* **94**, 531–535 (1935).]

technique, introduced by Gray and Wills, employed projection operators to project out of the determinantal state the desired angular momentum eigenstate. The disadvantage of such a procedure, as Racah pointed out, lies in the fact that it foregoes use of tensor operator methods, and is in consequence often unnecessarily cumbersome.[26]

An alternative approach, the *genealogical method* introduced by Bacher and Goudsmit, considers the problem recursively; the n particle state is built up by adding one particle to the $(n-1)$ particle *parent* states. The coefficients expressing this construction are called the coefficients of fractional parentage (cfp). One of the major advances contained in Racah's papers is the beginning of a systematic treatment of these coefficients.

To categorize the essential ideas in this problem consider a shell containing k degenerate states. The one particle states are thus characterized by the unitary group SU_k; the elementary n particle states are the (reducible) direct products of n one-particle states. The construction of *irreducible* n particle states is equivalent to the conceptually simpler problem of reducing the Kronecker product of an arbitrary m particle state by a one particle state—which is precisely the problem studied for the simply reducible (S.R.) groups in Wigner's paper (p. 87). The coefficients of fractional parentage are those Wigner coefficients involved in this reduction which refer to antisymmetric parent and final states. Thus we see that the fundamental problem for spectroscopy is the appropriate generalization of the concept of Wigner coefficients to arbitrary unitary groups, a topic of current research. Because of the present lack of a fully developed theoretical basis, much of the work on the cfp is of an incomplete nature, and purely numerical results are often the rule. For practical applications such numerical results are extremely valuable; the difficulty is primarily interpretative, in that a comprehensive view of the *structure* of the problem remains to be found. An essential element of this structure has been given by Racah (IV, 3) in the form of a corollary to Schur's lemma. Racah's result shows the conditions under which the cfp may be decomposed into a sum of factorized simpler structures. Specifically, if the group G contains a subgroup H, then the cfp may be written in the form $\sum AB$, where the elements A are the cfp for the group H, and the elements B are dependent only upon the invariants of H (i.e., independent of the "magnetic" quantum numbers).

This clearly suggests that one should look for a series of subgroups such that the cfp are completely factorized into simpler structures; it is this that Racah terms the "search for new quantum numbers." An excellent example is furnished by Racah's introduction of the *seniority quantum number* (IV,

[26] Projection operator methods have been greatly developed further, however, and are quite important in quantum chemistry, as extensively discussed, for example, by Löwdin (see bibliography).

4.2, p. 189); this is the quantum number implied by the subgroup inclusion $R_n \subset SU_n$ for n odd, or by the subgroup inclusion $Sp_n \subset SU_n$ for n even.

Probably the first to appreciate the significance of the work of Racah for nuclear structure was H. A. Jahn, who (with van Wieringen, Elliott, and others) embarked (in 1950) on an extensive program of calculations for the nuclear shell model, beginning with the p, d, and f orbital shells. (See bibliography, particularly Jahn's lectures at the Institut Poincaré.) For the j-j coupling model (whose importance was recognized at this time), the program was extended by the researches of Flowers and Edmonds (the reduced isotopic spin quantum number results from this work). Racah contributed quite essentially in these applications to nuclear structure; his Princeton lectures (1951) are a classic. Besides the account of Elliott and Lane ("Encyclopedia of Physics", 1957) let us note that an extensive treatment of nuclear spectroscopy has recently been given by de Shalit and Talmi (1963). Group theoretical techniques in both nuclear spectroscopy and in the many body problem are currently of great interest (see bibliography). We should like to mention especially the book by Hamermesh (1962), which contains not only a survey of nuclear spectroscopy but also the first extension of the $3n$-j symbols to a non-S.R. group.

Let us note, too, that a recent exposition of Racah's techniques, designed specifically for atomic spectroscopy, has been given by Judd (1962).

Returning to angular momentum theory—as distinct from applications—let us note that the symmetries of the Wigner and Racah coefficients were considered in the original work of Wigner (this volume pp. 92, 97) and Racah (this volume p. 152); the subject was generally taken to be closed. It was all the more of a surprise to learn from Regge's work in 1958 (included here on p. 296) that the symmetry group of the Wigner coefficient was not the twelve element group (six permutations combined with reflection) but rather the larger group of 72 elements which is, in fact, the symmetry group of the $9j$ symbol.[27] [The symmetry of the $9j$-symbol was discussed by Jahn and Hope (1954, p. 280)]. This wholly unsuspected symmetry has an implication for the symmetry of the Racah coefficient: instead of the previously known 24 element symmetry group (S_4) the actual symmetry group is $S_3 \times S_4$ which has 144 elements

[27] An alternative (but fully equivalent) form for the Regge symmetry of the $3j$-symbol is given by:

$$\begin{pmatrix} a & b & c \\ \alpha & \beta & \gamma \end{pmatrix} \to \begin{pmatrix} a & b & c \\ \dfrac{b+c-\alpha}{2} & \dfrac{a+c-\beta}{2} & \dfrac{a+b-\gamma}{2} \\ \dfrac{b+c+\alpha}{2} & \dfrac{a+c+\beta}{2} & \dfrac{a+b+\gamma}{2} \end{pmatrix}$$

(The editors are indebted to Professor W. T. Sharp for this point.)

(cf. p. 298). A deeper view of the nature of these extra symmetries remains an outstanding problem—a strong hint that angular momentum theory itself should not be considered a closed subject.

We conclude this selection of papers with the recent review of angular momentum theory contributed by Bargmann to the issue of the *Reviews of Modern Physics* honoring Wigner's 60th birthday; a more appropriate summary of the field than this beautiful exposition is hard to imagine. We shall let this paper speak for itself, noting only that Bargmann's treatment encompasses angular momentum theory in a far wider frame—indicative both of the possibilities for further development, and of the depth of the basic ideas of the quantum theory of angular momentum.

Zur Quantenmechanik des magnetischen Elektrons.

Von **W. Pauli jr.** in Hamburg.

(Eingegangen am 3. Mai 1927.)

Es wird gezeigt, wie man zu einer Formulierung der Quantenmechanik des magnetischen Elektrons nach der Schrödingerschen Methode der Eigenfunktionen ohne Verwendung zweideutiger Funktionen gelangen kann, indem man, gestützt auf die allgemeine Dirac-Jordansche Transformationstheorie, neben den Ortskoordinaten jedes Elektrons, um seinen rotatorischen Freiheitsgraden Rechnung zu tragen, die Komponente seines Eigenimpulsmomentes in einer festen Richtung als weitere unabhängige Veränderliche einführt. Im Gegensatz zur klassischen Mechanik kann diese Variable jedoch, ganz unabhängig von irgend einer speziellen Art der äußeren Kraftfelder, nur die Werte $+\frac{1}{2}\frac{h}{2\pi}$ und $-\frac{1}{2}\frac{h}{2\pi}$ annehmen. Das Hinzutreten der genannten neuen Variable bewirkt daher bei einem Elektron einfach ein Aufspalten der Eigenfunktion in zwei Ortsfunktionen ψ_α, ψ_β und allgemeiner bei N Elektronen in 2^N Funktionen, die als die „Wahrscheinlichkeitsamplituden" dafür zu betrachten sind, daß in einem bestimmten stationären Zustand des Systems nicht nur die Lagenkoordinaten der Elektronen in vorgegebenen infinitesimalen Intervallen liegen, sondern auch die Komponenten ihrer Eigenmomente in der festgewählten Richtung bei ψ_α zu $+\frac{1}{2}\frac{h}{2\pi}$, bei ψ_β zu $-\frac{1}{2}\frac{h}{2\pi}$ vorgegebene Werte haben. Es werden Methoden angegeben, um bei gegebener Hamiltonscher Funktion des Systems ebenso viele simultane Differentialgleichungen für die ψ-Funktionen aufzustellen, als ihre Anzahl beträgt (also 2 bzw. 2^N). Diese Gleichungen sind in ihren Folgerungen mit den Matrizengleichungen von Heisenberg und Jordan völlig äquivalent. Ferner wird im Fall mehrerer Elektronen diejenige Lösung der Differentialgleichungen, die der „Äquivalenzregel" genügt, im Anschluß an Heisenberg und Dirac durch ihre Symmetrieeigenschaften bei Vertauschung der Variablenwerte zweier Elektronen in einfacher Weise charakterisiert.

§ 1. Allgemeines über die Einordnung des Elektronenmagnetismus in die Schrödingersche Form der Quantenmechanik. Die zuerst von Goudsmit und Uhlenbeck zur Erklärung der Komplexstruktur der Spektren und ihrer anomalen Zeemaneffekte herangezogene Hypothese, gemäß welcher dem Elektron ein Eigenimpulsmoment von der Größe $\frac{1}{2}\frac{h}{2\pi}$ und ein magnetisches Moment von einem Magneton zukommt, ist durch Heisenberg und Jordan[1]) mit Hilfe der Methode der Matrizenrechnung in die Quantenmechanik eingegliedert und hierdurch quantitativ präzisiert worden. Während sonst die Matrizenmethode mathematisch völlig äquivalent ist der von Schrödinger entdeckten Methode der Eigenfunktionen in mehrdimensionalen Räumen,

---

[1]) ZS. f. Phys. **37**, 263, 1926.

stößt man bei einem Versuch, auch die durch das Eigenmoment des Elektrons bedingten Kräfte und Drehmomente, die dieses in äußeren Feldern erfährt, nach einer entsprechenden Methode zu behandeln, auf eigentümliche formale Schwierigkeiten. Bei Einführung eines weiteren Freiheitsgrades, welcher der Orientierung des Eigenimpulses des Elektrons im Raum entspricht, äußert sich nämlich die empirisch feststehende Tatsache der zwei quantenmäßig möglichen Lagen dieses Momentes in einem äußeren Magnetfeld darin, daß man zunächst auf Eigenfunktionen geführt wird, die in dem betreffenden Drehwinkel, z. B. dem Azimut des Impulses um eine raumfeste Achse, mehrdeutig, und zwar zweideutig sind. Man hat daher vielfach vermutet, daß diese zwar formal mögliche Darstellung mittels zweideutiger Eigenfunktionen dem wahren physikalischen Sachverhalt nicht gerecht wird und hat die Lösung des Problems in einer anderen Richtung gesucht. So hat kürzlich Darwin[1]) versucht, die in der Annahme des Elektronenimpulses zusammengefaßten Tatsachen ohne Einführung einer den Kreiselfreiheitsgraden des Elektrons entsprechenden neuen Dimension des Konfigurationsraumes dadurch zu erfassen, daß er die Amplituden der de Broglieschen Wellen als gerichtete Größen, das heißt die Schrödingersche Eigenfunktion als vektoriell betrachtet. Bei einem Versuch, diesen auf den ersten Anblick scheinbar verheißungsvollen Weg konsequent zu Ende zu denken, ergeben sich jedoch Schwierigkeiten, die gerade wieder mit der Zahl 2 der Lagen des Elektrons in einem äußeren Feld zusammenhängen und von denen ich nicht glaube, daß sie sich überwinden lassen werden. Andererseits ist eine Darstellung des quantenmechanischen Verhaltens des magnetischen Elektrons nach der Methode der Eigenfunktionen namentlich im Falle eines Atoms mit mehreren Elektronen deshalb sehr erwünscht, weil die Auswahl der in der Natur allein realisierten, die „Äquivalenzregel“ erfüllenden Lösung der quantenmechanischen Gleichungen aus allen nach der jetzigen Theorie möglichen Lösungen nach Heisenberg[2]) und Dirac[2]) am übersichtlichsten mit Hilfe der Symmetrieeigenschaften der Eigenfunktionen bei Vertauschen der zu zwei Elektronen gehörenden Variablenwerte erfolgt.

Wir wollen hier nun zeigen, daß durch eine geeignete Benutzung der von Jordan[3]) und Dirac[3]) aufgestellten Formulierung der Quanten-

[1]) Nature **119**, 282, 1927.

[2]) W. Heisenberg, ZS. f. Phys. **38**, 411, 1926; **39**, 499, 1926; **41**, 239, 1927. P. A. M. Dirac, Proc. Roy. Soc. **112**, 661, 1926.

[3]) P. Jordan, ZS. f. Phys. **40**, 809, 1927; Gött. Nachr. 1926, S. 161; P. A. M. Dirac, Proc. Roy. Soc. (A) **113**, 621, 1927; vgl. auch F. London, ZS. f. Phys. **40**, 193, 1926.

mechanik, welche allgemeine kanonische Transformationen der Schrödingerschen Funktionen ψ zu verwerten gestattet, eine quantenmechanische Darstellung des Verhaltens des magnetischen Elektrons nach der Methode der Eigenfunktionen in der Tat möglich ist, ohne daß mehrdeutige Funktionen herangezogen werden. Dies gelingt nämlich dadurch, daß man zu den Lagenkoordinaten q der Elektronenschwerpunkte die Komponenten des Eigenimpulses jedes Elektrons in einer festen Richtung (statt der zu diesen konjugierten Drehwinkel) als neue unabhängige Variable hinzufügt. Wie im folgenden § 2 zunächst im Spezialfall eines einzigen Elektrons ausgeführt wird, spaltet sich dann (bei Fehlen von Entartung) in jedem Quantenzustand die Eigenfunktion im allgemeinen in zwei Funktionen $\psi_\alpha(q_k)$ und $\psi_\beta(q_k)$, von denen die Quadrate der Absolutbeträge, mit $dq_1 \ldots dq_f$ multipliziert, die Wahrscheinlichkeit dafür angeben, daß in diesem Zustand nicht nur die q_k in dem vorzugebenden infinitesimalen Intervall $(q_k, q_k + dq_k)$ liegen, sondern außerdem noch die Komponente des Eigenimpulses in der fest gewählten Richtung den Wert $+\frac{1}{2}\frac{h}{2\pi}$ bzw. $-\frac{1}{2}\frac{h}{2\pi}$ annimmt. Es wird dann weiter gezeigt, wie durch Wahl geeigneter linearer Operatoren für die Komponenten s_x, s_y, s_z des Eigenmomentes nach einem vorzugebenden Koordinatenachsenkreuz für die Eigenfunktionen des magnetischen Elektrons in äußeren Kraftfeldern Differentialgleichungen aufgestellt werden können, die den Matrizengleichungen von Heisenberg und Jordan äquivalent sind. Für den Fall eines ruhenden Elektrons in einem äußeren Magnetfeld und für ein wasserstoffähnliches Atom wird dies in § 4 näher ausgeführt. Ferner wird untersucht, wie die Eigenfunktionen ψ_α, ψ_β sich bei Änderung der Koordinatenachsen transformieren (§ 3).

Die in der vorliegenden Arbeit angegebenen Differentialgleichungen der Eigenfunktionen des magnetischen Elektrons können nur als provisorisch und approximativ betrachtet werden, weil sie ebenso wie die Heisenberg-Jordansche Matrizenformulierung nicht relativistisch invariant geschrieben sind und im Wasserstoffatom nur in derjenigen Näherung gelten, in der das dynamische Verhalten des Eigenmomentes als säkulare Störung (in der klassischen Theorie: Mitteln über den Bahnumlauf) betrachtet werden kann. Insbesondere ist es also noch nicht möglich, die zu höheren Potenzen von $\alpha^2 Z^2$ proportionalen Korrektionen $\left(\alpha = \frac{2\pi e^2}{hc} = \text{Feinstrukturkonstante}\right)$ in der Größe der Wasserstoff-Feinstrukturaufspaltung quantenmechanisch zu berechnen, deren bei den

Röntgenspektren empirisch festgestellte Beträge durch die Sommerfeldsche Formel gut wiedergegeben werden. Die Schwierigkeiten, die der Lösung dieses Problems zurzeit noch entgegenstehen, werden in § 4 kurz diskutiert.

Obwohl also die hier mitgeteilte Formulierung der Quantenmechanik des magnetischen Elektrons in dieser Hinsicht noch gänzlich unbefriedigend ist, bietet sie andererseits den Vorteil, daß sie, wie in § 5 dargelegt wird, im Falle mehrerer Elektronen (im Gegensatz zur Darwinschen Formulierung) zu keinerlei neuen Schwierigkeiten Anlaß gibt und auch die nach Heisenberg zur Erfüllung der „Äquivalenzregel" notwendigen Symmetrieeigenschaften der Eigenfunktionen leicht zu formulieren erlaubt. Namentlich aus diesem Grunde schien mir bereits im gegenwärtigen Zeitpunkt eine Mitteilung der hier vorgeschlagenen Methode gerechtfertigt, und vielleicht kann man sogar hoffen, daß sie auch bei dem noch ungelösten Problem der Berechnung der Wasserstoff-Feinstruktur in höheren Näherungen sich als nützlich erweisen wird.

§ 2. Einführung der Komponente des Eigenmomentes des Elektrons in einer festen Richtung als unabhängige Variable in die Eigenfunktion. Definition der den Komponenten des Eigenmomentes entsprechenden Operatoren. In der klassischen Mechanik kann das dynamische Verhalten des Elektronenmomentes durch die folgenden Paare von kanonischen Variablen beschrieben werden: Der Betrag s des gesamten Eigenmomentes des Elektrons und der Drehwinkel χ um dessen Achse; zweitens die Komponente s_z dieses Momentes in einer festen Richtung z und das von der (xz)-Ebene aus gezählte Azimut φ des Momentvektors um die z-Achse. Da der Quotient s_z/s den Kosinus des Winkels zwischen diesem Vektor und der z-Achse angibt, sind dann dessen x- und y-Komponenten gegeben durch

$$s_x = \sqrt{s^2 - s_z^2}\cos\varphi, \quad s_y = \sqrt{s^2 - s_z^2}\sin\varphi.$$

Da der Drehwinkel χ stets zyklisch ist, in der Hamiltonschen Funktion also nicht auftritt, bleibt s konstant und kann als feste Zahl angesehen werden, so daß als eigentliches, das dynamische Verhalten des Elektronenmomentes bestimmendes kanonisches Variablenpaar nur (s_z, φ) verbleibt.

Bei Anwendung der ursprünglichen Schrödingerschen Methode hätte man also bei Vorhandensein eines einzigen Elektrons in jedem Quantenzustand (der bei Aufhebung der Entartung in äußeren Kraftfeldern durch einen bestimmten Energiewert E bereits eindeutig gekennzeichnet ist) eine Eigenfunktion ψ, die außer von den drei Ortskoordinaten des Elektronenschwerpunktes (kurz mit q_k oder auch q bezeichnet) noch vom Winkel φ abhängt. Es gibt dann

$$|\psi_E(q, \varphi)|^2 dq_1\, dq_2\, dq_3\, d\varphi$$

die Wahrscheinlichkeit an, daß in dem betreffenden Quantenzustand der Energie E sowohl die Ortskoordinaten in den Intervallen q_k, $q_k + dq_k$ als auch der Winkel φ in $(\varphi, \varphi + d\varphi)$ liegt. Wenn in irgend einer dynamischen Funktion die zu φ konjugierte Impulskoordinate s_z auftritt, so wäre sie dann zu ersetzen durch den Operator $\frac{h}{2\pi i}\frac{\partial}{\partial \varphi}$, angewandt auf die Eigenfunktion ψ, ebenso wie die zu q_k konjugierte Impulskomponente p_k der Translationsbewegung durch den Operator $\frac{h}{2\pi i}\frac{\partial}{\partial q_k}$ vertreten wird. Wie bekannt, hat jedoch der Umstand, daß die Zahl der quantenmäßig erlaubten Orientierungen des Elektronenmomentes 2 beträgt, zur Folge, daß die so definierte Funktion $\psi_E(q, \varphi)$ bei stetigem Fortschreiten von φ vom Wert 0 bis 2π nicht zu ihrem Ausgangswert zurückkehrt, sondern ihre Vorzeichen verändert.

Indessen kann man das Auftreten solcher Zweideutigkeiten, wie überhaupt die explizite Verwendung irgendwelcher Polarwinkel dadurch vermeiden, daß man an Stelle von φ die Impulskomponente s_z als unabhängige Variable in die Eigenfunktion einführt. Hierbei tritt in der Quantenmechanik noch ein besonderer vereinfachender Umstand auf: In der klassischen Mechanik wird im allgemeinen s_z, abgesehen von dem Sonderfall, wo s_z gerade ein Integral der Bewegungsgleichungen ist, bei bestimmter Energie eines Kontinuums von Werten fähig sein (z. B. wenn der Momentvektor um eine von der z-Achse verschiedene Richtung präzessiert). In der Quantenmechanik kann aber s_z, als zu einer Winkelkoordinate konjugiert, nur die charakteristischen Werte $+\frac{1}{2}\frac{h}{2\pi}$ und $-\frac{1}{2}\frac{h}{2\pi}$ annehmen; dies soll heißen, die Funktion $\psi_E(q_k, s_z)$ zerfällt in die beiden Funktionen

$$\psi_{\alpha, E}(q_k) \quad \text{und} \quad \psi_{\beta, E}(q_k),$$

die den Werten $s_z = +\frac{1}{2}\frac{h}{2\pi}$ und $s_z = -\frac{1}{2}\frac{h}{2\pi}$ entsprechen. Es gibt

$$|\psi_{\alpha, E}(q_k)|^2 dq_1\, dq_2\, dq_3$$

die Wahrscheinlichkeit dafür an, daß in dem betrachteten stationären Zustand gleichzeitig damit, daß q_k in $(q_k, q_k + dq_k)$ liegt, s_z den Wert $+\frac{1}{2}\frac{h}{2\pi}$ hat, und

$$|\psi_{\beta, E}(q_k)|^2 \, dq_1 \, dq_2 \, dq_3$$

die Wahrscheinlichkeit dafür, daß bei gleichem Wert der q_k die Impulskomponente s_z den Wert $-\frac{1}{2}\frac{h}{2\pi}$ annimmt. Jeder Versuch, die Größe von s_z in einem bestimmten stationären Zustand zu messen, wird immer nur die beiden Werte $+\frac{1}{2}\frac{h}{2\pi}$ und $-\frac{1}{2}\frac{h}{2\pi}$ ergeben, auch dann, wenn s_z kein Integral der Bewegungsgleichungen vorstellt. Dieser Sonderfall (z. B. starkes Magnetfeld in der z-Richtung) ist vielmehr dadurch ausgezeichnet, daß hier bei bestimmter Energie E stets nur eine der beiden Funktionen $\psi_{\alpha, E}$ oder $\psi_{\beta, E}$ von Null verschieden ist. Bei bestimmter Wahl des Koordinatensystems sind ψ_α und ψ_β in jedem stationären Zustand bei Normierung gemäß

$$\int (|\psi_\alpha|^2 + |\psi_\beta|^2) \, dq_1 \, dq_2 \, dq_3 = 1 \tag{1a}$$

bis auf einen gemeinsamen Phasenfaktor völlig bestimmt. Auch wird die Orthogonalitätsrelation gelten müssen

$$\int (\psi_{\alpha, n} \psi^*_{\alpha, m} + \psi_{\beta, n} \psi^*_{\beta, m}) \, dq_1 \, dq_2 \, dq_3 = 0, \text{ für } n \neq m. \tag{1b}$$

Hierin bezeichnen die Indizes n, m zwei voneinander verschiedene Quantenzustände und der beigefügte * (hier wie stets im folgenden) den konjugiert komplexen Wert[1]).

Um weiterhin die Differentialgleichungen aufstellen zu können, denen die Funktionen ψ_α, ψ_β bei gegebener Hamiltonscher Funktion genügen, könnte man so vorgehen, daß man diese als Funktion von (p_k, q_k) und (s_z, φ) ausdrückt und dann p_k durch den Operator $\frac{h}{2\pi i}\frac{\partial}{\partial q_k}$, φ durch den Operator $-\frac{h}{2\pi i}\frac{\partial}{\partial s_z}$ ersetzt. Der Gesamtoperator wäre dann auf $\psi(q_k, s_z)$ anzuwenden und schließlich hätte man zur Grenze überzugehen, wo ψ

[1]) Es sei an dieser Stelle nebenbei erwähnt, daß gemäß der Dirac-Jordanschen Transformationstheorie die früher erwähnte Funktion $\psi(q, \varphi)$ mit den Funktionen ψ_α, ψ_β gemäß den Formeln

$$\psi(q, \varphi) = \psi_\alpha(q) \, e^{\frac{i\varphi}{2}} + \psi_\beta(q) \, e^{-\frac{i\varphi}{2}}$$

zusammenhängt.

nur für $s_z = +\frac{1}{2}\frac{h}{2\pi}$ und $s_z = -\frac{1}{2}\frac{h}{2\pi}$ von Null verschieden ist.

Indessen wäre ein solches Verfahren unübersichtlich und wenig zweckmäßig. Die tatsächlich vorkommenden Hamilton-Funktionen enthalten zunächst immer die Drehimpulskomponenten s_x, s_y, s_z als Variable und es ist daher zweckmäßig, für diese direkt ohne den Umweg über den Polarwinkel φ geeignete Operatoren einzuführen.

Diese Operatoren müssen (abgesehen von einem Vorzeichen, vgl. unten) denselben Vertauschungsrelationen genügen, wie die betreffenden Matrizen, nämlich

$$[\mathbf{s}\mathbf{s}] = -\frac{h}{2\pi i}\,\mathbf{s};\quad \mathbf{s}^2 = \left(\frac{h}{2\pi}\right)^2 s(s+1) \text{ mit } s = {}^1/_2,$$

worin $\mathbf{s}$ eine Vektormatrix mit den Komponenten $\mathbf{s}_x$, $\mathbf{s}_y$, $\mathbf{s}_z$ bedeutet[1]). Messen wir $\mathbf{s}$ im folgenden der Einfachheit halber in der Einheit $\frac{1}{2}\frac{h}{2\pi}$ $\left(\text{d. h. man ersetze } \mathbf{s} \text{ durch } \frac{1}{2}\frac{h}{2\pi}\mathbf{s}\right)$ und schreiben die Vektorgleichungen in Komponenten aus, so erhalten wir

$$\left.\begin{aligned} \mathbf{s}_x\mathbf{s}_y - \mathbf{s}_y\mathbf{s}_x &= 2i\,\mathbf{s}_z, \cdots, \\ \mathbf{s}_x^2 + \mathbf{s}_y^2 + \mathbf{s}_z^2 &= 3, \end{aligned}\right\} \tag{2}$$

worin durch ... die Gleichungen angedeutet sind, die aus der angeschriebenen durch zyklische Vertauschung der Koordinaten hervorgehen[2]).

---

[1]) Vgl. W. Heisenberg und P. Jordan, l. c., Gl. (10). — Matrizen und Operatoren (oder „q-Zahlen") werden im folgenden stets durch Fettdruck gekennzeichnet.

[2]) Infolge des besonderen Umstandes, daß die Zahl der quantenmäßig erlaubten Lagen von $\mathbf{s}$ den Wert 2 hat (daß es sich also um zweizeilige Matrizen handelt), gelten außer (2) noch die weiteren verschärften Relationen

$$\left.\begin{aligned} \mathbf{s}_x\mathbf{s}_y &= -\mathbf{s}_y\mathbf{s}_x = i\,\mathbf{s}_z, \cdots, \\ \mathbf{s}_x^2 &= \mathbf{s}_y^2 = \mathbf{s}_z^2 = 1. \end{aligned}\right\} \tag{2a}$$

Man sieht dies am einfachsten ein, wenn man $\mathbf{s}_z$ als Diagonalmatrix wählt (die Relationen gelten aber allgemein). Bei mehrgliedrigen Matrizen, die (2) erfüllen (wobei der Wert 3 durch $r^2 - 1$ mit r = Zeilenzahl der Matrix zu ersetzen ist), würden dagegen $\mathbf{s}_x\mathbf{s}_y$ und $\mathbf{s}_x^2$ nicht verschwindende Matrixelemente an denjenigen Stellen haben, deren Zeilenindex sich vom Kolonnenindex um 2 unterscheidet (die also Übergängen der zu s_z gehörenden Quantenzahl um zwei Einheiten korrespondieren), so daß die Gleichungen (2a) nicht zu Recht bestehen könnten.

Auf das Bestehen der Relationen (2a) wurde ich von Herrn P. Jordan freundlichst hingewiesen, wofür ich ihm auch an dieser Stelle meinen Dank aus-

Es liegt nun nahe, für die Operatoren s_x, s_y, s_z, die den Relationen (2) genügen, den Ansatz von **linearen Transformationen** der ψ_α und ψ_β zu machen, und zwar ist der einfachst mögliche Ansatz der folgende:

$$\left.\begin{aligned} s_x(\psi_\alpha) &= \psi_\beta, & s_x(\psi_\beta) &= \psi_\alpha; \\ s_y(\psi_\alpha) &= -i\,\psi_\beta, & s_y(\psi_\beta) &= i\,\psi_\alpha; \\ s_z(\psi_\alpha) &= \psi_\alpha, & s_z(\psi_\beta) &= -\psi_\beta. \end{aligned}\right\} \quad (3)$$

Man kann diese Relationen auch in der symbolischen Matrizenform schreiben:

$$s_x(\psi) = \begin{pmatrix} 0, & 1 \\ 1, & 0 \end{pmatrix} \cdot \psi; \quad s_y(\psi) = \begin{pmatrix} 0, & -i \\ i, & 0 \end{pmatrix} \cdot \psi; \quad s_z(\psi) = \begin{pmatrix} 1, & 0 \\ 0, & -1 \end{pmatrix} \cdot \psi. \quad (3')$$

Die Relationen (2) sind hierbei so zu verstehen, daß die Matrizen (3') in (2) eingesetzt, bei Anwendung der gewöhnlichen Vorschrift zur Multiplikation der Matrizen[1]) diesen Relationen genügen. Die **entsprechenden Operatoren genügen jedoch Gleichungen, die aus (2) durch Vertauschung der Reihenfolge aller Multiplikationen hervorgehen**[2]). Die Rechtfertigung für diese Vorschrift wird sich uns aus dem allgemeinen Zusammenhang von Operator- und Matrixkalkül ergeben. Die letzte der Relationen (3) ist offenbar physikalisch notwendig, wenn ψ_α und ψ_β die Wahrscheinlichkeitsamplitude dafür bedeuten sollen, daß s_z $\left(\text{in der Einheit } \frac{1}{2}\frac{h}{2\pi} \text{ gemessen}\right)$ den Wert $+1$ oder -1 annimmt, weil der Operator s_z dann einfach Multiplikation der Eigenfunktion mit dem Zahlwert von s_z bedeuten muß. Daß die in der speziellen Wahl von s_x, s_y enthaltenen, über die Forderungen der Relationen (2) hinausgehen-

---

sprechen will. Er machte mich auch auf folgenden Zusammenhang mit der Quaternionentheorie aufmerksam. Schreibt man eine Quaternion Q in der Form

$$Q = k_1 A + k_2 B + k_3 C + D,$$

so genügen die „Einheiten" k_1, k_2, k_3 den Relationen

$$k_1 k_2 = -k_2 k_1 = k_3, \cdots,$$
$$k_1^2 = k_2^2 = k_3^2 = -1.$$

Diese sind mit den Relationen (2a) äquivalent, wenn man setzt

$$s_x = i\,k_1, \quad s_y = i\,k_2, \quad s_z = i\,k_3.$$

[1]) Vgl. Anm. 1, S. 612.

[2]) Die Notwendigkeit, an dieser Stelle zwischen Operatorrelation und Matrizenrelation zu unterscheiden, ergab sich mir erst nachträglich auf Grund einer brieflichen Mitteilung von Herrn C. G. Darwin betreffend den Vergleich der von ihm aufgestellten Gleichungen mit den meinen. (Siehe unten Anm. 2, S. 618.) Ich möchte auch an dieser Stelle Herrn Darwin für seine Anregung meinen besten Dank aussprechen.

den Normierungen keine Beschränkung der Allgemeinheit bedeuten, wird aus dem folgenden Paragraphen ersichtlich werden, wo das Verhalten der Funktionen ψ_α, ψ_β bei Verlagerung der Achsen des sie definierenden Koordinatensystems untersucht wird. [Vgl. unten S. 614, Gl. (3'').]

Ist nun irgend eine Hamiltonsche Funktion

$$H(p_k, q_k, s_x, s_y, s_z) = E$$

eines speziellen, ein magnetisches Elektron enthaltenden mechanischen Systems vorgegeben, so sind durch

$$\left.\begin{aligned} &H\left(\frac{h}{2\pi i}\frac{\partial}{\partial q_k}, q_k, \mathbf{s}_x, \mathbf{s}_y, \mathbf{s}_z\right)\psi_{E,\alpha} = E\psi_\alpha \\ \text{und}\quad &H\left(\frac{h}{2\pi i}\frac{\partial}{\partial q_k}, q_k, \mathbf{s}_x, \mathbf{s}_y, \mathbf{s}_z\right)\psi_{E,\beta} = E\psi_\beta, \end{aligned}\right\} \tag{4}$$

worin für $\mathbf{s}_x$, $\mathbf{s}_y$, $\mathbf{s}_z$ die Operatoren (3) einzusetzen sind, zwei simultane Differentialgleichungen für ψ_α und ψ_β gegeben, die zugleich die Eigenwerte E bestimmen.

Die Matrixkomponenten irgend einer Funktion $f(p, q, s_x, s_y, s_z)$, von der wir zunächst annehmen wollen, daß sie die Größen s_x, s_y, s_z entweder gar nicht oder nur linear enthält, sind definiert durch die simultanen Gleichungen

$$\mathbf{f}(\psi_{m\alpha}) = \sum_n f_{nm}\psi_{n\alpha}; \quad \mathbf{f}(\psi_{m\beta}) = \sum_n f_{nm}\psi_{n\beta}, \tag{5}$$

wenn unter $\mathbf{f}$ der Operator $f\left(\frac{h}{2\pi i}\frac{\partial}{\partial q}, q, \mathbf{s}_x, \mathbf{s}_y, \mathbf{s}_z\right)$ verstanden wird. Insbesondere gilt also

$$\mathbf{s}_x(\psi_{m\alpha}) = \psi_{m\beta} = \sum_n (s_x)_{nm}\psi_{n\alpha}; \quad \mathbf{s}_x(\psi_{m\beta}) = \psi_{m\alpha} = \sum_n (s_x)_{nm}\psi_{n\alpha} \tag{6}$$

und entsprechende Gleichungen für y und z. Daß auf der rechten Seite von (5) und (6) über den ersten Index der Matrix summiert wird, ist wesentlich, um zwischen der aufeinanderfolgenden Anwendung zweier Operatoren $\mathbf{f}$ und $\mathbf{g}$ und der Multiplikationsvorschrift der Matrizen Übereinstimmung herzustellen. Vermöge der Orthogonalitätsrelationen (1a) und (1b) folgt aus (6) leicht

$$f_{nm} = \int [\mathbf{f}(\psi_{m\alpha})\psi^*_{n\alpha} + \mathbf{f}(\psi_{m\beta})\psi^*_{n\beta}]\, dq_1\, dq_2\, dq_3 \ldots. \tag{5'}$$

Insbesondere ist also

$$\begin{aligned} (s_x)_{nm} &= \int [(\mathbf{s}_x\psi_{m\alpha})\psi^*_{n\alpha} + (\mathbf{s}_x\psi_{m\beta})\psi^*_{n\beta}]\, dq = \int (\psi_{m\beta}\psi^*_{n\alpha} + \psi_{m\alpha}\psi^*_{n\beta})\, dq, \\ (s_y)_{nm} &= \int [(\mathbf{s}_y\psi_{m\alpha})\psi^*_{n\alpha} + (\mathbf{s}_y\psi_{m\beta})\psi^*_{n\beta}]\, dq = \int i(-\psi_{m\beta}\psi^*_{n\alpha} + \psi_{m\alpha}\psi^*_{n\beta})\, dq, \\ (s_z)_{nm} &= \int [(\mathbf{s}_z\psi_{m\alpha})\psi^*_{n\alpha} + (\mathbf{s}_z\psi_{m\beta})\psi^*_{n\beta}]\, dq = \int (\psi_{m\alpha}\psi^*_{n\alpha} - \psi_{m\beta}\psi^*_{n\beta})\, dq. \end{aligned} \tag{6'}$$

Faßt man die allgemeine Eigenfunktion

$$\psi_\alpha = \sum c_n \psi_{n\alpha}, \quad \psi_\beta = \sum c_n \psi_{n\beta}$$

mit unbestimmten Faktoren c_n ins Auge, so spielen also die Ausdrücke

$$\left.\begin{aligned} d_x &= \psi_\beta \psi_\alpha^* + \psi_\alpha \psi_\beta^*, \\ d_y &= -i(\psi_\beta \psi_\alpha^* - \psi_\alpha \psi_\beta^*), \\ d_z &= (\psi_\alpha \psi_\alpha^* - \psi_\beta \psi_\beta^*) \end{aligned}\right\} \qquad (6'')$$

formal die Rolle von Volumdichten des Eigenmomentes des Elektrons.

Wir haben nun noch den Nachweis zu erbringen, daß die gemäß (6′) berechneten Matrizen allgemein den Relationen (2) von Heisenberg und Jordan genügen. Wenn wir mit i und k irgendwelche der Indizes x, y, z bezeichnen, bilden wir also

$$(s_i s_k)_{nm} = \sum_l (s_i)_{nl} (s_k)_{lm}.$$

Setzen wir hierin für $(s_k)_{lm}$ seinen aus (6′) folgenden Wert ein, so ergibt sich

$$(s_i s_k)_{nm} = \int \left\{ \left[\sum_l (s_i)_{nl} \psi_{l\alpha}^* \right] \mathbf{s}_k(\psi_{m\alpha}) + \sum_l \left[(s_i)_{nl} \psi_{l\beta}^* \right] \mathbf{s}_k(\psi_{m\beta}) \right\} dq.$$

Nun ist $(s_i)_{nl} = (s_i)_{ln}^*$, da die Matrizen $\mathbf{s}_i$ (wie man auf Grund von (6′) übrigens leicht bestätigt) hermitisch sind, also gilt gemäß (6)

$$\sum_l (s_i)_{nl} \psi_{l\alpha}^* = \sum (s_i)_{ln}^* \psi_{l\alpha}^* = [\mathbf{s}_i(\psi_{n\alpha})]^*,$$

und ebenso

$$\sum_l (s_i)_{nl} \psi_{l\beta}^* = [\mathbf{s}_i(\psi_{n\beta})]^*.$$

Das Endresultat ist also

$$(s_i s_k)_{nm} = \int \left\{ \left[\mathbf{s}_i(\psi_{n\alpha}) \right]^* \mathbf{s}_k(\psi_{m\alpha}) + \left[\mathbf{s}_i(\psi_{n\beta}) \right]^* \mathbf{s}_k(\psi_{m\beta}) \right\} dq.$$

Auf Grund dieser Relation bestätigt man durch Einsetzen der Operatoren (3) durch Vergleichen mit (6′) leicht alle Relationen (2), wenn diese als Matrizenrelationen aufgefaßt werden. Z. B. ergibt sich für $i = x$, $k = z$

$$(s_x s_y - s_y s_x)_{nm} = 2i \int (-\psi_{n\beta}^* \psi_{m\beta} + \psi_{n\alpha}^* \psi_{m\alpha}) = 2i (s_z)_{nm}$$

gemäß (6′). Ebenso verifiziert man die übrigen Relationen (2). Damit ist zugleich die Wahl der Operatoren (3) gerechtfertigt.

Beispiele für Gleichungen der Form (4) werden in § 4 gegeben werden.

§ 3. Verhalten der Funktionen ψ_α, ψ_β bei Drehungen des Koordinatensystems. In der Theorie von Dirac-Jordan wird allgemein die Frage beantwortet, wie bei Übergang von einem System kanonischer Variablen (p, q) zu einem neuen System P, Q sich die Eigen funktionen ψ transformieren. Ist $\boldsymbol{S}$ ein Operator, der die Operatoren $\boldsymbol{q}$ (Multiplikation mit q) und $\boldsymbol{p} = \frac{h}{2\pi i}\frac{\partial}{\partial q}$ gemäß

$$\boldsymbol{P} = \boldsymbol{S}\boldsymbol{p}\boldsymbol{S}^{-1}, \quad \boldsymbol{Q} = \boldsymbol{S}\boldsymbol{q}\boldsymbol{S}^{-1} \tag{7}$$

in die den neuen Variablen entsprechenden Operatoren $\boldsymbol{P}$, $\boldsymbol{Q}$ überführt, so erhält man die zu Q gehörige Eigenfunktion $\psi_E(Q)$ aus der zu q gehörigen Eigenfunktion $\psi_E(q)$ einfach durch Anwendung des Operators $\boldsymbol{S}$:

$$\psi_E(Q) = \boldsymbol{S}[\psi_E(Q)]. \tag{8}$$

Es stellt dann

$$|\psi_E(Q)|^2 \, d\,Q$$

wieder die Wahrscheinlichkeit dafür dar, daß bei bestimmter Energie E und beliebigem Wert von P die Variable Q zwischen Q und $Q + d\,Q$ liegt[1]).

In unserem Falle werden wir allerdings nicht mit den kanonischen Veränderlichen (s_z, φ) selbst rechnen, sondern mit den Komponenten s_x, s_y, s_z des Eigenmomentes, für welche die Vertauschungsrelationen die nicht kanonische Form (2) haben. Wir werden sodann die Frage zu beantworten haben, wie aus den gegebenen Eigenfunktionen ψ_α, ψ_β und Operatoren $\boldsymbol{s}_x$, $\boldsymbol{s}_y$, $\boldsymbol{s}_z$ in bezug auf ein bestimmtes Achsenkreuz (x, y, z) die Eigenfunktionen ψ'_α, ψ'_β und Operatoren $\boldsymbol{s}_{x'}$, $\boldsymbol{s}_{y'}$, $\boldsymbol{s}_{z'}$ in bezug auf ein neues Achsenkreuz (x', y', z') berechnet werden können. Die Quadrate der Absolutbeträge der neuen ψ'_α, ψ'_β bestimmen dann die Wahrscheinlichkeiten dafür, daß (bei gewissen Werten der Ortskoordinaten q des Elektrons) bei beliebigem Wert des Winkels φ' um die z'-Achse der Impuls $s_{z'}$ $\left(\text{in der Einheit } \frac{1}{2}\frac{h}{2\pi} \text{ gemessen}\right)$ die Werte $+1$ bzw. -1 hat.

Nun ist es für die Operatorgleichung (7) nicht wesentlich, daß die Vertauschungsrelationen zwischen $\boldsymbol{p}$ und $\boldsymbol{q}$ sowie zwischen $\boldsymbol{P}$ und $\boldsymbol{Q}$ gerade die kanonische Form haben. Es kommt vielmehr nur darauf an, daß die Vertauschungsrelationen bei der Transformation ihre Form be-

---

[1]) Daß wir gerade die Energie E als festen Parameter wählen, ist nur ein Sonderfall der von Dirac und Jordan betrachteten Transformationen. Diese Verfasser untersuchten auch noch näher den Zusammenhang zwischen zwei verschiedenen Darstellungen des Operators S: 1. der Differentialdarstellung, bei der $\boldsymbol{S} = \boldsymbol{S}\left(\frac{h}{2\pi i}\frac{\partial}{\partial x}, x\right)$ aus den Operatoren der Differentiation nach einer Variablen x und Multiplikation mit x zusammengesetzt gedacht wird, und 2. der Integraldarstellung von S, bei der gesetzt wird

$$\boldsymbol{S}[f(q)] = \int S(x, q)\, f(x)\, d\,x,$$

worin $S(x, q)$ eine gewöhnliche Funktion ist.

wahren, d. h. daß sie richtig bleiben, wenn man einfach an Stelle der alten Variablen die neuen schreibt. In unserem Falle ist es nun in der Tat bekannt, daß die Relationen (2) bei orthogonalen Koordinatentransformationen unverändert bestehen bleiben, so daß auch für die gestrichenen Größen gilt.

$$\left.\begin{aligned} s_{x'} s_{y'} - s_{y'} s_{x'} &= 2 i s_{z'}, \ldots, \\ s_{x'}^2 + s_{y'}^2 + s_{z'}^2 &= 3. \end{aligned}\right\} \tag{2'}$$

Also wird es erlaubt sein, zu setzen

$$s_{x'} = S s_x S^{-1}, \quad s_{y'} = S s_y S^{-1}, \quad s_{z'} = S s_z S^{-1}. \tag{9}$$

Die bequemste formale Darstellung der Operatoren, die wir immer auf das Eigenfunktionpaar ψ_α, ψ_β anzuwenden haben werden, ist die Matrixdarstellung, die schon oben in (3') benutzt wurde. Führt der Operator S das Paar $(\psi_\alpha, \psi_\beta)$ über in $(S_{11}\psi_\alpha + S_{12}\psi_\beta, S_{21}\psi_\alpha + S_{22}\psi_\beta)$, worin $S_{11}, S_{12}, S_{21}, S_{22}$ gewöhnliche Zahlkoeffizienten sind, so schreiben wir S als Matrix

$$S = \begin{pmatrix} S_{11}, & S_{12} \\ S_{21}, & S_{22} \end{pmatrix}.$$

Damit die Relationen (1 a) und (1 b) auch für das neue Paar $(S\psi_\alpha, S\psi_\beta)$ gelten, muß S der bekannten Orthogonalitätsrelation

$$S \bar{S}^* = 1 \tag{10}$$

genügen, worin der * Übergang zu konjugiert komplexem Wert und das Überstreichen Vertauschen von Zeilen und Kolonnen in der Matrix bedeutet. Also ausgeschrieben [1])

$$\begin{pmatrix} S_{11}, & S_{12} \\ S_{21}, & S_{22} \end{pmatrix} \begin{pmatrix} S_{11}^*, & S_{21}^* \\ S_{12}^*, & S_{22}^* \end{pmatrix} \equiv \begin{pmatrix} S_{11} S_{11}^* + S_{12} S_{12}^*, & S_{11} S_{21}^* + S_{12} S_{22}^* \\ S_{21} S_{11}^* + S_{22} S_{12}^*, & S_{21} S_{21}^* + S_{22} S_{22}^* \end{pmatrix} = \begin{pmatrix} 1, & 0 \\ 0, & 1 \end{pmatrix}. \tag{10'}$$

Andererseits folgt aus der Definition der Komponenten des Eigenmomentes, daß sich die ihnen entsprechenden Operatoren genau so transformieren müssen wie die Koordinaten, also bei Einführung der Eulerschen Winkel Θ, Φ, Ψ gemäß den Formeln [2])

$$\left.\begin{aligned} s_x &= (\cos\Phi\cos\Psi - \sin\Phi\sin\Psi\cos\Theta)\, s_{x'} \\ &\quad + (-\sin\Phi\cos\Psi - \cos\Phi\sin\Psi\cos\Theta)\, s_{y'} + \sin\Psi\sin\Theta\, s_{z'}, \\ s_y &= (\cos\Phi\sin\Psi + \sin\Phi\cos\Psi\cos\Theta)\, s_{x'} \\ &\quad + (-\sin\Phi\sin\Psi + \cos\Phi\cos\Psi\cos\Theta)\, s_y' - \cos\Psi\sin\Theta\, s_{z'}, \\ s_z &= \sin\Phi\sin\Theta\, s_{x'} + \cos\Phi\sin\Theta\, s_{y'} + \cos\Theta\, s_{z'}. \end{aligned}\right\} \tag{11}$$

[1]) Wir erinnern daran, daß man das an der Stelle (n, m) stehende Element im Produkt zweier Matrizen durch gliedweises Multiplizieren der n-ten Zeile der ersten Matrix mit der m-ten Kolonne der zweiten Matrix erhält.

[2]) Vgl. für das Folgende A. Sommerfeld und F. Klein, Theorie des Kreisels, I, § 2 bis 4, insbesondere die Definition der Parameter α, β, γ, δ. Auf deren Bedeutung für unser Problem hat mich Herr P. Jordan aufmerksam gemacht.

Unser Ziel wird es nun sein, die Matrix $\boldsymbol{S}$ so zu bestimmen, daß (9) und (11) übereinstimmen. Gelingt uns dies, dann ist unsere Frage nach der Transformation der $(\psi_\alpha, \psi_\beta)$ bei Drehungen des Koordinatensystems durch die Gleichungen

$$(\psi'_\alpha, \psi'_\beta) = \boldsymbol{S}(\psi_\alpha, \psi_\beta) \tag{12}$$

oder

$$\left.\begin{aligned} \psi'_\alpha &= S_{11}\psi_\alpha + S_{12}\psi_\beta, \\ \psi'_\beta &= S_{21}\psi_\alpha + S_{22}\psi_\beta \end{aligned}\right\} \tag{12a}$$

beantwortet.

Um nun (9) und (11) miteinander in Übereinstimmung zu bringen, ist es zweckmäßig, wie in der Kreiseltheorie üblich, die folgenden Bezeichnungen einzuführen:

$$\left.\begin{aligned} \xi &= s_x + i s_y, & \eta &= -s_x + i s_y, & \zeta &= -s_z, \\ \xi' &= s_{x'} + i s_{y'}, & \eta' &= -s_{x'} + i s_{y'}, & \zeta' &= -s_{z'}, \end{aligned}\right\} \tag{13}$$

$$\left.\begin{aligned} \alpha &= \cos\frac{\Theta}{2} e^{i\frac{\Phi-\Psi}{2}}, & \beta &= i\sin\frac{\Theta}{2} e^{i\frac{-\Phi+\Psi}{2}}, \\ \gamma &= i\sin\frac{\Theta}{2} e^{i\frac{\Phi-\Psi}{2}}, & \delta &= \cos\frac{\Theta}{2} e^{i\frac{-\Phi-\Psi}{2}}. \end{aligned}\right\} \tag{14}$$

Die Größen $\alpha, \beta, \gamma, \delta$ sind die Cayley-Kleinschen Drehungsparameter; zwischen ihnen bestehen die Relationen

$$\delta = \alpha^*, \quad \gamma = -\beta^*, \quad \alpha\delta - \beta\gamma = 1. \tag{14'}$$

Es ist dann (11) äquivalent mit[1])

$$\left.\begin{aligned} \xi &= \alpha^2\xi' + \beta^2\eta' + 2\alpha\beta\zeta', \\ \eta &= \gamma^2\xi' + \delta^2\eta' + 2\gamma\delta\zeta', \\ \zeta &= \alpha\gamma\xi' + \beta\delta\eta' + (\alpha\delta + \beta\gamma)\zeta', \end{aligned}\right\} \tag{11'}$$

(9) äquivalent mit

$$\xi = \boldsymbol{S}^{-1}\xi'\boldsymbol{S}, \quad \eta = \boldsymbol{S}^{-1}\eta'\boldsymbol{S}, \quad \zeta = \boldsymbol{S}^{-1}\zeta'\boldsymbol{S}. \tag{9'}$$

Wir behaupten nun, daß wir, um (9') mit (11') in Übereinstimmung zu bringen, einfach die Matrix $\boldsymbol{S}$ mit der Matrix $\begin{pmatrix}\alpha^*, & \beta^* \\ \gamma^*, & \delta^*\end{pmatrix}$ der konjugierten Werte der Cayley-Kleinschen Parameter identifizieren können:

$$\boldsymbol{S} = \begin{pmatrix}\alpha^*, & \beta^* \\ \gamma^*, & \delta^*\end{pmatrix} \text{ oder } S_{11} = \alpha^*, \; S_{12} = \beta^*, \; S_{21} = \gamma^*, \; S_{22} = \delta^*. \tag{15}$$

Dies ist zunächst erlaubt, weil die Relation (10) vermöge (14') gerade erfüllt ist:

$$\begin{pmatrix}\alpha^*, & \beta^* \\ \gamma^*, & \delta^*\end{pmatrix}\begin{pmatrix}\alpha & \gamma \\ \beta & \delta\end{pmatrix} = \begin{pmatrix}\delta, & -\gamma \\ -\beta, & \alpha\end{pmatrix}\begin{pmatrix}\alpha, & \gamma \\ \beta, & \delta\end{pmatrix} = \begin{pmatrix}1, & 0 \\ 0, & 1\end{pmatrix}.$$

[1]) Theorie des Kreisels, Gleichung (9), S. 21.

Setzen wir ferner in (9') und (11') für ξ', η', ζ' die aus (3') gemäß (13) folgenden Matrizen

$$\xi' = \begin{pmatrix} 0, & 1 \\ 1, & 0 \end{pmatrix} + i \begin{pmatrix} 0, & -i \\ i, & 0 \end{pmatrix} = \begin{pmatrix} 0, & 2 \\ 0, & 0 \end{pmatrix},$$

$$\eta' = -\begin{pmatrix} 0, & 1 \\ 1, & 0 \end{pmatrix} + i \begin{pmatrix} 0, & -i \\ i, & 0 \end{pmatrix} = \begin{pmatrix} 0, & 0 \\ -2, & 0 \end{pmatrix}$$

und

$$\zeta' = \begin{pmatrix} -1, & 0 \\ 0, & 1 \end{pmatrix}$$

ein, so erhalten wir aus beiden Gleichungen übereinstimmend:

$$\xi = \begin{pmatrix} -2\alpha\beta, & 2\alpha^2 \\ -2\beta^2, & 2\alpha\beta \end{pmatrix}, \quad \eta = \begin{pmatrix} -2\gamma\delta, & 2\gamma^2 \\ -2\delta^2, & 2\gamma\delta \end{pmatrix},$$

$$\zeta = \begin{pmatrix} -\alpha\delta - \beta\gamma, & 2\alpha\gamma \\ -2\beta\delta, & \alpha\delta + \beta\gamma \end{pmatrix}.$$

Hiermit ist der gewünschte Nachweis erbracht.

Wir haben nur noch einige ergänzende Bemerkungen hinzuzufügen. Die eine betrifft den Spezialfall einer Drehung des Koordinatensystems um die z-Achse, so daß $\Theta = 0$, $\beta = \gamma = 0$ und mit $\Phi + \Psi = \omega$, $\alpha = e^{\frac{i\omega}{2}}$, $\delta = e^{-\frac{i\omega}{2}}$ wird. Man erhält in diesem Falle

$$s_x = \begin{pmatrix} 0, & e^{-i\omega} \\ e^{i\omega}, & 0 \end{pmatrix}, \quad s_y = \begin{pmatrix} 0, & -i e^{-i\omega} \\ i e^{i\omega}, & 0 \end{pmatrix}, \quad s_z = \begin{pmatrix} 1, & 0 \\ 0, & -1 \end{pmatrix}. \qquad (3'')$$

Dies sind zugleich, wie leicht nachzurechnen ist, die allgemeinsten Matrizen (bzw. linearen Transformationen der ψ_α, ψ_β), die hermitisch sind, die Vertauschungsrelationen (2) erfüllen und bei denen außerdem noch s_z seine Normalform $\begin{pmatrix} 1, & 0 \\ 0, & -1 \end{pmatrix}$ hat. Man sieht hieraus, daß die Funktionen (ψ_α, ψ_β) durch Angabe der z-Richtung allein noch nicht eindeutig bestimmt sind (Willkür der Phase ω), sondern erst, wenn das ganze (x, y, z)-Achsenkreuz vorgegeben ist. Schon aus diesem Grunde scheint es kaum möglich, dem magnetischen Elektron gerichtete (vektorielle) Eigenfunktionen zuzuordnen.

Die zweite Bemerkung bezieht sich auf die Frage nach den allgemeinsten (hermitischen) linearen Transformationen der (ψ_α, ψ_β), die den Relationen (2) genügen. Es ist leicht zu sehen, daß diese allgemeinsten s_x, s_y, s_z stets durch eine Transformation der Form (9) [worin S die Relation (10) erfüllt] auf die Normalform (3') gebracht werden können. Wir wollen hier den Beweisgang nur kurz andeuten. Zunächst zeigt man, daß das allgemeinste (10) befriedigende S stets in der Form (14),

(15) durch Winkel Θ, Φ, Ψ ausgedrückt werden kann. Sodann kann jedenfalls zunächst s_z durch eine Transformation (9) in eine Diagonalmatrix verwandelt werden. Aus den Relationen (2) folgt dann bereits, daß s_z die gewünschte Normalform hat. Sodann muß nur noch durch eine geeignete Drehung um die z-Achse die Phase ω in den s_x, s_y zu Null gemacht werden.

Zusammenfassend können wir sagen, daß trotz Auszeichnung eines bestimmten Koordinatensystems durch die Wahl (3) der Operatoren s_x, s_y, s_z infolge der Invarianz der quantenmechanischen Gleichungen gegenüber Substitutionen der Form (9) und infolge des geschilderten Verhaltens der $(\psi_\alpha, \psi_\beta)$ bei Drehungen des auszeichnenden Achsenkreuzes die Unabhängigkeit aller endgültigen Resultate von einer speziellen Wahl des Achsenkreuzes garantiert ist.

§ 4. Differentialgleichungen der Eigenfunktionen eines magnetischen Elektrons in speziellen Kraftfeldern. a) Ruhendes Elektron im homogenen Magnetfeld. Bereits in Gleichung (3), (4) wurde angegeben, wie bei gegebener Hamiltonscher Funktion H die Differentialgleichungen für das Eigenfunktionenpaar $(\psi_\alpha, \psi_\beta)$ des magnetischen Elektrons aufgestellt werden können. Betrachten wir zunächst den Fall des ruhenden Elektrons in einem homogenen Magnetfeld, dessen Feldstärke die Komponenten H_x, H_y, H_z besitzen möge. Da das Elektron ruht, hängen hier die Eigenfunktionen von den Ortskoordinaten des Elektrons nicht ab. Bezeichnet e und m_0 Ladung und Masse des Elektrons,

$$\mu_0 = \frac{eh}{4\pi m_0 c}$$

die Größe des Bohrschen Magnetons, so lautet die Hamiltonfunktion hier

$$H = \mu_0 (H_x s_x + H_y s_y + H_z s_z),$$

wenn wir die konstante Translationsenergie fortlassen und $s_x, \ldots$ wieder in der Einheit $\frac{1}{2}\frac{h}{2\pi}$ gemessen werden. Ersetzt man s_x, s_y, s_z durch die Operatoren (3) (während natürlich μ_0, H_x, H_y, H_z gewöhnliche Zahlen bleiben), so ergibt sich für $(\psi_\alpha, \psi_\beta)$ das Gleichungssystem

$$\left.\begin{aligned} \mu_0 [(H_x - i H_y)\psi_\beta + H_z \psi_\alpha] &= E\psi_\alpha, \\ \mu_0 [(H_x + i H_y)\psi_\alpha - H_z \psi_\beta] &= E\psi_\beta. \end{aligned}\right\} \tag{16}$$

Wir haben absichtlich nicht von vornherein die Richtung des Magnetfeldes mit der [durch die Wahl der Operatoren (3) ausgezeichneten] z-Achse zusammenfallen lassen, um die physikalische Bedeutung unserer

Größen ψ_α, ψ_β und ihre im vorigen Paragraphen abgeleiteten Transformationseigenschaften an einem Beispiel erläutern zu können.

Aus (16) folgen zunächst die Eigenwerte E mittels der Determinantenbedingung

$$\begin{vmatrix} \mu_0 H_z - E, & \mu_0 (H_x - i H_y), \\ \mu_0 (H_x + i H_y), & -(\mu_0 H_z + E) \end{vmatrix} = 0$$

oder

$$-(\mu_0^2 H_z^2 - E^2) - \mu_0^2 (H_x^2 + H_y^2) = 0$$

zu

$$E = \pm \mu_0 \sqrt{H_x^2 + H_y^2 + H_z^2} = \pm \mu_0 |H|,$$

wie es für diesen Fall von vornherein zu fordern ist. Ferner folgt aus (16), wenn man den Winkel zwischen der Feldrichtung und der z-Achse mit Θ bezeichnet und $(\psi_\alpha, \psi_\beta)$ gemäß $|\psi_\alpha|^2 + |\psi_\beta|^2 = 1$ normiert, für $E = +\mu_0 |H|$:

$$|\psi_\alpha|^2 = \frac{\sin^2 \Theta}{\sin^2 \Theta + (1 - \cos \Theta)^2} = \frac{\sin^2 \Theta}{2(1 - \cos \Theta)} = \cos^2 \frac{\Theta}{2},$$

$$|\psi_\beta|^2 = \frac{(1 - \cos \Theta)^2}{2(1 - \cos \Theta)} = \sin^2 \frac{\Theta}{2},$$

analog für $E = -\mu_0 |H|$:

$$|\psi_\alpha|^2 = \sin^2 \frac{\Theta}{2}, \quad |\psi_\beta|^2 = \cos^2 \frac{\Theta}{2}.$$

Dieses Ergebnis ist auch im Einklang mit den Transformationseigenschaften (12), (14), (15) von $(\psi_\alpha, \psi_\beta)$. Es kann z. B. folgendermaßen physikalisch gedeutet werden: Es habe ursprünglich das äußere Magnetfeld die durch H_x, H_y, H_z angegebene Richtung und es seien nur parallel zum Felde gerichtete Elektronen vorhanden, jedoch keine antiparallelen; dann drehe man das Feld plötzlich in die z-Richtung. Man wird sodann finden, daß der Bruchteil $\cos^2 \frac{\Theta}{2}$ aller Elektronen parallel zur z-Achse gerichtete Momente, der Bruchteil $\sin^2 \frac{\Theta}{2}$ antiparallel zur z-Achse gerichtete Momente haben wird; und umgekehrt, wenn ursprünglich nur antiparallel zur Feldrichtung orientierte Elektronen vorhanden waren.

b) Ein magnetisches Elektron im Coulombschen Felde (wasserstoffähnliches Atom). Wenn wir nun dazu übergehen, die Gleichungen für das Eigenfunktionenpaar ψ_α, ψ_β des magnetischen Elektrons im Kernatom aufzustellen, wollen wir uns hier konsequent auf den Standpunkt stellen, bei dem die höheren Relativitäts- und magnetischen Korrektionen vernachlässigt und die von der Relativitätstheorie und dem Eigenmoment des Elektrons herrührenden Glieder als Störungsfunktion aufgefaßt werden. Analog wie in dem vorigen Beispiel nehmen wir sogleich ein äußeres homogenes Magnetfeld mit den Komponenten H_x, H_y, H_z als vorhanden an, um die Theorie des anomalen Zeemaneffekts mit zu umfassen. Wir betonen noch ausdrücklich, daß die hier aufgestellten Gleichungen mit den von Heisenberg und Jordan[1]) angegebenen Matrizengleichungen mathematisch und physikalisch völlig äquivalent sind. Von diesen Verfassern übernehmen wir auch die Form der Hamiltonschen Funktion.

Zunächst hat man die Hamiltonsche Funktion des ungestörten Kernatoms mit einem Elektron:

$$H_0 = \frac{1}{2m_0}(p_x^2 + p_y^2 + p_z^2) - \frac{Ze^2}{r}$$

(p_x, p_y, p_z = Translationsimpuls, Z = Kernladungszahl) oder als Operator geschrieben:

$$\boldsymbol{H}_0(\psi) = -\frac{1}{2m_0}\frac{h^2}{4\pi^2}\Delta\psi - \frac{Ze^2}{r}\psi, \tag{17}$$

worin wie üblich $\Delta = \frac{\partial^2}{\partial x^2} + \frac{\partial^2}{\partial y^2} + \frac{\partial^2}{\partial z^2}$ gesetzt ist. Sodann kommen die Terme, die schon bei einem Elektron ohne Eigenmoment infolge Wirkung des äußeren Magnetfeldes und infolge der Relativitätskorrektion hinzutreten:

$$H_1 = -\frac{1}{2m_0c^2}\left(E_0^2 + 2E_0Ze^2\frac{1}{r} + Z^2e^4\frac{1}{r^2}\right) + \frac{e}{2m_0c}(\mathfrak{H}[\mathfrak{r}\mathfrak{p}]),$$

worin E_0 den ungestörten Eigenwert, $\mathfrak{H}$ den Vektor des äußeren Magnetfeldes, $\mathfrak{p}$ den des Translationsimpulses und $\mathfrak{r}$ den vom Kern zum Elektron führenden Radiusvektor bedeutet.

Als Operator geschrieben gibt dies:

$$\boldsymbol{H}_1(\psi) = -\frac{1}{2m_0c^2}\left(E_0^2 + 2E_0Ze^2\frac{1}{r} + Z^2e^4\frac{1}{r^2}\right)\psi - i\mu_0(\mathfrak{H}[\mathfrak{r}\,\mathrm{grad}\,\psi]). \tag{18}$$

Die Operatoren $\boldsymbol{H}_0$ und $\boldsymbol{H}_1$ gelten in gleicher Weise für ψ_α und ψ_β, sie verändern den Index α oder β nicht. Es kommen nun noch die für das Eigenmoment des Elektrons charakteristischen Terme hinzu, die erstens den bereits im vorigen Beispiel angeschriebenen Wechselwirkungsgliedern des Eigenmomentes mit dem äußeren Magnetfeld und zweitens den gemäß der Relativitätstheorie daraus folgenden Wechselwirkungsgliedern eines

[1]) ZS. f. Phys., l. c., vgl. insbesondere Gleichung (2), (3), (4) dieser Arbeit.

bewegten Elektrons mit Eigenmoment mit dem Coulombschen elektrischen Felde entsprechen. Letztere übernehmen wir hier ohne neue Begründung von Thomas[1]) und Frenkel[1]), insbesondere was den Faktor $^1/_2$ betrifft. Beide Terme zusammen geben, gleich als Operator geschrieben:

$$H_2(\psi) = \frac{1}{4}\frac{h^2}{4\pi^2}\frac{Ze^2}{m_0^2c^2}\frac{1}{r^3}\frac{1}{i}(k_x s_x + k_y s_y + k_z s_z)(\psi)$$
$$+ \mu_0(H_x s_x + H_y s_y + H_z s_z)(\psi), \qquad (19)$$

worin k_x, k_y, k_z als Abkürzung für die $\left(\text{mit } \frac{2\pi i}{h} \text{ multiplizierten}\right)$ zum Bahnimpulsmoment gehörigen Operatoren geschrieben ist, die gegeben sind durch

$$k_x = y\frac{\partial}{\partial z} - z\frac{\partial}{\partial y}, \quad k_y = z\frac{\partial}{\partial x} - x\frac{\partial}{\partial z}, \quad k_z = x\frac{\partial}{\partial y} - y\frac{\partial}{\partial x}. \qquad (20)$$

Setzen wir endlich für s_x, s_y, s_z die durch (3) gegebenen Operatoren ein, so erhalten wir gemäß der allgemeinen Vorschrift (4) für $\psi_\alpha(x, y, z)$ und $\psi_\beta(x, y, z)$ in unserem Falle die simultanen Differentialgleichungen

$$\left.\begin{aligned}
(H_0 + H_1)(\psi_\alpha) + \frac{1}{4}\frac{h^2}{4\pi^2}\frac{Ze^2}{m_0^2c^2}\frac{1}{r^3}[-(i k_x + k_y)\psi_\beta - i k_z \psi_\alpha] \\
+ \mu_0[(H_x - iH_y)\psi_\beta + H_z\psi_\alpha] = E\psi_\alpha, \\
(H_0 + H_1)(\psi_\beta) + \frac{1}{4}\frac{h^2}{4\pi^2}\frac{Ze^2}{m_0^2c^2}\frac{1}{r^3}[(-i k_x + k_y)\psi_\alpha + i k_z \psi_\beta] \\
+ \mu_0[(H_x + iH_y)\psi_\alpha - H_z\psi_\beta] = E\psi_\beta,
\end{aligned}\right\} \qquad (21)$$

in denen also H_0, H_1 und k_x, k_y, k_z durch (17), (18) und (20) gegeben sind. Setzt man hierin speziell $H_x = H_y = 0$, so gehen diese Gleichungen in solche über, die bereits von Darwin[2]) aufgestellt worden sind. Im Gegensatz zu Darwin sehen wir aber als die Quelle dieser Gleichungen letzten Endes die Vertauschungsrelationen (2) [bzw. die verschärften Relationen (2a)] an, nicht aber die Vorstellung, daß die Amplituden der de Broglie-Wellen gerichtete Größen sind. Es ist ferner zu bemerken, daß die Gleichungen (21) gegenüber Drehungen des Koordinatensystems invariant sind, wenn hierbei das Funktionspaar (ψ_α, ψ_β) nach den Vorschriften des vorigen Paragraphen transformiert wird. Auf die Integration der Differentialgleichungen (21) brauchen wir nicht einzugehen, weil sie nach den Methoden von Heisenberg und Jordan ohne Schwierigkeit

[1]) L. H. Thomas, Nature **117**, 514, 1926; Phil. Mag. **3**, 1, 1927; J. Frenkel, ZS. f. Phys. **37**, 243, 1926.

[2]) C. G. Darwin, l. c., Gleichung (3).

durchgeführt werden kann und gegenüber den Ergebnissen dieser Verfasser zu nichts Neuem führt. Es sei auch noch kurz erwähnt, daß die Gleichungen (21) auch aus einem Variationsprinzip abgeleitet werden können, in welchem die durch (16) definierten Größen d_x, d_y, d_z eine Rolle spielen. Da sich eine neue physikalische Einsicht hieraus jedoch nicht ergibt, soll dies hier nicht näher ausgeführt werden.

Wie bereits in der Einleitung erwähnt, ist die hier formulierte Theorie nur als provisorisch anzusehen, da man von einer endgültigen Theorie verlangen muß, daß sie von vornherein relativistisch invariant formuliert ist und auch die höheren Korrektionen zu berechnen erlaubt. Nun bietet es keine Schwierigkeiten, den Drehimpulsvektor $\mathfrak{s}$ zu einem schiefsymmetrischen Tensor (Sechservektor) in der vierdimensionalen Raum-Zeit-Welt mit den Komponenten s_{ik} zu ergänzen und für diese gegenüber Lorentztransformationen invariante Vertauschungsrelationen aufzustellen, die als natürliche Verallgemeinerung von (2) [oder auch von (2a)] anzusehen sind. Man stößt dann jedoch auf eine andere Schwierigkeit, die bereits in den oben erwähnten, auf der klassischen Elektrodynamik basierenden Theorien von Thomas und Frenkel auftritt. In diesen Theorien braucht man in den höheren Näherungen besondere Zwangskräfte, um zu erreichen, daß in einem Koordinatensystem, wo das Elektron momentan ruht, dessen elektrisches Dipolmoment verschwindet; und zwar sind diese Zwangskräfte in den sukzessiven Näherungen jeweils höheren räumlichen Differentialquotienten der am Elektron angreifenden Feldstärken proportional. Es scheint, daß in der Quantenmechanik diese Schwierigkeit bestehen bleibt, und es ist mir aus diesem Grunde bisher nicht gelungen, zu einer relativistisch invarianten Formulierung der Quantenmechanik des magnetischen Elektrons zu gelangen, die als hinreichend naturgemäß und zwangsläufig angesehen werden kann. Man wird sogar, sowohl auf Grund des geschilderten Verhaltens der Zwangskräfte wie auch noch aus anderen Gründen, zu Zweifeln geführt, ob eine solche Formulierung der Theorie überhaupt möglich ist, solange man an der Idealisierung des Elektrons durch einen unendlich kleinen magnetischen Dipol (mit Vernachlässigung von Quadrupol- und höheren Momenten) festhält, ob nicht vielmehr für eine solche Theorie ein genaueres Modell des Elektrons erforderlich sein dürfte. Doch soll auf diese noch ungelösten Probleme hier nicht näher eingegangen werden.

§ 5. Der Fall mehrerer Elektronen. Der Fall, daß mehrere, sagen wir N Elektronen mit Eigenmoment im betrachteten mechanischen System vorhanden sind, bietet bei unserem physikalischen Ausgangspunkt der Methode der Eigenfunktionen gegenüber dem Falle eines einzigen Elektrons keine neuen Schwierigkeiten mehr.

Wir haben hier nach der Wahrscheinlichkeit zu fragen, daß in einem bestimmten, durch den Wert E der Gesamtenergie charakterisierten stationären Zustande des Systems die Lagenkoordinaten der Elektronen in bestimmten infinitesimalen Intervallen liegen und gleichzeitig die Komponenten ihrer Eigenmomente in einer fest zu wählenden z-Richtung, in der Einheit $\frac{1}{2}\frac{h}{2\pi}$ gemessen, entweder die Werte $+1$ oder -1 haben. Wir bezeichnen die Elektronen durch einen von 1 bis N fortlaufenden Index k, die Lagenkoordinaten des k-ten Elektrons kurz mit dem einen Buchstaben q_k (für x_k, y_k, z_k) und ihr infinitesimales Volumelement mit dq_k (für $dx_k\, dy_k\, dz_k$), ferner soll durch den Index α_k oder β_k angemerkt werden, ob für das k-te Elektron die Komponente seines Eigenmoments in der z-Richtung positiv oder negativ ist. Wir haben dann den Zustand des Systems zu charakterisieren durch die 2^N Funktionen

$$\psi_{\alpha_1..\alpha_N}(q_1\cdots q_N), \psi_{\beta_1\alpha_2..\alpha_N}(q_1\cdots q_N), \psi_{\alpha_1\beta_2\alpha_3..\alpha_N}(q_1\cdots q_N)\cdots\psi_{\alpha_1\alpha_2..\beta_N}(q_1\cdots q_N),$$
$$\psi_{\beta_1\beta_2\alpha_3..\alpha_N}(q_1\cdots q_N), \quad \cdots\cdots\cdots\cdots, \quad \psi_{\alpha_1..\alpha_{N-2}\beta_{N-1}\beta_N}(q_1\cdots q_N),$$
$$\cdots\cdots\cdots\cdots\cdots\cdots\cdots\cdots\cdots, \quad \psi_{\beta_1..\beta_N}(q_1\cdots q_N).$$

Es gibt dann z. B.

$$|\psi_{\beta_1\beta_2\alpha_3..\alpha_N}(q\cdots q_N)|^2\, dq\ldots dq_N$$

die Wahrscheinlichkeit dafür an, daß für das erste Elektron s_z gleich -1 und q in $(q_1, q_1 + dq)$, für das zweite Elektron s_z gleich -1 und q in $(q_2, q_2 + dq_2)\ldots$ und für das dritte bis N-te Elektron s_z gleich $+1$ und q bzw. in $[q_3, q_3 + dq_3 \cdots (q_N, q_N + dq_N)]$. Die Reihenfolge, mit der die Suffixe α_k oder β_k angeschrieben sind, soll belanglos sein, während die Variablen q ebenso wie der Index $k = 1\ldots N$ in einer bestimmten Reihenfolge auf die Elektronen bezogen sein sollen. Für die Komponenten s_{kx}, s_{ky}, s_{kz} des Eigenmoments des k-ten Elektrons können wir die Operatoren (3) direkt übernehmen, wenn wir die Festsetzung treffen, daß nur die Indizes α_k oder β_k dieses k-ten Elektrons an den Funktionen ψ durch diesen Operator verändert werden sollen, die der übrigen Elektronen $\alpha_{k'}$ oder $\beta_{k'}$ für $k' \neq k$) aber unverändert bleiben. Wir haben dann also z. B.

$$\left.\begin{aligned}
&s_{kx}(\psi_{\alpha_1..\alpha_k..\beta_N}(q_1\cdots q_N)) = \psi_{\alpha_1..\beta_k..\beta_N}, && s_{kx}(\psi_{..\beta_k..}) = \psi_{..\alpha_k..},\\
&s_{ky}(\psi_{..\alpha_k..}) = -i\psi_{..\beta_k..}, && s_{ky}(\psi_{..\beta_k..}) = i\psi_{..\alpha_k..},\\
&s_{kz}(\psi_{..\alpha_k..}) = \psi_{..\alpha_k..}, && s_{kz}(\psi_{..\beta_k..}) = -\psi_{..\beta_k..}.
\end{aligned}\right\} \qquad (22)$$

Ordnet man wie üblich den Impulskoordinaten p_k den Operator $\frac{h}{2\pi i}\frac{\partial}{\partial q_k}$ zu, so entspricht jetzt jeder Funktion

$$f(p_1 \dots p_N, \quad q_1 \dots q_N, \quad s_{1x}, s_{1y}, s_{1z} \dots s_{Nx}, s_{Ny}, s_{Nz})$$

ein Operator

$$\mathbf{f}\left(\frac{h}{2\pi i}\frac{\partial}{\partial q_1}, \dots \frac{h}{2\pi i}\frac{\partial}{\partial q_N}, q_1 \dots q_N, \mathbf{s}_{1x}, \mathbf{s}_{1y}, \mathbf{s}_{1z} \dots \mathbf{s}_{Nx}, \mathbf{s}_{Ny}, \mathbf{s}_{Nz}\right).$$

Insbesondere ergibt der Operator der Hamiltonschen Funktion H, angewandt auf die 2^N Funktionen $\psi \dots$, die 2^N simultanen Differentialgleichungen

$$H\left(\frac{h}{2\pi i}\frac{\partial}{\partial q_1} \dots \frac{h}{2\pi i}\frac{\partial}{\partial q_N}, q_1 \dots q_N, \mathbf{s}_{1x}, \mathbf{s}_{1y}, \mathbf{s}_{1z} \dots \mathbf{s}_{Nx} \mathbf{s}_{Ny} \mathbf{s}_{Nz}\right) \psi_{i_1 \dots i_N}$$
$$= E\psi_{i_1 \dots i_N} \text{ mit } i_k = \alpha_k \text{ oder } \beta_k. \tag{23}$$

Beziehen sich die Indizes n oder m auf die verschiedenen stationären Zustände, so gilt die Orthogonalitätsrelation

$$\int \sum_{i_k = \alpha_k \text{ oder } \beta_\alpha} (\psi_{n, i_1 \dots i_N} \psi^*_{m, i_1 \dots i_N})\, dq_1 \dots dq_N = \delta_{nm} \tag{24}$$

$$(\delta_{nm} = 0 \text{ für } n \neq m \text{ und } = 1 \text{ für } n = m)$$

und jeder Funktion f der oben beschriebenen Art entsprechen die Matrizen

$$f_{nm} = \int \sum_{i_k = \alpha_k \text{ oder } \beta_k} \{\mathbf{f}(\psi_{m, i_1 \dots i_N}) \cdot \psi^*_{n, i_1 \dots i_N}\}\, dq_1 \dots dq_N. \tag{25}$$

Es bedeutet hier $\mathbf{f}$ den oben definierten zu f gehörigen Operator und sowohl in (24) wie in (25) steht im Integranden eine Summe von 2^N Posten.

Die in Wirklichkeit vorkommenden Hamiltonschen Funktionen, ebenso wie alle zur Matrizendarstellung gelangenden Funktionen f, die tatsächliche physikalische Reaktionen des Systems beschreiben, haben nun wegen der Gleichheit der Elektronen die Eigenschaft, ihren Wert nicht zu ändern, wenn die Koordinaten zweier Elektronen, und zwar sowohl q_k als auch $\mathfrak{s}_k$ miteinander vertauscht werden; H und f können symmetrisch in den N Variablensystemen $(q_k, s_{kx}, s_{ky}, s_{kz})$ angenommeu werden. Dies hat nun nach Heisenberg und Dirac zur Folge, daß die Terme in verschiedene nicht miteinander kombinierende Gruppen zerfallen, die durch die Symmetrieeigenschaften der Eigenfunktionen bei Vertauschen zweier Elektronen charakterisiert sind. Dabei ist wesentlich zu beachten, daß sich das Vertauschen zweier Elektronen, etwa des ersten und zweiten, in der gleichzeitigen Vertauschung der Koordinatenwerte q_1 und q_2 und der zu den Indizes 1 und 2 gehörigen Suffixe α oder β, d. h. ja der Werte von s_{z_1} und s_{z_2}, bemerkbar macht.

Insbesondere gibt es eine symmetrische Lösung; für irgend zwei Indizes k und j bei unveränderten q und Suffixen der übrigen Indizes gilt:

$$\left.\begin{aligned}
\psi^{\text{sym.}} \ldots \alpha_k \alpha_j \ldots (\ldots q_k \ldots q_j \ldots) &= \psi^{\text{sym.}} \ldots \alpha_k \alpha_j \ldots (\ldots q_j \ldots q_k \ldots), \\
\psi^{\text{sym.}} \ldots \alpha_k \beta_j \ldots (\ldots q_k \ldots q_j \ldots) &= \psi^{\text{sym.}} \ldots \beta_k \alpha_j \ldots (\ldots q_j \ldots q_k \ldots), \\
\psi^{\text{sym.}} \ldots \beta_k \beta_j \ldots (\ldots q_k \ldots q_j \ldots) &= \psi^{\text{sym.}} \ldots \beta_k \beta_j \ldots (\ldots q_j \ldots q_k \ldots)
\end{aligned}\right\} \quad (26)$$

ferner eine antisymmetrische Lösung, bei der für irgend ein Indexpaar (Elektronenpaar) k und j bei Vertauschung Vorzeichenwechsel eintritt:

$$\left.\begin{aligned}
\psi^{\text{antis.}} \ldots \alpha_k \alpha_j \ldots (\ldots q_k \ldots q_j \ldots) &= -\psi^{\text{antis.}} \ldots \alpha_k \alpha_j \ldots (\ldots q_j \ldots q_k \ldots), \\
\psi^{\text{antis.}} \ldots \alpha_k \beta_j \ldots (\ldots q_k \ldots q_j \ldots) &= -\psi^{\text{antis.}} \ldots \beta_k \alpha_j \ldots (\ldots q_j \ldots q_k \ldots), \\
\psi^{\text{antis.}} \ldots \beta_k \beta_j \ldots (\ldots q_k \ldots q_j \ldots) &= -\psi^{\text{antis.}} \ldots \beta_k \beta_j \ldots (\ldots q_j \ldots q_k \ldots).
\end{aligned}\right\} \quad (27)$$

Es folgt dies einfach daraus, daß symmetrische Operatoren f den Symmetriecharakter der Funktionen, auf die sie ausgeübt werden, unverändert lassen. Auch das Nichtkombinieren der symmetrischen und der unsymmetrischen Klasse folgt einfach aus (25).

Es wäre interessant, die gruppentheoretische Untersuchung von Wigner[1]) für den Fall von N Elektronen ohne Eigenmoment auf solche mit Eigenmoment zu übertragen und zugleich festzustellen, wie die Terme, die den verschiedenen Symmetrieklassen entsprechen, die man bei Vernachlässigung des Eigenmoments erhält, sich auf die Symmetrieklassen der Elektronen mit Eigenmoment verteilen. Im Falle von 2 Elektronen gibt es nur die symmetrische und die schiefsymmetrische Klasse, die also nach (26), (27) in diesem Falle ($N = 2$) charakterisiert sind durch die Gleichungen

$$\left.\begin{aligned}
&\psi^{\text{sym.}} \alpha_1 \alpha_2 (q_1, q_2) = \psi^{\text{sym.}} \alpha_1 \alpha_2 (q_2, q_1), \quad \psi^{\text{sym.}} \beta_1 \beta_2 (q_1, q_2) = \psi^{\text{sym.}} \beta_1 \beta_2 (q_2, q_1), \\
&\psi^{\text{sym.}} \alpha_1 \beta_2 (q_1, q_2) = \psi^{\text{sym.}} \beta_1 \alpha_2 (q_2, q_1), \\
&\qquad\qquad \psi^{\text{sym.}} \beta_1 \alpha_2 (q_1, q_2) = \psi^{\text{sym.}} \alpha_1 \beta_2 (q_2, q_1),
\end{aligned}\right\} \quad (26')$$

$$\left.\begin{aligned}
&\psi^{\text{antis.}} \alpha_1 \alpha_2 (q_1, q_2) = -\psi^{\text{antis.}} \alpha_1 \alpha_2 (q_2, q_1), \\
&\qquad\qquad \psi^{\text{antis.}} \beta_1 \beta_2 (q_1, q_2) = -\psi^{\text{antis.}} \beta_1 \beta_2 (q_2, q_1), \\
&\psi^{\text{antis.}} \alpha_1 \beta_2 (q_1, q_2) = -\psi^{\text{antis.}} \beta_1 \alpha_2 (q_2, q_1), \\
&\qquad\qquad \psi^{\text{antis.}} \beta_1 \alpha_2 (q_1, q_2) = -\psi^{\text{antis.}} \alpha_1 \beta_2 (q_2, q_1).
\end{aligned}\right\} \quad (27')$$

Dagegen besteht im allgemeinen keine einfache Beziehung zwischen den Funktionswerten $\psi_{\alpha_1, \beta_2}(q_1, q_2)$ und $\psi_{\alpha_1, \beta_2}(q_2, q_1)$; denn diesen entsprechen zwei Konfigurationen verschiedener potentieller Energie; nämlich einmal hat das Elektron mit positivem s_z die Lagenkoordinaten q_1 und das Elektron mit negativem s_z die Lagenkoordinaten q_2; das andere Mal ist umgekehrt das Elektron mit positivem s_z im Raumpunkt, der q_2 entspricht, und das Elektron mit negativem s_z im Raumpunkt, der q_1 entspricht.

[1]) E. Wigner, ZS. f. Phys. **40**, 883, 1927.

Die schiefsymmetrische Lösung ist auch im allgemeinen Falle von N Elektronen diejenige, welche die „Äquivalenzregel" erfüllt und in der Natur allein vorkommt[1]). Es scheint mir ein Vorzug der Methode der Eigenfunktionen, daß diese Lösung in so einfacher Weise charakterisiert werden kann, und gerade deshalb schien mir die formale Ausdehnung dieser Methode auf Elektronen mit Eigenmoment nicht ohne Bedeutung, obwohl sie gegenüber den Heisenbergschen Matrizenmethoden zu keinen neuen Resultaten führen kann. Auch dürften sich die Intensitäten der Interkombinationslinien zwischen Singulett- und Triplettermen, worüber neue Resultate von Ornstein und Burger[2]) vorliegen, nach diesen Methoden in übersichtlicher Weise quantenmechanisch berechnen lassen.

---

[1]) Bei dieser Gelegenheit möchte ich gern betonen, daß das alleinige Vorkommen der schiefsymmetrischen Lösung zunächst nur bei Elektronen, und zwar bei Berücksichtigung ihres Eigenmomentes von der Erfahrung gefordert wird. In einer früheren Mitteilung (ZS. f. Phys. **41**, 81, 1927) wird die Fermische Statistik ebenfalls nur für das Elektronengas beim Vergleich mit der Erfahrung herangezogen. Die Möglichkeit anderer Arten von Statistik bei anderen materiellen Gasen bleibt immer noch offen, was in dieser Mitteilung leider nicht genügend hervorgehoben wurde. Vgl. hierzu auch F. Hund, ZS. f. Phys. **42**, 93, 1927.

[2]) L. S. Ornstein und H. C. Burger, ZS. f. Phys. **40**, 403, 1926.

Einige Folgerungen aus der Schrödingerschen Theorie für die Termstrukturen.

Von **E. Wigner** in Berlin.

Mit 2 Abbildungen. (Eingegangen am 5. Mai 1927.)

Es wird versucht, aus der Form der Schrödingerschen Differentialgleichung einige strukturelle Eigenschaften der Spektren abzuleiten. Es wird die Aufspaltung im elektrischen und magnetischen Feld, das Aufbauprinzip der Serienspektren und einiges Verwandte behandelt. Es ergibt sich — soweit das rotierende Elektron nicht in Betracht zu ziehen ist — Übereinstimmung mit der Erfahrung.

1. Die einfache Gestalt der Schrödingerschen Differentialgleichung gestattet die Anwendung einiger Methoden der Gruppen, genauer gesagt, der Darstellungstheorie. Diese Methoden haben den Vorteil, daß man mit ihrer Hilfe beinahe ganz ohne Rechnung Resultate erhalten kann, die nicht nur für das Einkörperproblem (Wasserstoffatom), sondern auch für beliebig komplizierte Systeme exakt gültig sind. Der Nachteil der Methode ist, daß sie keine Näherungsformeln abzuleiten gestattet. Es ist auf diese Weise möglich, einen großen Teil unserer qualitativen spektroskopischen Erfahrung zu erklären. Die Methode ist dabei so allgemein, daß sie vielfach gar nicht an die spezielle Gestalt der Differentialgleichung gebunden ist. So z. B. kann die Frage nach der Anzahl der Aufspaltungskomponenten im Magnetfeld behandelt werden, ohne daß man mehr voraussetzen müßte, als daß die Terme Eigenwerte einer linearen homogenen Differentialgleichung sind, in die physikalisch gleichwertige Dinge (z. B. Richtungen im Raume, solange keine äußeren Felder da sind) in gleicher Weise eingehen. Die relativen Intensitäten der Komponenten können noch bei schwachen Feldern berechnet werden, ohne daß man eine Annahme über die Differentialgleichung mit Feld machen müßte.

Viele hier abgeleitete Beziehungen sind schon bekannt. Andererseits kenne ich außer der Londonschen Ableitung[1]) des Kuhn-Reiche-Thomasschen Summensatzes keine streng gültige Beziehung, die hier nicht vorkäme, es sei denn, daß sie sich auf das Wasserstoffatom bezieht.

Um das Resultat vorwegzunehmen, sei schon hier gesagt, daß in allen Fällen Übereinstimmung mit den spektroskopischen Erfahrungen besteht [wie sie zuletzt von F. Hund[2]) so schön zusammengefaßt worden

---

[1]) ZS. f. Phys. **39**, 322, 1926.

[2]) F. Hund, Linienspektren und periodisches System der Elemente. Berlin 1927. Dortselbst auch weitere Literatur.

sind], soweit hierbei das „rotierende Elektron" keine Rolle spielt. Wenn man also aus dem Hundschen Buche alle Kapitel über das rotierende Elektron streichen könnte, so würde das übrigbleibende Material sich unseren Ergebnissen fügen.

In der vorliegenden Arbeit ist das rotierende Elektron zumeist nicht mit berücksichtigt und die Resultate sind dementsprechend vielfach nur für Singulettsysteme gültig.

Es wird allerdings bei vielen Ableitungen gar nicht angenommen, daß die Differentialgleichung nur die Lagenkoordinaten der Elektronen enthält; sie könnte noch beliebig viele, sich auf die Rotation der Elektronen beziehende Koordinaten enthalten. Da bei Nichtsingulettsystemen auch in diesen Fällen keine Übereinstimmung mit der Erfahrung erzielt werden konnte, so würde ich glauben, daß man ohne einen neuen Gedanken, lediglich durch Einführung neuer Koordinaten zur Beschreibung der Elektronenmagnete, nicht auskommen kann.

Im folgenden allgemeinen Teil werden die mathematischen Hilfsmittel bereitgestellt, die dann im speziellen Teile verwertet werden.

Allgemeiner Teil.

2. Im folgenden ist die Determinante aller Matrizen ungleich Null, alle Integrationen, wo nichts anderes angegeben ist, sind über den gesamten Bereich der Variablen zu erstrecken, $\delta_{jk} = 0, 1$, je nachdem $j \neq k$ oder $j = k$.

Ist $\psi(x_1, x_2, \ldots x_n)$ eine Funktion, R eine lineare Substitution in n Variablen

$$\left.\begin{aligned} x_1' &= \alpha_{11} x_1 + \alpha_{12} x_2 + \cdots \alpha_{1n} x_n, \\ x_2' &= \alpha_{21} x_1 + \alpha_{22} x_2 + \cdots \alpha_{2n} x_n, \\ &\cdot\;\cdot\;\cdot\;\cdot\;\cdot \\ x_n' &= \alpha_{n1} x_1 + \alpha_{n2} x_2 + \cdots \alpha_{nn} x_n, \end{aligned}\right\} \tag{1}$$

so verstehe ich unter der Funktion $\psi(R(x_1, x_2 \ldots x_n))$ die Funktion, für die identisch in den $x_1, x_2 \ldots x_n$ gilt

$$\psi(R(x_1, x_2 \ldots x_n)) \equiv \psi(x_1', x_2' \ldots x_n'), \tag{2}$$

wobei für $x_1', x_2' \ldots x_n'$ die Werte aus (1) eingesetzt gedacht werden müssen. Wir werden für $\psi(R(x_1, x_2 \ldots x_n))$ auch kurz $\psi(R)$ schreiben. Da E immer die Einheitssubstitution ist ($\alpha_{ik} = \delta_{ik}$), ist $\psi(E) = \psi$.

Es sei nun eine lineare homogene Differentialgleichung, ein Eigenwertproblem gegeben: $H(\psi, \varepsilon) = 0$. Hat die Differentialgleichung, bzw.

die Substitution R die Eigenschaft, daß mit jeder $\psi(x_1, x_2 \ldots x_n)$ auch $\psi(R(x_1, x_2 \ldots x_n))$ eine Lösung der Differentialgleichung bei demselben ε ist, so sagen wir „R ist in der Substitutionsgruppe der Differentialgleichung $H(\psi, \varepsilon)$ enthalten". Es ist nämlich klar, daß die Substitutionen, für die dies gilt, eine Gruppe bilden.

Zumeist ist ihre Substitutionsgruppe der Differentialgleichung direkt anzusehen. Zum Beispiel wenn ich in der Schrödingerschen Gleichung unter x_i, y_i, z_i die Lagenkoordinaten des i-ten Teilchens verstehe, unter R dagegen die Substitution

$$\left.\begin{array}{lll} x_1' = \alpha_1 x_1 + \beta_1 y_1 + \gamma_1 z_1, & y_1' = \alpha_2 x_1 + \beta_2 y_1 + \gamma_2 z_1, & z_1' = \alpha_3 x_1 + \beta_3 y_1 + \gamma_3 z_1, \\ \vdots & \vdots & \vdots \\ x_i' = \alpha_1 x_i + \beta_1 y_i + \gamma_1 z_i, & y_i' = \alpha_2 x_i + \beta_2 y_i + \gamma_2 z_i, & z_i' = \alpha_3 x_i + \beta_3 y_i + \gamma_3 z_i, \\ \vdots & \vdots & \vdots \end{array}\right\} \quad (A)$$

wobei die Matrix

$$\begin{pmatrix} \alpha_1 & \beta_1 & \gamma_1 \\ \alpha_2 & \beta_2 & \gamma_2 \\ \alpha_3 & \beta_3 & \gamma_3 \end{pmatrix}$$

orthogonal ist, so verifiziert man leicht, daß R in der Substitutionsgruppe der Schrödingerschen Gleichung enthalten ist. In einer vorangehenden Mitteilung[1]) wurde zur Ableitung des Heisenberg-Diracschen Termzerfalls nur die Tatsache benutzt, daß z. B. bei dem He die Substitution

$$x_1' = x_2, \ y_1' = y_2, \ z_1' = z_2, \ x_2' = x_1, \ y_2' = y_1, \ z_2' = z_1$$

in die Substitutionsgruppe gehört.

3. Ist mir umgekehrt die Substitutionsgruppe $E, R_2, R_3, \ldots$ einer Differentialgleichung bekannt, so kann ich aus jeder Lösung sofort andere angeben, nämlich aus $\psi(x_1, x_2 \ldots x_n)$, die weiteren $\psi(R_2(x_1, x_2 \ldots x_n))$, $\psi(R_3(x_1, x_2 \ldots x_n))$ usw., die alle zum Eigenwert ε gehören. Diese Funktionen werden im allgemeinen nicht alle linear unabhängig voneinander sein. $\psi_1, \psi_2, \ldots \psi_l$ sei ein linear unabhängiges Aggregat, durch welches sich alle zu diesem Eigenwert ε gehörende Eigenfunktionen ausdrücken lassen. Da auch $\psi_\lambda(R_i)$ eine solche Eigenfunktion ist, ist

$$\psi_\lambda(R_i) = \sum_{\varkappa=1}^{l} a_{\lambda\varkappa}^{R_i} \psi_\varkappa(E). \tag{3}$$

[1]) E. Wigner, Über nichtkombinierende Terme in der neueren Quantentheorie, 2. Teil, ZS. f. Phys. **40**, 883, 1927.

Führen wir in diese Gleichung durch die Substitution R_j neue Variable ein, so erhalten wir

$$\psi_\lambda(R_i R_j) = \sum_{\varkappa=1}^{l} a_{\lambda\varkappa}^{R_i} \psi_\varkappa(R_j) = \sum_{\varkappa=1}^{l} \sum_{\mu=1}^{l} a_{\lambda\varkappa}^{R_i} a_{\varkappa\mu}^{R_j} \psi_\mu(E).$$

Andererseits ist

$$\psi_\lambda(R_i R_j) = \sum_{\mu=1}^{l} a_{\lambda\mu}^{R_i R_j} \psi_\mu(E).$$

Wegen der linearen Unabhängigkeit der ψ_μ folgt hieraus

$$a_{\lambda\mu}^{R_i R_j} = \sum_{\varkappa=1}^{l} a_{\lambda\varkappa}^{R_i} a_{\varkappa\mu}^{R_j}. \tag{4}$$

Mit anderen Worten: die Matrizen (a_{ik}^{E}), $(a_{ik}^{R_2})$, $(a_{ik}^{R_3}) \ldots$ usw. bilden eine zur Substitutionsgruppe E, R_2, $R_3 \ldots$ der Differentialgleichung isomorphe Substitutionsgruppe, oder wie man das in der Gruppentheorie sagt, sie bilden eine Darstellung dieser Gruppe [1]).

Diese Darstellung nennen wir dann zur Kürze die Darstellung des betreffenden Terms ε.

Durch eine andere Wahl der linear unabhängigen $\psi_1, \psi_2 \ldots \psi_l$, wenn wir z. B.

$$\psi_1' = \alpha_{11}\psi_1 + \alpha_{12}\psi_2 + \cdots \alpha_{1l}\psi_l$$
$$\vdots$$
$$\psi_l' = \alpha_{l1}\psi_1 + \alpha_{l2}\psi_2 + \cdots \alpha_{ll}\psi_l$$

gewählt hätten, erfährt die Darstellung nur eine „Ähnlichkeitstransformation“, indem aus den Matrizen (a_{ik}^{E}), $(a_{ik}^{R_2})$, $(a_{ik}^{R_3})$ usw. die Matrizen $(\alpha_{ik})(\alpha_{ik}^{E})(\alpha_{ik})^{-1}$, $(\alpha_{ik})(a_{ik}^{R_2})(\alpha_{ik})^{-1}$, $(\alpha_{ik})(a_{ik}^{R_3})(\alpha_{ik})^{-1}$ usw. entstehen. Solche Darstellungen, die sich nur durch eine Ähnlichkeitstransformation unterscheiden, wollen wir als nicht verschieden voneinander ansehen. In diesem Sinne hat also jeder Term e i n e Darstellung.

Umgekehrt entspricht natürlich einer Ähnlichkeitstransformation der Darstellung lediglich eine andere Wahl der linear unabhängigen $\psi_1, \psi_2 \ldots \psi_l$.

---

[1]) Siehe z. B. A. Speiser, Theorie der Gruppen von endlicher Ordnung. Berlin, 1923. Eine sehr schöne und leicht verständliche Darstellung findet man bei J. Schur, Neue Begründung der Theorie der Gruppencharaktere. Berl. Ber. 1905, S. 406 in den ersten 6 bis 8 Seiten. Die hier gebrauchten Sätze sind entweder hier abgeleitet oder wenigstens [(6), I, II und Punkt 11], ausführlich besprochen.

4. Nehmen wir insbesondere die $\psi_1, \psi_2 \ldots \psi_l$ orthogonal normiert, bzw. im komplexen Falle hermetisiert an, so daß $\int \psi_i \widetilde{\psi_j} = \delta_{ij}$, so folgt

$$\int \psi_\varkappa(R_i) \overline{\psi_\lambda(R_i)} = \sum_{\mu\nu} \int a^{R_i}_{\varkappa\mu} \psi_\mu(E) \widetilde{a^{R_i}_{\lambda\nu}} \overline{\psi_\nu(E)} = \sum_{\mu\nu} a^{R_i}_{\varkappa\mu} \widetilde{a^{R_i}_{\lambda\nu}} \delta_{\mu\nu}.$$

Führen wir nun auf der linken Seite die Transformation R_i^{-1} der Integrationsvariablen aus, so können wir dafür einfach $\int \psi_\varkappa(E) \overline{\psi_\lambda(E)} = \delta_{\varkappa\lambda}$ schreiben, und wir sehen unmittelbar, daß

$$\sum_\mu a^{R_i}_{\varkappa\mu} \widetilde{a^{R_i}_{\lambda\mu}} = \delta_{\varkappa\lambda}, \tag{5}$$

d. h. die Darstellung hat eine orthogonale, bzw. hermitische Form. Es folgt hieraus bekanntlich, da $\left(a^{R_i^{-1}}_{jk}\right) = \left(a^{R_i}_{jk}\right)^{-1}$ ist, daß

$$\widetilde{a^{R_i}_{jk}} = a^{R_i^{-1}}_{kj}. \tag{5a}$$

Eine Umkehrung dieses Satzes — die für „irreduzible" Darstellungen gilt — wird uns später begegnen.

5. Irreduzibel nennt man eine Darstellung, wenn es nicht möglich ist, die Matrix (t_{ik}) der Ähnlichkeitstransformation so zu wählen, daß alle Matrizen $(t_{ik})\,(a^E_{ik})\,(t_{ik})^{-1}$, $(t_{ik})\,(a^{R_2}_{ik})\,(t_{ik})^{-1}$, $(t_{ik})\,(a^{R_3}_{ik})\,(t_{ik})^{-1}$ usw. die Gestalt haben

$$\begin{pmatrix} & & 0 & 0\ldots 0 \\ & & \vdots & \vdots \\ & & 0 & 0\ldots 0 \\ 0\ldots 0 & & & \\ 0\ldots 0 & & & \\ \vdots\quad \vdots & & & \\ 0\ldots 0 & & & \end{pmatrix}$$

wobei die leer gelassenen quadratförmigen Stellen bei allen $(t_{ik})\,(a^R_{ik})\,(t_{ik})^{-1}$ beliebig besetzt sein können, während die ausgeschriebenen Nullen bei allen dieselben Stellen einnehmen. Es würden etwa bei allen in den ersten l'-Zeilen nur die ersten l'-Spalten besetzt sein, in den letzten $l-l'$-Zeilen nur die letzten $l-l'$-Spalten. In den ersten l'-Zeilen stehen in den letzten $l-l'$-Spalten, in den letzten $l-l'$-Zeilen in den ersten l'-Spalten lauter Nullen. Ist eine solche Bestimmung des (t_{ik}) nicht möglich, ist also die Darstellung des Terms eine irreduzible, so sagen wir, die Entartung (d. h. das Zugehören mehrerer Eigenfunktionen zu einem Eigenwert) ist normal. Sonst sprechen wir von einer zufälligen Entartung.

In dem Falle der zufälligen Entartung, wo also die Darstellung des Terms keine irreduzible ist, können wir diese durch eine zweckmäßig gewählte Ähnlichkeitstransformation $(t_{ik}) \cdot\cdot (t_{ik})^{-1}$ „ausreduzieren", d. h. alle Matrizen (a_{ik}^{E}), $(a_{ik}^{R_2})$, $(a_{ik}^{R_3}) \ldots$ in die Form bringen:

$$\left(\begin{array}{cccc}
J_1 & \begin{matrix}0\cdots0\\ \vdots\quad\vdots\\ 0\cdots0\end{matrix} & \cdots\cdots & \begin{matrix}0\\ \vdots\\ 0\end{matrix} \\
\begin{matrix}0\cdots0\\ \vdots\quad\vdots\\ 0\cdots0\end{matrix} & J_2 & \cdots\cdots & \begin{matrix}0\\ \vdots\\ 0\end{matrix} \\
\vdots\quad\vdots & 0\cdots0 & \ddots & \begin{matrix}\vdots\\ 0\cdots0\end{matrix} \\
\begin{matrix}\vdots\quad\vdots\\ 0\cdots0\end{matrix} & \begin{matrix}\vdots\quad\vdots\\ 0\cdots0\end{matrix} & \begin{matrix}0\\ \vdots\\ \cdots0\end{matrix} & J_s
\end{array}\right) \tag{6}$$

wo an den quadratförmigen Stellen $J_1, J_2, \ldots J_s$ Matrizen irreduzibler Darstellungen, sonst lauter Nullen stehen. Führen wir nun mit Hilfe der t_{ik} neue Eigenfunktionen ein

$$\psi_i' = \sum_k t_{ik} \psi_k,$$

so zerfallen diese Eigenfunktionen in s-Gruppen. Die $\psi_i'(R)$ für $R = E_1, R_2, R_3 \ldots$ usw. der einen Gruppe lassen sich durch lineare Kombination der ψ' dieser selben Gruppe allein ausdrücken. Dies geht aus der Gestalt (6) der Darstellung hervor, die für die ψ' gilt. Daß der Eigenwert der einen dieser Gruppen mit dem Eigenwert der anderen Gruppen übereinstimmt, ist quasi zufällig, woher auch das Wort zufällige Entartung genommen wurde. Im entgegengesetzten Falle der normalen Entartung sind — bei beliebiger Wahl des ψ — alle linear unabhängigen Eigenfunktionen, die zu diesem Eigenwert gehören, unter den $\psi(E)$, $\psi(R_2)$, $\psi(R_3) \ldots$ usw. enthalten. Im Falle der zufälligen Entartung können wir auch sagen, daß die s-Terme normaler Entartung (mit den Darstellungseigenschaften $J_1, J_2, \ldots J_s$) zusammenfallen. Wir werden einen solchen Eigenwert s-fach zählen, indem wir jeder der s-Gruppen von ψ' einen Eigenwert zuordnen. In diesem Sinne ist also die Darstellung eines jeden Terms eine irreduzible. Wir wollen an dieser Nomenklatur festhalten.

6. Für die Matrixelemente einer irreduziblen Darstellung gelten zwei wichtige Sätze. Sind (a_{ik}) und (b_{ik}) zwei nicht ähnliche irreduzible Darstellungen derselben Gruppe, so gilt[1])

$$\sum_R a^R_{k\varkappa} b^{R^{-1}}_{\lambda l} = 0 \tag{II}$$

und

$$\sum_R a^R_{k\varkappa} a^{R^{-1}}_{\lambda l} = \delta_{kl}\delta_{\varkappa\lambda} c, \tag{I}$$

wobei die Summation über alle Elemente der Gruppe zu erstrecken ist und c eine Zahl ist, deren Größe nur von den Eigenschaften der Gruppe und der Darstellung abhängt, von k, l, $\varkappa$, λ dagegen unabhängig ist.

Kombinieren wir diese Gleichung mit (5a), indem wir annehmen daß die Form der Darstellung eine hermitische ist, so erhalten wir

$$\sum_R a^R_{k\varkappa} \widetilde{b^R_{l\lambda}} = 0 \tag{II a}$$

und

$$\sum_R a^R_{k\varkappa} a^R_{l\lambda} = \delta_{kl}\delta_{\varkappa\lambda} c. \tag{I a}$$

Für kontinuierliche Gruppen (Drehgruppe usw.) sind an Stelle der Summen Integrale zu setzen[2]).

$$\int_R a^R_{k\varkappa} \widetilde{b^R_{l\lambda}} = 0, \tag{II b}$$

$$\int_R a^R_{k\varkappa} \widetilde{a^R_{l\lambda}} = \delta_{kl}\delta_{\varkappa\lambda} c. \tag{I b}$$

Jede Gruppe, also auch die Substitutionsgruppe unserer Differentialgleichung, besitzt offenbar eine Darstellung — die „identische Darstellung“ — bei der jedem R die Matrix (1) zugeordnet ist. Funktionen, die diese Darstellungseigenschaft haben, für die also gilt

$$f(R) = f(E)$$

nennen wir in Anlehnung an Dirac[3]) symmetrische Funktionen.

Haben wir nun zwei Eigenwerte mit den Darstellungseigenschaften (a^R_{ik}) bzw. (b^R_{ik}) (die voneinander verschieden sind), und den Eigenfunktionen $\psi_1, \psi_2, \ldots$ bzw. $\psi'_1, \psi'_2, \ldots$, die wir unter sich orthogonalhermitisch annehmen wollen und ist die Funktion f symmetrisch, so gilt

$$\int \psi_k \widetilde{\psi'_l} f = A_{kl} = 0. \tag{7}$$

---

[1]) J. Schur, Berl. Ber. 1905, S. 411.

[2]) Vgl. J. Schur, Neue Anwendungen der Integralrechnung auf Probleme der Invariantentheorie. Berl. Ber. S. 199, 1924.

[3]) Proc. Roy. Soc. **112**, 661, 1926.

Es ist nämlich

$$A_{kl} = \int \psi_k(E)\,\overline{\psi'_l(E)}\,f(E) = \int \psi_k(R)\,\overline{\psi'_l(R)}\,f(R),$$

wie bei (5) ausgeführt ist. Wir haben nun

$$A_{kl} = \int \psi_k(R)\,\overline{\psi'_l(R)}\,f(R)$$

$$= \sum_{\varkappa\lambda} a^R_{k\varkappa}\,\widetilde{b^R_{l\lambda}} \int \psi_\varkappa(E)\,\overline{\psi'_\lambda(E)}\,f E = \sum_{\varkappa\lambda} a^R_{k\varkappa}\,\widetilde{b^R_{l\lambda}}\,A_{\varkappa\lambda}.$$

Summieren wir diese Gleichung über alle R der Substitutionsgruppe der Differentialgleichung, so erhalten wir wegen (IIa) $A_{kl} = 0$.

7. Wir haben im Punkt 4 folgendes gesehen: Nehmen wir die Eigenfunktionen $\psi_1, \psi_2, \psi_3, \ldots$, die zu einem Eigenwert gehören, orthogonal-hermitisch an, so ist die zugehörige Form der Darstellung des Eigenwertes ebenfalls orthogonal-hermitisch. Es gilt nun für irreduzible Darstellungen auch die Umkehrung: Ist die Form der Darstellung eine hermitische, was durch eine Ähnlichkeitstransformation immer zu erreichen ist, ist also

$$\sum_\mu a^R_{\varkappa\mu}\,\widetilde{a^R_{\lambda\mu}} = \delta_{\varkappa\lambda}\ (R = E, R_2, R_3 \ldots) \tag{5}$$

oder, was damit gleichbedeutend ist,

$$a^R_{\varkappa\lambda} = \overline{a^{R^{-1}}_{\lambda\varkappa}}\quad (R = E, R_2, R_3 \ldots), \tag{5a}$$

so sind die Eigenfunktionen $\psi_1, \psi_2, \psi_3, \ldots$ auch orthogonal-hermitisch. Es ist sogar allgemeiner (f wieder eine symmetrische Funktion)

$$\int \psi_k(E)\,\overline{\psi_l(E)}\,f(E) = \delta_{kl}\,C.$$

Setzen wir nämlich

$$\int \psi_k(E)\,\overline{\psi_l(E)}\,f(E) = \int \psi_k(R)\,\overline{\psi_l(R)}\,f(R)$$

und

$$\int \psi_k(R)\,\overline{\psi_l(R)}\,f(R) = \sum_{\varkappa,\lambda} a^R_{k\varkappa}\,\widetilde{a^R_{l\lambda}} \int \psi_k(E)\,\overline{\psi_l(E)}\,f(E),$$

dies ergibt über R summiert wegen (Ia)

$$\int \psi_k(E)\,\overline{\psi_l(E)}\,f(E) = \delta_{kl}\,C, \tag{7a}$$

wo C eine von k, l unabhängige, von der Funktion f dagegen abhängige Konstante ist.

Man verifiziert auch leicht folgende Verallgemeinerung: Gehören zu zwei Termen derselben Darstellungseigenschaft die Eigenfunktionen $\psi_1, \psi_2, \psi_3, \ldots$ bzw. $\psi'_1, \psi'_2 \ldots$, wobei diese so gewählt sind, daß auch die Form der Darstellung dieselbe ist, so gilt ebenso wie vorher

$$\int \psi_k\,\widetilde{\psi'_l}\,f = C\,\delta_{kl}. \tag{7b}$$

8. Wir haben nun alles in der Hand, um die Störungstheorie von Schrödinger[1]) zu behandeln. Aus seinen Formeln in § 2 und den Formeln (7), (7 a), (7 b) geht ohne weiteres hervor, daß eine normale Entartung, solange die Störung eine symmetrische Funktion ist, also die Substitutionsgruppe der Differentialgleichung sich nicht ändert — niemals aufgehoben wird und der Term seine Darstellung beibehält. Halten wir bei zufälligen Entartungen an der Nomenklatur des Punktes 5 fest, so ändert sich hieran auch in diesem Falle nichts: Es rücken höchsten die zufällig zusammengefallenen Terme etwas auseinander, oder es rücken einige bisher getrennte zusammen, um sich eventuell bei einer weiteren Störung wieder zu trennen.

Dieses Bild ändert sich ganz, wenn die Störung nicht „symmetrisch" ist, die neue Differentialgleichung also nicht mehr die Substitutionsgruppe der alten hat. Hier interessiert uns namentlich der Fall, daß die Substitutionsgruppe der gestörten Differentialgleichung eine Untergruppe der ungestörten ist. Dies ist z. B. der Fall, wenn man eine elektrische oder magnetische Kraft auf das System einwirken läßt.

Die Behandlung dieses Falles ist auch sehr einfach. Die Substitutionsgruppe der alten Differentialgleichung bezeichne ich mit $\mathfrak{R}$, die der neuen mit $\mathfrak{R}'$, die Darstellung des betrachteten Terms (a_{ik}^{R}) sei eine irreduzible Darstellung von $\mathfrak{R}$. Sie ist selbstverständlich auch eine Darstellung ihrer Untergruppe $\mathfrak{R}'$, aber als solche wahrscheinlich nicht irreduzibel. Wir können sie aber mit Hilfe einer Ähnlichkeitstransformation $(t_{ik})\,(a_{ik}^{R})\,(t_{ik})^{-1}$ ausreduzieren. Sind E, R_2', R_3', ... die Elemente von $\mathfrak{R}'$ (sie sind unter den E, R_2, R_3, ... enthalten), so haben die $(t_{ik})\,(a_{ik}^{R'})\,(t_{ik})^{-1}$ die Form (Punkt 5)

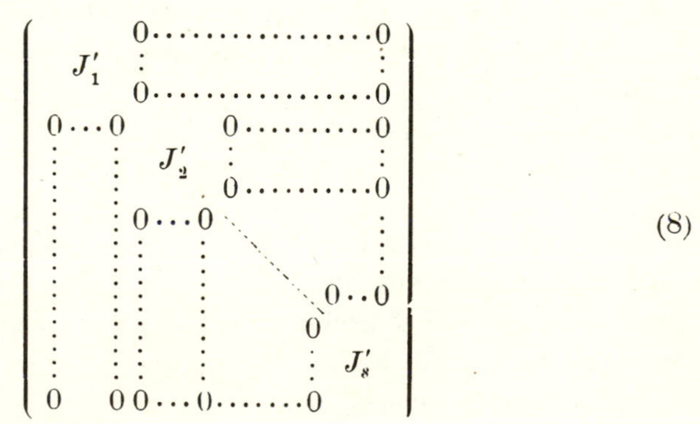

(8)

[1]) Ann. d. Phys. **80**, 437, 440—457, 1926.

für alle Elemente R' der Gruppe $\mathfrak{R}'$, wo $J'_1, J'_2, \ldots J'_s$ irreduzible Darstellungen von $\mathfrak{R}'$ sind. Bei den anderen Elementen von $\mathfrak{R}$ werden natürlich andere Formen auftreten und insbesondere nicht überall, wo hier, lauter Nullen stehen.

Wir werden dieser Form der Matrix entsprechend auch neue Eigenfunktionen einführen $\left(\psi'_i = \sum_k t_{ik} \psi_k\right)$, für die die Darstellung die Form (8) hat. Wenn wir dann der ursprünglichen Differentialgleichung auch nur die Gruppe $\mathfrak{R}'$ als Substitutionsgruppe zuschreiben, ist der betrachtete Term s-fach „zufällig" entartet und enthält eigentlich einen Term der Darstellungseigenschaft J'_1, einen weiteren der Darstellungseigenschaft J'_2 usw. bis J'_s. Verändern wir etwas die Differentialgleichung, wobei nun die jetzt betrachtete Substitutionsgruppe nicht mehr verändert wird, so wird die zufällige Entartung im allgemeinen aufgehoben, und es entstehen die s-Terme mit den Darstellungseigenschaften $J'_1, J'_2, \ldots J'_s$ aus den betrachteten.

Um die Aufspaltung eines Terms mit der Darstellung (a^R_{ik}) bei einer Störung, die die Substitutionsgruppe $\mathfrak{R}$ der Differentialgleichung ändert, zu erhalten, muß man also (a^R_{ik}) als Darstellung der neuen Substitutionsgruppe $\mathfrak{R}'$ ausreduzieren. Hat sie dann die Gestalt (8), so spaltet der ursprüngliche Term in s Terme mit den Darstellungseigenschaften J'_1, $J'_2, \ldots J'_s$ auf.

Natürlich hätten wir all dies auch durch direkte Rechnung mit Hilfe der Formeln von Schrödinger erhalten können.

9. Im Punkte 5 haben wir jedem Term eine irreduzible Darstellung der Gruppe $\mathfrak{R}$ zugeordnet. Es erhebt sich die Frage, ob auch umgekehrt alle irreduziblen Darstellungen der Gruppe $\mathfrak{R}$ gebraucht werden.

Es sei $H\,[\psi(x_1, x_2, \ldots x_n), \varepsilon]$ eine Differentialgleichung und $\left(a^R_{\varkappa\lambda}\right)$ eine Darstellung ihrer Substitutionsgruppe. Existieren nun beliebige Funktionen $f_1(x_1, x_2, \ldots x_n), f_2(x_1, x_2, \ldots x_n), \ldots f_l(x_1, x_2, \ldots x_n)$, die diese Darstellungseigenschaft haben, daß also

$$f_\varkappa(R) = \sum_\lambda a^R_{\varkappa\lambda} f_\lambda(E),$$

so gibt es (im allgemeinen unendlich viele) Eigenwerte mit der Darstellung $(a^R_{\varkappa\lambda})$. (Im entgegengesetzten Falle gibt es natürlich keinen, da ja sonst schon seine Eigenfunktionen $\psi_1, \psi_2, \ldots \psi_l$ die für $f_1, f_2, \ldots f_l$ verlangte Eigenschaft hätten, diese also nicht unerfüllbar wäre.)

Um dies einzusehen, muß man nur eine der Funktionen $f_\varkappa$ nach dem vollständigen System von Eigenfunktionen der Differentialgleichung $H[\psi(x_1, x_2, \ldots x_n), \varepsilon]$ entwickeln. Die Entwicklungskoeffizienten nach

Eigenfunktionen, deren Darstellung von $(a_{\varkappa\lambda}^{R})$ verschieden ist, verschwinden hierbei wegen (7). Hieraus folgt unser Satz.

Die Voraussetzungen dieses Satzes sind — wie wir sehen werden — im Falle von wenigstens vier Teilchen für alle irreduziblen Darstellungen der in Betracht kommenden Substitutionsgruppen erfüllt.

10. Die Substitutionsgruppen, die für uns in Betracht kommen, sind im wesentlichen Verknüpfungen von symmetrischen Gruppen, zwei- oder dreidimensionalen Dreh- oder Drehspiegelungsgruppen, eventuell noch mit einer Spiegelung. Die Darstellungen dieser Gruppen einzeln sind bekannt[1]), wir brauchen nur einen Satz für die Verknüpfung.

Unter der Dimension einer Darstellung versteht man die Anzahl Zeilen oder Spalten in ihren Matrizen, diese ist also gleich der Anzahl der linear unabhängigen Eigenfunktionen, die zu diesem Term gehören.

Habe ich eine Gruppe $\mathfrak{A}$ vom Grade $\mathfrak{a}$ mit den Elementen $E, A_2, A_3, \ldots A_{\mathfrak{a}}$ und eine andere $\mathfrak{B}$ vom Grade $\mathfrak{b}$ mit den Elementen $E, B_2, B_3, \ldots B_{\mathfrak{b}}$ so, daß sie außer E kein gemeinsames Element haben, so sei das direkte Produkt $\mathfrak{C} = \mathfrak{A} . \mathfrak{B} = \mathfrak{B} . \mathfrak{A}$ der Gruppen $\mathfrak{A}$ und $\mathfrak{B}$ vom Grade $\mathfrak{a}\,\mathfrak{b}$ und enthalte die Elemente $E, A_2, A_3, \ldots A_{\mathfrak{a}}, B_2, B_2 A_2, B_2 A_3, \ldots B_2 A_{\mathfrak{a}}, \ldots B_{\mathfrak{b}}, B_{\mathfrak{b}} A_2, B_{\mathfrak{b}} A_3, \ldots B_{\mathfrak{b}} A_{\mathfrak{a}}$. Alle A sind mit allen B vertauschbar, $A_n B_m = B_m A_n$. Es ist klar, daß $\mathfrak{C}$ die Gruppeneigenschaften besitzt.

Sind nun die α irreduziblen Darstellungen der Gruppe $\mathfrak{A}$ die Matritzen

$$\left({}^{1}a_{ik}^{A_n}\right),\ \left({}^{2}a_{ik}^{A_n}\right), \ldots \left({}^{\alpha}a_{ik}^{A_n}\right)$$

mit den Dimensionen $u_1, u_2, \ldots u_\alpha$; die β irreduziblen Darstellungen der Gruppe $\mathfrak{B}$ die Matrizen

$$\left({}^{1}b_{ik}^{B_m}\right),\ \left({}^{2}b_{ik}^{B_m}\right), \ldots \left({}^{\beta}b_{ik}^{B_m}\right)$$

mit den Dimensionen $v_1, v_2, \ldots v_\beta$, so hat die Gruppe $\mathfrak{C}$ genau $\alpha\beta$ irreduzible Darstellungen, und zwar sind diese

$$\begin{matrix} \left({}^{1,1}c_{i_1,i_2;k_1,k_2}^{A_n B_m}\right), & \left({}^{1,2}c_{i_1,i_2;k_1,k_2}^{A_n B_m}\right), & \ldots & \left({}^{1,\beta}c_{i_1,i_2;k_1,k_2}^{A_n B_m}\right) \\ \left({}^{2,1}c_{i_1,i_2;k_1,k_2}^{A_n B_m}\right), & \left({}^{2,2}c_{i_1,i_2;k_1,k_2}^{A_n B_m}\right), & \ldots & \left({}^{2,\beta}c_{i_1,i_2;k_1,k_2}^{A_n B_m}\right) \\ \vdots & \vdots & & \\ \left({}^{\alpha,1}c_{i_1,i_2;k_1,k_2}^{A_n B_m}\right), & \left({}^{\alpha,2}c_{i_1,i_2;k_1,k_2}^{A_n B_m}\right), & \ldots & \left({}^{\alpha,\beta}c_{i_1,i_2;k_1,k_2}^{A_n B_m}\right) \end{matrix} \tag{9}$$

dabei ist

$${}^{\nu,\mu}c_{i_1,i_2;k_1,k_2}^{A_n B_m} = {}^{\nu}a_{i_1,k_1}^{A_n} \cdot {}^{\mu}b_{i_2,k_2}^{B_m} \tag{9a}$$

[1]) Vgl. A. Speiser, Theorie der Gruppen usw. Für die symmetrische Gruppe insbesondere J. Schur, Berl. Ber. 1908, S. 664. Für die Drehgruppen: J. Schur, Neue Anwendungen der Integralrechnung auf Probleme der Invariantentheorie I., II. und III. Berl. Ber. 1924, S. 189, 297 und 346. H. Weyl, Zur Theorie der Darstellung der einfachen kontinuierlichen Gruppen. Berl. Ber. 1924, S. 338. Für den Hinweis auf diese Arbeiten bin ich wiederum Herrn J. v. Neumann zu Danke verpflichtet.

mit den Dimensionen $u_1 v_1$, $u_1 v_2$, ... $u_1 v_\beta$, $u_2 v_1$, $u_2 v_2$, ... $u_2 v_\beta$, ... $u_\alpha v_1$, $u_\alpha v_2$, ... $u_\alpha v_\beta$. i_1 und i_2 beziehen sich auf die Zeilen, k_1 und k_2 beziehen sich auf die Spalten.

Man beweist zunächst leicht, daß diese wirklich Darstellungen von $\mathfrak{C}$ sind, dann mit Hilfe der Gleichung (I) von J. Schur[1]), daß alle Funktionen der y

$$\sum_{B_m = E}^{B_\mathfrak{b}} \sum_{A_n = E}^{A_\mathfrak{a}} {}^{\nu,\mu} C^{A_n\, B_m}_{i_1, i_2;\, k_1\, k_2,} y_{n,\, m}$$

linear unabhängig sind, woraus die Irreduzibilität folgt. Die Anzahl der Klassen konjugierter Elemente von $\mathfrak{A}$ ist α, von $\mathfrak{B}$ ist β[2]). Die von $\mathfrak{C}$ bestimmt sich zu $\alpha\beta$. Dies muß auch die Anzahl der irreduziblen Darstellungen sein, folglich existieren außer diesen keine weiteren.

Wir werden im folgenden die Terme nach ihren Darstellungen unterscheiden und bezeichnen. Unser (trivialer) Satz sagt also folgendes aus: besteht die Substitutionsgruppe einer Differentialgleichung aus dem direkten Produkt mehrerer Substitutionsgruppen (die eine Gruppe sei z. B. die Vertauschung der Elektronen untereinander, die andere etwa die der Wasserstoffkerne), läßt also die Differentialgleichung zwei vertauschbare Substitutionsgruppen zu, so definiert jedes Paar von irreduziblen Darstellungen der beiden Gruppen eine irreduzible Darstellung der ganzen Gruppe. Diese Darstellung, sowie die Terme mit diesen Darstellungen, benennt man dann zweckmäßig so, daß man die Zeichen beider Darstellungen nebeneinander fügt.

Aus den Formeln (9), (9a) liest man nun auch leicht ab — da alle $({}^\nu a^E_{ik})$ sowie $({}^\mu b^E_{ik})$ Einheitsmatrizen sind —, daß die ψ sich bei einer Substitution der einen Untergruppe so transformieren, wie wenn diese Untergruppe allein da wäre.

Spezieller Teil.

11. Wir werden jetzt die Substitutionsgruppe unseres Eigenwertproblems aufsuchen.

Es gehören folgende Substitutionen in diese Gruppe:

I. Auf alle Fälle das Vertauschen aller Koordinaten aller Elektronen.

II. Ebenfalls auf alle Fälle das Vertauschen anderer gleichwertiger Teilchen.

---

[1]) l. c.

[2]) Ebenda, Satz XIII.

Diese beiden Gruppen sind Permutationsgruppen oder, wie man sie auch nennt, symmetrische Gruppen. Ihre irreduziblen Darstellungen sind bekannt [1]).

III a [1]). Im feldlosen Falle ist der Raum isotrop. Ich kann also das ganze System von einem anderen, zum ursprünglichen beliebig verdrehten, Koordinatensystem betrachten und kann dieses Koordinatensystem auch noch spiegeln. Diese Gruppe ist also die dreidimensionale Drehspiegelungsgruppe, ihre Substitutionen stehen in (A). Ihre irreduziblen Darstellungen sind ebenfalls bekannt; es existieren zwei Darstellungen von der Dimension $2l + 1$ (wo l eine nicht negative ganze Zahl ist), die in den Matrizen, die reinen Drehungen zugeordnet sind, übereinstimmen. Der reinen Spiegelung durch den Ursprungspunkt entspricht bei der einen Darstellung eine Diagonalmatrix mit lauter Gliedern -1^l, wir nennen diese die „normale" Darstellung, bei der zweiten eine Diagonalmatrix mit lauter Gliedern -1^{l+1}, wir nennen diese die „gespiegelte" Darstellung [2]).

III b. Im Falle des elektrischen Feldes wird die Isotropie des Raumes aufgehoben. Das Koordinatensystem kann bloß um die Achse des elektrischen Feldes gedreht werden und außerdem in einer zum Felde parallelen Ebene gespiegelt. Es bleibt uns also nur eine zweidimensionale Drehspiegelungsgruppe. Sie hat zwei Darstellungen der Dimension 1, unendlich viele der Dimension 2. Der reinen Verdrehung um einen Winkel α entsprechen dabei die Matrizen

$$(1)^{3)},\quad (1)^{4)},\quad \begin{pmatrix} e^{i\alpha} & 0 \\ 0 & e^{-i\alpha} \end{pmatrix}, \begin{pmatrix} e^{2i\alpha} & 0 \\ 0 & e^{-2i\alpha} \end{pmatrix}, \begin{pmatrix} e^{3i\alpha} & 0 \\ 0 & e^{-3i\alpha} \end{pmatrix} \ldots,$$

der Spiegelung in der Ebene durch die Feldachse

$$(1)^{3)},\quad (-1)^{4)},\quad \begin{pmatrix} 0 & 1 \\ 1 & 0 \end{pmatrix}, \quad \begin{pmatrix} 0 & 1 \\ 1 & 0 \end{pmatrix}, \quad \begin{pmatrix} 0 & 1 \\ 1 & 0 \end{pmatrix} \ldots$$

III c. Im Falle des magnetischen Feldes kann man das Koordinatensystem um die Feldachse drehen und in der Ebene senkrecht zu dieser spiegeln. Dies ist ein direktes Produkt einer zweidimensionalen Drehgruppe mit einer Spiegelung.

In den irreduziblen Darstellungen der zweidimensionalen Drehgruppe entsprechen einer Verdrehung um den Winkel α die Matrizen

$$(1),\ (e^{i\alpha}),\ (e^{-i\alpha}),\ (e^{2i\alpha}),\ (e^{-2i\alpha}) \ldots$$

---

[1]) Vgl. J. Schur, l. c. H. Weyl, l. c.

[2]) Diese Unterscheidung wird sich als zweckmäßig erweisen.

[3]) Diese Darstellung werden wir normal nennen.

[4]) Diese Darstellung werden wir gespiegelt nennen.

In den beiden Darstellungen der Spiegelung entsprechen einer solchen die Matrizen

$$(1),\ (-1).$$

In den Darstellungen der Gruppe IIIc entsprechen also (Punkt 10) einer Drehung um den Winkel α

$$(1),\ (1),\ (e^{i\alpha}),\ (e^{i\alpha}),\ (e^{-i\alpha}),\ (e^{-i\alpha}),\ (e^{2i\alpha}),\ (e^{2i\alpha}),\ (e^{-2i\alpha})\ldots,$$

der Spiegelung in der zur Achse des Feldes senkrechten Ebene die Matrizen

$$(1),\ (-1),\ (1),\ (-1),\ (1),\ (-1),\ (1),\ (-1),\ (1)\ldots$$

Der Fall, daß beide Felder vorhanden sind, wird sich als nicht mehr wesentlich erweisen.

Da nun die Gruppen I und II sowohl untereinander, wie mit jeder der Gruppen IIIa, IIIb, IIIc vertauschbar sind, können wir die Substitutionsgruppe der Differentialgleichung erhalten, indem wir das direkte Produkt der Gruppen I und II mit derjenigen Gruppe III bilden, die in Frage kommt.

12. Zunächst betrachten wir den Fall gekreuzter elektrischer und magnetischer Felder, um die Wirkung der Gruppen III ganz auszuschließen.

Wäre nur die Gruppe I vorhanden, so würde im Falle von n Elektronen jeder Aufteilung der Zahl n in ganzzahlige Summanden (wobei jeder Summand größer oder gleich dem vorangehenden sein muß) eine Darstellung entsprechen. Sind außerdem n_1 Teilchen der Sorte (1) (etwa Protonen), n_2 Teilchen der Sorte (2) usw. da, so muß man im Sinne des Punktes 10 jede Aufteilung der Zahl n mit jeder Aufteilung der Zahl n_1 und jeder Aufteilung der Zahl n_2 usw. kombinieren, um alle irreduziblen Darstellungen zu erhalten. Wegen 9. werden unendlich viele Eigenwerte jeder Darstellungseigenschaft existieren. Daß keine Interkombinationen zwischen Termen mit verschiedenen Darstellungen vorkommen, ist einfach eine Folge von (7).

Es werden aber nicht alle Eigenwerte in der Natur als Terme realisiert sein. Vielmehr ist z. B. bei Elektronen am einfachsten zu fordern, daß bei adiabatischem Verschwindenlassen der Wechselwirkung zwischen den Elektronen nur solche Zustände entstehen sollen, die höchstens zwei Elektronen auf einer Bahn enthalten. Dies sind diejenigen Termsysteme, bei denen die Zahl n in lauter 1 und 2 zerlegt wird [1]).

[1]) W. Heisenberg (ZS. f. Phys. **41**, 239, 1927) leitet diese Termsysteme ab, indem er davon ausgeht, daß die Wirkung der Elektronenrotation durch neue

Wir entnehmen nun mit Heisenberg[1]) der Erfahrung, daß ein Term bei einem Atom zu einer Multiplizität gehört, die um 1 größer ist als die Anzahl der 1 in der Zerlegung von n in Summanden.

Bei	H	$1 = 1$	Dublett
„	He	$2 = 2$	Singulett
„	„	$2 = 1 + 1$	Triplett
„	Li	$3 = 1 + 2$	Dublett
„	„	$3 = 1 + 1 + 1$	Quartett
„	Be	$4 = 2 + 2$	Singulett
„	„	$4 = 1 + 1 + 2$	Triplett
„	„	$4 = 1 + 1 + 1 + 1$	Quintett

usw.

Wie die Auswahl der in der Natur vorkommenden Termsysteme bei Protonen usw. ist, kann vorläufig nur durch Vergleich mit der Erfahrung entschieden werden, wie insbesondere in der Diskussion dieses Themas bei F. Hund[2]) auseinandergesetzt ist[3]).

13. Wir gehen nun zu dem Falle über, wo auf unser System keine Kraft wirkt. Dann sind die Darstellungen des IIIa mit den Darstellungen I und II zu kombinieren.

Wir müssen nun die Darstellungen der dreidimensionalen Drehgruppe etwas genauer betrachten. Dazu führen wir die Parameterdarstellung von Euler ein. Die Verdrehung, die den Bogen ▬ in den

---

Eigenfunktionen, und zwar entsprechend den zwei Einstellungsmöglichkeiten des Elektrons, zwei Eigenfunktionen beschrieben wird. Dann sucht er diejenigen Termsysteme aus, die durch zwei Funktionen antisymmetrisiert werden können. Diese Vorschrift führt genau zu den im Text angegebenen Termsystemen. Ich kann davon absehen näheres hierüber mitzuteilen, da Herr F. Hund unabhängig von mir nach einer etwas anderen Methode zu demselben Ergebnis gekommen ist und seine Resultate bald in dieser Zeitschrift erscheinen.

[1]) l. c.

[2]) ZS. f. Phys. **42**, 93, 1927.

[3]) Ich möchte an dieser Stelle noch eine Bemerkung zur Frage der Bose-Einsteinschen Statistik machen, soweit sie die Gasentartung betrifft. Ob in einem Gase die Bose-Einsteinsche oder die Fermi-Diracsche Theorie zu Recht besteht, hängt lediglich davon ab (W. Pauli jr., ZS. f. Phys. **41**, 81, 1927; siehe auch R. H. Fowler, Proc. Roy. Soc. (A) **113**, 432, 1926), ob bei der Vertauschung zweier Atome bzw. Moleküle die Eigenfunktion ungeändert bleibt oder ihr Vorzeichen umkehrt. Dies läßt sich mit großer Wahrscheinlichkeit für Moleküle, die aus zwei gleichen Atomen bestehen, angeben. Wie immer nämlich das allgemeine Gesetz sein mag, das den Vorzeichenwechsel bei dem Austausch von Elektronen und etwa von Stickstoffkernen regelt, wenn man zwei N_2 miteinander vertauscht, kann sich am Ende das Vorzeichen nicht geändert haben, weil ja jede Operation zweimal ausgeführt worden ist. In solchen Fällen wenigstens besteht also die Einsteinsche Theorie der Gasentartung höchstwahrscheinlich zu Recht. Natürlich gilt dies nicht für Elektronen.

Bogen $-\cdot-\cdot-$ überführt, kennzeichnen wir mit den drei Eulerschen Winkeln α, β, γ. Sie ist aus einer Drehung um die Z-Achse mit dem Winkel γ, um die X-Achse mit β und aus noch einer Drehung um die Z-Achse mit dem Winkel α zusammengesetzt. Das Matrizenelement, das in der $2l+1$-dimensionalen Darstellung in der j-ten Zeile und der k-ten Spalte steht, bezeichnen wir mit D^l_{jk} (j und k laufen von $-l$ bis $+l$). Wir betrachten diejenige Form der Darstellung, die als Darstellung der zweidimensionalen Drehgruppe um die Z-Achse ausreduziert und außerdem hermitisch ist. D^l_{jk} hat die Form

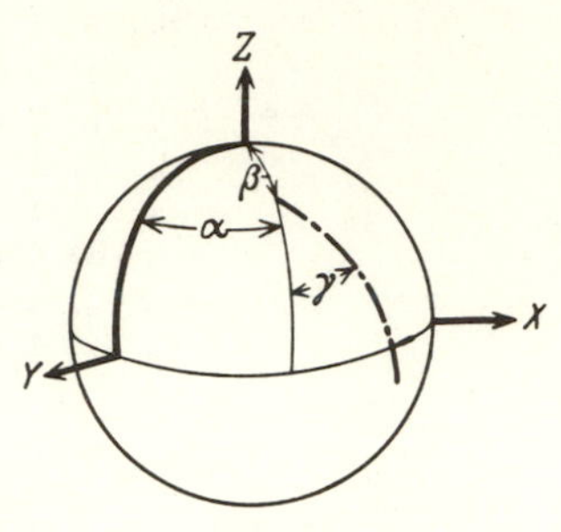

Fig. 1.

$$D^l_{jk}(\alpha, \beta, \gamma) = e^{ij\alpha}\, d^l_{jk}(\beta)\, e^{ik\gamma}, \tag{10}$$

$d^l_{jk}(\beta)$ ist reell, und es gilt

$$\begin{aligned} d^l_{jk}(\beta) &= (-1)^{j+k} d^l_{jk}(-\beta) = (-1)^{j+k} d^l_{kj}(\beta) \\ &= d^l_{kj}(-\beta) = (-1)^{j+k} d_{-j,-k}(\beta). \end{aligned} \tag{11}$$

Für später merken wir uns die Rekursionsformeln an:

$$D^{l+1}_{jk} \frac{\sqrt{(l+1)^2 - j^2}\sqrt{(l+1)^2 - k^2}}{(2l+1)(l+1)} - \\ - D^l_{jk}\cos\beta + D^{l-1}_{jk} \frac{\sqrt{l^2 - j^2}\sqrt{l^2 - k^2}}{(2l+1)l} = 0, \tag{12}$$

$$D^{l+1}_{j-1,k} \frac{\sqrt{l-j+1}\sqrt{l-j+2}\sqrt{(l+1)^2 - k^2}}{(2l+1)(l+1)} - \\ - D^l_{jk}\sin\beta\, e^{-i\alpha} - D^{l-1}_{j-1,k} \frac{\sqrt{l+j}\sqrt{l+j-1}\sqrt{l^2 - k^2}}{(2l+1)l} = 0, \tag{12 a}$$

$$D^{l+1}_{j+1,k} \frac{\sqrt{l+j+1}\sqrt{l+j+2}\sqrt{(l+1)^2 - k^2}}{(2l+1)(l+1)} + \\ + D^l_{jk}\sin\beta\, e^{i\alpha} - D^{l-1}_{j+1,k} \frac{\sqrt{l-j-1}\sqrt{l-j}\sqrt{l^2 - k^2}}{(2l+1)l} = 0. \tag{12 b}$$

Durch (11) gehen aus ihnen weitere hervor.

Es seien noch die ersten zwei Darstellungen für reine Drehung explizite hergeschrieben. Die „normalen“ und „gespiegelten“ Darstellungen der Drehspiegelungsgruppe ergeben sich daraus gemäß Punkt 11.

$$\text{(1)},\quad \begin{pmatrix} e^{-i\alpha}\dfrac{1+\cos\beta}{2}e^{-i\gamma}, & e^{-i\alpha}\dfrac{\sin\beta}{\sqrt{2}}, & e^{-i\alpha}\dfrac{1-\cos\beta}{2}e^{i\gamma}, \\ -\dfrac{\sin\beta}{\sqrt{2}}e^{-i\gamma}, & \cos\beta, & \dfrac{\sin\beta}{\sqrt{2}}e^{i\gamma}, \\ e^{i\alpha}\dfrac{1-\cos\beta}{2}e^{-i\gamma}, & e^{i\alpha}\dfrac{\sin\beta}{\sqrt{2}}, & e^{i\alpha}\dfrac{1+\cos\beta}{2}e^{i\gamma}. \end{pmatrix}$$

14. Ich will nun $2l+1$ Funktionen finden:

$$f_{-l}(x_1, y_1, z_1 \ldots x_n, y_n, z_n),\quad f_{-l+1}(x_1, y_1, z_1 \ldots x_n, y_n, z_n)\ \ldots,$$
$$f_l(x_1, y_1, z_1 \ldots x_n, y_n, z_n),$$

die sich bei der Substitution (A) nach Maßgabe der $2l+1$-dimensionalen irreduziblen Darstellung der dreidimensionalen Drehgruppe transformieren, für die also gilt

$$f_\varkappa(x_1', y_1', z_1' \ldots x_n', y_n', z_n') = \sum_\lambda D^l_{\varkappa\lambda} f_\lambda(x_1, y_1, z_1 \ldots x_n, y_n, z_n), \quad (13)$$

worin für $x_1', y_1', z_1' \ldots x_n', y_n', z_n'$ die Werte aus (A) zu entnehmen sind.

Ich betrachte die Funktionen an der Überfläche $x_1 = y_1 = x_2 = 0$. An dieser Überfläche sind die Funktionen durch die gegenseitige Konfiguration der Teilchen bestimmt, während die Lage der ganzen Konfiguration im Raume vorgeschrieben ist, ich nenne sie die Lage B. Für die Koordinaten, die die gegenseitige Konfiguration beschreiben, schreibe ich g. Die Werte der Funktionen f_{-l} bis f_l an dieser Überfläche bezeichne ich mit $\chi_{-l}(g)$, $\chi_{-l+1}(g)$, $\chi_{-l+2}(g) \ldots \chi_l(g)$. Die Funktion $f_\varkappa(x_1, y_2, z_1 \ldots x_n, y_n, z_n)$ ist dann wegen (13)

$$f_\varkappa(x_1, y_1, z_1 \ldots x_n, y_n, z_n) = \sum_\lambda D^l_{\varkappa\lambda}(\alpha, \beta, \gamma)\chi_\lambda(g), \quad (13\text{a})$$

wo α das Azimut des ersten Elektrons, β sein Polabstand und γ der Winkel zwischen den Ebenen Zr_1 und $r_1 r_2$ ist[1]). Umgekehrt bestimme ich die $\chi_\lambda(g)$ ganz beliebig, so haben die in (13a) definierten Funktionen $f_\varkappa$ die verlangte Transformationseigenschaft (13).

Sollen die $f_\varkappa$ die normale Darstellungseigenschaft der Drehspiegelungsgruppe haben, so muß

$$f_\varkappa(-x_1, -y_1, -z_1 \cdots -x_n, -y_n, -z_n) = (-1)^l f_\varkappa(x_1, y_1, z_1 \cdots x_n, y_n, z_n) \quad (14)$$

sein, soll die Darstellungseigenschaft die gespiegelte sein, so ist

$$f_\varkappa(-x_1, -y_1, -z_1 \cdots -x_n, -y_n, -z_n) = -(-1)^l f_\varkappa(x_1, y_1, z_1 \cdots x_n, y_n, z_n). \quad (14\text{a})$$

---

[1]) Es geht hieraus hervor, daß die $D(\alpha, \beta, \gamma)$ ein vollständiges Funktionensystem bilden müssen.

Ist die Anzahl der Teilchen, die in die Differentialgleichung eingehen, mindestens gleich 3 (die gesamte Anzahl der Teilchen also mindestens gleich 4, da ein Teilchen wegen des Schwerpunktsatzes absepariert werden muß: z. B. gehen in die Gleichung des H-Atoms nur drei Koordinaten einem Teilchen entsprechend ein. Vgl. z. B. E. Fues, Ann. d. Phys. **80**, 367, 1926), so kann man beides erreichen, indem man etwa die χ in der Lage B nur dort willkürlich definiert, wo $x_3 > 0$ und die Werte für $x_3 < 0$ durch die Gleichung (14) bzw. (14a) bestimmt. Im Falle von nur drei Teilchen (z. B. He) ist dies jedoch nicht möglich; dort kommen nur die normalen Darstellungen in Betracht.

Will man Funktionen bilden [1]), die sich bei der Vertauschung von Elektronen so benehmen, wie eine bestimmte in 11. I beschriebene Darstellung dies verlangt, so hat man lediglich die χ nur dort willkürlich zu definieren, wo $r_1 < r_2 < \cdots < r_n$ ist (r_i Abstand des i-ten Teilchens vom Schwerpunkt) und an den anderen Stellen die Funktionen mit Hilfe ihrer Darstellungseigenschaften (9), (9a) zu bestimmen.

Wegen 9. können wir hieraus schließen: Vom Li ab kommen in jedem Multiplettsystem Terme jeder in Punkt 11, IIIa beschriebenen Darstellungseigenschaft vor. Bei He und H kommen nur normale Terme vor.

Die Eigenfunktionen sind durch die Ausdrücke (13a) gegeben, die $\chi_\lambda(g)$ hängen nur von der gegenseitigen Konfiguration der Teilchen ab.

15. In diesem Punkte möchte ich die Auswahlregeln für die azimutale Quantenzahl ableiten [2]). Unter Intensität ist hier die „Stärke des Oszillators" zu verstehen, also die lineare Polarisation

$$\left|\int x\, \psi_i\, \psi_k\right|,$$

die höheren Momente von Heisenberg [3]) sind nicht in Betracht gezogen.

---

[1]) Es gilt dies nur, wenn ein Teilchen (bei uns der Kern) da ist, von dem nur ein Exemplar vorhanden ist und dessen Koordinaten man mit Hilfe des Schwerpunktsatzes absepariert.

[2]) M. Born, W. Heisenberg und P. Jordan erhielten bereits in ihrer Arbeit (ZS. f. Phys. **35**, 557, 1926) Resultate, die sich in vielen Punkten mit den hier folgenden decken. Insbesondere findet sich dort die Auswahlregel für die magnetische Quantenzahl m und auch die Intensitätsformel (18). Auch bei P. A. M. Dirac (Proc. Roy. Soc. **111**, 281, 1926) finden sich diese Formeln. Unsere Betrachtungen führen insofern etwas weiter, als wir auch einen Übergang mit $l_1 = l_2$ (dort $j_1 = j_2$) ausschließen können, wodurch die Deutung des l als azimutale Quantenzahl gesichert erscheint. Außerdem kann l (j) dort ganz- oder halbzahlige Werte annehmen; bei uns sind nur ganzzahlige Werte möglich. Für den Fall eines einzigen Elektrons findet sich dies auch bei L. Brillouin, Journ. de phys. et le Rad. **8**, 74, 1927.

[3]) ZS. f. Phys. **38**, 411, 1926.

Da wir die Wirkung des rotierenden Elektrons nicht berücksichtigen konnten, werden alle unsere „Multiplettsysteme" (siehe Punkt 12) die Eigenschaften haben, die experimentell bei Singulettsystemen bekannt sind.

Die Dimensionen der Darstellungen der Drehgruppe sind $2l + 1$. Die Zahl l nennen wir azimutale Quantenzahl, da folgendes gilt: l ändert sich bei einem Übergang um $+1$ oder -1. Alle anderen Übergänge sind verboten.

Es genügt, nachzuweisen, daß bei solchen verbotenen Übergängen die Polarisation, die vom ersten Elektron herrührt, verschwindet. Bezeichnen wir mit r_1, r_2 den Radiusvektor des ersten bzw. zweiten Elektrons, mit α, β Azimut und Polabstand von der Z-Achse des ersten, γ ist der Winkel der Ebenen $Z r_1$, $r_1 r_2$, so ist die $\varkappa$-te Eigenfunktion, die zu einem Term mit der azimutalen Quantenzahl l gehört, wegen (13a)

$$\sum_{\lambda} D^{l}_{\varkappa\lambda}(\alpha, \beta, \gamma)\, \chi_{\lambda}(g),$$

die $\varkappa'$-te Eigenfunktion mit der azimutalen Quantenzahl l'

$$\sum_{\lambda'} D^{l'}_{\varkappa'\lambda'}(\alpha, \beta, \gamma)\, \chi'_{\lambda'}(g).$$

Die Polarisation in der Z-Achse, die vom ersten Elektron herrührt,

$$\sum_{\lambda, \lambda'} \int D^{l}_{\varkappa\lambda}(\alpha, \beta, \gamma)\, \chi_{\lambda}(g)\, \overline{D^{l'}_{\varkappa'\lambda'}(\alpha, \beta, \gamma)}\, \overline{\chi'_{\lambda'}(g)}\, r_1 \cos\beta$$

oder wegen (12)

$$\sum_{\lambda, \lambda'} \int (D^{l+1}_{\varkappa\lambda}\, u_{\varkappa\lambda} + D^{l-1}_{\varkappa\lambda}\, v_{\varkappa\lambda})\, \chi_{\lambda}(g)\, \overline{D^{l'}_{\varkappa'\lambda'}}\, \overline{\chi'_{\lambda'}(g)}\, r_1,$$

$$u_{\varkappa\lambda} = \frac{\sqrt{(l+1)^2 - \varkappa^2}\,\sqrt{(l+1)^2 - \lambda^2}}{(2l+1)(l+1)}, \quad v_{\varkappa\lambda} = \frac{\sqrt{l^2 - \varkappa^2}\,\sqrt{l^2 - \lambda^2}}{(2l+1)\,l},$$

dies verschwindet schon bei der Integration über α, β, γ wegen (II b), wenn weder $l + 1 = l'$ noch $l - 1 = l'$. Ähnlich die Polarisation in der X- und Y-Richtung.

Terme, zu denen eine normale Darstellung der Drehspiegelungsgruppe gehört, bezeichnen wir als normal, Terme, zu denen eine gespiegelte Darstellung gehört, als gespiegelt. Übergänge zwischen normalen und gespiegelten Termen kommen nicht vor.

Es genügt, dies für eine Änderung von l um ± 1 nachzuweisen, da andere Übergänge überhaupt nicht vorkommen. Die Polarisation in der Z-Richtung vom ersten Elektron ist (ψ^l normal, ψ^{l+1} gespiegelt)

$$\int \psi^l\, \overline{\psi^{l+1}}\, z_1.$$

Führen wir für x_i, y_i, z_i neue Integrationsvariable $-x_i$, $-y_i$, $-z_i$ ein, so ändert sich ψ^l mit dem Faktor $(-1)^l$; ψ^{l+1} mit dem Faktor $-(-1)^{l+1}$, z mit -1. Es ist also

$$\int \psi^l \overline{\psi^{l+1}} z_1 = -\int \psi^l \overline{\psi^{l+1}} z_1 = 0.$$

Dieser Satz wird sich als identisch mit der Laporteschen Regel erweisen. [Punkt 25][1]).

16. Wirkung des elektrischen Feldes. Durch ein elektrisches Feld in der Z-Achse wird die Substitutionsgruppe unserer Differentialgleichung verkleinert. Wir müssen also im Sinne des achten Punktes verfahren und die Darstellung der dreidimensionalen Drehgruppe als Darstellung der zweidimensionalen (um die Z-Achse) ausreduzieren. In (11) ist die Darstellung schon in dieser Form angenommen worden. [Es gilt nämlich für $\beta = 0$

$$d^l_{\varkappa\lambda}(0) = \delta_{\varkappa\lambda},$$

wie schon daraus hervorgeht, daß $D^l_{\varkappa\lambda}(\alpha, \beta, \gamma)$ für $\alpha = \beta = \gamma = 0$ die Einheitsmatrix sein muß.] In den irreduziblen Bestandteilen von $D^l_{\varkappa\lambda}(\alpha, 0, 0)$, als Darstellung der zweidimensionalen Drehspiegelungsgruppe um die Z-Achse, sind also

$$(1), \quad \begin{pmatrix} e^{i\alpha} & 0 \\ 0 & e^{-i\alpha} \end{pmatrix}, \quad \begin{pmatrix} e^{2i\alpha} & 0 \\ 0 & e^{-2i\alpha} \end{pmatrix}, \ldots, \quad \begin{pmatrix} e^{il\alpha} & 0 \\ 0 & e^{-il\alpha} \end{pmatrix}$$

die Matrizen für reine Drehung um den Winkel α; für die reine Spiegelung in der YZ-Ebene (nicht mehr durch den Koordinatenursprungspunkt!) haben wir, wenn die Darstellung normal war,

$$(1), \quad \begin{pmatrix} 0 & 1 \\ 1 & 0 \end{pmatrix}, \quad \begin{pmatrix} 0 & 1 \\ 1 & 0 \end{pmatrix}, \ldots, \quad \begin{pmatrix} 0 & 1 \\ 1 & 0 \end{pmatrix}$$ [2]),

bei gespiegelter Darstellung

$$(-1), \quad \begin{pmatrix} 0 & 1 \\ 1 & 0 \end{pmatrix}, \quad \begin{pmatrix} 0 & 1 \\ 1 & 0 \end{pmatrix}, \ldots, \quad \begin{pmatrix} 0 & 1 \\ 1 & 0 \end{pmatrix}$$ [2]) [3]).

Wir finden also: im homogenen elektrischen Felde spaltet ein Term mit der azimutalen Quantenzahl l in $l + 1$ Terme auf. Ein neuer Term

---

[1]) Vgl. F. Hund, Linienspektren usw., S. 132. Die Laportesche Regel soll also vom Li ab gelten.

[2]) Vgl. Fußnote 3 und 4, S. 636.

[3]) Die Matrizen $\begin{pmatrix} e^{i\varkappa\alpha} & 0 \\ 0 & e^{-i\varkappa\alpha} \end{pmatrix}$ und $\begin{pmatrix} 0 & -1 \\ -1 & 0 \end{pmatrix}$ gehen durch eine Ähnlichkeitstransformation in die Matrizen $\begin{pmatrix} e^{i\varkappa\alpha} & 0 \\ 0 & e^{-i\varkappa\alpha} \end{pmatrix}$ und $\begin{pmatrix} 0 & 1 \\ 1 & 0 \end{pmatrix}$ über.

ist einfach, und je nachdem der ursprüngliche Term normal oder gespiegelt war, normal oder gespiegelt, die anderen haben noch zwei Eigenfunktionen und unterscheiden sich in den beiden Fällen nicht in ihren Darstellungseigenschaften.

17. Größe der Aufspaltung und Auswahlregeln. Ein Term mit der Darstellung $\begin{pmatrix} e^{im\alpha} & 0 \\ 0 & e^{-im\alpha} \end{pmatrix}$ hat die elektrische Quantenzahl m. Bei einer Strahlung, bei der der elektrische Vektor parallel zum Felde ist, ändert sich m nicht (ein normaler Term geht in einen normalen, ein gespiegelter in einen gespiegelten über). Ist der elektrische Vektor der Strahlung senkrecht zum Feld, so kann sich m nur mit $+1$ oder -1 ändern. Wir sehen, daß die Auswahlregeln für schwache Felder durchbrochen sind: es kommt gar nicht mehr darauf an, aus was für einem Term der neue entstanden ist, nur die elektrische Quantenzahl m spielt eine Rolle.

18. Mehr Interesse beanspruchen die Verhältnisse in schwachen Feldern[1]), wo wir noch die Störungstheorie von Schrödinger anwenden können. Das Störungspotential ist

$$\sum_{i=1}^{n} \mathfrak{E} e z_i.$$

Ist keine zufällige Entartung vorhanden, so wird sich der Starkeffekt immer als quadratisch erweisen; wir müssen also, um zu einer Aufspaltung zu gelangen, bis zur zweiten Näherung vorschreiten.

Als linear unabhängige unter den $2l+1$ Eigenfunktionen mit der azimutalen Quantenzahl l nehmen wir diejenigen heraus, deren Darstellungseigenschaft sich in der Form (11) schreiben läßt: sie haben die Gestalt (13a). Die Eigenfunktionen numerieren wir so, daß wir die azimutale Quantenzahl und die laufende Nummer des Terms mit dieser azimutalen Quantenzahl als oberen Index, die Nummer der Eigenfunktion, die zu diesem Term gehört, als unteren Index hinzufügen. Die oberen Indizes sind also bei allen Eigenfunktionen, die zu demselben Term gehören, dieselben, während der untere Index von $-l$ bis $+l$ läuft.

Für die Störungstheorie brauchen wir Integrale der Gestalt

$$(l_1 s_1 m_1; l_2 s_2 m_2)_1 = \int z_1 \Psi_{m_1}^{l_1 s_1} \overline{\Psi_{m_2}^{l_2 s_2}} = \int r_1 \cos\beta \, \Psi_{m_1}^{l_1 s_1} \overline{\Psi_{m_2}^{l_2 s_2}} \tag{15}$$

(da das i-te Elektron von dem ersten in keiner Weise bevorzugt ist, rechnen wir immer mit dem ersten).

---

[1]) Vgl. auch A. Unsöld, Beiträge zur Quantenmechanik der Atome. Ann. d. Phys. **82**, 355, 1927, § 10.

Zur Berechnung setzen wir (13a) in (15) ein:

$$(l_1 s_1 m_1;\ l_2 s_2 m_2)_1 = \sum_{\lambda_1, \lambda_2} \int D^{l_1}_{m_1 \lambda_1} \chi^{l_1 s_1}_{\lambda_1}(g) \overline{D^{l_2}_{m_2 \lambda_2} \chi^{l_2 s_2}_{\lambda_2}}(g)\, r_1 \cos\beta \qquad (15\text{a})$$

und beachten (12), wodurch wir $D^{l_2}_{m_2 \lambda_2}(\alpha, \beta, \gamma) \cos\beta$ durch $D^{l_2+1}_{m_2 \lambda_2}(\alpha, \beta, \gamma)$ und $D^{l_2-1}_{m_2 \lambda_2}(\alpha, \beta, \gamma)$ ausdrücken. Dann führen wir die Integration über α, β und γ (d. h. über alle Lagen im Raume bei gleicher gegenseitiger Konfiguration der Teilchen) mit Hilfe von Ib und IIb aus, indem wir

$$\int D^{l_1}_{m_1 \lambda_1}(\alpha, \beta, \gamma) D^{k}_{m_2 \lambda_2}(\alpha, \beta, \gamma) = c_{l_1} \delta_{l_1 k} \delta_{m_1 m_2} \delta_{\lambda_1 \lambda_2}$$

setzen und erhalten für (15a)

$$\sum_{\lambda_1, \lambda_2} \{u c_{l_1} \delta_{l_1, l_2+1} \delta_{m_1 m_2} \delta_{\lambda_1 \lambda_2} + v c_{l_1} \delta_{l_1, l_2-1} \delta_{m_1 m_2} \delta_{\lambda_1 \lambda_2}\} \chi^{l_1 s_1}_{\lambda_1}(g) \overline{\chi^{l_2 s_2}_{\lambda_2}}(g)\, r_1\, dg,$$

$$u = \frac{\sqrt{(l_2+1)^2 - m_2^2}\sqrt{(l_2+1)^2 - \lambda_2^2}}{(2 l_2 + 1)(l_2 + 1)}, \quad v = \frac{\sqrt{l_2^2 - m_2^2}\sqrt{l_2^2 - \lambda_2^2}}{(2 l_2 + 1) l_2}.$$

Dies ist für $l_1 = l_2 + 1$, $m_1 = m_2$

$$(l_2 + 1, s_1, m_2;\ l_2 s_2 m_2)_1 = \sum_{\lambda_2} u\, c_{l_2+1} \int \chi^{l_2+1, s_1}_{\lambda_2}(g) \overline{\chi^{l_2 s_2}_{\lambda_2}(g)}\, r_1\, dg.$$

für $l_1 = l_2 - 1$, $m_1 = m_2$

$$(l_2 - 1, s_1 m_2;\ l_2 s_2 m_2)_1 = \sum_{\lambda_2} v\, c_{l_2-1} \int \chi^{l_2-1, s_1}_{\lambda_2}(g) \overline{\chi^{l_2 s_2}_{\lambda_2}(g)}\, r_1\, dg.$$

Wir können noch die Abhängigkeit von m — auf die es uns allein ankommt — besser zum Ausdruck bringen, indem wir für $l_1 = l_2 + 1$, $m_2 = m_1$

$$(l_1 s_1 m_1;\ l_1 - 1, s_2, m_1)_1 = \sqrt{l_1^2 - m_1^2}\, C_{l_1 s_1 s_2} \qquad (16\text{a})$$

für $l_1 = l_2 - 1$, $m_2 = m_1$

$$(l_1 s_1 m_1;\ l_1 + 1, s_2, m_1)_1 = \sqrt{(l_1+1)^2 - m_1^2}\, C'_{l_1 s_1 s_2} \qquad (16\text{b})$$

für $m_1 \neq m_2$ und für $|l_1 - l_2| \neq 1$

$$(l_1 s_1 m_1;\ l_2 s_2 m_2)_1 = 0, \quad m_1 \neq m_2 \qquad (16\text{c})$$

$$(l_1 s_1 m_1;\ l_2 s_2 m_1)_1 = 0, \quad |l_1 - l_2| \neq 1. \qquad (16\text{d})$$

Ist der eine Term normal, der andere gespiegelt, so gilt offenbar auf alle Fälle

$$(l_1 s_1 m_1;\ l_2, s_2, m_2)_1 = 0. \qquad (16\text{e})$$

In diesen Formeln hängt $C_{l_1 s_1 s_2}$ und $C'_{l_1 s_1 s_2}$ nicht mehr von m_1 ab. Da diese Formeln (16a), (16b), (16c), (16d) und (16e) nicht nur für das erste, sondern für alle Elektronen gelten, gelten sie auch für die Summe $\sum_i (l_1 s_1 m_1;\ l_2 s_2 m_2)_i$, nur daß die Zahlen $C_{l_1 s_1 s_2}$ und $C'_{l_1 s_1 s_2}$ eine etwas andere Bedeutung haben.

Jetzt können wir die Schrödingersche Störungstheorie anwenden. Für den linearen Effekt kommen hierbei — solange keine zufällige Entartung auftritt — nur Ausdrücke der Art

$$(l_1 s_1 m_1; l_1 s_2 m_2)$$

in Frage. Diese verschwinden aber wegen (16 d). Es ist nur ein quadratischer Effekt zu erwarten.

Die Existenz eines linearen Effekts hängt also (z. B. bei dem Wasserstoffatom) mit einer zufälligen Entartung zusammen. Auch in der älteren Theorie war dies ähnlich der Fall.

Wir können noch das quadratische Glied berechnen. Die Verschiebung des Eigenwerts, der zu der Eigenfunktion $\Psi_{m_1}^{l_1 s_1}$ gehört, ist

$$\Delta E_{l_1 s_1 m_1} = \mathfrak{E}^2 e^2 \sum_{l_2 s_2 m_2} \frac{(l_1 s_1 m_1; l_2 s_2 m_2)^2}{E_{l_1 s_1} - E_{l_2 s_2}}$$

oder wegen (16 c), (16 d) und (16 a), (16 b)

$$\Delta E_{l_1 s_1 m_1} = \mathfrak{E}^2 \{(l_1^2 - m_1^2) A' + ((l_1 + 1)^2 - m_1^2) B'\}$$

wo A' und B' nicht mehr von m abhängen. Anders geschrieben

$$\Delta E_{l s m} = \mathfrak{E}^2 (A_{l s} - B_{l s} m^2). \tag{17}$$

Die Zusatzenergie ist für die m-te und $-m$-te Eigenfunktion dieselbe, was ja auch natürlich ist, da diese Entartung selbst bei einem starken Feld (Punkt 16) nicht aufgehoben wird. Die absolute Größe der Verschiebung und Aufspaltung (A und B) läßt sich nicht allgemein angeben, jedoch gibt unsere Formel (17) das Bild der Aufspaltung an.

Dieselbe Abhängigkeit von m findet sich auch bei R. Becker[1]), dessen Theorie auch eine sehr gute Abschätzung von A und B ermöglicht. Auf Grund der Quantenmechanik hat seine Rechnung A. Unsöld[2]) wiederholt, der im wesentlichen zu derselben Formel gelangt.

Unsere Formel (17) muß — wenigstens für Singulettsysteme — etwa ebenso genau gelten wie z. B. die Formeln für die Aufspaltung bei dem Zeemaneffekt. Leider ist sie am experimentellen Material vorläufig nicht zu prüfen, da man bisher eine wirkliche Aufspaltung (außer natürlich bei H) kaum erreicht hat.

In der Fig. 2 ist das Bild eines solchen Effekts schematisch aufgezeichnet. Bei den vier aufgezeichneten Termen (ein S-, P-, D- und B-Term) sind die Linien mit $m = 0$ untereinander gezeichnet, und die $B_{l s}$ sind bei allen vieren gleich groß angenommen, was in Wirklichkeit

[1]) ZS. f. Phys. **9**, 332, 1922.

[2]) l. c.

natürlich nicht der Fall sein wird. Die gestrichelten Linien bedeuten Übergänge, die Zahlen auf ihnen sind Intensitäten.

19. Ich möchte noch kurz auf die Verhältnisse im magnetischen Felde eingehen. Zu diesem Zwecke muß man wieder die Überlegungen des Punktes 8 heranziehen.

Ein Term habe die azimutale Quantenzahl l und gehöre zu einem Multiplettsystem M. Bis kein Feld da ist, ist die Substitutionsgruppe der Differentialgleichung (vgl. Punkt 11) das direkte Produkt der dreidimensionalen Drehspiegelungsgruppe IIIa mit der symmetrischen Gruppe I. Die Darstellung unseres Termes entsteht, indem wir die Darstellung, die zum Multiplettsystem M gehört, mit der $2l+1$ dimensionalen Darstellung der Drehgruppe kombinieren. Ich nehme die letztere Darstellung in der Form (11) an und das entstehende magnetische Feld in der Richtung der Z-Achse. Unsere Darstellung ist dann als Darstellung des direkten Produkts der symmetrischen Gruppe I mit der Gruppe IIIc bereits ausreduziert. Sie hat $2l+1$ irreduzible Bestandteile. Diese sind zusammengesetzt aus derjenigen Darstellung der Gruppe I, die zu unserem Multiplettsystem gehört, mit einer der folgenden Darstellungen von IIIc

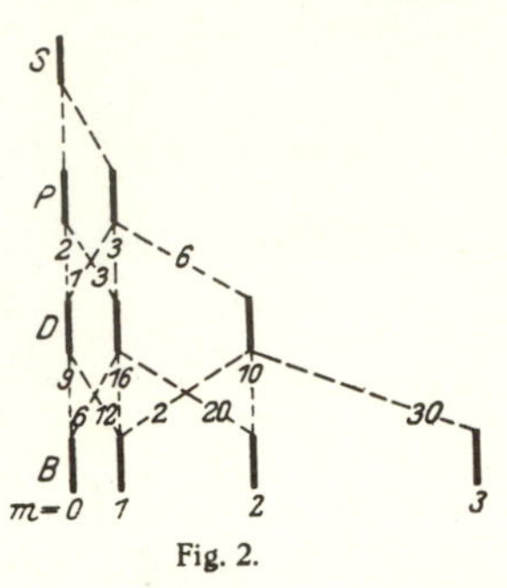

Fig. 2.

$$(e^{-il\alpha}),\ (e^{-i(l-1)\alpha})\ \ldots\ (e^{-i\alpha}),\ (1),\ (e^{i\alpha})\ \ldots\ (e^{il\alpha})$$

für eine Drehung. Diese $2l+1$ Darstellungen sind die Darstellungseigenschaften der entstandenen $2l+1$ Terme.

Dieses Resultat ist von der speziellen Gestalt der Schrödingerschen Gleichung sowie von der Darstellungseigenschaft der Gruppe I, die zu unserem Multiplettsystem M gehört, wesentlich unabhängig. Trotzdem zeigt es sich, daß die Aufspaltung immer in eine ungerade Anzahl von Komponenten eintritt, was bekanntlich der Erfahrung bei Dublett-, Quartett- usw. Systemen widerspricht. Diese Schwierigkeit tritt natürlich nicht nur an dieser Stelle auf; sie ist aber hier am krassesten zu formulieren. Es scheint mir, daß hier zur Erklärung der Verhältnisse noch ein wesentlicher Punkt fehlt.

20. Wir könnten nun — ebenso wie bei dem Starkeffekt — durch Berücksichtigung der speziellen Gestalt der Schrödingerschen Gleichung (in der allerdings das rotierende Elektron auf keine Weise

berücksichtigt ist) eine Aufspaltungsformel erhalten. Es würde sich[1]) — wie zu erwarten — die für Singulettsysteme gültige Formel des normalen Zeemaneffekts ergeben (Aufspaltungsfaktor $g = 1$). Ebenso könnten wir die Auswahlregel im beliebig starken Feld berechnen, was allerdings ein etwas komplizierteres Resultat ergibt. Für schwache Felder stehen die Integrale für die Intensitäten der π-Komponenten fertig in (16a) oder (16b) da, die Intensitäten der σ-Komponenten ergeben sich ebenso einfach durch Anwendung der Formel (12a) und (12b). Wenn wir die magnetische Quantenzahl mit m bezeichnen, so gilt für einen Übergang

$$\left.\begin{array}{lll} & & \text{Relative Intensitäten} \\ l,\, m \rightarrow l+1.\, m & (\pi\text{-Komponente}) & (l+1)^2 - m^2 \\ l,\, m \rightarrow l+1,\, m+1 & (\sigma\text{-Komponente}) & \frac{1}{2}(l+1+m)(l+2+m) \\ l,\, m \rightarrow l+1,\, m-1 & (\sigma\text{-Komponente}) & \frac{1}{2}(l+1-m)(l+2-m) \end{array}\right\} \quad (18)$$

Auch diese Formeln gelten für Singulettsysteme.

21. Als vorletzten Punkt unserer Überlegungen behandeln wir das sukzessive Einfangen von Elektronen von einem Ion. Wir werden hierbei von dem im wesentlichen von Heisenberg (l. c.) gegebenen Schema der Zuordnung der Darstellungen zu Multiplettsystemen ausgehen, die in Punkt 12 beschrieben ist.

Unsere Aufgabe besteht eigentlich aus zwei Teilen. Im ersten Teil[2]) gehen wir von einem „ungestörten" System aus, in dem keine Wechselwirkung zwischen den Elektronen vorhanden ist und diese nur unter der Einwirkung des Kerns stehen. Dann sind die Eigenfunktionen Produkte von Funktionen der Koordinaten nur je eines Elektrons, oder lineare Kombinationen solcher Produkte. Es läßt sich dann jedem Elektron i eine Hauptquantenzahl N_i und eine Azimutalquantenzahl l_i zuordnen. Wir fragen nun, was für Terme (Multiplettsystem und Azimutalquantenzahl l des Atoms) aus diesem Term entstehen, wenn wir die Wechselwirkung zwischen den Elektronen langsam, als Störung angewachsen denken. Wir haben also zu jedem Elektron eine Azimutalquantenzahl zugeordnet und fragen, wie sich diese zu der gesamten azimutalen Quantenzahl des Atoms zusammensetzen. Dieser Teil der Aufgabe läßt sich auch ganz formal erledigen; ich werde indessen hier eine etwas andere Behandlung wählen.

---

[1]) Siehe auch M. Born und P. Jordan, ZS. f. Phys. **34**, 858, 1925, und E. Schrödinger, Ann. d. Phys. **81**, 139, 1926.

[2]) Vgl. F. Hund, Linienspektren usw. §§ 24, 25.

Die zweite Frage[1]) betrifft die Zuordnung der einzelnen Seriengrenzen des Atoms zu Termen des Ions. Es ist mir leider nicht gelungen, diese Aufgabe in dieser Allgemeinheit zu lösen, obwohl es mit den hier entwickelten Methoden durchaus möglich sein muß. Ich verzichte daher auf die Behandlung dieser Frage.

22. Wir müssen hier eine kleine Hilfsbetrachtung einschalten. Es sei ein System von n gleichen Teilchen gegeben, die alle unter der Wirkung einer Kraft stehen (man denke an die n Elektronen des Atoms, wobei die Wechselwirkung vernachlässigt ist). Die Eigenfunktionen dieses Systems sind Produkte von Eigenfunktionen der einzelnen Teilchen, oder Summen solcher Produkte. Insbesondere betrachten wir einen Zustand, in dem d Bahnen doppelt besetzt sind, alle anderen einfach. Zu einem Term gehören dann $\frac{n!}{2^d}$ Eigenfunktionen. Lassen wir nun die Wechselwirkung zwischen den Teilchen eintreten, so spalten die Terme im allgemeinen auf, und es entstehen solche, deren Darstellungen für Substitutionsgruppen der Art I irreduzibel sind. Die Frage ist nun, wieviel Terme einer bestimmten Darstellungseigenschaft aus einem bestimmten, „ungestörten" Term entstehen. Das Problem wurde für den Fall $d = 0$ bereits in einer vorangehenden Mitteilung gelöst[2]), die $N_{\lambda_1 \lambda_2 \ldots \lambda_\varrho}$ geben die Zahlen der Terme an, die zu der Darstellung λ_1 $\lambda_2, \ldots \lambda_\varrho$ gehören ($\lambda_1 + \lambda_2 + \cdots + \lambda_\varrho = n$). Im allgemeinen ist das Problem etwas verwickelter, glücklicherweise interessieren uns aber nur diejenigen Termsysteme (vgl. Punkt 12), bei denen in der partitio nur die Zahlen 1 und 2 vorkommen. Die Anzahl der Zweier bezeichnen wir mit z, die Anzahl der Einser ist dann $n - 2z$, die Multiplizität des Terms $n - 2z + 1$.

Das Resultat der Rechnung, die ich nicht anführen möchte (sie ist mit der für $d = 0$ vollkommen analog, nur erfordert sie etwas mehr Rechenarbeit), ist folgendes: Aus einem ungestörten Term mit d doppelt besetzten Bahnen entstehen

$$\binom{n-2d}{z-d} - \binom{n-2d}{z-d-1} \tag{19}$$

Terme der Multiplizität $n - 2z + 1$.

---

[1]) Vgl. F. Hund, Linienspektren usw. § 39.
[2]) E. Wigner, ZS. f. Phys. **40**, 883, 1927.

Diese Formel gilt natürlich auch für $d = 0$ und ist in diesem Falle – wie man sich leicht überzeugt — mit der Formel für $N_{\lambda_1 \lambda_2 \ldots \lambda_\varrho}$ identisch, wenn nämlich alle λ gleich 1 oder 2 sind.

23. Wir betrachten nunmehr ein „ungestörtes" Atom mit n Elektronen mit den Hauptquantenzahlen $N_1, N_2, \ldots, N_n$ und den Azimutalquantenzahlen $l_1, l_2, \ldots, l_n$. Die direkte Berechnung der Aufspaltung der Eigenwerte ist zwar durchaus möglich, doch kommt man einfacher zum Ziele, wenn man sich ein schwaches magnetisches Feld angelegt denkt und so den $2l_i + 1$fachen Term des $i$ten Elektrons in $2l_i + 1$ Terme zerlegt. Dann ist unsere Aufgabe einfach die: Ein Elektron ist auf einer der $2l_1 + 1$ Bahnen N_1, l_1, m_1 (m_1 läuft von $-l_1$ bis $+l_1$), ein zweites auf einer der $2l_2 + 1$ Bahnen N_2, l_2, m_2 usw. und schließlich ein Elektron auf einer der $2l_n + 1$ Bahnen N_n, l_n, m_n (m_n läuft von $-l_n$ bis $+l_n$). Es ist nun zu beachten, daß die magnetische Quantenzahl m des Atoms aufgebaut aus Elektronen mit den magnetischen Quantenzahlen $m_1, m_2, \ldots, m_n$ gleich $m_1 + m_2 + \cdots m_n$ ist. Die Eigenfunktion des Atoms ist nämlich das Produkt der Eigenfunktionen der Elektronen, diese multiplizieren sich bei einer Drehung mit dem Winkel α um die magnetische Feldachse mit $e^{im_1\alpha}, e^{im_2\alpha}, \ldots, e^{im_n\alpha}$, die Eigenfunktion des Atoms also mit $e^{i(m_1 + m_2 + \cdots m_n)\alpha}$.

Hiernach können wir alle etwa $(2l_1 + 1)(2l_2 + 1) \ldots (2l_n + 1)$ Zustände separat betrachten, die durch spezielle Wahl der $m_1, m_2, \ldots, m_n$ entstehen und auf alle separat das Ergebnis des vorangehenden Punktes anwenden. Dieses genügt offenbar, da Fälle, wo mehr als zwei Elektronen dieselben drei Zahlen N, l, m haben, für uns überhaupt ausscheiden (Punkt 12).

24. Ich möchte dies an einem Beispiel illustrieren. Es seien ein $1p$- und zwei $2p$-Elektronen vorhanden ($N_1 = 1$, $N_2 = N_3 = 2$, $l_1 = l_2 = l_3 = 1$). Die erste Spalte in Tabelle 1 repräsentiert durch die drei Striche ||| die drei Bahnen $N = 1$, $l = 1$, $m = -1, 0, +1$, in der zweiten Spalte sind die drei Striche die drei Zustände $N = 2$, $l = 1$, $m = -1, 0, +1$. Die Punkte auf den Strichen bedeuten Elektronen, die in diesem Zustand sind, d die Anzahl der doppelt besetzten Bahnen, m sind magnetische Quantenzahlen des Atoms, also $m_1 + m_2 + m_3$. Darunter stehen die Ausdrücke

$$\binom{n-2d}{z-d} - \binom{n-2d}{z-d-1}, \tag{19}$$

Tabelle 1.

	$N=1,\ l=1$			$N=2,\ l=1$			d	$z=0$, Quartett					$z=1$, Dublett						
$m=$	-1	0	1	-1	0	1		-2	-1	0	1	2	-3	-2	-1	0	1	2	3
				••	|	|	1						1						
				•	•	|	0	1						2					
				•	|	•	0		1						2				
	•	|	|	|	••	|	1								1				
				|	•	•	0			1						2			
				|	|	••	1										1		
				••	|	|	1							1	1				
				•	•	|	0		1	1					2	2			
	|	•	|	•	|	•	0			1	1					2	2		
		und		|	••	|	1									1	1		
	|	|	•	|	•	•	0				1	1					2	2	
				|	|	••	1												1

die angeben, wieviel solche Terme entstehen. Es entsteht also ein Quartetterm mit $m = -2$, zwei mit $m = -1$, drei mit $m = 0$, zwei mit $m = 1$ und einer mit $m = 2$. Diese kann man nur in ein ^4D, ^4P, ^4S zusammenfassen. Ähnlich erhalten wir für die Dubletterme ^2B, ^2D, ^2D, ^2P, ^2P, ^2P, ^2S. Es sind dies offenbar durchweg dieselben Terme, die nach dem vorhandenen empirischen Material erwartet werden müssen. (Vgl. z. B. F. Hund, l. c. § 24, 25, wo hierfür viele Beispiele gegeben sind.)

Auf die empirische Bedeutung dieses Aufbauprinzipes hinzuweisen, ist kaum notwendig, kann man doch bekanntlich mit seiner Hilfe einen großen Teil des Baues der Serienspektren sowie des periodischen Systems erklären.

25. Wir untersuchen noch die Frage, welche Terme normal und welche gespiegelt sind. Da die Eigenfunktionen eines Elektrons immer normal sind (Punkt 14), so geht die Eigenfunktion des $i$ten Elektrons bei der Spiegelung durch den Koordinatenursprungspunkt in den $(-1)^{l_i}$fachen Wert über. Die ganze Eigenfunktion des Atoms geht in den $(-1)^{l_1 + l_2 + \cdots l_n}$fachen Wert über. Ist also l die azimutale Quantenzahl des ganzen Atoms, so ist der zugehörige Term normal oder gespiegelt, je nachdem $(-1)^{l_1 + l_2 + \cdots l_n - l}$ gleich $+1$ oder -1 ist. Nun kommen Übergänge nur von normalen in normale, von gespiegelten in gespiegelte Terme vor. $l_1 + l_2 + \cdots l_n - l$ kann sich also nur um eine gerade Zahl ändern. Da sich l nur um $+1$ oder -1 ändert, darf sich $l_1 + l_2 + \cdots l_n$ nur um eine ungerade Zahl ändern. Dies ist der Inhalt

der Laporteschen Regel. Es ist merkwürdig, wie wenig anschaulich diese Deutung ist, so daß man die in der älteren Quantentheorie von W. Heisenberg[1]) gegebene Deutung kaum wiedererkennen kann.

26. Zum Schluß möchte ich noch darauf hinweisen, daß es leicht ist — nicht nur die Kugelsymmetrie aller S-Terme [die ja von unserem Standpunkte aus trivial ist][2]) — zu beweisen, sondern das Verschwinden der Dipolmomente aller stationären Zustände zu zeigen. Auch die Quadrupolmomente lassen sich berechnen. Ich möchte aber hierauf nicht eingehen, da die vorliegende Note ohnehin schon allzu lang geworden ist. Die Bedeutung dieser Beziehungen besteht hauptsächlich in der zuerst von A. Unsöld[3]) bemerkten Tatsache, daß man bei jeder Störungsrechnung in erster Näherung so rechnen kann, wie wenn die elektrische Ladung nach der Schrödingerschen Formel über den ganzen Raum „verschmiert" wäre und nach den Regeln der klassischen Elektrostatik wirken würde.

27. Zweck der vorliegenden Arbeit war, zu zeigen, daß man durch ganz einfache Symmetriebetrachtungen über die Schrödingersche Differentialgleichung schon einen wesentlichen Teil der rein qualitativen spektroskopischen Erfahrung erklären kann.

Wie die in Punkt 19 gestreifte Schwierigkeit zu beheben ist, vermag ich nicht zu sagen. Es erscheint einem zunächst recht schwierig, mit einer Differentialgleichung mit eindeutigem ψ eine Aufspaltung in eine gerade Anzahl von Komponenten im magnetischen Felde herauszuholen. Auch zeigen sich noch einige andere Schwierigkeiten in solchen Fällen, wo die rotierenden Elektronen eine Rolle spielen.

Berlin, Institut für theoretische Physik der Techn. Hochschule.

---

[1]) ZS. f. Phys. **32**, 841, 1925.

[2]) Zu diesem Ergebnis kommt auch A. Unsöld (l. c.) für S-Terme wasserstoffähnlicher Spektren und für abgeschlossene Schalen (die natürlich in unserem Sinne auch S-Terme sind), indem er sich für die Wechselwirkung der Elektronen auf die erste bzw. zweite Näherung beschränkt. Vgl. auch A. Sommerfeld, Phys. ZS. **28**, 231, 1927.

[3]) l. c.

Berichtigung zu der Arbeit: Einige Folgerungen aus der Schrödingerschen Theorie für die Termstrukturen.

Von **E. Wigner** in Berlin.

(Eingegangen am 8. September 1927.)

1. Bei der Ableitung der Rekursionsformeln (12), (12a), (12b) in zitierter Arbeit[1] ist mir ein Fehler unterlaufen. Genannte Formeln lauten richtigerweise:

$$D^l_{jk}\cos\beta = u_j\, u_k D^{l-1}_{j,k} + v_j\, v_k D^l_{j,k} + w_j\, w_k D^{l+1}_{j,k},$$
$$D^l_{jk}\sin\beta\, e^{-i\alpha} = u^-_j\, u_k D^{l-1}_{j-1,k} + v^-_j\, v_k D^l_{j-1,k} + w^-_j\, w_k D^{l+1}_{j-1,k},$$
$$D^l_{jk}\sin\beta\, e^{i\alpha} = u^+_j\, u_k D^{l-1}_{j+1,k} + v^+_j\, v_k D^l_{j+1,k} + w^+_j\, w_k D^{l+1}_{j+1,k}$$

mit

$$u_\lambda = \frac{\sqrt{l^2-\lambda^2}}{\sqrt{l}\sqrt{2l+1}}, \quad v_\lambda = \frac{\lambda}{\sqrt{l}\sqrt{l+1}}, \quad w_\lambda = \frac{\sqrt{(l+1)^2-\lambda^2}}{\sqrt{l+1}\sqrt{2l+1}},$$
$$u^-_\lambda = -\frac{\sqrt{l+\lambda}\sqrt{l+\lambda-1}}{\sqrt{l}\sqrt{2l+1}}, \quad v^-_\lambda = \frac{\sqrt{l-\lambda+1}\sqrt{l+\lambda}}{\sqrt{l}\sqrt{l+1}}, \quad w^-_\lambda = \frac{\sqrt{l-\lambda+1}\sqrt{l-\lambda+2}}{\sqrt{l+1}\sqrt{2l+1}},$$
$$u^+_\lambda = \frac{\sqrt{l-\lambda-1}\sqrt{l-\lambda}}{\sqrt{l}\sqrt{2l+1}}, \quad v^+_\lambda = \frac{\sqrt{l+\lambda+1}\sqrt{l-\lambda}}{\sqrt{l}\sqrt{l+1}}, \quad w^+ = -\frac{\sqrt{l+\lambda+1}\sqrt{l+\lambda+2}}{\sqrt{l+1}\sqrt{2l+1}}.$$

L. c. fehlten die Glieder mit den v. Man leitet diese Formeln ab, indem man die Darstellung $D^l_{jk}\, D^1_{\iota\varkappa}$ ausreduziert.

2. Hierdurch ändert sich Punkt 15. Außer den dort erwähnten Übergängen sind nämlich noch Übergänge mit $\Delta l = 0$ zwischen normalen und gespiegelten Termen erlaubt.

Die Auswahlregel für die azimutale Quantenzahl lautet also:

Ist ein Zustand aufgebaut aus einem l_1-, einem l_2- usw. und einem l_n-Elektron und hat er dabei die azimutale Quantenzahl l, so ist er normal oder gespiegelt, je nachdem $l_1 + l_2 + \cdots l_n - l$ eine gerade oder ungerade Zahl ist[2].

Es kommen Übergänge vor: mit $\Delta l = \pm 1$ von normalen in normale, von gespiegelten in gespiegelte Terme; mit $\Delta l = 0$, wenn $l \neq 0$ ist,

[1] ZS. f. Phys. **43**, 624, 1927.
[2] l. c. Punkt 25.

von normalen in gespiegelte, gespiegelten in normale Terme[1]. Alle anderen Übergänge sind verboten.

Man beweist dies ganz nach Punkt 15. Diese Auswahlregel (wie auch das Verbot von Interkombination zwischen verschiedenen Multiplettsystemen) gilt zunächst nur — wie l. c. auseinandergesetzt — unter Vernachlässigung der Elektronenrotation, d. h. bei kleiner Multiplettaufspaltung. Die Elemente niedriger Ordnungszahl (z. B. Stickstoff) fügen sich diesen Verboten noch gut, während sie bei den höheren Elementen durchbrochen werden.

3. Auch in Punkt 18 muß man noch einige Glieder in Betracht ziehen, wodurch jedoch das Resultat nicht geändert wird.

4. In Punkt 20 kommen zu den Formeln (18) noch folgende hinzu:

	Relative Intensität
$l, m \rightarrow l, m$ (π-Komponente)	m^2
$l, m \rightarrow l, m+1$ (σ-Komponente) . .	$\frac{1}{2}(l+1+m)(l-m)$

Auch diese Formeln stehen bei Born, Heisenberg und Jordan[2]. Es ist zu beachten, daß solche Übergänge nur von normalen in gespiegelte Terme oder umgekehrt vorkommen.

Alles andere bleibt ungeändert.

Berlin, Inst. f. theoret. Phys. d. Techn. Hochschule.

[1] Diese Auswahlregel ist empirisch zuerst von O. Laporte (ZS. f. Phys. **23**, 135, 1924) und H. N. Russell (Science **59**, 512, 1924) mit Bezug auf „ungestrichene" und „gestrichene" Terme gefunden worden, welche Einteilung nach W. Heisenberg (ZS. f. Phys. **32**, 841, 1925) mit der Einteilung in „normale" und „gespiegelte" Terme identisch ist. Hierauf machte mich freundlichst Herr W. Pauli jr. aufmerksam.

[2] ZS. f. Phys. **35**, 557, 1926.

Zur Hyperfeinstruktur von Li^+.

Teil II.

Von **P. Güttinger** und **W. Pauli** in Zürich.

Mathematischer Anhang.

I. Die relativen Intensitäten der Komponenten eines Multipletts.

Eine Herleitung der bekannten Formeln für die Multiplettintensitäten aus der Quantenmechanik findet sich in der Literatur nur bei Dirac* angegeben. Seine Methode beruht auf der Einführung von Wirkungs- und Winkelvariablen als q-Zahlen, was uns jedoch eine unnötige Komplikation zu sein scheint. Ferner hat uns Herr Kramers freundlicherweise eine andere, bisher noch nicht veröffentlichte wellenmechanische Methode zur Herleitung dieser Formeln brieflich mitgeteilt**. Da diese aber die genaue Kenntnis der Transformationseigenschaften der Spinvektoren bei räumlichen Drehungen voraussetzt, schien uns die Durchführung und Mitteilung einer elementaren, matrizenmethodischen Ableitung der in Rede stehenden Formeln nicht überflüssig zu sein, zumal für die Rechnungen des Textes die Kenntnis der absoluten Beträge der Matrixelemente nicht genügt, sondern auch die ihrer relativen Vorzeichen erforderlich ist.

* P. A. M. Dirac, Proc. Roy. Soc. London (A) **111**, 281, 1926.
** Inzwischen erschienen. Proc. Amst. Akad. **23**. 953, 1930.

Für die Intensitäten der Zeemankomponenten (Abhängigkeit der Matrixelemente von der Quantenzahl m) liegt eine solche Herleitung bereits vor* und wir haben daher ihre Resultate verwendet.

§ 1. *Formulierung des Problems. Abspaltung der Quantenzahl m.* Wir haben zwei Drehimpulsvektoren, das Spinmoment $\mathfrak{s}$ und das Bahnmoment $\mathfrak{l}$, beide in der Einheit $h/2\pi$ gemessen, die sich zu einem Vektor $\mathfrak{j}$ zusammensetzen gemäß

$$\mathfrak{j} = \mathfrak{s} + \mathfrak{l}. \tag{1}$$

Es gelten die Vertauschungsrelationen (V.-R.)

$$\left.\begin{aligned} [j_x, A_x] &= 0, \\ [j_x, A_y] &= [A_x, j_y] = iA_z, \end{aligned}\right\} \tag{2}$$

und

$$[j_y, B] = 0, \tag{3}$$

wenn A_x, A_y, A_z die Komponenten eines *beliebigen Vektors* sind und B ein beliebiger Skalar ist, wobei diese sonst irgendwelche Funktionen der Spinmomente, Orts- und Impulskoordinaten der Elektronen des Atoms sind**. Insbesondere kann für $\mathfrak{A}$ der Koordinatenvektor $\mathfrak{r}$ eines Elektrons oder sein Impulsvektor $\mathfrak{p}$, oder auch der Vektor des gesamten elektrischen Moments oder $\mathfrak{s}$, $\mathfrak{l}$ oder $\mathfrak{j}$ selbst genommen werden. Im letzteren Falle erhalten wir durch Spezialisierung

$$[j_x, j_y] = ij_z. \tag{1'}$$

Wir interessieren uns hier speziell für den Fall, daß j_z und j^2 Diagonalmatrizen sind. Bekanntlich gehören dann zu dem Eigenwert $j(j+1)$ von j^2 die Eigenwerte $-j \leqq m \leqq +j$ von j_z. Weiter folgt aus (3) für $B = j^2$

$$[j^2, j_x] = 0, \ldots, \tag{3'}$$

so daß j_x, j_y, j_z diagonal in bezug auf den Betrag j sind. Für den beliebigen Vektor $\mathfrak{A}$ folgt nach Dirac*** aus (1)

$$\begin{aligned} \left[j^2 [j^2, \mathfrak{A}]\right] &\equiv j^4 \mathfrak{A} - 2 j^2 \mathfrak{A} j^2 + \mathfrak{A} j^4 \\ &= 2(j^2 \mathfrak{A} + \mathfrak{A} j^2) - 4(\mathfrak{A}\mathfrak{j})\mathfrak{j}. \end{aligned} \tag{4}$$

---

* Vgl. M. Born und P. Jordan, Elementare Quantenmechanik, Berlin 1930, 4. Kap., § 25—29.

** Nur dürfen diese Funktionen keine c-Zahlvektoren wie äußere Feldstärken usw. explizit enthalten.

*** P. A. M. Dirac, Die Prinzipien der Quantentheorie, Leipzig 1930, insbesondere § 47. — Es ist zu beachten, daß wir hier mit $[a, b]$ den Ausdruck $ab - ba$ bezeichnet haben, während Dirac $[a, b] = i(ab - ba)$ definiert. Ferner setzt Dirac $k^2 = j^2 + {}^1/_4$.

Betrachtet man hierin zunächst die in bezug auf j diagonalen Matrixelemente, so gibt das Klammerprodukt auf der linken Seite hierzu keinen Beitrag und es folgt, da $j(j+1)$ der Eigenwert von $\mathfrak{j}^2$ ist,

$$j(j+1)\mathfrak{A}_{j,m'}^{j,m} = (\mathfrak{A}\mathfrak{j})_{j,m}^{j,m} \cdot \mathfrak{j}_{j,m'}^{j,m}. \tag{5}$$

Das skalare Produkt $(\mathfrak{A}\mathfrak{j})$ ist nämlich als Skalar gemäß (3) mit allen Komponenten von $\mathfrak{j}$ vertauschbar, also diagonal in bezug auf j und m. Betrachten wir andererseits ein Nichtdiagonalelement $(j' \neq j'')$ in bezug auf j, so gibt also der letzte Term der rechten Seite von (4) keinen Beitrag und man erhält

$$[j'^2(j'+1)^2 - 2j'(j'+1)j''(j''+1) + j''^2(j''+1)^2] \cdot \mathfrak{A}_{j'',m''}^{j',m'}$$
$$= 2[j'(j'+1) + j''(j''+1)] \cdot \mathfrak{A}_{j'',m''}^{j',m}.$$

Die Klammer auf der linken Seite ist

$$[j'(j'+1) - j''(j''+1)]^2 = (j'-j'')^2(j'+j''+1)^2,$$

während

$$2[j'(j'+1) + j''(j''+1)] = (j'+j''+1)^2 + (j'-j'')^2 - 1,$$

also folgt

$$[(j'+j''+1)^2 - 1][(j'-j'')^2 - 1]\,\mathfrak{A}_{j'',m''}^{j',m'} = 0.$$

Das Matrixelement verschwindet also für alle $\mathfrak{A}$, außer wenn einer der Faktoren verschwindet. Da die j', j'' beide positiv oder mindestens Null sind und da $j' \neq j''$, also nicht beide Null sind, kann der erste Faktor nicht verschwinden, also muß es die zweite Klammer, was für $j' - j'' = \pm 1$ der Fall ist. Wir haben also für jeden Vektor $\mathfrak{A}$ die Auswahlregel

$$j' - j'' = \pm 1 \text{ oder } 0, \tag{6}$$

wie es auf diesem Wege Dirac gezeigt hat. Insbesondere gilt dies für die Koordinatenmatrizen sowie für $\mathfrak{s}$ und $\mathfrak{l}$.

Die Auflösung der Gleichungen (2) hinsichtlich der Abhängigkeit der Matrizen von m hat ergeben:

$$\left.\begin{aligned}
(A_x \pm iA_y)_{j,m\mp1}^{j,m} &= A_j^j \cdot \sqrt{(j\pm m)(j\mp m+1)},\\
A_{z\,j,m}^{\,j,m} &= A_j^j \cdot m,\\
(A_x \pm iA_y)_{j+1,m\mp1}^{j,m} &= A_{j+1}^j \cdot \pm\sqrt{(j\mp m+1)(j\mp m+2)},\\
A_{z\,j+1,m}^{\,j,m} &= A_{j+1}^j \cdot \sqrt{(j+1)^2 - m^2},\\
(A_x \pm iA_y)_{j-1,m\mp1}^{j,m} &= A_{j-1}^j \cdot \mp\sqrt{(j\pm m)(j\pm m-1)},\\
A_{z\,j-1,m}^{\,j,m} &= A_{j-1}^j \cdot \sqrt{j^2 - m^2},
\end{aligned}\right\} \tag{7}$$

worin die $A_{j'}^j$ von m unabhängig sind. Speziell für j_x, j_y, j_z ist $A_j^j = 1$ und $A_{j\pm1}^j = 0$ zu setzen.

Außer der Anwendung der V.-R. (2) auf $\mathfrak{s}$, $\mathfrak{l}$ und die Koordinatenmatrix $\mathfrak{r}$ müssen wir hier noch andere V.-R. in Betracht ziehen, nämlich

$$[s_x, s_y] = i s_z, \ldots, \tag{8}$$

aus der mit Hilfe von (1) und (2) folgt

$$[s_x, \mathfrak{l}] = 0, \ldots, [l_x, l_y] = i l_z \ldots \tag{8'}$$

und

$$[s_x \cdot \mathfrak{r}] = 0 \ldots. \tag{9}$$

aus der nach (1) und (2) folgt

$$[l_x, x] = 0, \ldots, [l_x, y] = [x, l_y] = i l_z \ldots \tag{9'}$$

Ferner haben wir zu benutzen, daß die Eigenwerte von $\mathfrak{s}^2$ und $\mathfrak{l}^2$ bzw. $s(s+1)$ und $l(l+1)$ betragen. Der erstere Betrag s kann in allen folgenden Betrachtungen als c-Zahl angesehen werden, da alle betrachteten Größen mit ihm vertauschbar sind. In bezug auf l ist zwar $\mathfrak{s}$ diagonal, nicht aber $\mathfrak{r}$. Wir nennen $s_{l,j'}^{l,j}$ bzw. $a_{l',j'}^{l,j}$ den aus den Matrixelementen von $\mathfrak{s}$ bzw. $\mathfrak{r}$ gemäß (7) abgespaltenen, von m unabhängigen Faktor. Wir haben bereits gesehen, daß $j' = j+1$, j oder $j-1$ sein muß. Ebenso folgt aus (9'), daß $l' = l+1$, l oder $l-1$ sein muß. Es wird nun unsere Aufgabe sein, die Größen $s_{l,j'}^{l,j}$ und $a_{l',j'}^{l,j}$ in ihrer Abhängigkeit von j, j' zu bestimmen. Wir beginnen mit der erstgenannten Größe.

§ 2. *Bestimmung der Impulsmomentmatrizen.* Der Fall $j' = j''$ erledigt sich sofort gemäß (5), angewandt auf $\mathfrak{A} = \mathfrak{s}$, was

$$j(j+1) s_j^j = (\mathfrak{s}\mathfrak{j})_j^j$$

ergibt. Es ist aber

$$\mathfrak{l}^2 = (\mathfrak{s} - \mathfrak{j})^2 = \mathfrak{s}^2 - 2(\mathfrak{s}\mathfrak{j}) + \mathfrak{j}^2,$$

also

$$(\mathfrak{s}\mathfrak{j}) = \tfrac{1}{2}(\mathfrak{s}^2 - \mathfrak{l}^2 + \mathfrak{j}^2).$$

Die Eigenwerte der auf der rechten Seite stehenden Größen sind aber bekannt und man erhält

$$s_j^j = \frac{s(s+1) - l(l+1) + j(j+1)}{2j(j+1)}. \tag{10}$$

Um $s_{j+1}^j = s_j^{j+1}$ und $s_{j-1}^j = s_j^{j-1}$ zu bestimmen, benutzen wir zunächst die Gleichung*

$$[s_x - i s_y, s_z]_{j, m+1}^{j, m} = (s_x - i s_y)_{j, m+1}^{j, m}$$

* Welche Komponenten von $\mathfrak{s}$ und welche Matrixel mente in bezug auf m genommen werden, ist unwesentlich.

und erhalten schließlich mit Benutzung von (7) und Fortheben der m enthaltenden Faktoren

$$|s^j_{j-1}|^2(2j-1)-|s^j_{j+1}|^2(2j+3)+(s^j_j)^2 = s^j_j.$$

Es ist nach (10)

$$s^j_j-(s^j_j)^2 = s^j_j(1-s^j_j)$$

$$= \frac{[s(s+1)-l(l+1)+j(j+1)]}{[2j(j+1)]^2}[-s(s+1)+l(l+1)+j(j+1)],$$

und mit der Abkürzung

$$A = s(s+1)-l(l+1) = (s-l)(s+l+1) \qquad \text{(a)}$$

$$|s^j_{j-1}|^2(2j-1)-|s^j_{j+1}|^2(2j+3) = \frac{j^2(j+1)^2-A^2}{[2j(j+1)]^2}. \qquad (11)$$

Als zweite Gleichung finden wir aus

$$(\mathbf{s}^2)^{j,\,m}_{j,\,m} = s(s+1)$$

$$|s^j_{j-1}|^2 j(2j-1)+|s^j_{j+1}|^2(j+1)(2j+3)+(s^j_j)^2 j(j+1) = s(s+1)$$

und mit Einsetzen von (10) und Einführung der Abkürzung

$$B = s(s+1)+l(l+1) \qquad \text{(b)}$$

$$|s^j_{j-1}|^2 j(2j-1)+|s^j_{j+1}|^2(j+1)(2j+3) = \frac{-A^2+2Bj(j+1)-j^2(j+1)^2}{4j(j+1)}. \qquad (12)$$

Nun kann man (11) und (12) nach $|s^j_{j-1}|^2$ und $|s^j_{j+1}|^2$ auflösen. Man findet

$$|s^j_{j-1}|^2(2j+1)(2j-1) = \frac{-j^2(j-1)(j+1)^2-A^2(j+1)+2Bj^2(j+1)}{4j^2(j+1)}$$

$$= \frac{-j^2(j^2-1)-A^2+2Bj^2}{4j^2} = \frac{-j^4+(2B+1)j^2-A^2}{4j^2}.$$

Da nach (b)

$$2B+1 = 2(l^2+s^2+l+s)+1 = (l-s)^2+(l+s+1)^2,$$

wird nach (a)

$$|s^j_{j-1}|^2 = \frac{[j^2-(l-s)^2][(l+s+1)^2-j^2]}{4j^2(4j^2-1)}. \qquad (13)$$

Löst man (11) und (12) andererseits nach $|s^j_{j+1}|^2$ auf, so erhält man einen Ausdruck, der aus (13) mittels Ersetzen von j durch $j+1$ hervorgeht, wie es sein muß, wenn die Gleichungen (11) und (12) verträglich sein sollen. Ferner verschwindet (13) „am Rand“, d. h. für $j = j_{\max}+1$ und $j = j_{\min}$, worin bekanntlich $j_{\max} = l+s$ und $j_{\min} = |l-s|$.

§ 3. *Bestimmung der Koordinatenmatrizen.* Zur Bestimmung der Koordinatenmatrizen, d. h. des Faktors $a_{l',j'}^{l,j}$, der aus dieser nach Abspaltung der in (7) angegebenen, die Quantenzahl m enthaltenden Wurzelfaktoren entsteht, verwenden wir die in (9) enthaltene Relation

$$[\mathfrak{s}_x - i\,\mathfrak{s}_y,\ \mathfrak{z}]_{l',j'',m+1}^{l,j,m} = 0,$$

und zwar zuerst an den Stellen $j' = j - 1$, $j'' = j + 1$, sodann an den Stellen $j' = j$, $j'' = j + 1$ und $j' = j$, $j'' = j - 1$. Wir erhalten unter Berücksichtigung, daß $\mathfrak{s}$ diagonal in Bezug auf l ist, und nach Fortheben gemeinsamer Faktoren die Gleichungen

$$s_{l,j}^{l,j-1}\, a_{l',j+1}^{l,j} = s_{l',j+1}^{l',j}\, a_{l'j}^{l,j-1}. \tag{14}$$

Diese Formel ermöglicht eine rekursive Bestimmung von $a_{l',j+1}^{l,j}$, da sie bei bekannten s den Quotienten dieses Ausdrucks für aufeinanderfolgende Werte von j zu berechnen gestattet. Wir werden sehen, daß es leicht ist, diese Gleichung analytisch nach $a_{l',j+1}^{l,j}$ aufzulösen. Für die Stelle $j' = j$, $j'' = j + 1$ erhält man zunächst

$$\begin{aligned} &+ s_{l,j}^{l,j}\, a_{l',j+1}^{l,j}(j - m) - s_{l,j+1}^{l,j}\, a_{l',j+1}^{l,j+1}(m + 1) \\ &+ s_{l',j+1}^{l',j}\, a_{l',j}^{l,j}\cdot m - s_{l',j+1}^{l',j+1}\, a_{l',j+1}^{l,j}(j + 1 - m) = 0. \end{aligned}$$

Da diese Gleichung für alle m gelten muß, spaltet sie in zwei Gleichungen, die einzeln ausdrücken, daß die von m unabhängigen und die zu m proportionalen Terme für sich verschwinden.

$$\left.\begin{aligned} &\left[+ s_{l,j}^{l,j}\, j - s_{l',j+1}^{l',j+1}(j + 1)\right] a_{l',j+1}^{l,j} - s_{l,j+1}^{l,j}\, a_{l',j+1}^{l,j+1} = 0, \\ &\left[- s_{l,j}^{l,j} + s_{l',j+1}^{l',j+1}\right] a_{l',j+1}^{l,j} - s_{l,j+1}^{l,j}\, a_{l',j+1}^{l,j+1} + s_{l',j+1}^{l',j}\, a_{l',j}^{l,j} = 0. \end{aligned}\right\} \tag{15}$$

Subtraktion der zweiten von der ersten Gleichung gibt

$$\left[+ s_{l,j}^{l,j}(j + 1) - s_{l',j+1}^{l',j+1}(j + 2)\right] a_{l',j+1}^{l,j} - s_{l',j+1}^{l',j}\, a_{l',j}^{l,j} = 0. \tag{15'}$$

Analog ergibt die Betrachtung der Stelle $j' = j$, $j'' = j - 1$ die beiden Gleichungen

$$\left.\begin{aligned} &\left[s_{l,j}^{l,j}(j + 1) - s_{l',j-1}^{l',j-1}\, j\right] a_{l',j-1}^{l,j} + s_{l,j-1}^{l,j}\, a_{l',j-1}^{l,j-1} = 0, \\ &\left[s_{l,j}^{l,j} - s_{l',j-1}^{l',j-1}\right] a_{l',j-1}^{l,j} + s_{l,j-1}^{l,j}\, a_{l',j-1}^{l,j-1} - s_{l',j-1}^{l',j}\, a_{l',j}^{l,j} = 0. \end{aligned}\right\} \tag{16}$$

Subtraktion der ersten Gleichung von der zweiten gibt

$$\left[- s_{l,j}^{l,j}\, j + s_{l',j-1}^{l',j-1}(j - 1)\right] a_{l',j-1}^{l,j} - s_{l',j-1}^{l',j}\, a_{l',j}^{l,j} = 0. \tag{16'}$$

Hat man aus (14) die $a_{l',j+1}^{l,j}$ bestimmt, so folgen die $a_{l',j}^{l,j}$ und $a_{l',j-1}^{l,j}$ aus (15), (16), oder aus (15'), (16').

Für den Fall $l = l'$, folgt aus (14) zunächst

$$a^{l,j}_{l,j+1} = \mathrm{const}\, s^{l,j}_{l,j+1},$$

und zwar wollen wir setzen

$$a^{l,j}_{l,j+1} = -a^l_l s^{l,j}_{l,j+1}, \tag{17}$$

wenn in letzterem Ausdruck die positive Wurzel aus (13) gezogen wird. Dann kommt

$$a^{l,j}_{l,j} = a^l_l \frac{l(l+1) - s(s+1) + j(j+1)}{2j(j+1)}, \tag{18}$$

$$a^{l,j}_{l,j-1} = -a^l_l s^{l,j}_{l,j-1}. \tag{19}$$

Für $l' = l - 1$ dagegen erhält man aus (14) durch Einsetzen von (13) und Fortheben gemeinsamer Faktoren

$$a^{l,j}_{l-1,j+1} \frac{\sqrt{(j-l+s)(l+s+1-j)}}{2j\sqrt{(2j-1)(2j+1)}} = a^{l,j-1}_{l-1,j} \frac{\sqrt{(j-l+s+2)(l+s-1-j)}}{2(j+1)\sqrt{(2j+1)(2j+3)}}.$$

Das Prinzip der Auflösung dieser Formel besteht darin, daß man sie durch Hinzufügen gleicher Faktoren auf beiden Seiten auf die Form bringt

$$a^{l,j}_{l-1,j+1} f(j-1) = a^{l,j-1}_{l-1,j} f(j),$$

woraus dann

$$a^{l,j}_{l-1,j+1} = \mathrm{const}\, f(j)$$

folgt. Dieser Multiplikator ist nun in unserem Falle $\sqrt{(j-l+s+1)(l+s-j)}$. Man erhält so

$$a^{l,j}_{l-1,j+1} = -a^l_{l-1} \frac{\sqrt{(j-l+s+2)(j-l+s+1)(l+s-j)(l+s-1-j)}}{2(j+1)\sqrt{(2j+1)(2j+3)}}. \tag{20}$$

Durch Einsetzen in (15'), (16') ergibt dann weiter die Ausrechnung

$$a^{l,j}_{l-1,j-1} = a^l_{l-1} \frac{\sqrt{(j+l-s-1)(j+l-s)(l+s+1+j)(l+s+j)}}{2j\sqrt{(2j-1)(2j+1)}}, \tag{21}$$

$$a^{l,j}_{l-1,j} = a^l_{l-1} \frac{\sqrt{(j+l-s)(j-l+s+1)(l+s+1+j)(l+s-j)}}{2j(j+1)}. \tag{22}$$

Der Fall $l' = l + 1$ ergibt wegen der Symmetrie der Formeln [Vertauschbarkeit der oberen und unteren Indizes in den Gleichungen (20) bis (22)] nichts Neues. Die gewonnenen Ausdrücke (17) bis (22) stimmen mit den bekannten Formeln für die Intensitäten der Multiplettkomponenten überein.

§ 4. *Summenregeln. Normierung.* Es ist leicht, aus den gewonnenen Formeln die Summenregeln zu bestätigen. Zunächst erhält man aus (7) allgemein für die Summen der Quadrate der Matrixelemente eines

Vektors $\mathfrak{A}$ von einem bestimmten Teilniveau m, j nach einem anderen Teilniveau mit bestimmtem j', aber irgendeinem m', also für

$$\sum_{j'}^{j} = \sum_{m'} \left(\left| A_{x\,j',\,m'}^{\;j,\,m} \right|^2 + \left| A_{y\,j',\,m'}^{\;j,\,m} \right|^2 + \left| A_{z\,j',\,m'}^{\;j,\,m} \right|^2 \right)$$

in bekannter Weise die Ausdrücke

$$\left. \begin{aligned} \sum_{j}^{j} = |A_j^j|^2 j(j+1), \quad \sum_{j+1}^{j} &= |A_{j+1}^j|^2 (j+1)(2j+3), \\ \sum_{j-1}^{j} &= |A_{j-1}^j|^2 j(2j-1). \end{aligned} \right\} \tag{23}$$

(Es ist zu beachten, daß die $\sum_{j'}^{j}$, im Gegensatz zu den $A_{j'}^{j}$, **nicht** im oberen und unteren Index symmetrisch sind.)

Die weitere Summation über die Werte j' des Endzustandes bei festem Teilniveau j, m des Anfangszustandes gibt (bei den Matrixelementen der Koordinaten)

$$\sum_{l'}^{l} = \sum_{l',\,j}^{l,\,j} + \sum_{l',\,j+1}^{l,\,j} + \sum_{l',\,j-1}^{l,\,j}$$

$$\left. \begin{aligned} &\text{für } l' = l & \sum_{l}^{l} &= |a_l^l|^2 \,.\, l(l+1), \\ &\text{für } l' = l-1 & \sum_{l-1}^{l} &= |a_{l-1}^l|^2 \, l(2l-1), \\ &\text{für } l' = l+1 & \sum_{l+1}^{l} &= |a_{l+1}^l|^2 (l+1)(2l+3). \end{aligned} \right\} \tag{24}$$

Diese Summen sind also von den Quantenzahlen j, m des Anfangsniveaus unabhängig. Eine analoge Regel mit ähnlichen Formeln gilt bekanntlich auch, wenn man bei festem Endniveau über die Anfangsniveaus summiert.

Die Kenntnis der Werte der Summen $\sum_{l'}^{l}$, gestattet einen Vergleich der Matrixelemente der Koordinaten (und Impulsmomente) für den Fall eines diagonalen $\boldsymbol{j}^2$ (und $\boldsymbol{j}_z$) mit dem eines diagonalen $\boldsymbol{s}_z$ (und $\boldsymbol{j}_z$, also auch $\boldsymbol{l}_z$). Letzteres entspricht beim Zeemaneffekt dem Falle starker Felder. Sind m_s, m_l, $m = m_l + m_s$ [mit $-s \leqq |m_s| \leqq s$, $-l \leqq m_l \leqq l$] in bekannter Weise die Eigenwerte von $\boldsymbol{s}_z$, $\boldsymbol{l}_z$, $\boldsymbol{j}_z$, so sind in diesem Falle die (in Bezug auf m_s notwendig diagonalen) Matrixelemente der Koordinaten direkt durch Formeln vom Typus (7) gegeben:

$$\left. \begin{aligned} (\boldsymbol{x} \pm i\boldsymbol{y})_{l,\,m_l \mp l}^{l,\,m_l} &= a^l \sqrt{(l \pm m_l)(l \mp m_l + 1)}, \\ \boldsymbol{z}_{l,\,m_l}^{l,\,m_l} &= a_l^l \, m \;, \\ (\boldsymbol{x} \pm i\boldsymbol{y})_{l-1,\,m_l \mp 1}^{l,\,m_l} &= a_{l-1}^l \cdot \mp \sqrt{(l \pm m_l)(l \pm m_l - 1)}, \\ \boldsymbol{z}_{l-1,\,m_l}^{l,\,m_l} &= a_{l-1}^l \sqrt{l^2 - m_l^2} \\ (\boldsymbol{x} \pm i\boldsymbol{y})_{l+1,\,m_l \mp 1}^{l,\,m} &= a_{l+1}^l \cdot \pm \sqrt{(l \mp m_l + 1)(l \mp m_l + 2)}, \\ \boldsymbol{z}_{l+1,\,m_l}^{l,\,m_l} &= a_{l+1}^l \sqrt{(l+1)^2 - m_l^2}. \end{aligned} \right\} \tag{25}$$

Die Formeln für die Matrixelemente $\boldsymbol{l}_x$, $\boldsymbol{l}_y$, $\boldsymbol{l}_z$ erhält man hieraus, in dem man $a_l^l = 1$, $a_{l+1}^l = a_{l-1}^l = 0$ setzt, die Matrixelemente von $\boldsymbol{s}_x$, $\boldsymbol{s}_y$, $\boldsymbol{s}_z$, aus denen von $\boldsymbol{l}_x$, $\boldsymbol{l}_y$, $\boldsymbol{l}_z$, indem man überall l, m_l durch s, m_s ersetzt.

Die Summen der Quadrate der Beträge dieser Matrixelemente von einem festen Anfangszustand l, m_l nach allen Endzuständen mit festem l' kann analog zu (23) berechnet werden (wobei nur j, m durch l, m_l zu ersetzen ist) und stimmt mit den Summen (24) genau überein. Deshalb sind wir berechtigt, die Größen $a_{l'}^l$ in (25) gleichzusetzen den Größen $a_{l'}^l$ in den Ausdrücken (17) bis (22) für die Koordinatenmatrizen bei diagonalem j. Tun wir dies, so müssen alle angegebenen Matrixelemente in schwachen Feldern ($\boldsymbol{j}^2$ diagonal) aus denen in starken Feldern ($\boldsymbol{l}_z$ diagonal) durch eine unitäre Transformation, vermittelt durch eine unitäre Matrix $S_{m_l}^j$ *, hervorgehen.

II. Die Matrixelemente bei Zusammensetzung von drei Impulsmomenten. Im Text werden gewisse Matrixelemente benötigt für den Fall, daß drei Impulsmomente, nämlich Elektronenspin $\mathfrak{s}$, Bahnmoment $\mathfrak{l}$ und Kernmoment $\mathfrak{i}$ zusammengesetzt werden. Die Resultierende heiße $\mathfrak{f} = \mathfrak{l} + \mathfrak{s} + \mathfrak{i} = \mathfrak{j} + \mathfrak{i}$, wenn wieder $\mathfrak{j} = \mathfrak{l} + \mathfrak{s}$ gesetzt wird. Es ist für den Zweck der vorliegenden Arbeit angemessen, neben $\boldsymbol{i}^2$, $\boldsymbol{s}^2$, $\boldsymbol{l}^2$, $\mathfrak{f}^2$ und $\mathfrak{f}_z$ [Eigenwerte $i(i+1)$, $s(s+1)$, $l(l+1)$, $f(f+1)$ und m (letztere Quantenzahl bezieht sich jetzt also nicht mehr auf $\boldsymbol{j}_z$); $\boldsymbol{i}^2$ und $\boldsymbol{s}^2$ können als c-Zahlen angesehen werden] noch $\boldsymbol{j}^2$ mit den Eigenwerten $j(j+1)$ auf Diagonalform zu bringen. Die Angabe der Matrixelemente für diesen Fall erfordert keine neue Rechnung, sondern ergibt sich aus einer zweimaligen Anwendung der Formeln in I. Die V.-R. von $\mathfrak{s}$ und $\mathfrak{r}$ mit $\mathfrak{f}$, $\mathfrak{i}$, $\mathfrak{j}$; $\mathfrak{f}^2$, $\boldsymbol{i}^2$, $\boldsymbol{j}^2$ sind nämlich analog zu der V.-R. von $\mathfrak{r}$ mit $\mathfrak{j}$, $\mathfrak{s}$, $\mathfrak{l}$; $\boldsymbol{j}^2$, $\boldsymbol{s}^2$, $\boldsymbol{l}^2$. Also sind die unitären Transformationen

$$\mathfrak{s}_{(l),\, j',\, m'_j}^{(l),\, j,\, m_j} \rightarrow \mathfrak{s}_{(l),\, j',\, f',\, m'_f}^{(l),\, j,\, f,\, m_f} \quad \text{und} \quad \mathfrak{r}_{(l'),\, j',\, m'_j}^{(l),\, j,\, m_j} \rightarrow \mathfrak{r}_{(l'),\, j',\, f',\, m'_f}^{(l),\, j,\, f,\, m_f}$$

beide analog zu

$$\mathfrak{r}_{l',\, m'_l}^{l,\, m_l} \rightarrow \mathfrak{r}_{l',\, j'\, m'_j}^{l,\, j,\, m_j}.$$

Die schließlichen Ergebnisse mögen hier noch zum Gebrauch zusammengestellt werden.

---

* Es ist bisher nicht gelungen, geschlossene analytische Formeln für die Elemente dieser Matrix aufzustellen.

$$\mathfrak{i}_{z\,l,j,f,m}^{\,l,j,f,m} / i_{l,j,f}^{l,j,f} = \mathfrak{s}_{z\,l,j',f,m}^{\,l,j,f,m} / s_{l,j',f}^{l,j,f} = \mathfrak{z}_{l',j',f,m}^{l,j,f,m} / a_{l',j',f}^{l,j,f} = m,$$

$$(\mathfrak{i}_x \pm i\,\mathfrak{i}_y)_{l,j,f,m\mp 1}^{l,j,f,m} / i_{l,j,f}^{l,j,f} = (\mathfrak{s}_x \pm i\,\mathfrak{s}_y)_{l,j,f,m\mp 1}^{l,j,f,m} / s_{l,j',f}^{l,j,f}$$
$$= (\mathfrak{x} \pm i\,\mathfrak{y})_{l',j',f,m\mp 1}^{l,j,f,m} / a_{l',j',f}^{l,j,f} = \sqrt{(f \pm m)(f \mp m + 1)}.$$

$$\mathfrak{i}_{z\,l,j,f+1,m}^{\,l,j,f,m} / i_{l,j,f+1}^{l,j,f} = \mathfrak{s}_{z\,l,j',f+1,m}^{\,l,j,f,m} / s_{l,j',f+1}^{l,j,f} = \mathfrak{z}_{l',j',f+1,m}^{l,j,f,m} / a_{l',j',f+1}^{l,j,f}$$
$$= \sqrt{(f+1)^2 - m^2},$$

$$(\mathfrak{i}_x \pm i\,\mathfrak{i}_y)_{l,j,f+1,m\mp 1}^{l,j,f,m} / i_{l,j,f+1}^{l,j,f} = (\mathfrak{s}_x \pm i\,\mathfrak{s}_y)_{l,j',f+1,m\mp 1}^{l,j,f,m} / s_{l,j',f+1}^{l,j,f}$$
$$= (\mathfrak{x} \pm i\,\mathfrak{y})_{l',j',f+1,m\mp 1}^{l,j,f,m} / a_{l',j',f+1}^{l,j,f} = \pm\sqrt{(f \mp m + 1)(f \mp m + 2)}.$$

$$\mathfrak{i}_{z\,l,j,f-1,m}^{\,l,j,f,m} / i_{l,j,f-1}^{l,j,f} = \mathfrak{s}_{z\,l,j',f-1,m}^{\,l,j,f,m} / s_{l,j',f-1}^{l,j,f} = \mathfrak{z}_{l',j',f-1,m}^{l,j,f,m} / a_{l',j',f-1}^{l,j,f}$$
$$= \sqrt{f^2 - m^2},$$

$$(\mathfrak{i}_x \pm i\,\mathfrak{i}_y)_{l,j,f-1,m\mp 1}^{l,j,f,m} / i_{l,j,f-1}^{l,j,f} = (\mathfrak{s}_x \pm i\,\mathfrak{s}_y)_{l,j',f-1,m\mp 1}^{l,j,f,m} / s_{l,j',f-1}^{l,j,f}$$
$$= (\mathfrak{x} \pm i\,\mathfrak{y})_{l',j',f-1,m\mp 1}^{l,j,f,m} / a_{l',j',f-1}^{l,j,f} = \mp\sqrt{(f \pm m)(f \pm m - 1)}.$$

$$i_{l,j,f}^{l,j,f} = \frac{i(i+1) - j(j+1) + f(f+1)}{2f(f+1)},$$

$$i_{l,j,f+1}^{l,j,f} = \frac{\sqrt{(f+j-i+1)(f-j+i+1)(j+i+2+f)(j+i-f)}}{2(f+1)\sqrt{(2f+1)(2f+3)}},$$

$$i_{l,j,f-1}^{l,j,f} = \frac{\sqrt{(f+j-i)(f-j+i)(j+i+1+f)(j+i+1-f)}}{2f\sqrt{(2f-1)(2f+1)}}.$$

$$s_{l,j,f}^{l,j,f} / s_{l,j}^{l,j} = a_{l',j,f}^{l,j,f} / a_{l',j}^{l,j} = \frac{j(j+1) - i(i+1) + f(f+1)}{2f(f+1)},$$

$$s_{l,j,f+1}^{l,j,f} / s_{l,j}^{l,j} = a_{l',j,f+1}^{l,j,f} / a_{l',j}^{l,j} = -\frac{\sqrt{(f+j-i+1)(f-j+i+1)(j+i+2+f)(j+i-f)}}{2(f+1)\sqrt{(2f+1)(2f+3)}},$$

$$s_{l,j,f-1}^{l,j,f} / s_{l,j}^{l,j} = a_{l',j,f-1}^{l,j,f} / a_{l',j}^{l,j} = -\frac{\sqrt{(f+j-i)(f-j+i)(j+i+1+f)(j+i+1-f)}}{2f\sqrt{(2f-1)(2f+3)}}.$$

$$s^{l,j,f}_{l,j-1,f} \Big/ s^{l,j}_{l,j-1} = a^{l,j,f}_{l',j-1,f} \Big/ a^{l,j}_{l',j-1} = \frac{\sqrt{(f+j-i)(f-j+i+1)(j+i+1+f)(j+i-f)}}{2f(f+1)}$$

$$s^{l,j,f}_{l,j-1,f+1} \Big/ s^{l,j}_{l,j-1} = a^{l,j,f}_{l',j-1,f+1} \Big/ a^{l,j}_{l',j-1} = -\frac{\sqrt{(f-j+i+2)(f-j+i+1)(j+i-f)(j+i-1-f)}}{2(f+1)\sqrt{(2f+1)(2f+3)}},$$

$$s^{l,j,f}_{l,j-1,f-1} \Big/ s^{l,j}_{l,j-1} = a^{l,j,f}_{l',j-1,f-1} \Big/ a^{l,j}_{l',j-1} = \frac{\sqrt{(f+j-i-1)(f+j-i)(j+i+1+f)(j+i+f)}}{2f\sqrt{(2f-1)(2f+1)}}.$$

$$s^{l,j,f}_{l,j+1,f} \Big/ s^{l,j}_{l,j+1} = a^{l,j,f}_{l',j+1,f} \Big/ a^{l,j}_{l',j+1} = \frac{\sqrt{(f+j-i+1)(f-j+i)(j+i+2+f)(j+i+1-f)}}{2f(f+1)},$$

$$s^{l,j,f}_{l,j+1,f+1} \Big/ s^{l,j}_{l,j+1} = a^{l,j,f}_{l',j+1,f+1} \Big/ a^{l,j}_{l',j+1} = \frac{\sqrt{(f+j-i+1)(f+j-i+2)(j+i+3+f)(j+i+2+f)}}{2(f+1)\sqrt{(2f+1)(2f+3)}},$$

$$s^{l,j,f}_{l,j+1,f-1} \Big/ s^{l,j}_{l,j+1} = a^{l,j,f}_{l',j+1,f-1} \Big/ a^{l,j}_{l',j+1} = -\frac{\sqrt{(f-j+i)(f-j+i-1)(j+i+2-f)(j+i+1-f)}}{2f\sqrt{(2f+1)(2f-1)}}.$$

$$s^{l,j}_{l,j} = \frac{s(s+1)-l(l+1)+j(j+1)}{2j(j+1)}, \quad a^{l,j}_{l,j} = a_l \frac{l(l+1)-s(s+1)+j(j+1)}{2j(j+1)},$$

$$s^{l,j}_{l,j-1} = -a^{l,j}_{l,j-1} \Big/ a_l = \frac{\sqrt{(j+l-s)(j-l+s)(l+s+1+j)(l+s+1-j)}}{2j\sqrt{(2j-1)(2j+1)}},$$

$$s^{l,j}_{l,j+1} = -a^{l,j}_{l,j+1} \Big/ a_l = \frac{\sqrt{(j+l-s+1)(j-l+s+1)(l+s+2+j)(l+s-j)}}{2(j+1)\sqrt{(2j+1)(2j+3)}}.$$

$$a^{l,j}_{l-1,j} \Big/ a_{l-1} = \frac{\sqrt{(j+l-s)(j-l+s+1)(l+s+1+j)(l+s-j)}}{2j(j+1)},$$

$$a^{l,j}_{l-1,j-1} \Big/ a_{l-1} = \frac{\sqrt{(j+l-s-1)(j+l-s)(l+s+1+j)(l+s+j)}}{2j\sqrt{(2j-1)(2j+1)}},$$

$$a^{l,j}_{l-1,j+1} \Big/ a_{l-1} = -\frac{\sqrt{(j-l+s+2)(j-l+s+1)(l+s-j)(l+s-1-j)}}{2(j+1)\sqrt{(2j+1)(2j+3)}}.$$

$$a^{l,j}_{l+1,j} \Big/ a_{l+1} = \frac{\sqrt{(j+l-s+1)(j-l+s)(l+s+2+j)(l+s+1-j)}}{2j(j+1)},$$

$$a^{l,j}_{l+1,j-1} \Big/ a_{l+1} = -\frac{\sqrt{(j-l+s)(j-l+s-1)(l+s+2+j)(l+s+1-j)}}{2j\sqrt{(2j-1)(2j+1)}},$$

$$a^{l,j}_{l+1,j+1} \Big/ a_{l+1} = \frac{\sqrt{(j+l-s+1)(j+l-s+2)(l+s+3+j)(l+s+2+j)}}{2(j+1)\sqrt{(2j+1)(2j+3)}}.$$

Im Text werden ferner die Matrixelemente des skalaren Produktes aus $\mathfrak{i}$ und $\mathfrak{s}$, also von $(\mathfrak{i}\,\mathfrak{s})$ benötigt, weil die Störungsenergie zwischen Kern und Elektron zu dieser Größe proportional ist. Die V.-R. (2) und (3) von Anhang I, bleiben hier bestehen, wenn man $\mathfrak{j}$ durch $\mathfrak{f}$ ersetzt. Also ist $(\mathfrak{i}\,\mathfrak{s})$ als Skalar in bezug auf f und m diagonal. Da ferner $\mathfrak{i}$ diagonal in bezug auf j ist, erhält man zunächst

$$(\mathfrak{i}\,\mathfrak{s})^{j,\,f}_{j',\,f} = i^{j,\,f}_{j,\,f}\, s^{j,\,f}_{j',\,f}\, f(f+1) + i^{j,\,f}_{j,\,f+1}\, s^{j,\,f+1}_{j',\,f}\,(f+1)\,(2f+3) \\ + i^{j,\,f}_{j,\,f-1}\, s^{j,\,f-1}_{j',\,f}\, f\,(2f-1). \qquad (26)$$

Das in bezug auf j diagonale Matrixelement $(j' = j)$ erhält man am einfachsten aus I, Gleichung (5), wenn man sie auf $\mathfrak{s}^{j,\,m}_{j,\,m'}$ anwendet*; die übrigen Matrixelemente $(j' = j - 1$ und $j' = j + 1)$ kann man durch Einsetzen und Ausrechnen finden. Es ergibt sich schließlich

$$\left.\begin{aligned} (\mathfrak{i}\,\mathfrak{s})^{j,\,f}_{j-1,\,f} &= s^{l,\,j}_{l,\,j-1} \cdot \tfrac{1}{2}\sqrt{(f+j-i)(f-j+i+1)(j+i+1-f)(j+i-f)}, \\ (\mathfrak{i}\,\mathfrak{s})^{j,\,f}_{j+1,\,f} &= s^{l,\,j}_{l,\,j+1} \cdot \tfrac{1}{2}\sqrt{(f+j-i+1)(f-j+i)(j+i+2+f)(j+i+1-f)}, \\ (\mathfrak{i}\,\mathfrak{s})^{j,\,f}_{j,\,f} &= s^{l,\,j}_{l,\,j} \cdot \tfrac{1}{2}[f(f+1) - i(i+1) - j(j+1)]. \end{aligned}\right\} \qquad (27)$$

* Vgl. l. c. A, S. 753.

Reprinted from the *American Journal of Mathematics*,
Vol. LXIII, No. 1, January 1941

ON REPRESENTATIONS OF CERTAIN FINITE GROUPS.*

By EUGENE P. WIGNER.

1. The purpose of this paper is the derivation of a classification of the representations of finite groups with special reference to groups which satisfy the following two conditions:

a. Every element is equivalent to its reciprocal, i. e., all classes are ambivalent.

b. The Kronecker (or "direct") product of any two irreducible representations of the group contains no representation more than once.

Groups of this character will be called S. R. groups (simply reducible). The symmetric permutation groups of the third and fourth degree, the quaternion group, the three dimensional rotation group, the two dimensional unimodular unitary group are S. R. groups. The significance of condition *a* is that every representation is equivalent to the conjugate imaginary representation. One sees this most easily by assuming the representation to be unitary. Then, the traces of reciprocal elements are conjugate complex. They are, on the other hand, equal, since they belong to the same class. Thus all traces are real and conjugate complex representations are equivalent.

The groups of most eigen-value problems occurring in quantum theory are S. R. This is important for the following reason.

Let us assume that we have two eigen-value problems $H_1\psi_1 = \lambda_1\psi_1$ and $H_2\psi_2 = \lambda_2\psi_2$ which allow the same group; ψ_1 and ψ_2 shall be defined in different spaces. One often considers then [1] the "united system" the wave functions Ψ of which are defined in the product space of the spaces of ψ_1 and ψ_2. The "unperturbed" eigen-value equation is $(H_1 + H_2)\Psi = \Lambda\Psi$. The multiplicity of the eigen-value $\Lambda = \lambda_1 + \lambda_2$ is the product of the multiplicities of λ_1 and λ_2. The eigen-value Λ splits up if one introduces a small perturbation term into the last equation. If this perturbation allows the same group as the original two problems and if this group satisfies the above condition *b*, the characteristic functions of the eigen-values into which Λ splits can be determined in "first approximation" by the invariance of the eigen-value problem under the group. The properties of S. R. groups to be derived here

* Received May 1, 1940.

[1] For a more complete discussion cf. e. g. E. Wigner: *Gruppentheorie*, etc., Braunschweig, 1931.

give a basis for a suitable normalization of (and numerous relations between) these eigen-functions which will be dealt with elsewhere.

We shall denote the different irreducible representations of a group by letters j, k, l, etc. The identical representation (in which the matrix (1) corresponds to every element) by 0. The elements of the group will be P, Q, R, S, T, etc. The rows and columns of the representations will be designated by small Greek letters κ, λ, μ, ν, etc. The $\kappa\lambda$ element of the matrix which corresponds in the j-th representation to the element R will be denoted by $\begin{bmatrix} jR \\ \kappa\lambda \end{bmatrix}$ so that one has

$$\sum_{\lambda} \begin{bmatrix} jR \\ \kappa\lambda \end{bmatrix} \begin{bmatrix} jS \\ \lambda\mu \end{bmatrix} = \begin{bmatrix} j\,RS \\ \kappa\ \mu \end{bmatrix} \tag{1}$$

The character will be abbreviated to

$$\sum_{\kappa} \begin{bmatrix} jR \\ \kappa\kappa \end{bmatrix} = [j; R]. \tag{1a}$$

The summation over the indices κ, λ etc. referring to the rows or columns of the representations will always run over all values. The unit element of the group will be E, the degree of the representations j is

$$[j; E] = [j]. \tag{1b}$$

A star will denote the conjugate complex. The Kronecker product of the representations j and k coordinates to the group element R the matrix $M_{\kappa\mu;\,\lambda\nu}$ the rows and columns of which are denoted by double indices $\kappa\mu$ and $\lambda\nu$ respectively. We set

$$M_{\kappa\mu;\,\lambda\nu} = \begin{bmatrix} jR \\ \kappa\lambda \end{bmatrix} \begin{bmatrix} kR \\ \mu\nu \end{bmatrix}. \tag{2}$$

The character corresponding to the element R is

$$\sum_{\kappa\mu} M_{\kappa\mu;\kappa\mu} = \sum_{\kappa\mu} \begin{bmatrix} jR \\ \kappa\kappa \end{bmatrix} \begin{bmatrix} kR \\ \mu\mu \end{bmatrix} = [j; R][k; R]. \tag{2a}$$

The significance of condition b for the groups under consideration becomes evident if one reduces the Kronecker product of two representations, i. e. brings it into the form in which it appears as the sum of irreducible representations. The matrix by which this transformation can be carried out is—apart from some phase factors—uniquely determined.

It may be useful to write down the well known orthogonality and completeness relations for irreducible representations. These are

$$\sum_{R} \begin{bmatrix} jR \\ \kappa\lambda \end{bmatrix}^* \begin{bmatrix} kR \\ \mu\nu \end{bmatrix} = \frac{h}{[j]} \delta_{jk}\delta_{\kappa\mu}\delta_{\lambda\nu} \tag{3}$$

(3a) $$\sum_R [j;R]^*[k;R] = \sum_C n_c[j;C][k;C] = h\delta_{jk}.$$

The summation is to be extended in this and all similar formulas over all group elements;

(4) $$h = \sum_R 1$$

is the order of the group. The summation over C in the second part of (3a) is to be extended over all different classes, n_c is the number of elements of the class C. The completeness relations yield

(5) $$\sum_{j\kappa\lambda} \frac{[j]}{h} \begin{bmatrix} jR \\ \kappa\lambda \end{bmatrix}^* \begin{bmatrix} jS \\ \kappa\lambda \end{bmatrix} = \delta_{R,S}$$

(5a) $$\sum_j [j;R]^*[j;S] = h\, \Delta_{R,S}/n_R = \sum_Q \delta_{RQ,QS}.$$

The summation over j is to be extended over all different irreducible representations, $\delta_{R,S}$ is 1 for $R = S$, zero otherwise, $\Delta_{R,S}$ is 1 if R and S are in the same class, zero otherwise, n_R is the number of the elements of the class of R. All representations are assumed to be in the unitary form, i. e.

(6) $$\sum_\lambda \begin{bmatrix} jR \\ \kappa\lambda \end{bmatrix}^* \begin{bmatrix} jR \\ \mu\lambda \end{bmatrix} = \delta_{\kappa\mu};\ \sum_\kappa \begin{bmatrix} jR \\ \kappa\lambda \end{bmatrix}^* \begin{bmatrix} jR \\ \kappa\nu \end{bmatrix} = \delta_{\lambda\nu}.$$

The irreducible representations can be classified,[2] in general, into three groups: those which can be transformed into a real form, those which cannot but are equivalent to the conjugate complex representation, and those which are not equivalent to the conjugate complex representation. In analogy to the notation customarily used for the two dimensional unimodular unitary group, we shall call the representations of the first kind *integer* representations. Correspondingly $c_j = 1$ will hold for representations j which can be transformed into a real form, $c_j = -1$ will hold for *half integer* representations j which cannot be transformed into a real form but are equivalent to the conjugate complex representation. Finally $c_j = 0$ if the representation j is not equivalent to the conjugate complex of j. According to G. Frobenius and I. Schur [2]

(7) $$\sum_R [j;R^2] = c_j h.$$

2. The number of square roots of an element R will be denoted by $\zeta(R)$

(8) $$\zeta(R) = \sum_S \delta_{R,S^2}.$$

We have

$$\sum_R \zeta(R)^2 = \sum_{R,S} \zeta(R)\delta_{R,S^2} = \sum_S \zeta(S^2) = \sum_{S,T} \delta_{S^2,T^2}$$

[2] G. Frobenius and I. Schur, Berl. Ber. 1906, p. 186.

One can replace S by TR in the last summation and obtain

$$\sum_R \zeta(R)^2 = \sum_{R,T} \delta_{TRTR,T^2} = \sum_{R,T} \delta_{R,TR^{-1}T^{-1}}$$

as $TRTR = T^2$ if and only if $R = TR^{-1}T^{-1}$. For a given R, there will be a T such that $R = TR^{-1}T^{-1}$ only if R and R^{-1} are in the same class, i. e. if R is in an ambivalent class. In this case, the number of T satisfying $R = TR^{-1}T^{-1}$ is equal to h/n_R, since each of the n_R members of the class of R is obtained h/n_R times when T runs over all h elements of the group. Hence

$$\sum_R \zeta(R)^2 = \sum_R{}' h/n_R = h \times (\text{number of ambivalent classes}). \tag{9}$$

The second summation is to be extended only over the elements of the ambivalent classes. The result thus obtained [3] holds for every finite group:

THEOREM 1. *The sum of the squares of the numbers of square roots of all elements of a finite group is equal to the order of the group, multiplied by the number of ambivalent classes.*

All classes are ambivalent in the S. R. groups. Hence

$$\sum_R \zeta(R)^2 = hn \tag{9a}$$

holds for these, where n is the number of all classes.

The number of times the representation i is contained in the Kronecker product of the representations j and k is given by the equation

$$(i, j, k) = \sum_C [i; C]^* [j; C] [k; C] n_c/h \tag{10}$$

where the summation has to be extended, as in (3a) over all classes. Multiplying (7) by $[j; S]$ and summing over j gives, for (5a)

$$\begin{aligned} \sum_j c_j h [j; S] &= \sum_{R,j} [j; R^2][j; S] \\ &= \sum_R \Delta_{R^2,S}\, h/n_S = h\zeta(S). \end{aligned} \tag{11}$$

The $R^2 = S$ equation is satisfied for $\zeta(S)$ group elements R but R^2 is in the class of S for $n_S\zeta(S)$ group elements.

LEMMA 1. *The Kronecker product of two integer representations or of two half integer representations of a S. R. group contains only integer representations; the Kronecker product of an integer and a half integer represen-*

[3] This must have been known to the authors of Reference 2 since it follows immediately from a comparison of the last sentence of § 4 with the sentence in italics on page 201.

tation contains only half integer representations. The unitary matrix which transforms an integer representation into the conjugate complex form is symmetric, that which transforms a half integer representation into the conjugate complex form is skew symmetric.[2] Hence the unitary matrix S which transforms the Kronecker product M of two integer or two half integer representations into the conjugate complex form is symmetric:

$$SM = M^*S; \qquad S = S'. \tag{12}$$

If the unitary U brings M into the reduced form $UMU^{-1} = M_r$

$$S_rM_r = M^*{}_rS_r; \qquad S_r = U^*SU^{-1} \tag{12a}$$

and S_r is again symmetric. Since the corresponding parts of M_r and $M^*{}_r$ are equivalent and since M_r does not contain any irreducible representation more than once, S_r is a step matrix just as M_r is and every submatrix of S_r is symmetric on account of the symmetry of S_r. Hence, every submatrix of M_r, i. e. all irreducible parts of M, can be transformed into the conjugate complex form by a symmetric matrix and are integer representations.

If M is the product of an integer and a half integer representation, S will be skew symmetric and the same will hold for S_r and its submatrices. Consequently, all the irreducible parts of M will be half integer representations.

For a S. R. group $c_ic_jc_k = 1$ if (ijk) is different from zero.

Since the (ijk) are positive integers or zero,

$$(ijk)^2 \geq c_ic_jc_k(ijk). \tag{13}$$

The equality sign can hold only if either $(ijk) = 0$, or $(ijk) = 1$ and $c_ic_jc_k = 1$. Hence

$$\sum_{ijk} (ijk)^2 \geq \sum_{ijk} c_ic_jc_k(ijk) \tag{13a}$$

and the equality sign can hold only if for all i, j, k either $(ijk) = 0$ or $(ijk) = 1$ and $c_ic_jc_k = 1$. This is the case, according to the definition of S. R. groups and Lemma 1, for S. R. groups and conversely, if the equality sign holds in (13a), the group must be a S. R. group.

Because of (11), we have

$$\begin{aligned}\sum_{ijk} c_ic_jc_k(i, j, k) &= \sum_{ijk} c_ic_jc_k[i; C]^*[j; C][k; C]n_c/h \\ &= \sum_{ijk} c_ic_jc_k[i; R]^*[j; R][k; R]/h = \sum_R \zeta(R)^3/h\end{aligned}$$

For the left side of (13a) we have, because of (5a)

$$\begin{aligned}
&\sum_{ijk} (i,j,k)^2 \\
&\quad = \sum_{ijk} \sum_{CC'} [i;C]^*[j;C][k;C][i;C'][j;C']^*[k;C']^* n_c n_{c'}/h^2 \\
(14a) \qquad &\quad = \sum_{ik} \sum_{C} [i;C]^2[k;C]^2 n_c/h = \sum_{C} h/n_c \\
&\quad = \sum_{R} h/n_R{}^2 = \sum_{R} v_R{}^2/h
\end{aligned}$$

where $v_R = h/n_R$ is the number of elements which commute with R. Hence (13a) is equivalent to

THEOREM 2. *The inequality*

$$\sum_{R} \zeta(R)^3 \leq \sum_{R} v_{R^2} \tag{15}$$

holds for every finite group. The equality sign in (15) *holds for all finite S. R. groups and only these.*

3. The Kronecker product of a representation with itself can be decomposed into a symmetric part

$$B_{\kappa\mu;\lambda\nu} = \frac{1}{2}\begin{bmatrix} jR \\ \kappa\lambda \end{bmatrix}\begin{bmatrix} jR \\ \mu\nu \end{bmatrix} + \frac{1}{2}\begin{bmatrix} jR \\ \kappa\nu \end{bmatrix}\begin{bmatrix} jR \\ \mu\lambda \end{bmatrix} \tag{16}$$

and an antisymmetric part

$$A_{\kappa\mu;\lambda\nu} = \frac{1}{2}\begin{bmatrix} jR \\ \kappa\lambda \end{bmatrix}\begin{bmatrix} jR \\ \mu\nu \end{bmatrix} - \frac{1}{2}\begin{bmatrix} jR \\ \kappa\nu \end{bmatrix}\begin{bmatrix} jR \\ \mu\lambda \end{bmatrix} \tag{16a}$$

It is easy to see that both B and A form a representation of the group. The irreducible parts of both B and A are integer representations in case of $S. R.$ groups. The irreducible parts of the B for integer j and the irreducible parts of the A for half integer j will be called even representations. Conversely, the irreducible parts of the B for integer j will be called odd representations. This notation is taken again from the theory of representations of the two dimensional unitary group.

THEOREM 3. *In S. R. groups no representation can be both even and odd.*

The trace of the symmetric part of the square of the representation j is

$$X_{js}(R) = \tfrac{1}{2}[j;R]^2 + \tfrac{1}{2}[j;R^2] \tag{17}$$

and the trace of the antisymmetric part is

$$X_{ja}(R) = \tfrac{1}{2}[j;R]^2 - \tfrac{1}{2}[j;R^2] \tag{17a}$$

The condition that two representations have no common part is that the sum of the products of their characters vanish. Theorem 3 is equivalent therefore with the validity of

$$(18) \qquad \tfrac{1}{4} \sum_R ([j;R]^2 + c_j[j;R^2])([k;R]^2 - c_k[k;R^2]) = 0$$

for all j and k. Since the left side of (18) by its nature cannot be negative (it is $\Sigma n_i m_i$ where n_i and m_i are the numbers of times the representation i is contained in the first and second representation of (18)), the validity of (18) for all j and k is equivalent with the vanishing of

$$(18a) \qquad \sum_R \sum_{jk} ([j;R]^2 + c_j[j;R^2])([k;R]^2 - c_k[k:R^2])$$
$$= \sum_R (h/n_R + \zeta(R^2))(h/n_R - \zeta(R^2))$$
$$= \sum_R v_R{}^2 - \sum_R \zeta(R^2)^2$$

where (5a) and (11) have been utilized. Now evidently

$$\sum_S \zeta(S)^3 = \sum_{S,R} \zeta(S)^2 \delta_{S,R^2} = \sum_R \zeta(R^2)^2$$

so that (18a) vanishes on account of Theorem 2. Hence Theorem 3 is valid.

Of course, the Kronecker product of an even and an odd representation e. g., contains, in general, both even and odd representations. It has not been shown, either, that every integer representation is either even, or odd. In fact, one can easily find a group which has an integer representation which does not occur in the square of any representation. A group of this character is formed by the elements $1, -1, x, -x, y, -y, z, -z$, with the multiplication rules $x^2 = y^2 = 1$, $z^2 = -1$, $xy = -yx = z$, $xz = -zx = y$, $zy = -yz = x$.

PRINCETON UNIVERSITY.

ON THE MATRICES WHICH REDUCE THE KRONECKER PRODUCTS OF REPRESENTATIONS OF S.R. GROUPS*

Eugene P. Wigner

The present article deals with simply reducible groups which are finite or compact and with their unitary representations (more precisely: with their representations which are in unitary form) in general. A simply reducible group satisfies two criteria. First, all classes are ambivalent, i.e., contain with an element X also its reciprocal, X^{-1}. Since the characters of all elements of a class are equal to each other in every representation, it follows that the characters of X and X^{-1} are equal. If the representation is in unitary form, it is at once evident that the characters of reciprocals are conjugate complex. It therefore follows from the ambivalent nature of all classes that all characters are real in every representation of a simply reducible group. As a

* *Editors' note*: The original version of Professor Wigner's paper differed in only minor ways from the later version given here. The changes are entirely of a clarifying nature, except that references to a few articles which have appeared since 1940 have been added, and the more modern terminology (Racah coefficients, multiplicity free representations, etc.) occasionally mentioned. Also, the left-handed coordinate system used in the original manuscript was changed into a right-handed one by interchanging X and Y. We feel that the usefulness of these changes outweighs that of historical accuracy.

result, every representation is equivalent to the conjugate complex representation.

The second condition for a group to be called simply reducible is that the irreducible representations of the group be, in the terminology of G. R. Mackey, multiplicity free. This means that the Kronecker (direct) product of two irreducible representations, j_1 and j_2, does not contain any irreducible representation more than once. If the direct product of j_1 and j_2 contains j_3, we set the symbol $(j_1j_2j_3) = 1$; if it does not contain j_3, we set $(j_1j_2j_3) = 0$. Since all representations to be considered are equivalent to their conjugate complex, if the product of j_1 and j_2 contains j_3, the product of j_2 and j_3 contains j_1. Hence, the symbol $(j_1j_2j_3)$ is symmetric in its arguments. The multiplicity free nature of the representations is the condition for the validity of what is often called the Eckart-Wigner theorem in quantum mechanics, and it was this circumstance which attracted this writer's attention to the subject.

The present article, which is published here for the first time, is the continuation of the article[6] in which simply reducible groups were defined. A summary of the relevant results of this article is given in Reference 6, and its notation is followed in the present article. Group elements are denoted by capitals, R, S, T, ... Summation over these is always extended over all elements in case of finite groups; the invariant integral is meant in the case of continuous groups. In the former case, h is the order of the group; in the latter case, h is the volume thereof. Other symbols to be used are:

$[j]$ dimension of the irreducible representation j; in the case of the three-dimensional rotation group $[j] = 2j + 1$.

$\begin{bmatrix} j & R \\ \kappa & \lambda \end{bmatrix}$ $\kappa\lambda$ matrix element of the matrix which the unitary irreducible representation j coordinates to the group element R. The customary notation is $\begin{bmatrix} j & R \\ \kappa & \lambda \end{bmatrix} = D^{(j)}(R)_{\kappa\lambda}$.

$\begin{pmatrix} j_1 & j_2 & j_3 \\ \kappa_1 & \kappa_2 & \kappa_3 \end{pmatrix}$ coefficient of the vector addition model. An equivalent quantity was calculated for the three-dimensional rotation group in the writer's book.[3] The relation between the two notations is, explicitly,

$$s^{(l\bar{l})}_{\mathscr{L}\mu\nu} = (-)^{l-\bar{l}+\mu+\nu}(2\mathscr{L}+1)^{1/2} \begin{pmatrix} l & \bar{l} & \mathscr{L} \\ \mu & \nu & -\mu-\nu \end{pmatrix}.$$

$\begin{Bmatrix} j_1 & j_2 & j_3 \\ l_1 & l_2 & l_3 \end{Bmatrix}$ is called, in the case of the three-dimensional rotation group, the Racah coefficient[7] $W(j_1l_1, j_2l_2, j_3l_3)$.

Much more extensive tables for both coefficients than are contained in the present article have been published by H. A. Jahn[8] and L. C. Biedenharn.[9] and also others.*

The star denotes the conjugate imaginary.

* Editors' note: See Bibliography, Section E, p. 321.

I. Simply Reducible Groups

1. *Definition of Simply Reducible Groups*

The chief purpose of the present paper is to derive some formulas for the irreducible representations of the three-dimensional rotation or the two-dimensional unimodular unitary group. Many of the formulas and relations to be derived were given already in the books quoted in References 3 and 5. It seems appropriate, however, to give a more systematic and exhaustive treatment to this subject than has been done hitherto. The results of the first five sections will be valid not only for the above continuous groups, but for all finite S.R. (simply reducible) groups because the special properties of the above groups will not be utilized in the first five sections.

Equations derived in earlier parts of the paper will be quoted in the customary manner when used again. Some equations expressing symmetry properties of certain symbols form an exception under this rule and will be used without being quoted. The numbers of these equations are set in bold face type where they are derived.

The S.R. groups were defined[6] as groups for which (a) the direct product of two irreducible representations contains no irreducible representation more than once, and (b) every element is equivalent with (contained in the class of) its reciprocal. The notation adopted in the first part of Reference 6 will be used here also. All equations of Sections 2–5 are valid for finite S.R. groups and also for the above continuous groups if the summation over the group elements is replaced by the invariant group integral and h is replaced by the volume of the group, i.e., the integral of 1 over the group. In order to avoid the necessity of distinguishing between the two kinds of groups, the necessary results in Reference 6 (Lemma 1, Theorem 3), will be established for the above continuous groups by direct enumeration in Section 7.

2. *Invariant Vectors in Representation Space (Three-j-Symbols or Clebsch Gordan Coefficients); Covariant and Contravariant Components*

Let us denote the matrix elements of the unitary matrix which brings the Kronecker product of the representations j_1 and j_2 into the reduced form by

$$U_{j_3\kappa_3;\,\kappa_1\kappa_2} = [j_3]^{1/2}\begin{pmatrix} j_1 & j_2 & j_3 \\ \kappa_1 & \kappa_2 & \kappa_3 \end{pmatrix}. \tag{1}$$

All representations are assumed to be in unitary form. In U, j_3 and κ_3 specify the row, κ_1 and κ_2 the column into which the matrix element (1) belongs. The second factor on the right side of (1) will be called a "three-j-symbol." Because of (1), we have

$$\begin{bmatrix} j_1 & R \\ \kappa_1 & \lambda_1 \end{bmatrix} \begin{bmatrix} j_2 & R \\ \kappa_2 & \lambda_2 \end{bmatrix} = \sum_{j_3} [j_3] \begin{pmatrix} j_1 & j_2 & j_3 \\ \kappa_1 & \kappa_2 & \kappa_3 \end{pmatrix}^* \begin{bmatrix} j_3 & R \\ \kappa_3 & \lambda_3 \end{bmatrix}^* \begin{pmatrix} j_1 & j_2 & j_3 \\ \lambda_1 & \lambda_2 & \lambda_3 \end{pmatrix} \tag{2}$$

or

$$\begin{pmatrix} j_1 & j_2 & j_3 \\ \kappa_1 & \kappa_2 & \kappa_3 \end{pmatrix} \begin{bmatrix} j_1 & R \\ \kappa_1 & \lambda_1 \end{bmatrix} \begin{bmatrix} j_2 & R \\ \kappa_2 & \lambda_2 \end{bmatrix} = \begin{bmatrix} j_3 & R \\ \kappa_3 & \lambda_3 \end{bmatrix}^* \begin{pmatrix} j_1 & j_2 & j_3 \\ \lambda_1 & \lambda_2 & \lambda_3 \end{pmatrix}. \tag{2a}$$

The left side is the $j_3\ \kappa_3;\ \lambda_1\ \lambda_2$ matrix element of the product of U with the Kronecker product of the representations j_1 and j_2, the right side is the same matrix element of the product of the reduced form of the above Kronecker product with U. *The summation convention is adopted in these and all following formulas for the greek letters*, i.e., a summation must be carried out over all greek letters occurring twice in the same expression. In order to make it unnecessary to give the limits of summation, we shall set the three-j-symbols equal to zero if the last j does not occur in the Kronecker product of the first two. Then all summations can be extended over all possible values of the variables. This holds, naturally, for the greek letters also, the possible values of which serve to denote the rows and columns of the corresponding representation.

Equation (2) can be given, in addition to (2a), further particular forms. Multiplying this equation by $\begin{bmatrix} j_3 & R \\ \kappa_3 & \lambda_3'' \end{bmatrix}$ and summing over κ_3, one obtains because of the unitary nature of the representation

$$\begin{bmatrix} j_1 & R \\ \kappa_1 & \lambda_1 \end{bmatrix} \begin{bmatrix} j_2 & R \\ \kappa_2 & \lambda_2 \end{bmatrix} \begin{bmatrix} j_3 & R \\ \kappa_3 & \lambda_3 \end{bmatrix} \begin{pmatrix} j_1 & j_2 & j_3 \\ \kappa_1 & \kappa_2 & \kappa_3 \end{pmatrix} = \begin{pmatrix} j_1 & j_2 & j_3 \\ \lambda_1 & \lambda_2 & \lambda_3 \end{pmatrix}. \tag{2b}$$

This expression can be said to express the fact that the three-j-symbols are invariant under the operations of the group. The conjugate imaginary of (2b) can be written, again on account of the unitary nature of the representations

$$\begin{bmatrix} j_1 & R \\ \lambda_1 & \kappa_1 \end{bmatrix} \begin{bmatrix} j_2 & R \\ \lambda_2 & \kappa_2 \end{bmatrix} \begin{bmatrix} j_3 & R \\ \lambda_3 & \kappa_3 \end{bmatrix} \begin{pmatrix} j_1 & j_2 & j_3 \\ \kappa_1 & \kappa_2 & \kappa_3 \end{pmatrix}^* = \begin{pmatrix} j_1 & j_2 & j_3 \\ \lambda_1 & \lambda_2 & \lambda_3 \end{pmatrix}^* \tag{2c}$$

where R^{-1} has been replaced by R. This equation shows that the conjugate imaginaries of the three-j-symbols are also invariant under the group operations though the role of rows and columns is interchanged.

Another form of (2) expresses the fact that while $U^{-1} \ldots U$ transforms the reduced form of the Kronecker product into the Kronecker product $U \ldots U^{-1}$ transforms the Kronecker product into its reduced form. This equation reads

$$\begin{pmatrix} j_1 & j_2 & j_3 \\ \kappa_1 & \kappa_2 & \kappa_3 \end{pmatrix} \begin{bmatrix} j_1 & R \\ \kappa_1 & \lambda_1 \end{bmatrix} \begin{bmatrix} j_2 & R \\ \kappa_2 & \lambda_2 \end{bmatrix} \begin{pmatrix} j_1 & j_2 & j_3' \\ \lambda_1 & \lambda_2 & \lambda_3 \end{pmatrix}^* = [j_3]^{-1}\, \delta_{j_3 j_3'} \begin{bmatrix} j_3 & R \\ \kappa_3 & \lambda_3 \end{bmatrix}^*. \tag{2c}$$

A last form of (2) is given by (5).

If we again adopt the notation that $(-1)^{2j}=1$ if j is an integer representation, $(-1)^{2j}=-1$ if j is a half integer representation, the above convention allows us to write

$$\begin{pmatrix} j_1 & j_2 & j_3 \\ \lambda_1 & \lambda_2 & \lambda_3 \end{pmatrix} = (-1)^{2j_1+2j_2+2j_3} \begin{pmatrix} j_1 & j_2 & j_3 \\ \lambda_1 & \lambda_2 & \lambda_3 \end{pmatrix} \tag{3}$$

because the three-j-symbol in (3) vanishes on account of the lemma in Reference 6 unless $(-1)^{2j_1+2j_2+2j_3}=1$. We shall write, furthermore $(-1)^j=+1$, if j is an even representation; $(-1)^j=-1$ if j is an odd representation (cf. Theorem 3 of reference 6). If j is integer but neither even nor odd, the value of $(-1)^j$ is immaterial for the following and can be taken to be $+1$. If j is half integer, $(-1)^j$ can be assumed to be i or $-i$ arbitrarily, provided that its value, once adopted, is maintained throughout.

The unitary character of the matrix (1) is expressed by

$$\sum_{j_3} [j_3] \begin{pmatrix} j_1 & j_2 & j_3 \\ \kappa_1 & \kappa_2 & \kappa_3 \end{pmatrix} \begin{pmatrix} j_1 & j_2 & j_3 \\ \lambda_1 & \lambda_2 & \kappa_3 \end{pmatrix}^* = \delta_{\kappa_1\lambda_1}\delta_{\kappa_2\lambda_2} \tag{4}$$

$$\begin{pmatrix} j_1 & j_2 & j \\ \lambda_1 & \lambda_2 & \lambda \end{pmatrix} \begin{pmatrix} j_1 & j_2 & j' \\ \lambda_1 & \lambda_2 & \lambda' \end{pmatrix}^* = [j]^{-1}(j_1 \quad j_2 \quad j)\delta_{jj'}\delta_{\lambda\lambda'}. \tag{4a}$$

Here again $(j_1 \quad j_2 \quad j)=1$, if the Kronecker product of j_1 and j_2 contains j; it is zero otherwise. The normalization of the three-j-symbols adopted in (1) is such that

$$\begin{pmatrix} j_1 & j_2 & j_3 \\ \lambda_1 & \lambda_2 & \lambda_3 \end{pmatrix} \begin{pmatrix} j_1 & j_2 & j_3 \\ \lambda_1 & \lambda_2 & \lambda_3 \end{pmatrix}^* = (j_1 \quad j_2 \quad j_3). \tag{4b}$$

The three-j-symbols are not completely determined by the condition that they satisfy (2). The reduced form of the Kronecker product commutes with all diagonal matrices in which the diagonal elements corresponding to one representation are the same (and only with these). Hence, the transforming matrix (1) can be multiplied by a diagonal matrix of this character. This corresponds to multiplying the three-j-symbol of (1) by a quantity depending only on j_3 (and $j_1 j_2$), but independent of κ_1, κ_2, κ_3. If we require that (1) shall continue to be unitary, the modulus of the indeterminate factor, $\omega(j_1, j_2, j_3)$, must be 1.

One can bring, by means of the orthogonality relations, the representation coefficient from the right side of (2) to the left side. This gives

$$\sum_R \begin{bmatrix} j_1 & R \\ \kappa_1 & \lambda_1 \end{bmatrix} \begin{bmatrix} j_2 & R \\ \kappa_2 & \lambda_2 \end{bmatrix} \begin{bmatrix} j_3 & R \\ \kappa_3 & \lambda_3 \end{bmatrix} = h \begin{pmatrix} j_1 & j_2 & j_3 \\ \kappa_1 & \kappa_2 & \kappa_3 \end{pmatrix}^* \begin{pmatrix} j_1 & j_2 & j_3 \\ \lambda_1 & \lambda_2 & \lambda_3 \end{pmatrix}. \tag{5}$$

The purpose of the factor $[j]^{1/2}$ in (1) was to make this expression symmetric in the representation coefficients. It follows from (5) at once that

$$\left|\begin{pmatrix} j_1 & j_2 & j_3 \\ \kappa_1 & \kappa_2 & \kappa_3 \end{pmatrix}\right|^2 = \left|\begin{pmatrix} j_2 & j_1 & j_3 \\ \kappa_2 & \kappa_1 & \kappa_3 \end{pmatrix}\right|^2 = \left|\begin{pmatrix} j_1 & j_3 & j_2 \\ \kappa_1 & \kappa_3 & \kappa_2 \end{pmatrix}\right|^2 \text{ etc.} \tag{5a}$$

i.e., that the interchange of the columns does not affect the absolute value of the three-j-symbols. It will be shown now that it is possible to normalize the latter in such a way that

$$\begin{pmatrix} j_1 & j_2 & j_3 \\ \kappa_1 & \kappa_2 & \kappa_3 \end{pmatrix} = (-1)^{j_1+j_2+j_3} \begin{pmatrix} j_2 & j_1 & j_3 \\ \kappa_2 & \kappa_1 & \kappa_3 \end{pmatrix} = (-1)^{j_1+j_2+j_3} \begin{pmatrix} j_1 & j_3 & j_2 \\ \kappa_1 & \kappa_3 & \kappa_2 \end{pmatrix} \tag{6}$$

and hence

$$\begin{pmatrix} j_1 & j_2 & j_3 \\ \kappa_1 & \kappa_2 & \kappa_3 \end{pmatrix} = \begin{pmatrix} j_2 & j_3 & j_1 \\ \kappa_2 & \kappa_3 & \kappa_1 \end{pmatrix} = \begin{pmatrix} j_3 & j_1 & j_2 \\ \kappa_3 & \kappa_1 & \kappa_2 \end{pmatrix}, \tag{6a}$$

i.e., that the three-j-symbols remain unchanged under an even permutation of their columns and are multiplied by $(-1)^{j_1+j_2+j_3}$ for an odd permutation. According to the lemma of Reference 6 and our convention for the values of $(-1)^j$ the $(-1)^{j_1+j_2+j_3} = \pm 1$, unless the three-$j$-symbols in (6) and (6a) vanish.

In order to obtain (6) and (6a), let us first choose a triple $\kappa_{10}, \kappa_{20}, \kappa_{30}$ of κ for which the three-j-symbol does not vanish. Let us multiply then the three-j-symbols of $j_1 j_2 j_3$ by a factor of modulus 1 which makes

$$\begin{pmatrix} j_1 & j_2 & j_3 \\ \kappa_{10} & \kappa_{20} & \kappa_{30} \end{pmatrix} \tag{*}$$

real and positive. We can then multiply all the symbols

$$\begin{pmatrix} j_2 & j_1 & j_3 \\ \kappa_2 & \kappa_1 & \kappa_3 \end{pmatrix} \tag{†}$$

with a second factor of modulus 1 so that the first part of (6) shall be correct for the triple $\kappa_{10}, \kappa_{20}, \kappa_{30}$. This is evidently possible if j_2 and j_1 are different, because in this case the factors by which (*) and (†) can be multiplied are entirely independent of each other. For $j_1 = j_2$ on the other hand, (6) only expresses the fact that the representation j_3 is in the symmetric or antisymmetric part of the Kronecker product of j_1 with itself, for even and odd values of $j_3 - 2j_1$, respectively. In a similar way, all the equations (6) and (6a) can be made to hold for a single triple $\kappa_1 = \kappa_{10}, \kappa_2 = \kappa_{20}, \kappa_3 = \kappa_{30}$ for which (*) does not vanish. It follows then from (5)—its left side is clearly unaltered if one interchanges the indices 1 and 2—that (6) and (6a) hold for all $\kappa_1, \kappa_2, \kappa_3$. In the following, we shall not necessarily retain the normalization which makes (*) real but may multiply all three-j-symbols with factors of modulus unity which depend symmetrically on j_1, j_2, j_3. This will leave equations (6) and (6a) valid. In the case of the three-dimensional rotation group, it will turn out to be possible to choose all three-j-symbols real and this will be the normalization adopted.

An important special case of (2) is obtained by taking for j_2 the identical representation $j_2 = 0$. The Kronecker product of this with j_1 is of course j_1 itself so that j_3 must be j_1 also. Since j_2 has only one row, no summation over κ_2 is necessary. We shall denote the only row of $j_2 = 0$ by $\kappa_2 = 0$. It then follows from (2) that

$$[j]^{1/2}\begin{pmatrix} j & 0 & j \\ \kappa & 0 & \mu \end{pmatrix} = \begin{pmatrix} & j & \\ \kappa & & \mu \end{pmatrix} \tag{7}$$

is the unitary matrix which transforms the representation j into the conjugate complex form

$$\begin{bmatrix} j & R \\ \kappa & \lambda \end{bmatrix} = \begin{pmatrix} & j & \\ \kappa & & \mu \end{pmatrix}^* \begin{bmatrix} j & R \\ \mu & \nu \end{bmatrix}^* \begin{pmatrix} & j & \\ \lambda & & \nu \end{pmatrix}. \tag{8}$$

The unitary condition (4) and (4a) becomes

$$\begin{pmatrix} & j & \\ \kappa & & \mu \end{pmatrix}^* \begin{pmatrix} & j & \\ \kappa & & \mu' \end{pmatrix} = \delta_{\mu\mu'}; \qquad \begin{pmatrix} & j & \\ \kappa' & & \mu \end{pmatrix}^* \begin{pmatrix} & j & \\ \kappa & & \mu \end{pmatrix} = \delta_{\kappa'\kappa}, \tag{9}$$

and instead of (6) we have

$$\begin{pmatrix} & j & \\ \kappa & & \mu \end{pmatrix} = (-1)^{2j} \begin{pmatrix} & j & \\ \mu & & \kappa \end{pmatrix} \tag{10}$$

a result well known from the investigations of Frobenius and Schur.[1] Since $(-1)^{4j} = 1$ for all j, it is permissible to interchange the indices on *both* "one-j-symbols" in (8).

The one-j-symbol defined in (7) can be made to play the role of a metric tensor by which to raise and lower indices. Thus we can write

$$\begin{pmatrix} j_1 & j_2 & j_3 \\ \kappa_1 & \kappa_2 & \kappa_3 \end{pmatrix} \begin{pmatrix} & j_3 & \\ \kappa_3 & & \lambda_3 \end{pmatrix}^* = \begin{pmatrix} j_1 & j_2 & \lambda_3 \\ \kappa_1 & \kappa_2 & j_3 \end{pmatrix} \tag{10a}$$

and

$$\begin{pmatrix} j_1 & j_2 & j_3 \\ \lambda_1 & \lambda_2 & \lambda_3 \end{pmatrix} \begin{pmatrix} & j & \\ \lambda_1 & & \kappa_1 \end{pmatrix}^* \begin{pmatrix} & j & \\ \lambda_2 & & \kappa_2 \end{pmatrix}^* \begin{pmatrix} & j & \\ \lambda_3 & & \kappa_3 \end{pmatrix}^* = \begin{pmatrix} \kappa_1 & \kappa_2 & \kappa_3 \\ j_1 & j_2 & j_3 \end{pmatrix}. \tag{10b}$$

If one wants to use the notation of (10a) and (10b), defining covariant (straight) and contravariant (time inverted) three-j-symbols, one has to be careful when substituting numbers for the j and κ. It is necessary then to use a different type of number for the designation of the representations $j = 0, 1, 2, \ldots$ and the rows and columns of the representations $\lambda = 0, 1, 2, \ldots$.

Because of (9), one can lower an index by the conjugate complex of the metric

$$\begin{pmatrix} & j_3 & \\ \kappa_3 & & \lambda_3 \end{pmatrix} \begin{pmatrix} j_1 & j_2 & \lambda_3 \\ \kappa_1 & \kappa_2 & j_3 \end{pmatrix} = \begin{pmatrix} j_1 & j_2 & j_3 \\ \kappa_1 & \kappa_2 & \kappa_3 \end{pmatrix}. \tag{10c}$$

If one wanted to adopt this notation consistently, it would be appropriate to write

$$\begin{pmatrix} & j & \\ \kappa & & \lambda \end{pmatrix}^* = \begin{pmatrix} \kappa & & \lambda \\ & j & \end{pmatrix} \tag{10d}$$

whence the raising of the index in (10a) and (10b) would follow the natural rule. It should be remembered though that the metric tensor (7) is symmetric only for integer j; for half integer j it is skew symmetric. Hence the raising of an index must be done by using the left index of the metric tensor as summation index; in the lowering process the right index must be used as summation index. Since the three-j-symbols will be all real in the case of the three-dimensional rotation group, the same will hold also for the metric or one-j-symbol. Hence raising and lowering of indices is done by the same tensor in this case; nevertheless, the order of the indices remains to be observed. If the indices refer to a half-integer j, the scalar products $a_\mu b^\mu$ and $a^\mu b_\mu$ have opposite signs.

The idea of using the contravariant and covariant notation for three-j-symbols is due to C. Herring. It will not be made full use of in the present article.

It follows from (8) and (9) that

$$\begin{bmatrix} j_1 & R \\ \kappa_1 & \lambda_1 \end{bmatrix}\begin{bmatrix} j_2 & R \\ \kappa_2 & \lambda_2 \end{bmatrix}\begin{bmatrix} j_3 & R \\ \kappa_3 & \lambda_3 \end{bmatrix}\begin{pmatrix} & j_1 & \\ \lambda_1 & & \nu_1 \end{pmatrix}^*\begin{pmatrix} & j_2 & \\ \lambda_2 & & \nu_2 \end{pmatrix}\begin{pmatrix} & j_3 & \\ \lambda_3 & & \nu_3 \end{pmatrix}^*$$

$$= \begin{pmatrix} & j_1 & \\ \kappa_1 & & \mu_1 \end{pmatrix}^*\begin{pmatrix} & j_2 & \\ \kappa_2 & & \mu_2 \end{pmatrix}^*\begin{pmatrix} & j_3 & \\ \kappa_3 & & \mu_3 \end{pmatrix}^*\begin{bmatrix} j_1 & R_1 \\ \mu_1 & \nu_1 \end{bmatrix}^*\begin{bmatrix} j_2 & R_2 \\ \mu_2 & \nu_2 \end{bmatrix}^*\begin{bmatrix} j_3 & R_3 \\ \mu_3 & \nu_3 \end{bmatrix}^*$$

and this equation gives with the aid of (5) when summed over all R

$$\begin{pmatrix} j_1 & j_2 & j_3 \\ \kappa_1 & \kappa_2 & \kappa_3 \end{pmatrix}^*\begin{pmatrix} j_1 & j_2 & j_3 \\ \lambda_1 & \lambda_2 & \lambda_3 \end{pmatrix}\begin{pmatrix} & j_1 & \\ \lambda_1 & & \nu_1 \end{pmatrix}^*\begin{pmatrix} & j_2 & \\ \lambda_2 & & \nu_2 \end{pmatrix}^*\begin{pmatrix} & j_3 & \\ \lambda_3 & & \nu_3 \end{pmatrix}^*$$

$$= \begin{pmatrix} & j_1 & \\ \kappa_1 & & \mu_1 \end{pmatrix}^*\begin{pmatrix} & j_2 & \\ \kappa_2 & & \mu_2 \end{pmatrix}^*\begin{pmatrix} & j_3 & \\ \kappa_3 & & \mu_3 \end{pmatrix}^*\begin{pmatrix} j_1 & j_2 & j_3 \\ \mu_1 & \mu_2 & \mu_3 \end{pmatrix}\begin{pmatrix} j_1 & j_2 & j_3 \\ \nu_1 & \nu_2 & \nu_3 \end{pmatrix}^*.$$

Dividing this by the first factor on the left side one sees that

$$\begin{pmatrix} j_1 & j_2 & j_3 \\ \lambda_1 & \lambda_2 & \lambda_3 \end{pmatrix}\begin{pmatrix} & j_1 & \\ \lambda_1 & & \nu_1 \end{pmatrix}^*\begin{pmatrix} & j_2 & \\ \lambda_2 & & \nu_2 \end{pmatrix}^*\begin{pmatrix} & j_3 & \\ \lambda_3 & & \nu_3 \end{pmatrix}^* = c\begin{pmatrix} j_1 & j_2 & j_3 \\ \nu_1 & \nu_2 & \nu_3 \end{pmatrix}^* \tag{11}$$

where c does not depend on the ν. One can bring by means of the orthogonality relations (9), the last three factors of (11) to the right side. The conjugate complex of the resulting equation is

$$(c^*)^{-1}\begin{pmatrix} j_1 & j_2 & j_3 \\ \lambda_1 & \lambda_2 & \lambda_3 \end{pmatrix}^* = \begin{pmatrix} & j_1 & \\ \lambda_1 & & \nu_1 \end{pmatrix}^*\begin{pmatrix} & j_2 & \\ \lambda_2 & & \nu_2 \end{pmatrix}^*\begin{pmatrix} & j_3 & \\ \lambda_3 & & \nu_3 \end{pmatrix}^*\begin{pmatrix} j_1 & j_2 & j_3 \\ \nu_1 & \nu_2 & \nu_3 \end{pmatrix}.$$

Interchanging herein the λ with the ν, one obtains (11) back again and

$$|c|^2 = (-1)^{2j_1+2j_2+2j_3} = 1. \tag{12}$$

One sees from (11) furthermore that c is a symmetric function of the j. Hence one can make $c = 1$ by multiplying the three-j-symbols by factors of modulus unity which depend only on the j and are symmetric functions of these. The only equation which could be changed by this renormalization is (7). One convinces oneself easily, however, that $c = 1$ automatically holds if $j_2 = 0$, provided that

$$\begin{pmatrix} & 0 & \\ 0 & & 0 \end{pmatrix} = 1 \tag{13}$$

i.e., if we choose the unit matrix for the matrix which transforms the unit matrix into its conjugate complex. One then has, instead of (11)

$$\begin{pmatrix} j_1 & j_2 & j_3 \\ \nu_1 & \nu_2 & \nu_3 \end{pmatrix}^* = \begin{pmatrix} j_1 & j_2 & j_3 \\ \lambda_1 & \lambda_2 & \lambda_3 \end{pmatrix} \begin{pmatrix} & j_1 & \\ \lambda_1 & & \nu_1 \end{pmatrix}^* \begin{pmatrix} & j_2 & \\ \lambda_2 & & \nu_2 \end{pmatrix}^* \begin{pmatrix} & j_3 & \\ \lambda_3 & & \nu_3 \end{pmatrix}^* . \tag{11a}$$

Because of (10) and (3), the order of the greek indices can be reversed in all one-j-symbols of (11a) simultaneously. It shows that *the fully covariant and fully contravariant components of a three-j-symbol are conjugate complex.* For the sake of future reference, we rewrite (2) by means of (8) and (11), (9), (10)

$$\begin{aligned}
&\begin{bmatrix} j_1 & R \\ \kappa_1 & \lambda_1 \end{bmatrix} \begin{bmatrix} j_2 & R \\ \kappa_2 & \lambda_2 \end{bmatrix} \\
&= \sum_{j_3} [j_3] \begin{pmatrix} j_1 & j_2 & j_3 \\ \mu_1 & \mu_2 & \mu_3 \end{pmatrix} \begin{pmatrix} & j_1 & \\ \mu_1 & & \kappa_1 \end{pmatrix}^* \begin{pmatrix} & j_2 & \\ \mu_2 & & \kappa_2 \end{pmatrix}^* \begin{pmatrix} & j_3 & \\ \kappa_3 & & \lambda_3 \end{pmatrix}^* \begin{bmatrix} j_3 & R \\ \mu_3 & \kappa_3 \end{bmatrix} \begin{pmatrix} j_1 & j_2 & j_3 \\ \lambda_1 & \lambda_2 & \lambda_3 \end{pmatrix} \\
&= \sum_{j_3} [j_3] \begin{pmatrix} \kappa_1 & \kappa_2 & j_3 \\ j_1 & j_2 & \mu_3 \end{pmatrix} \begin{bmatrix} j_3 & R \\ \mu_3 & \kappa_3 \end{bmatrix} \begin{pmatrix} & j_3 & \\ \kappa_3 & & \lambda_3 \end{pmatrix}^* \begin{pmatrix} j_1 & j_2 & j_3 \\ \lambda_1 & \lambda_2 & \lambda_3 \end{pmatrix}
\end{aligned} \tag{14}$$

which shows that the row-index of the representation should be a contravariant (upper) index.

The particular case of (4a) in which $j' = 0$, $j_1 = j_2$ is significant for many applications

$$\begin{pmatrix} j_1 & j & j_1 \\ \lambda_1 & \lambda & \lambda_2 \end{pmatrix} \begin{pmatrix} & j_1 & \\ \lambda_1 & & \lambda_2 \end{pmatrix}^* = [j_1]^{1/2} \delta_{j0} \delta_{\lambda 0} , \tag{15}$$

Again, this can be written as

$$\begin{pmatrix} j_1 & \lambda_1 & j \\ \lambda_1 & j_1 & \lambda \end{pmatrix} = [j_1]^{1/2} \delta_{j0} \delta_{\lambda 0} . \tag{15a}$$

It may be well to record here to what extent the indeterminate phases in our different symbols are fixed. Naturally, all the irreducible representations can be transformed by any unitary matrix, and this will affect the three-j-symbols and also (7). Even if the form of the representations is supposed to be fixed, all one-j-symbols [except (13)] can be multiplied by a factor depending on j. The three-j-symbols will then be multiplied by the $-\frac{1}{2}$ power of the product of the corresponding three factors. If, however, all one-j-symbols (7) are supposed to be fixed, only the signs of the three-j-symbols can be changed and these only simultaneously for all three-j-symbols containing the same three j's.

3. *The Six-j-Symbols* (*Racah Coefficients*)

It is clear that some connection must exist between the three-j-symbols which corresponds to the associative law of ordinary multiplication. One can expand by (14) the product of three representation coefficients in three different ways, one of which is

$$\begin{aligned}
&\begin{bmatrix} j_1 & R \\ \kappa_1 & \lambda_1 \end{bmatrix}\begin{bmatrix} j_2 & R \\ \kappa_2 & \lambda_2 \end{bmatrix}\begin{bmatrix} j_3 & R \\ \kappa_3 & \lambda_3 \end{bmatrix} \\
&= \sum_j [j] \begin{pmatrix} j_1 & j_2 & j \\ \mu_1 & \mu_2 & \kappa \end{pmatrix}\begin{pmatrix} j_1 & j_2 & j \\ \lambda_1 & \lambda_2 & \eta \end{pmatrix}\begin{pmatrix} & j_1 & \\ \mu_1 & & \kappa_1 \end{pmatrix}^* \\
&\quad \cdot \begin{pmatrix} & j_2 & \\ \mu_2 & & \kappa_2 \end{pmatrix}^*\begin{pmatrix} & j & \\ \lambda & & \eta \end{pmatrix}^*\begin{bmatrix} j & R \\ \kappa & \lambda \end{bmatrix}\begin{bmatrix} j_3 & R \\ \kappa_3 & \lambda_3 \end{bmatrix} \\
&= \sum_{jj_4} [j][j_4] \begin{pmatrix} j_1 & j_2 & j \\ \mu_1 & \mu_2 & \kappa \end{pmatrix}\begin{pmatrix} j_1 & j_2 & j \\ \lambda_1 & \lambda_2 & \eta \end{pmatrix}\begin{pmatrix} j & j_3 & j_4 \\ \mu & \mu_3 & \mu_4 \end{pmatrix}\begin{pmatrix} j & j_3 & j_4 \\ \lambda & \lambda_3 & \lambda_4 \end{pmatrix} \\
&\quad \cdot \begin{pmatrix} & j & \\ \lambda & & \eta \end{pmatrix}^*\begin{pmatrix} & j & \\ \mu & & \kappa \end{pmatrix}^*\begin{pmatrix} & j_1 & \\ \mu_1 & & \kappa_1 \end{pmatrix}^*\begin{pmatrix} & j_2 & \\ \mu_2 & & \kappa_2 \end{pmatrix}^*\begin{pmatrix} & j_3 & \\ \mu_3 & & \kappa_3 \end{pmatrix}^*\begin{pmatrix} & j_4 & \\ \eta_4 & & \lambda_4 \end{pmatrix}^*\begin{bmatrix} j_4 & R \\ \mu_4 & \eta_4 \end{bmatrix}.
\end{aligned} \tag{15b}$$

The left side of (15b) is symmetric in 1 and 3 and the right side must also remain unchanged if one interchanges these two indices. This gives an equation which is valid for all R. Since the representation coefficients are orthogonal, their coefficients must be equal on the two sides of the equation. After some obvious simplifications, this assumes the form

$$\begin{aligned}
&\sum_{\substack{j \\ \kappa\mu}} [j] \begin{pmatrix} j_1 & j_2 & j \\ \mu_1 & \mu_2 & \kappa \end{pmatrix}\begin{pmatrix} j_3 & j_4 & j \\ \mu_3 & \mu_4 & \mu \end{pmatrix}\begin{pmatrix} & j & \\ \mu & & \kappa \end{pmatrix}^* \\
&\qquad\qquad \cdot \sum_{\eta\lambda} \begin{pmatrix} j_1 & j_2 & j \\ \lambda_1 & \lambda_2 & \eta \end{pmatrix}\begin{pmatrix} j_3 & j_4 & j \\ \lambda_3 & \lambda_4 & \lambda \end{pmatrix}\begin{pmatrix} & j & \\ \lambda & & \eta \end{pmatrix}^* \\
&= \sum_{j\kappa\mu} [j] \begin{pmatrix} j_3 & j_2 & j \\ \mu_3 & \mu_2 & \kappa \end{pmatrix}\begin{pmatrix} j_1 & j_4 & j \\ \mu_1 & \mu_4 & \mu \end{pmatrix}\begin{pmatrix} & j & \\ \mu & & \kappa \end{pmatrix}^* \\
&\qquad\qquad \cdot \sum_{\eta\lambda} \begin{pmatrix} j_3 & j_2 & j \\ \lambda_3 & \lambda_2 & \eta \end{pmatrix}\begin{pmatrix} j_1 & j_4 & j \\ \lambda_1 & \lambda_4 & \lambda \end{pmatrix}\begin{pmatrix} & j & \\ \lambda & & \eta \end{pmatrix}^*.
\end{aligned} \tag{16}$$

All summations occurring in (16) are so indicated. Equations (16) form a set of linear equations for the quantities appearing on the left side after the dot. One can solve them by multiplying (16) with

$$\begin{pmatrix} j_1 & j_2 & j' \\ \mu_1 & \mu_2 & \kappa' \end{pmatrix}^* \begin{pmatrix} j_3 & j_4 & j' \\ \mu_3 & \mu_4 & \mu' \end{pmatrix}^* \begin{pmatrix} & j' & \\ \mu' & & \kappa' \end{pmatrix}$$

and summing over $\mu_1\mu_2\mu_3\mu_4\kappa'\mu'$. The quantities after the dot on the left will appear then as linear combinations of the expressions after the dot on the right side of (16), with coefficients which depend on all j but are independent of the λ. We can, further, interchange the lower indices of the one-j-symbols by means of (10) and write,

$$\begin{pmatrix} j_1 & j_2 & j' \\ \lambda_1 & \lambda_2 & \eta \end{pmatrix} \begin{pmatrix} j_3 & j_4 & j' \\ \lambda_3 & \lambda_4 & \lambda \end{pmatrix} \begin{pmatrix} & j' & \\ \eta & & \lambda \end{pmatrix}^*$$
$$= \sum_j (-1)^{2j_4}[j] \begin{Bmatrix} j_1 & j_2 & j' \\ j_3 & j_4 & j \end{Bmatrix} \begin{pmatrix} j_3 & j_2 & j \\ \lambda_3 & \lambda_2 & \eta \end{pmatrix} \begin{pmatrix} j_1 & j_4 & j \\ \lambda_1 & \lambda_4 & \lambda \end{pmatrix} \begin{pmatrix} & j & \\ \eta & & \lambda \end{pmatrix}^*, \quad (17)$$

where the summation convention is again used for the greek indices. The coefficients given by the curly brackets will be called six-j-symbols.

It is best now to bring the three-j-symbols of the right side of (17) over to the left side. This can be done by means of the orthogonality relations (4) and (9), multiplying (17) with

$$\begin{pmatrix} j_3 & j_2 & j'' \\ \lambda_3 & \lambda_2 & \eta'' \end{pmatrix}^* \begin{pmatrix} j_1 & j_4 & j'' \\ \lambda_1 & \lambda_4 & \lambda'' \end{pmatrix}^* \begin{pmatrix} & j'' & \\ \eta'' & & \lambda'' \end{pmatrix}.$$

This gives, after a slight change in notation and by means of (11a)

$$\begin{Bmatrix} j_1 & j_2 & j_3 \\ l_1 & l_2 & l_3 \end{Bmatrix} = \begin{pmatrix} j_1 & j_2 & j_3 \\ \iota_1 & \iota_2 & \iota_3 \end{pmatrix} \begin{pmatrix} j_1 & l_2 & l_3 \\ \kappa_1 & \lambda_2 & \mu_3 \end{pmatrix} \begin{pmatrix} l_1 & j_2 & l_3 \\ \mu_1 & \kappa_2 & \lambda_3 \end{pmatrix} \begin{pmatrix} l_1 & l_2 & j_3 \\ \lambda_1 & \mu_2 & \kappa_3 \end{pmatrix}$$
$$\cdot \begin{pmatrix} & j_1 & \\ \iota_1 & & \kappa_1 \end{pmatrix}^* \begin{pmatrix} & j_2 & \\ \iota_2 & & \kappa_2 \end{pmatrix}^* \begin{pmatrix} & j_3 & \\ \iota_3 & & \kappa_3 \end{pmatrix}^* \begin{pmatrix} & l_1 & \\ \lambda_1 & & \mu_1 \end{pmatrix}^* \begin{pmatrix} & l_2 & \\ \lambda_2 & & \mu_2 \end{pmatrix}^* \begin{pmatrix} & l_3 & \\ \lambda_3 & & \mu_3 \end{pmatrix}^*$$
$$= \begin{pmatrix} \kappa_1 & \kappa_2 & \kappa_3 \\ j_1 & j_2 & j_3 \end{pmatrix} \begin{pmatrix} j_1 & \mu_2 & l_3 \\ \kappa_1 & l_2 & \mu_3 \end{pmatrix} \begin{pmatrix} l_1 & j_2 & \mu_3 \\ \mu_1 & \kappa_2 & l_3 \end{pmatrix} \begin{pmatrix} \mu_1 & l_2 & j_3 \\ l_1 & \mu_2 & \kappa_3 \end{pmatrix}. \quad (18)$$

It follows from (18) and (6a) that

$$\begin{Bmatrix} j_1 & j_2 & j_3 \\ l_1 & l_2 & l_3 \end{Bmatrix} = \begin{Bmatrix} j_2 & j_3 & j_1 \\ l_2 & l_3 & l_1 \end{Bmatrix} = \begin{Bmatrix} j_3 & j_1 & j_2 \\ l_3 & l_1 & l_2 \end{Bmatrix} \quad \mathbf{(19)}$$

and with (6) also

$$\begin{Bmatrix} j_1 & j_2 & j_3 \\ l_1 & l_2 & l_3 \end{Bmatrix} = \begin{Bmatrix} j_2 & j_1 & j_3 \\ l_2 & l_1 & l_3 \end{Bmatrix} = \begin{Bmatrix} j_1 & j_3 & j_2 \\ l_1 & l_3 & l_2 \end{Bmatrix} \text{etc.} \quad \mathbf{(19a)}$$

so that the columns of a six-j-symbol can be interchanged arbitrarily. It follows in a similar way that

$$\begin{Bmatrix} j_1 & j_2 & j_3 \\ l_1 & l_2 & l_3 \end{Bmatrix} = \begin{Bmatrix} l_1 & l_2 & j_3 \\ j_1 & j_2 & l_3 \end{Bmatrix} = \begin{Bmatrix} l_1 & j_2 & l_3 \\ j_1 & l_2 & j_3 \end{Bmatrix} = \begin{Bmatrix} j_1 & l_2 & l_3 \\ l_1 & j_2 & j_3 \end{Bmatrix}, \qquad \textbf{(19b)}$$

i.e., that it is possible to interchange in two columns simultaneously the upper representation with the lower. The symmetry of the six-j-symbols corresponds to that of a regular tetrahedron; the representations in one column correspond to opposite edges. Those which occur in the same three-j-symbol in (18) meet at one corner. The six-j-symbol vanishes unless the Kronecker product of any two edges meeting at one corner contains the representation corresponding to the third edge of that corner.

By substituting (11a) into the conjugate complex of (18) one easily verifies that it is equal to (18) itself, i.e., the six-j-symbols are *real*.

Interchanging now the indices 1 and 3 in (17), the expressions on the right side of (17) will appear as linear combinations of the expressions on the left side. Since these are, as functions of $\lambda_1, \lambda_2, \lambda_3, \lambda_4$ evidently linearly independent, it is clear that the two systems of coefficients form reciprocal matrices:

$$\sum_j [j] \begin{Bmatrix} j_1 & j_2 & j' \\ j_3 & j_4 & j \end{Bmatrix} \cdot [j''] \begin{Bmatrix} j_3 & j_2 & j \\ j_1 & j_4 & j'' \end{Bmatrix} = \delta_{j'j''}. \qquad (20)$$

Since the first and last column in the second six-j-symbol can be turned upside down, (20) expresses the fact that

$$R_{j'j} = ([j][j'])^{1/2} \begin{Bmatrix} j_1 & j_2 & j' \\ j_3 & j_4 & j \end{Bmatrix} \qquad (20a)$$

is a real orthogonal matrix. The possible values of j' are those which are contained in both Kronecker products $j_1 \times j_2$ and $j_3 \times j_4$. The possible values of j are the common parts of $j_1 \times j_4$ and $j_3 \times j_2$. According to (20a), the number of the common irreducible parts of $j_1 \times j_2$ and $j_3 \times j_4$ must be equal with the number of common irreducible parts of $j_1 \times j_4$ and $j_2 \times j_3$—a fact which can be also directly verified.

One can interchange the first two columns of the first three-j-symbol on the right side of (17). This introduces a factor $(-1)^{j+j_2+j_3}$. After this, one can apply (17) again to the right side of the resulting equation which becomes hereupon

$$\sum_{jj''} (-1)^{j+j_2+j_3} [j][j''] \cdot \begin{Bmatrix} j_1 & j_2 & j' \\ j_3 & j_4 & j \end{Bmatrix} \begin{Bmatrix} j_2 & j_3 & j \\ j_1 & j_4 & j'' \end{Bmatrix} \begin{pmatrix} j_1 & j_3 & j'' \\ \lambda_1 & \lambda_3 & \eta \end{pmatrix} \begin{pmatrix} j_2 & j_4 & j'' \\ \lambda_2 & \lambda_4 & \lambda \end{pmatrix} \begin{pmatrix} & j'' & \\ \eta & & \lambda \end{pmatrix}^*.$$

On the other hand, one can first interchange the first two columns in the first three-j-symbol on the left side of (17) and then use (17). This gives

$$\sum_{j''}(-1)^{2j_4+j_1+j_2+j'}[j'']\begin{Bmatrix} j_2 & j_1 & j' \\ j_3 & j_4 & j'' \end{Bmatrix}\begin{pmatrix} j_3 & j_1 & j'' \\ \lambda_3 & \lambda_1 & \eta \end{pmatrix}\begin{pmatrix} j_2 & j_4 & j'' \\ \lambda_2 & \lambda_4 & \lambda \end{pmatrix}\begin{pmatrix} & j'' & \\ \eta & & \lambda \end{pmatrix}^*.$$

Comparing this with the preceding expression yields

$$\sum_{j}(-1)^{j+j'+j''}[j]\begin{Bmatrix} j_1 & j_2 & j' \\ j_3 & j_4 & j \end{Bmatrix}\begin{Bmatrix} j_2 & j_3 & j \\ j_1 & j_4 & j'' \end{Bmatrix} = \begin{Bmatrix} j_2 & j_1 & j' \\ j_3 & j_4 & j'' \end{Bmatrix}. \tag{21}$$

Bringing the right side over to the left by (20) gives a more symmetric form to this

$$\sum_{ll'}(-1)^{j+l+l'}[j][l][l']\begin{Bmatrix} j_1 & j_2 & j \\ j_3 & j_4 & l \end{Bmatrix}\begin{Bmatrix} j_1 & j_4 & l \\ j_2 & j_3 & l' \end{Bmatrix}\begin{Bmatrix} j_1 & j_3 & l' \\ j_4 & j_2 & j' \end{Bmatrix} = \delta_{jj'}. \tag{21a}$$

The arrangement of the j_k in (21a) is as follows: j_1 is always in the upper left corner. It would be equally possible, however, to bring any other j_k to that place. Otherwise, the arrangement of the j_k in all three six-j-symbols is different from each other but such that both six-j-symbols, in which a summation index (l or l') occurs, limit the values of this index to the same set. This is, for l, e.g., the common part $j_1 \times j_4$ and $j_2 \times j_3$. Similarly, j and j' are also limited to the same set. As special cases of the above formulas we may note the following ones. Writing 0 for l_1 in (18), i.e., substituting the unit representation therefor, gives zero, unless $l_3 = j_2$, $l_2 = j_3$. In this case the summation can be carried out by (4b), (7), (9), and (11a) and gives

$$\begin{Bmatrix} j_1 & j_2 & j_3 \\ 0 & j_3 & j_2 \end{Bmatrix} = (-1)^{j_1+j_2+j_3}([j_2][j_3])^{-1/2}(j_1 j_2 j_3). \tag{18b}$$

Writing now 0 for j'' in (20) we obtain in case of $j_1 = j_2$, $j_3 = j_4$

$$\sum_{j}(-1)^{j_1+j_3+j}[j]\begin{Bmatrix} j_1 & j_1 & j' \\ j_3 & j_3 & j \end{Bmatrix} = \delta_{j'0}([j_1][j_3])^{1/2}. \tag{20b}$$

Writing 0 for j'' in (21) gives in a similar way

$$\sum_{j}[j]\begin{Bmatrix} j_1 & j_2 & j' \\ j_1 & j_2 & j \end{Bmatrix} = (-1)^{2j_1+2j_2}. \tag{21b}$$

We finally note a formula which immediately follows from (17) and which will be useful for the calculation of the three-j-symbols. One obtains it by bringing a three-j-symbol in (17) from the right to the left side by (11a) and (4a). After changing the notation somewhat, one has

$$\begin{Bmatrix} j_1 & j_2 & j_3 \\ l_1 & l_2 & l_3 \end{Bmatrix}\begin{pmatrix} j_1 & j_2 & j_3 \\ \iota_1 & \iota_2 & \iota_3 \end{pmatrix}$$

$$= \begin{pmatrix} j_1 & l_2 & l_3 \\ \iota_1 & \lambda_2 & \eta_3 \end{pmatrix}\begin{pmatrix} l_1 & j_2 & l_3 \\ \eta_1 & \iota_2 & \lambda_3 \end{pmatrix}\begin{pmatrix} l_1 & l_2 & j_3 \\ \lambda_1 & \eta_2 & \iota_3 \end{pmatrix}\begin{pmatrix} & l_1 & \\ \lambda_1 & & \eta_1 \end{pmatrix}^*\begin{pmatrix} & l_2 & \\ \lambda_2 & & \eta_2 \end{pmatrix}^*\begin{pmatrix} & l_3 & \\ \lambda_3 & & \eta_3 \end{pmatrix}^*$$

$$= \begin{pmatrix} j_1 & \eta_2 & l_3 \\ \iota_1 & l_2 & \eta_3 \end{pmatrix}\begin{pmatrix} l_1 & j_2 & \eta_3 \\ \eta_1 & \iota_2 & l_3 \end{pmatrix}\begin{pmatrix} \eta_1 & l_2 & j_3 \\ l_1 & \eta_2 & \iota_3 \end{pmatrix}. \tag{22}$$

4. *The Metric in Representation Space*

If one subjects the representation j_1 to a unitary transformation

$$\left[\left[\begin{matrix} j_1 & R \\ \mu & \nu \end{matrix}\right]\right] = u^*_{\kappa\mu} \left[\begin{matrix} j_1 & R \\ \kappa & \lambda \end{matrix}\right] u_{\lambda\nu} \tag{23}$$

all the preceding formulae remain valid if the three-j-symbols are replaced at the same time by

$$\left(\left(\begin{matrix} j_1 & j_2 & j_3 \\ \mu & \kappa_2 & \kappa_3 \end{matrix}\right)\right) = \left(\begin{matrix} j_1 & j_2 & j_3 \\ \kappa & \kappa_2 & \kappa_3 \end{matrix}\right) u_{\kappa\mu} \qquad (j_2, j_3 \neq j_1). \tag{23a}$$

The one-j-symbols have to be replaced by

$$\left(\left(\begin{matrix} & j & \\ \mu & & \nu \end{matrix}\right)\right) = \left(\begin{matrix} & j & \\ \kappa & & \lambda \end{matrix}\right) u_{\kappa\mu} u_{\lambda\nu}. \tag{23b}$$

This shows that the index κ_1 of three-j-symbol $\left(\begin{matrix} j_1 & j_2 & j_3 \\ \kappa_1 & \kappa_2 & \kappa_3 \end{matrix}\right)$ has indeed the character of the index of a vector in the unitary representation space of the representation j_1. The same holds, naturally, of the other greek indices in three-j-symbols. The one-j-symbol on the other hand, is a tensor, symmetric or antisymmetric, depending on the integer or half integer character of j. The possibility to use the one-j-symbol as a metric tensor is connected with the possibility of introducing, in the spaces of the irreducible representations, the operation of the time inversion. In fact, if ψ_λ are the components of a vector in representation space then $\phi_\lambda = (\psi^\lambda)^*$ are the components of the vector obtained by time inversion where

$$\psi^\lambda = \psi_\kappa \left(\begin{matrix} & j & \\ \kappa & & \lambda \end{matrix}\right)^*. \tag{24}$$

Since the operation of time inversion is independent of the choice of coordinates in the representation space and invariant under the transformations of the underlying group, which are all spatial, the transition to the contravariant components must also have these properties. This can be verified also directly and, unless one is willing to accept the above arguments, such a verification is necessary to establish the fact that the calculus with co- and contravariant components is consistent. As far as the group transformations are concerned, this is a consequence of (8), but the verification will not be carried out in detail. If the components of the vector ψ in the second coordinate system are $\bar{\psi}_\nu$, one has

$$\bar{\psi}^\nu = \bar{\psi}_\mu \left(\left(\begin{matrix} & j & \\ \mu & & \nu \end{matrix}\right)\right)^* = \psi_{\kappa'} u_{\kappa'\mu} \left(\begin{matrix} & j & \\ \kappa & & \lambda \end{matrix}\right)^* u^*_{\kappa\mu} u^*_{\lambda\nu} = \psi_\kappa \left(\begin{matrix} & j & \\ \kappa & & \lambda \end{matrix}\right)^* u^*_{\lambda\nu} = \psi^\lambda u^*_{\lambda\nu}. \tag{24a}$$

This gives the transformation law for the contravariant components. One easily verifies now that the contraction of two vectors ψ and ϕ

$$\bar{\psi}^{\nu}\bar{\phi}_{\nu} = \psi^{\lambda}u^{*}_{\lambda\nu}\phi_{\kappa}u_{\kappa\nu} = \psi^{\kappa}\phi_{\kappa} \tag{25}$$

is independent of the coordinate system. This assures the consistency of the calculus. It should be noted, however, that for half integer j the antisymmetric nature of the one-j-symbol renders $\psi^{\kappa}\phi_{\kappa} = -\psi_{\kappa}\phi^{\kappa}$.

The question of invariants with respect to coordinate transformations in representation space naturally arises. One such invariant is clearly

$$\begin{pmatrix} j_1 & j_2 & j_3 \\ \kappa_1 & \kappa_2 & \kappa_3 \end{pmatrix}\begin{pmatrix} \kappa_1 & \kappa_2 & \kappa_3 \\ j_1 & j_2 & j_3 \end{pmatrix}.$$

However, the value of this is given by (4b) to be $(j_1j_2j_3)$. It is reasonable to try, next, combinations of the form

$$\begin{pmatrix} j_1 & \kappa_1 & j_2 \\ \kappa_1 & j_1 & \kappa_2 \end{pmatrix}\begin{pmatrix} \kappa_2 & j_3 & \kappa_3 \\ j_2 & \kappa_3 & j_3 \end{pmatrix}.$$

However, these vanish, on account of (15a), unless $j_2 = 0$, i.e., unless the three-j-symbol is actually a one-j-symbol. In this case, the expression is again trivial.

The simplest nontrivial invariants with respect to coordinate transformations in the representation spaces are made up of four three-j-symbols; (18) shows that the six-j-symbol is such an invariant. In view of (4a), one easily convinces oneself that it is the most general invariant consisting of only four three-j-symbols which cannot be fully contracted. This explains the significance of the six-j-symbols.

5. *Further Properties of the Six-j-Symbols*

One can derive a relation for the six-j-symbols which is similar both in its derivation and its applications to Eq. (17), valid for the three-j-symbols. For this purpose, one has to consider such combinations of three-j-symbols which can be expressed in different ways by a series of three-j-symbols. In (15b), we started from a combination of representation coefficients which we expressed in two ways as a series of representation coefficients. Comparison of coefficients of the two series gave us a relation for the three-j-symbols and will give us a relation for six-j-symbols in this section.

In

$$\begin{pmatrix} l_1 & k_2 & j_3 \\ \lambda_1 & \kappa_2 & \iota_3 \end{pmatrix}\begin{pmatrix} j_1 & l_2 & k_3 \\ \iota_1 & \lambda_2 & \kappa_3 \end{pmatrix}\begin{pmatrix} k_1 & j_2 & l_3 \\ \kappa_1 & \iota_2 & \lambda_3 \end{pmatrix}\begin{pmatrix} l_1 & l_2 & l_3 \\ \lambda_1' & \lambda_2' & \lambda_3' \end{pmatrix}\begin{pmatrix} k_1 & k_2 & k_3 \\ \kappa_1' & \kappa_2' & \kappa_3' \end{pmatrix}$$
$$\cdot\begin{pmatrix} & l_1 & \\ \lambda_1 & & \lambda_1' \end{pmatrix}^{*}\begin{pmatrix} & l_2 & \\ \lambda_2 & & \lambda_2' \end{pmatrix}^{*}\begin{pmatrix} & l_3 & \\ \lambda_3 & & \lambda_3' \end{pmatrix}^{*}\begin{pmatrix} & k_1 & \\ \kappa_1 & & \kappa_1' \end{pmatrix}^{*}\begin{pmatrix} & k_2 & \\ \kappa_2 & & \kappa_2' \end{pmatrix}^{*}\begin{pmatrix} & k_3 & \\ \kappa_3 & & \kappa_3' \end{pmatrix}^{*} \tag{26}$$

(17) can be applied to the product of the third and fourth and to the product of the second and fifth three-j-symbols (together with the corresponding one-j-symbols). One obtains in this way

$$\sum_{lk} \begin{pmatrix} l_1 & k_2 & j_3 \\ \lambda_1 & \kappa_2 & \iota_3 \end{pmatrix} (-1)^{2k_1+2k_2+2l_2}[l][k] \begin{Bmatrix} k_1 & j_2 & l_3 \\ l_1 & l_2 & l \end{Bmatrix} \begin{pmatrix} l_1 & j_2 & l \\ \lambda_1' & \iota_2 & \lambda \end{pmatrix} \begin{pmatrix} k_1 & l_2 & l \\ \kappa_1' & \lambda_2 & \lambda \end{pmatrix}^*$$

$$\cdot \begin{Bmatrix} j_1 & l_2 & k_3 \\ k_1 & k_2 & k \end{Bmatrix} \begin{pmatrix} k_1 & l_2 & k \\ \kappa_1' & \lambda_2 & \kappa \end{pmatrix} \begin{pmatrix} j_1 & k_2 & k \\ \iota_1 & \kappa_2' & \kappa' \end{pmatrix} \begin{pmatrix} & k & \\ \kappa & & \kappa' \end{pmatrix}^* \begin{pmatrix} & l_1 & \\ \lambda_1 & & \lambda_1' \end{pmatrix}^* \begin{pmatrix} & k_2 & \\ \kappa_2 & & \kappa_2' \end{pmatrix}^*$$

where (11a) also has been made use of. This expression can be contracted by means of the orthogonality relation (4a) after which only the summation over l remains. The resulting expression has two three-j-symbols in which the last upper index is l. The product of these can be transformed again by (17) which gives the expression

$$\sum_{lj} (-1)^{2k_1+2l_2}[l][j] \begin{Bmatrix} k_1 & j_2 & l_3 \\ l_1 & l_2 & l \end{Bmatrix} \begin{Bmatrix} j_1 & l_2 & k_3 \\ k_1 & k_2 & l \end{Bmatrix} \begin{Bmatrix} l_1 & j_2 & l \\ j_1 & k_2 & j \end{Bmatrix}$$

$$\cdot \begin{pmatrix} l_1 & k_2 & j_3 \\ \lambda_1 & \kappa_2 & \iota_3 \end{pmatrix} \begin{pmatrix} j_1 & j_2 & j \\ \iota_1 & \iota_2 & \iota \end{pmatrix} \begin{pmatrix} l_1 & k_2 & j \\ \lambda_1' & \kappa_2' & \iota' \end{pmatrix} \begin{pmatrix} & j & \\ \iota & & \iota' \end{pmatrix}^* \begin{pmatrix} & l_1 & \\ \lambda_1 & & \lambda_1' \end{pmatrix}^* \begin{pmatrix} & k_2 & \\ \kappa_2 & & \kappa_2' \end{pmatrix}^*.$$

This can be contracted again by (11a) and (4a) and the summation over j drops out hereupon. One obtains in this way for (26) the left side of (26a)

$$\sum_{l} (-1)^{2l}[l] \begin{Bmatrix} j_1 & l_2 & k_3 \\ k_1 & k_2 & l \end{Bmatrix} \begin{Bmatrix} k_1 & j_2 & l_3 \\ l_1 & l_2 & l \end{Bmatrix} \begin{Bmatrix} l_1 & k_2 & j_3 \\ j_1 & j_2 & l \end{Bmatrix} \begin{pmatrix} j_1 & j_2 & j_3 \\ \iota_1 & \iota_2 & \iota_3 \end{pmatrix}$$

$$= \sum_{l} (-1)^{2l}[l] \begin{Bmatrix} j_1 & l_2 & k_3 \\ l & j_2 & j_3 \end{Bmatrix} \begin{Bmatrix} k_1 & j_2 & l_3 \\ l & k_2 & k_3 \end{Bmatrix} \begin{Bmatrix} l_1 & k_2 & j_3 \\ l & l_2 & l_3 \end{Bmatrix} \begin{pmatrix} j_1 & j_2 & j_3 \\ \iota_1 & \iota_2 & \iota_3 \end{pmatrix}. \quad (26a)$$

The right side results from the remark that (26) is invariant under a cyclic permutation of the indices 1, 2, 3. The last factor in (26a) can be dropped, of course. If one brings one of the six-j-symbols over the left side to the right side by means of the orthogonality relation (20) one obtains after some rewriting

$$\begin{Bmatrix} j_1 & j_2 & j_3 \\ l_1 & l_2 & l_3 \end{Bmatrix} \begin{Bmatrix} j_1 & j_2 & j_3 \\ k_1 & k_2 & k_3 \end{Bmatrix}$$

$$= \sum_{kl} (-1)^{2j_1+2l}[k][l] \begin{Bmatrix} j_1 & l_2 & l_3 \\ k & k_2 & k_3 \end{Bmatrix} \begin{Bmatrix} l_1 & j_2 & l_3 \\ k & k_2 & l \end{Bmatrix} \begin{Bmatrix} k_1 & j_2 & k_3 \\ k & l_2 & l \end{Bmatrix} \begin{Bmatrix} k_1 & k_2 & j_3 \\ l_1 & l_2 & l \end{Bmatrix}. \quad (26b)$$

This equation is analogous to (22) and will be used for the calculation of the six-j-symbols. Note that the exponent of -1 could be replaced by $2j_3 + 2k$ since $2j_1 + 2l + 2j_3 + 2k$ is necessarily even, since this holds for $2j_1 + 2j_2 + 2j_3$, for $2l + 2j_2 + 2k$ and, of course, for $4j_2$.

6. *Expressions for Six-j-Symbols in Terms of Group Integrals*

It was mentioned before that if one subjects the representation j to a unitary transformation with the matrix u, the three-j-symbols will transform according to

$$\begin{pmatrix} j_1 & j_2 & j \\ \lambda_1 & \lambda_2 & \lambda \end{pmatrix} \to \begin{pmatrix} j_1 & j_2 & j \\ \lambda_1 & \lambda_2 & \eta \end{pmatrix} u_{\eta\lambda}.$$

On the other hand, the one-j-symbols transform according to

$$\begin{pmatrix} & j & \\ \kappa & & \lambda \end{pmatrix} \to \begin{pmatrix} & j & \\ \mu & & \nu \end{pmatrix} u_{\mu\kappa} u_{\nu\lambda}.$$

We saw that, as a result of these expressions, the six-j-symbols are invariant under a unitary transformation of the underlying representations. On the other hand, a change in the sign in one of the four three-j-symbols of (18a) (cf. the last paragraph of Section 2) will change the sign of the six-j-symbol also.

It appears reasonable, therefore, to consider such products of the six-j-symbols which remain unchanged if the sign of the three-j-symbols is changed. Such a combination is, first of all [cf. (18) and (11a)]

$$\begin{Bmatrix} j_1 & j_2 & j_3 \\ j_1' & j_2' & j_3' \end{Bmatrix}^2$$

$$= \begin{pmatrix} j_1 & j_2 & j_3 \\ \kappa_1 & \kappa_2 & \kappa_3 \end{pmatrix}^* \begin{pmatrix} j_1 & j_2' & j_3' \\ \kappa_1' & \mu_2 & \mu_3 \end{pmatrix}^* \begin{pmatrix} & j_1 & \\ \kappa_1 & & \kappa_1' \end{pmatrix} \begin{pmatrix} j_1' & j_2 & j_3' \\ \mu_1 & \kappa_2 & \mu_3 \end{pmatrix} \begin{pmatrix} j_1' & j_2' & j_3 \\ \mu_1' & \mu_2 & \kappa_3 \end{pmatrix} \begin{pmatrix} & j_1' & \\ \mu_1 & & \mu_1' \end{pmatrix}^*$$

$$\begin{pmatrix} j_1 & j_2 & j_3 \\ \iota_1 & \iota_2 & \iota_3 \end{pmatrix} \begin{pmatrix} j_1 & j_2' & j_3' \\ \iota_1' & \lambda_2 & \lambda_3 \end{pmatrix} \begin{pmatrix} & j_1 & \\ \iota_1 & & \iota_1' \end{pmatrix}^* \begin{pmatrix} j_1' & j_2 & j_3' \\ \lambda_1 & \iota_2 & \lambda_3 \end{pmatrix}^* \begin{pmatrix} j_1' & j_2' & j_3 \\ \lambda_1' & \lambda_2 & \iota_3 \end{pmatrix}^* \begin{pmatrix} & j_1' & \\ \lambda_1 & & \lambda_1' \end{pmatrix}$$

$$= \sum_{RSTU} h^{-4} \begin{bmatrix} j_1 & R \\ \kappa_1 & \iota_1 \end{bmatrix} \begin{bmatrix} j_2 & R \\ \kappa_2 & \iota^2 \end{bmatrix} \begin{bmatrix} j_3 & R \\ \kappa_3 & \iota_3 \end{bmatrix} \begin{bmatrix} j_1 & S \\ \kappa_1 & \iota_1 \end{bmatrix}^* \begin{bmatrix} j_2' & S \\ \mu_2 & \lambda_2 \end{bmatrix} \begin{bmatrix} j_3' & S \\ \mu_3 & \lambda_3 \end{bmatrix} \begin{bmatrix} j_1' & T \\ \lambda_1 & \mu_1 \end{bmatrix}$$

$$\cdot \begin{bmatrix} j_2 & T \\ \iota_2 & \kappa_2 \end{bmatrix} \begin{bmatrix} j_3' & T \\ \lambda_3 & \mu_3 \end{bmatrix} \begin{bmatrix} j_1' & U \\ \lambda_1 & \mu_1 \end{bmatrix}^* \begin{bmatrix} j_2' & U \\ \lambda_2 & \mu_2 \end{bmatrix} \begin{bmatrix} j_3 & U \\ \iota_3 & \kappa_3 \end{bmatrix}$$

$$= \sum_{RSTU} h^{-4} [j_1; RS^{-1}][j_2; RT][j_3; RU][j_1'; TU^{-1}][j_2'; SU][j_3'; ST]$$

where (5) and (8) have been used; $[j; X]$ is the character of the element X in the representation j. Changing the notation somewhat, this gives

$$\begin{Bmatrix} j_1 & j_2 & j_3 \\ j_1' & j_2' & j_3' \end{Bmatrix}^2 = h^{-3} \sum_{R_1 R_2 R_3} [j_1; R_1][j_2; R_2][j_3; R_3]$$

$$\cdot [j_1'; R_2 R_3^{-1}][j_2'; R_3 R_1^{-1}][j'; R_1 R_2^{-1}]. \quad (27)$$

The equations

$$\begin{Bmatrix} j_1 & j_2 & j' \\ j_1 & j_2 & j \end{Bmatrix} = \frac{(-1)^{2j}}{h^2} \sum_{RS} [j_1; SR][j_2; S^{-1}R][j; S][j'; R], \tag{27a}$$

$$\begin{Bmatrix} j_1 & j_2 & j' \\ j_3 & j_2 & j \end{Bmatrix} \begin{Bmatrix} j_1 & j_2 & j'' \\ j_3 & j_2 & j' \end{Bmatrix} \begin{Bmatrix} j_1 & j_2 & j \\ j_3 & j_2 & j'' \end{Bmatrix}$$

$$= \frac{(-1)^{2j_1}}{h^4} \sum_{RXYZ} [j_1; RY][j_2; RXYZR][j_3; RXZ][j; X][j'; Y][j''; Z] \tag{27b}$$

$$\begin{Bmatrix} j_1 & j_2 & j_3 \\ l_1 & l_2 & l_3 \end{Bmatrix} \begin{Bmatrix} j_1 & l_2 & l_3 \\ j_1 & j_2 & j_3 \end{Bmatrix} \begin{Bmatrix} l_1 & j_2 & l_3 \\ j_1 & j_2 & j_3 \end{Bmatrix} \begin{Bmatrix} l_1 & l_2 & j_3 \\ j_1 & j_2 & j_3 \end{Bmatrix}$$

$$= \frac{(-1)^{2l_1}}{h^5} \sum_{URVSW} [j_1; R][j_1; S][j_2; UVWRWS]$$

$$\cdot [j_3; R^{-1}UVWSV][l_1; U][l_2; V][l_3; W] \tag{27c}$$

follow in a similar way. The first of these formulas provides for finite groups a reasonably convenient way to determine the absolute value of the six-j-symbols. The remaining ones permit the determination of those signs of the six-j-symbols which cannot be chosen arbitrarily. In most cases, it will be easier to compare the six-j-symbols by the formulas of the preceding sections than by Eqs. (27), and use these for the evaluation of sums or integrals over the group. The left sides of (27b) and (27c) show a high degree of symmetry which is not manifest on the right sides. This shows that the character sums satisfy certain identities which, however, have not been explored.

II. The Three-Dimensional Rotation Group

7. *The Representations of the Three-Dimensional Rotation Group*

The following sections will be devoted to an application of the formulas of the preceding sections to the representations of the two-dimensional unimodular unitary group. For this purpose, we briefly summarize first the well known results concerning the representations of this group. We shall parametrize the group in the customary fashion, by writing the elements as a product of a diagonal, a real and another diagonal matrix. The group elements can be characterized in this way by three angles, α, β, γ, the variability domain of α extends from -2π to 2π, β varies from 0 to π, and γ from $-\pi$ to π. The group element corresponding to the angles α, β, γ is

$$Z_\alpha Y_\beta Z_\gamma = \begin{Vmatrix} e^{-i\alpha/2} & 0 \\ 0 & e^{i\alpha/2} \end{Vmatrix} \cdot \begin{Vmatrix} \cos \frac{1}{2}\beta & -\sin \frac{1}{2}\beta \\ \sin \frac{1}{2}\beta & \cos \frac{1}{2}\beta \end{Vmatrix} \cdot \begin{Vmatrix} e^{-i\gamma/2} & 0 \\ 0 & e^{i\gamma/2} \end{Vmatrix} . \tag{28}$$

The matrices in (28) form naturally a representation of their own group; it is the representation $j = \frac{1}{2}$. In conformity herewith, we denote the two rows and columns of (28) by $-\frac{1}{2}$ and $\frac{1}{2}$. The representation j is the symmetric part of the $2j$ power of (28), defined as[2]

$$\begin{bmatrix} j & R \\ \kappa & \lambda \end{bmatrix} = \frac{[(j+\kappa)!(j-\kappa)!(j+\lambda)!(j-\lambda)!]^{1/2}}{(2j)!} \cdot \sum_{\substack{\nu_1+\cdots+\nu_{2j}=\kappa \\ \mu_1+\cdots+\mu_{2j}=\lambda}} \begin{bmatrix} \frac{1}{2} & R \\ \nu_1 & \mu_1 \end{bmatrix} \begin{bmatrix} \frac{1}{2} & R \\ \nu_2 & \mu_2 \end{bmatrix} \cdots \begin{bmatrix} \frac{1}{2} & R \\ \nu_{2j} & \mu_{2j} \end{bmatrix}. \quad (29)$$

where κ and λ can assume the values $-j, -j+1, \ldots, j-1, j$, and

$$\frac{(2j)!}{(j+\kappa)!(j-\kappa)!}$$

is the number of ways the $\nu_1, \ldots, \nu_{2j}$ can be picked so as to have $j+\kappa$ of the ν to be $+\frac{1}{2}$ and $j-\kappa$ to be $-\frac{1}{2}$, which makes their sum equal to κ. One can easily convince oneself that (29) forms a unitary representation of the group, for all values of $j = 0, \frac{1}{2}, 1, \frac{3}{2}, 2, \ldots$.

In particular, (29) gives for the representatives of the diagonal matrices

$$\begin{bmatrix} j & Z_a \\ \kappa & \lambda \end{bmatrix} = \frac{[(j+\kappa)!(j-\kappa)!(j+\lambda)!(j-\lambda)!]^{1/2}}{(2j)!} \cdot \sum_{\substack{\nu_1+\cdots+\nu_{2j}=\kappa \\ \mu_1+\cdots+\mu_{2j}=\lambda}} e^{i(\nu_1+\cdots+\nu_{2j})a} \delta_{\nu_1\mu_1} \cdots \delta_{\nu_{2j}\mu_{2j}}$$

$$= \frac{(j+\kappa)!(j-\kappa)!}{(2j)!} \sum_{\nu_1+\cdots+\nu_{2j}=\kappa} e^{i\kappa a} \delta_{\kappa\lambda} = \delta_{\kappa\lambda} e^{i\kappa a} \quad \textbf{(30)}$$

again diagonal matrices and the representatives of real matrices are, evidently, real matrices. These can be calculated to be[3]

$$\begin{bmatrix} j & Y_\beta \\ \kappa & \lambda \end{bmatrix} = \sum_\eta (-1)^{j+\lambda-\eta} \frac{[(j+\kappa)!(j-\kappa)!(j+\lambda)!(j-\lambda)!]^{1/2}}{(\eta-\kappa-\lambda)!\eta!(j+\kappa-\eta)!(j+\lambda-\eta)!} \cdot \cos^{2\eta-\kappa-\lambda}\tfrac{1}{2}\beta \sin^{2j+\kappa+\lambda-2\eta}\tfrac{1}{2}\beta. \quad (30a)$$

The summation over η has to be extended over those integers for which none of the factorials in the denominator becomes negative. We shall note the special case of $\kappa = \pm j$ of (30a). Evidently, the sum over the ν in (29) reduces in this case to the single term $\nu_1 = \nu_2 = \ldots = \nu_{2j} = \frac{1}{2}$ and all terms of the summation are equal. Hence

$$\begin{bmatrix} j & R \\ j & \lambda \end{bmatrix} = \left(\frac{(2j)!}{(j-\lambda)!(j+\lambda)!}\right)^{1/2} e^{ija} \cos^{j+\lambda}\tfrac{1}{2}\beta \sin^{j-\lambda}\tfrac{1}{2}\beta e^{i\lambda\gamma}. \quad (30b)$$

We note here that those unitary matrices which have the same characteristic values $e^{\frac{1}{2}i\phi}$, $e^{-\frac{1}{2}i\phi}$ form a class ϕ. Every class is represented among the diagonal matrices Z_ϕ, in fact Z_ϕ and $Z_{-\phi}$ are in the same class. The character of the representation j can be easily obtained from (30)

$$[j;\phi] = \sum_{\kappa=-j}^{j} e^{i\kappa\phi}. \tag{31}$$

It follows from (31) easily that the Kronecker product of the representations j and j' contains as irreducible parts the representations

$$j \times j' = |j-j'|, |j-j'|+1, |j-j'|+2, \ldots, j+j'-1, j+j'. \tag{31a}$$

Evidently, the dimension of the representation j is

$$[j] = 2j+1. \tag{31b}$$

The matrix which transforms the representation into the conjugate complex form must be, apart from an arbitrary factor which we shall choose to be 1, equal to

$$\begin{pmatrix} & j & \\ \kappa & & \lambda \end{pmatrix} = \begin{bmatrix} j & Y_{-\pi} \\ \kappa & \lambda \end{bmatrix}. \tag{32}$$

This follows from the fact that this matrix must commute with all real matrices of the representation and that it must transform those of the form (30) into their reciprocals. Both conditions are evidently satisfied by the matrix corresponding to Y_π. One sees from (28) and (29) easily that

$$\begin{pmatrix} & j & \\ \kappa & & \lambda \end{pmatrix} = (-1)^{j-\lambda}\delta_{\kappa,-\lambda} = (-1)^{j+\kappa}\delta_{\kappa,-\lambda}. \tag{32a}$$

The possibility to assume the form (32a) for the one-j-symbols can be easily seen without reference to the preceding developments if one assumes that the representations with integer j can be brought into a real form, those with half integer j cannot (i.e., that a representation is integer or half integer according to the integer or half integer character of its j). For integer j, in fact, the representation can be first assumed to be real in which case the unit matrix transforms it into the conjugate complex form. If we now transform the representation by a unitary matrix u^{-1}, it will be brought into the conjugate complex form by a further transformation with $u'u$. Denoting the matrix (32a) (which is symmetric for integer j) by J and the unit matrix by I, we can choose

$$u = \tfrac{1}{2}(1-i)(iI+J) \tag{32b}$$

and we have on account of $J^2 = I$,

$$u'u = \tfrac{1}{4}(1-i)^2(-I+2iJ+I) = J.$$

For half integer j (32a) already has—apart from the order of rows and columns—the usual form.

From the fact that (32a) commutes with the matrices corresponding to Y_β it follows that

$$\begin{pmatrix} j & \\ \kappa & \lambda \end{pmatrix}\begin{bmatrix} j & Y_\beta \\ \lambda & -\mu \end{bmatrix} = \begin{bmatrix} j & Y_\beta \\ \kappa & \lambda \end{bmatrix}\begin{pmatrix} j & \\ \lambda & -\mu \end{pmatrix} \tag{33}$$

or

$$\begin{bmatrix} j & Y_\beta \\ -\kappa & -\mu \end{bmatrix} = (-1)^{\kappa-\mu}\begin{bmatrix} j & Y_\beta \\ \kappa & \mu \end{bmatrix}. \tag{33a}$$

The condition that Z_π transforms Y_β into its reciprocal gives, using the real orthogonal character of the matrix corresponding to Y,

$$\begin{bmatrix} j & Z_\pi \\ \kappa & \lambda \end{bmatrix}\begin{bmatrix} j & Y_\beta \\ \lambda & \mu \end{bmatrix} = \begin{bmatrix} j & Y_\beta \\ \lambda & \kappa \end{bmatrix}\begin{bmatrix} j & Z_\pi \\ \lambda & \mu \end{bmatrix}$$

or

$$\begin{bmatrix} j & Y_\beta \\ \kappa & \mu \end{bmatrix} = (-1)^{\kappa-\mu}\begin{bmatrix} j & Y_\beta \\ \mu & \kappa \end{bmatrix} \tag{33b}$$

and

$$\begin{bmatrix} j & Y_\beta \\ \kappa & \mu \end{bmatrix} = \begin{bmatrix} j & Y_\beta \\ -\mu & -\kappa \end{bmatrix}. \tag{33c}$$

One has, furthermore, for $Y_\beta = Y_\pi(Y_{\pi-\beta})^{-1}$

$$\begin{bmatrix} j & Y_\beta \\ \kappa & \lambda \end{bmatrix} = \begin{bmatrix} j & Y_\pi \\ \kappa & \mu \end{bmatrix}\begin{bmatrix} j & Y_{\pi-\beta} \\ \lambda & \mu \end{bmatrix}$$

or

$$\begin{bmatrix} j & Y_\beta \\ \kappa & \lambda \end{bmatrix} = (-1)^{j-\kappa}\begin{bmatrix} j & Y_{\pi-\beta} \\ \lambda & -\kappa \end{bmatrix} = (-1)^{j+\lambda}\begin{bmatrix} j & Y_{\pi-\beta} \\ -\kappa & \lambda \end{bmatrix} \tag{33d}$$

all these formulas are written down here only for the sake of easier references. They can be, of course, also deduced directly from (30a) and are further illustrated by the table at the end of Section 12, which gives some of the matrices corresponding to Y_β for $\beta = \frac{1}{2}\pi$.

8. *The Three-Dimensional Rotation Group is Simply Reducible*

We now wish to establish the validity of the results of the paper of Reference 6, which are mentioned in this reference, for the unimodular unitary group.

The validity of the Lemma follows from the fact that both the considerations of Frobenius and Schur which are used for the derivation of the Lemma and also the derivation itself can be carried out without assuming that the underlying group is finite. It then follows from the half integer character of the representation $j = \frac{1}{2}$ and the above-described structure of the Kronecker products that (as is well known) the representations with half integer numerical values of j (even dimensions) are half integer and the representations with integer numerical values of j (odd dimensions) are integer.

In order to determine the even and odd character of the representations, it is necessary to show, first, that the representations $j = 2j', 2j' - 2, 2j' - 4$, ... are in the symmetric part $j' \bar{\times} j'$ of the Kronecker product $j' \times j'$ of the representation j' with itself, while $j = 2j' - 1,\ 2j - 3$ ... are in the antisymmetric part $j' \underline{\times} j'$ of $j' \times j'$. This statement is evidently correct for $j' = 0$. It is also correct for $j' = \frac{1}{2}$ in which case $j' \bar{\times} j'$ must be three-dimensional (and is $j = 1$) while $j' \underline{\times} j'$ must be one-dimensional (and is $j = 0$). Assuming the above structure for all $j \leq j'$, it will be shown now for $j' + \frac{1}{2}$ which will make it generally valid.

Let us consider the symmetric part of the Kronecker product of $(j \times \frac{1}{2})$ with itself. This is[4]

$$(j' \times \tfrac{1}{2}) \bar{\times} (j' \times \tfrac{1}{2}) = [(j' + \tfrac{1}{2}) \dot{+} (j' - \tfrac{1}{2})] \bar{\times} [(j' + \tfrac{1}{2}) \dot{+} (j' - \tfrac{1}{2})]. \tag{34}$$

The trace of the right side can be easily expressed in terms of the traces of the Kronecker products of the representations involved. It is, according to (17) of Reference 6, identical with the trace of

$$(j' + \tfrac{1}{2}) \bar{\times} (j' + \tfrac{1}{2}) \dot{+} (j' + \tfrac{1}{2}) \times (j' - \tfrac{1}{2}) \dot{+} (j' - \tfrac{1}{2}) \bar{\times} (j' - \tfrac{1}{2}).$$

and is, according to our surmise, equivalent to the sum of the representations

$$\begin{array}{l} (j' + \tfrac{1}{2}) \bar{\times} (j' + \tfrac{1}{2}),\ 2j',\ 2j' - 1,\ 2j' - 2,\ 2j' - 3,\ 2j' - 4,\ 2j' - 5, \ldots \\ \qquad\qquad\qquad\quad 2j' - 1, \qquad\quad 2j' - 3, \qquad\quad 2j' - 5, \ldots . \end{array} \tag{34a}$$

On the other hand, one easily verifies by the same equation that the trace of $(j' \times \frac{1}{2}) \bar{\times} (j' \times \frac{1}{2})$ is also equal to the trace of

$$(j' \bar{\times} j') \times (\tfrac{1}{2} \bar{\times} \tfrac{1}{2}) \dot{+} (j' \underline{\times} j') \times (\tfrac{1}{2} \underline{\times} \tfrac{1}{2})$$

which is equal, according to our surmise, to

$$\begin{array}{l} [(2j') \dot{+} (2j' - 2) \dot{+} (2j' - 4) \ldots] \times (1) \dot{+} [(2j' - 1) \dot{+} (2j' - 3) \dot{+} \\ (2j' - 5) \ldots] \times (0) = [(2j' + 1) \dot{+} (2j') \dot{+} (2j' - 1)] \dot{+} [(2j' - 1) \dot{+} (2j' - 2) \\ \qquad\qquad\qquad\qquad\qquad\qquad\qquad\qquad\quad \dot{+} (2j' - 1) \\ \dot{+} (2j' - 3)] \dot{+} [(2j' - 3) \ldots \\ \dot{+} (2j' - 3) \ldots . \end{array} \tag{34b}$$

Comparing (34a) and (34b) one sees that $(j' + \frac{1}{2}) \bar{\times} (j' \times \frac{1}{2})$ contains the representations $j = (2j' + 1), (2j' - 1), (2j' - 3), \ldots$ which is the desired result.

It follows that if j' is integer the representations $2j',\ 2j' - 2,\ 2j' - 4$, are even while $2j' - 1,\ 2j' - 3, \ldots$ are odd representations. If j' is half integer, the first series consists of odd and the second set of even representations; an integer representation j is even or odd according to the even or odd character of j. Thus no representation is both even and odd, in accordance with Theorem 3 which had to be verified. The notation half integer, integer, odd and even for the representations of S.R. groups was chosen so that the numerical value of j shall have the character which the representation denotes.

For the definition of the $(-1)^j$ (cf. Section 2) we shall adopt the convention

$$(-1)^j = i^{2j}. \tag{35}$$

It may be remarked, however, that the convention $(-1)^j = (-i)^{2j}$ would lead to identical formulas, since the exponent of -1 is an integer in all equations.

9. *The Three-j-Symbols for the Three-Dimensional Rotation Group*

A few simple relations immediately follow from the above normalizations and the definition (2) and (2a) of the three-j-symbols. Writing in (2a) Z_a for R, we have, for (30), (2a)

$$\begin{pmatrix} j_1 & j_2 & j_3 \\ \lambda_1 & \lambda_2 & \lambda_3 \end{pmatrix} e^{i(\lambda_1+\lambda_2)a} = e^{-i\lambda_3 a} \begin{pmatrix} j_1 & j_2 & j_3 \\ \lambda_1 & \lambda_2 & \lambda_3 \end{pmatrix}$$

or

$$\begin{pmatrix} j_1 & j_2 & j_3 \\ \lambda_1 & \lambda_2 & \lambda_3 \end{pmatrix} = 0 \qquad \text{unless } \lambda_1 + \lambda_2 + \lambda_3 = 0. \tag{36}$$

Setting, on the other hand, $R = Y_{-\pi}$ in (2a), we have

$$\begin{pmatrix} j_1 & j_2 & j_3 \\ -\lambda_1 & -\lambda_2 & \kappa_3 \end{pmatrix} (-1)^{j_1-\lambda_1+j_2-\lambda_2} = (-1)^{j_3+\kappa_3} \begin{pmatrix} j_1 & j_2 & j_3 \\ \lambda_1 & \lambda_2 & -\kappa_3 \end{pmatrix}$$

or

$$\begin{pmatrix} j_1 & j_2 & j_3 \\ -\lambda_1 & -\lambda_2 & -\lambda_3 \end{pmatrix} = \begin{pmatrix} j_1 & j_2 & j_3 \\ \lambda_1 & \lambda_2 & \lambda_3 \end{pmatrix} (-1)^{j_1+j_2+j_3} \tag{36a}$$

since $j_3 + \kappa_3$ being an integer $(-1)^{j_3+\kappa_3} = (-1)^{-j_3-\kappa_3}$ and since the three-j-symbols vanishes anyway unless $\lambda_1 + \lambda_2 + \lambda_3 = 0$.

Finally, (11a) and (32a) show that

$$\begin{pmatrix} j_1 & j_2 & j_3 \\ \lambda_1 & \lambda_2 & \lambda_3 \end{pmatrix}^* = (-1)^{j_1+j_2+j_3+\lambda_1+\lambda_2+\lambda_3} \begin{pmatrix} j_1 & j_2 & j_3 \\ -\lambda_1 & -\lambda_2 & -\lambda_3 \end{pmatrix}$$

so that the normalization adopted makes the three-j-symbols *real*.

It follows from (30b) and the well known orthogonality relations for representation coefficients that

$$\int \left| \begin{bmatrix} j & R \\ j & \lambda \end{bmatrix} \right|^2 dR = \frac{(2j)!}{(j-\lambda)!(j+\lambda)!} \int \cos^{2j+2\lambda} \tfrac{1}{2}\beta \, \sin^{2j-2\lambda} \tfrac{1}{2}\beta \, dR = \frac{h}{2j+1}. \tag{37}$$

where $h = \int dR$ is the volume of the group. Hence, we have, for (5), (30b), (33a), and (33c)

$$h\begin{pmatrix} j_1 & j_2 & j_3 \\ j_1 & -j_2 & j_2-j_1 \end{pmatrix}\begin{pmatrix} j_1 & j_2 & j_3 \\ \lambda & -\lambda+j_3 & -j_3 \end{pmatrix}$$

$$=\int\begin{bmatrix} j_1 & R \\ j_1 & \lambda \end{bmatrix}\begin{bmatrix} j_2 & R \\ -j_2 & -\lambda+j_3 \end{bmatrix}\begin{bmatrix} j_3 & R \\ j_2-j_1 & -j_3 \end{bmatrix}dR$$

$$=(-1)^{j_2+j_3-\lambda}$$

$$\cdot\frac{[(2j_1)!(2j_2)!(2j_3)!]^{1/2}\int\cos^{2j_1+2\lambda}\tfrac{1}{2}\beta\,\sin^{2j_2+2j_3-2\lambda}\tfrac{1}{2}\beta\,dR}{[(j_1-\lambda)!(j_1+\lambda)!(j_2-j_3+\lambda)!(j_2+j_3-\lambda)!(j_3-j_2+j_1)!(j_3+j_2-j_1)!]^{1/2}}$$

$$=\frac{(-1)^{j_2+j_3-\lambda}[(2j_1)!(2j_2)!(2j_3)!]^{1/2}(j_1+\lambda)!(j_2+j_3-\lambda)!h}{[(j_1-\lambda)!(j_1+\lambda)!(j_2-j_3+\lambda)!(j_2+j_3-\lambda)!\cdot(j_3-j_2+j_1)!(j_3+j_2-j_1)!]^{1/2}(j_1+j_2+j_3+1)!}. \tag{37a}$$

For $\lambda=j_1$, this gives

$$\begin{pmatrix} j_1 & j_2 & j_3 \\ j_1 & -j_2 & j_2-j_1 \end{pmatrix}\begin{pmatrix} j_1 & j_3 & j_2 \\ j_1 & -j_3 & j_3-j_1 \end{pmatrix}$$

$$=\frac{(-1)^{2j_1}(2j_1)!}{(j_1+j_2+j_3+1)!}\left[\frac{(2j_2)!(2j_3)!}{(j_1+j_2-j_3)!(j_1-j_2+j_3)!}\right]^{1/2}. \tag{37b}$$

This equation shows that if we give to the first factor the sign of $(-1)^{2j_3}$ (which we are yet at liberty to do), the sign of the second factor will be automatically $(-1)^{2j_2}$ since $(-1)^{2j_1}=(-1)^{2j_2+2j_3}$. On account of (6) and (36a), the sign $(-1)^{2j_3}$ will hold also for the three-j-symbol obtained by interchanging j_1 and j_2 in the first factor of (37b). The equations obtained from (37b) by cyclic permutations of the j_1, j_2, j_3 then show that the above convention will hold true for all three-j-symbols obtained from those on the left side of (37b) by an arbitrary interchange of the j. This normalization of the sign of the three-j-symbols leaves, furthermore, (7) and (32a) valid and will be adopted henceforth. In order to obtain the value of the first factor of (37b) separately, one can make a cyclic permutation of the j and multiply the resulting equation with (37b). Dividing this product by the equation obtained from (37b) by the other cyclic permutation of the j, one obtains an equation for the square of one three-j-symbol of (37b) alone. This gives

$$\begin{pmatrix} j_1 & j_2 & j_3 \\ j_1 & -j_2 & j_2-j_1 \end{pmatrix}=(-1)^{2j_3}\left[\frac{(2j_1)!(2j_2)!}{(j_1+j_2+j_3+1)!(j_1+j_2-j_3)!}\right]^{1/2} \tag{37c}$$

and now (37a)

$$\begin{pmatrix} j_1 & j_2 & j_3 \\ \lambda & j_3-\lambda & -j_3 \end{pmatrix}$$

$$=\frac{(-1)^{j_2-j_3-\lambda}[(2j_3)!(j_1+j_2-j_3)!(j_1+\lambda)!(j_2+j_3-\lambda)!]^{1/2}}{[(j_1+j_2+j_3+1)!(-j_1+j_2+j_3)!(j_1-j_2+j_3)!(j_1-\lambda)!(j_2-j_3+\lambda)!]^{1/2}}. \tag{37d}$$

This equation gives immediately all three-j-symbols in which one j is $\frac{1}{2}$. We have

$$\begin{pmatrix} j & j+\frac{1}{2} & \frac{1}{2} \\ \lambda & \frac{1}{2}-\lambda & -\frac{1}{2} \end{pmatrix} = \frac{(-1)^{j-\lambda}(j+1-\lambda)^{1/2}}{[(2j+1)(2j+2)]^{1/2}} \tag{37e}$$

and from this, all others can be obtained by interchanging the first two columns, or using (36a), or both.

10. *Calculation of the Racah Coefficients*

We now go over to the calculation of the six-j-symbols. Since these do not depend on the particular form in which the representations are assumed, it should be possible to calculate them directly, i.e., without first computing the three-j-symbols. This is in fact the case and can be done, for instance, in the following way.

One can remark, first, that the orthogonality relations (20) directly give those six-j-symbols for which the orthogonal matrix (20a) has only one dimension. This is the case, e.g., if $j_2+j_4=\pm(j_1-j_3)$ or $\pm(j_2-j_4)=j_1+j_3$. If $j_3=\frac{1}{2}$, the orthogonal matrix (20a) has, if it is not actually one-dimensional, only two rows and two columns and its opposite elements are therefore equal, apart from the sign. This means, e.g.,

$$\begin{Bmatrix} j_1 & j_2 & j_4-\frac{1}{2} \\ \frac{1}{2} & j_4 & j_2-\frac{1}{2} \end{Bmatrix} (2j_2 \cdot 2j_4)^{1/2} = \pm \begin{Bmatrix} j_1 & j_2 & j_4+\frac{1}{2} \\ \frac{1}{2} & j_4 & j_2+\frac{1}{2} \end{Bmatrix} ((2j_2+2)(2j_4+2))^{1/2}$$

which, together with (19a) allows to reduce j_2 and j_4 simultaneously by $\frac{1}{2}$. This reduction can be carried to the point at which $j_2+j_4=j_1-\frac{1}{2}$, after which the resulting six-j-symbol is given by the first remark. The other six-j-symbols with $j_3=\frac{1}{2}$ can be calculated in a similar way. This being carried out, the calculation of the remaining six-j-symbols can proceed much in the same way as will be described toward the end of this section.

The reason that the above-outlined procedure will not be followed is that it yields only the absolute values of the six-j-symbols. The determination of the signs, on the other hand, becomes inelegant. Some of them can be chosen arbitrarily, as mentioned at the beginning of Section 6. This being done, the remaining signs can be obtained by the above orthogonality relations with an occasional reference to (21a). The whole discussion becomes, however, somewhat tedious. We shall make use therefore in the following calculation of the results of the preceding section.

If the six-j-symbol

$$\begin{Bmatrix} j_1 & j_2 & j_3 \\ l_1 & l_2 & l_3 \end{Bmatrix}$$

is different from zero, the following numbers must be integers:

$$\begin{aligned} J &= j_1+j_2+j_3 & J_1 &= j_1+l_2+l_3 \\ J_2 &= l_1+j_2+l_3 & J_3 &= l_1+l_2+j_3 \end{aligned} \tag{38}$$

and since they will frequently occur in the ensuing formulas, we shall use the above abbreviations for them.

A general formula for the six-j-symbols is contained in (22). By inserting j_1 for ι_1, $-j_2$ for ι_2, and j_2-j_1 for ι_3, one obtains a reasonably simple expression for the six-j-symbols. In the case $l_1+l_2=j_3$ one can set more advantageously $\iota_3=j_3$ and $\iota_1=-j_1$ or $\iota_2=-j_2$. Then, because of the last three-j-symbol on the right side, λ_1 and η_2 can assume only the values $-l_1$ and $-l_2$, respectively, and the right side of (22) reduces to a single term. One obtains with (37c) and (37d)

$$\begin{Bmatrix} j_1 & j_2 & j_3 \\ l_1 & l_2 & l_3 \end{Bmatrix} = (-1)^{j_1+j_2+j_3}$$

$$\cdot\left[\frac{(J+1)!(J-2j_1)!(J-2j_2)!(J_1-2l_2)!(J_2-2l_1)!(2l_1)!(2l_2)!}{(J_1+1)!(J_2+1)!(J-2j_3)!(J_1-2j_1)! \cdot (J_2-2j_2)!(J_1-2l_3)!(J_2-2l_3)!(2j_3+1)!}\right]^{1/2}$$

$$\text{for } l_1+l_2=j_3. \quad (39)$$

This equation immediately gives all six-j-symbols in which one representation is $j=0$ or $j=\frac{1}{2}$. For $l_1=0$, we must set $l_2=j_3$ and $l_3=j_2$; otherwise the six-j-symbol vanishes.

$$\begin{Bmatrix} j_1 & j_2 & j_3 \\ 0 & j_3 & j_2 \end{Bmatrix} = (-1)^{j_1+j_2+j_3}[(2j_2+1)(2j_3+1)]^{-1/2}. \quad (39a)$$

This is identical with (18b) and thus independent of the normalizations of Section 9.

For $l_1=\frac{1}{2}$, we have to distinguish between two cases:

$$\begin{Bmatrix} j_1 & j_2 & j_3 \\ \frac{1}{2} & j_3-\frac{1}{2} & j_2+\frac{1}{2} \end{Bmatrix} = (-1)^{j_1+j_2+j_3}\left[\frac{(J-2j_2)(J-2j_3+1)}{(2j_2+1)(2j_2+2)2j_3(2j_3+1)}\right]^{1/2}, \quad (39b)$$

$$\begin{Bmatrix} j_1 & j_2 & j_3 \\ \frac{1}{2} & j_3-\frac{1}{2} & j_2-\frac{1}{2} \end{Bmatrix} = (-1)^{j_1+j_2+j_3}\left[\frac{(J+1)(J-2j_1)}{2j_2(2j_2+1)2j_3(2j_3+1)}\right]^{1/2}. \quad (39c)$$

We may note finally

$$\begin{Bmatrix} j_1 & j_2 & j_3 \\ \frac{1}{2}J-j_1 & \frac{1}{2}J-j_2 & \frac{1}{2}J-j_3 \end{Bmatrix}$$

$$= (-1)^J\left[\frac{(J+1)!(J-2j_1)!(J-2j_2)!(J-2j_3)!}{(2j_1+1)!(2j_2+1)!(2j_3+1)!}\right]^{1/2}. \quad (40)$$

One can obtain a recursive formula for the six-j-symbols by setting $k_2=\frac{1}{2}$, $k_1=j_3-\frac{1}{2}$, $k_3=j_1-\frac{1}{2}$ in (26b). Both k and l will assume two values, $l_3\pm\frac{1}{2}$ and $l_1\pm\frac{1}{2}$, respectively, and the six-j-symbol will appear as a sum of four six-j-symbols in which both j_1 and j_3 are smaller by $\frac{1}{2}$ than in the original one:

$$\frac{(2l_1+1)(2l_3+1)[(J+1)(J-2j_2)]^{1/2}}{[(J_1-2l_2)(J_1-2l_3)(J_3-2l_1)(J_3-2l_2)]^{1/2}}\begin{Bmatrix} j_1 & j_2 & j_3 \\ l_1 & l_2 & l_3 \end{Bmatrix}$$
$$= -\left[\frac{J_1+1}{J_1-2l_3}(J_2+1)(J_2-2j_2)\frac{J_3+1}{J_3-2l_1}\right]^{1/2}\begin{Bmatrix} j_1-\frac{1}{2} & j_2 & j_3-\frac{1}{2} \\ l_1-\frac{1}{2} & l_2 & l_3-\frac{1}{2} \end{Bmatrix}$$
$$-\left[\frac{J_1-2j_1+1}{J_1-2l_2}(J_2-2l_1+1)(J_2-2l_3)\frac{J_3+1}{J_3-2l_1}\right]^{1/2}\begin{Bmatrix} j_1-\frac{1}{2} & j_2 & j_3-\frac{1}{2} \\ l_1-\frac{1}{2} & l_2 & l_3+\frac{1}{2} \end{Bmatrix}$$
$$-\left[\frac{J_1+1}{J_1-2l_3}(J_2-2l_1)(J_2-2l_3+1)\frac{J_3-2j_3+1}{J_3-2l_2}\right]^{1/2}\begin{Bmatrix} j_1-\frac{1}{2} & j_2 & j_3-\frac{1}{2} \\ l_1+\frac{1}{2} & l_2 & l_3-\frac{1}{2} \end{Bmatrix}$$
$$+\left[\frac{J_1-2j_1+1}{J_1-2l_2}(J_2+2)(J_2-2j_2+1)\frac{J_3-2j_3+1}{J_3-2l_2}\right]^{1/2}\begin{Bmatrix} j_1-\frac{1}{2} & j_2 & j_3-\frac{1}{2} \\ l_1+\frac{1}{2} & l_2 & l_3+\frac{1}{2} \end{Bmatrix}. \tag{41}$$

This formula is valid, of course, only if $J-2j_2=j_1-j_2+j_3>0$. It permits one to diminish the j occurring in a six-j-symbol until this assumes a form in which one of the equations (39a, b, or c) is applicable. The following tables were obtained in this way. The numerical ones illustrate, in particular, the orthogonality relations: every table, considered as a matrix, is orthogonal if the weight factor $((2j+1)(2l+1))^{1/2}$ is inserted. It may be mentioned that this fact also follows a successive calculation of all six-j-symbols.

$$\begin{Bmatrix} j_1 & j_2 & j_3 \\ 1 & j_3-1 & j_2-1 \end{Bmatrix} = (-1)^{j_1+j_2+j_3} \cdot\left[\frac{J(J+1)(J-2j_1-1)(J-2j_1)}{(2j_2-1)2j_2(2j_2+1)(2j_3-1)2j_3(2j_3+1)}\right]^{1/2} \tag{42a}$$

$$\begin{Bmatrix} j_1 & j_2 & j_3 \\ 1 & j_3-1 & j_2 \end{Bmatrix} = (-1)^{j_1+j_2+j_3} \cdot\left[\frac{2(J+1)(J-2j_1)(J-2j_2)(J-2j_3+1)}{2j_2(2j_2+1)(2j_2+2)(2j_3-1)2j_3(2j_3+1)}\right]^{1/2} \tag{42b}$$

$$\begin{Bmatrix} j_1 & j_2 & j_3 \\ 1 & j_3-1 & j_2+1 \end{Bmatrix} = (-1)^{j_1+j_2+j_3} \cdot\left[\frac{(J-2j_2-1)(J-2j_2)(J-2j_3+1)(J-2j_3+2)}{(2j_2+1)(2j_2+2)(2j_2+3)(2j_3-1)2j_3(2j_3+1)}\right]^{1/2} \tag{42c}$$

$$\begin{Bmatrix} j_1 & j_2 & j_3 \\ 1 & j_3 & j_2 \end{Bmatrix} = (-1)^{j_1+j_2+j_3} \cdot\frac{2(j_1^2+j_1-j_2^2-j_2-j_3^2-j_3)}{[2j_2(2j_2+1)(2j_2+2)2j_3(2j_3+1)(2j_3+2)]^{1/2}}. \tag{42d}$$

Formulas (42) give the values of all six-j-symbols in which one of the j is 1, some of them by means of the symmetry relations (19).

Table for $\begin{Bmatrix} 3/2 & 3/2 & j \\ 3/2 & 3/2 & l \end{Bmatrix}$

$l \backslash j$	0	1	2	3
0	−1/4	1/4	−1/4	1/4
1	1/4	−11/60	1/20	3/20
2	−1/4	1/20	3/20	1/20
3	1/4	3/20	1/20	1/140

Table for $\begin{Bmatrix} 3/2 & 3/2 & j \\ 2 & 2 & l \end{Bmatrix}$

$l \backslash j$	0	1	2	3
$\frac{1}{2}$	$\frac{5^{1/2}}{10}$	$-\frac{18^{1/2}}{20}$	$\frac{14^{1/2}}{20}$	$-\frac{2^{1/2}}{10}$
$\frac{3}{2}$	$-\frac{5^{1/2}}{10}$	$\frac{2^{1/2}}{10}$	0	$-\frac{2^{1/2}}{10}$
$\frac{5}{2}$	$\frac{5^{1/2}}{10}$	$-\frac{2^{1/2}}{60}$	$-\frac{14^{1/2}}{28}$	$-\frac{18^{1/2}}{70}$
$\frac{7}{2}$	$-\frac{5^{1/2}}{10}$	$-\frac{2^{1/2}}{10}$	$-\frac{14^{1/2}}{70}$	$-\frac{2^{1/2}}{140}$

Table for $\begin{Bmatrix} 3/2 & 2 & j \\ 3/2 & 2 & l \end{Bmatrix}$

$l \backslash j$	1/2	3/2	5/2	7/2
1/2	1/20	−1/10	3/20	−1/5
3/2	−1/10	3/20	−1/10	−1/10
5/2	3/20	−1/10	−47/420	−1/35
7/2	−1/5	−1/10	−1/35	−1/280

Table for $\begin{Bmatrix} 2 & 2 & l \\ 2 & 2 & j \end{Bmatrix}$

$l \backslash j$	0	1	2	3	4
0	1/5	−1/5	1/5	−1/5	1/5
1	−1/5	1/6	−1/10	0	2/15
2	1/5	−1/10	−3/70	4/35	2/35
3	−1/5	0	4/35	1/14	1/70
4	1/5	2/15	2/35	1/70	1/630

11. *Calculation of the Three-j-Symbols*

The three-j-symbols in which one of the j is equal to the sum of two others can be obtained rather simply from (2a). Writing therein $-\kappa_3 = j_3 = j_1 + j_2$ only the term with $\kappa_1 = j_1$, $\kappa_2 = j_2$ remains on the left side. The representation coefficients are all of the form (30b) or can be brought to this form by (33a). One obtains with (37c)

$$\begin{pmatrix} j_1 & j_2 & j_1+j_2 \\ \lambda_1 & \lambda_2 & -\lambda_1-\lambda_2 \end{pmatrix} = (-1)^{j_1-j_2+\lambda_1+\lambda_2} \cdot \left[\frac{(2j_1)!(2j_2)!(j_1+j_2+\lambda_1+\lambda_2)!(j_1+j_2-\lambda_1-\lambda_2)!}{(2j_1+2j_2+1)!(j_1+\lambda_1)!(j_1-\lambda_1)!(j_2+\lambda_2)!(j_2-\lambda_2)!}\right]^{1/2}. \tag{43}$$

This yields, together with (39d), (32a), and (22), a general formula for the three-j-symbols. For this purpose, we write $\frac{1}{2}J - j_1$ for l_1, $\frac{1}{2}J - j_2$ for l_2 and $\frac{1}{2}J - j_3$ for l_3 in (22) which gives to all three-j-symbols on the right side the form (43). The six-j-symbol is given by (40) and one obtains

$$\begin{pmatrix} j_1 & j_2 & j_3 \\ \lambda_1 & \lambda_2 & \lambda_3 \end{pmatrix} = \sum_{\kappa_1\kappa_2\kappa_3} \frac{(-1)^{\kappa_1+\kappa_2+\kappa_3-\frac{1}{2}J}[(J-2j_1)!(J-2j_2)!(J-2j_3)!]^{1/2}}{[(J+1)!]^{1/2}(\frac{1}{2}J-j_1-\kappa_1)!(\frac{1}{2}J-j_1+\kappa_1)!} \cdot \frac{[(j_1-\lambda_1)!(j_1+\lambda_1)!(j_2-\lambda_2)!(j_2+\lambda_2)!(j_3-\lambda_3)!(j_3+\lambda_3)!]^{1/2}}{(\frac{1}{2}J-j_2-\kappa_2)!(\frac{1}{2}J-j_2+\kappa_2)!(\frac{1}{2}J-j_3-\kappa_3)!(\frac{1}{2}J-j_3+\kappa_3)!} \tag{44}$$

$$\kappa_3 - \kappa_2 = \lambda_1, \; \kappa_1 - \kappa_3 = \lambda_2, \; \kappa_2 - \kappa_1 = \lambda_3.$$

The summation runs in reality only over one index because the last line determines all three κ if one is given. The summation has to be extended over all values of κ_1, κ_2, κ_3, compatible with the last line of (44), for which all factorials in the denominator become non-negative integers. Equation (44) is essentially identical with Eq. (27) in Chapter XVII of Reference 3. One can show that the connection between the s and our three-j-symbols is

$$s^{(ll)}_{\mathscr{L}\mu\nu} = (-1)^{l-l+\mu+\nu}(2\mathscr{L}+1)^{1/2}\begin{pmatrix} l & l & \mathscr{L} \\ \mu & \nu & -\mu-\nu \end{pmatrix}. \tag{44a}$$

One can obtain a recursive formula for the three-j-symbols from (22), which is similar to (41) but much more easily manageable. Setting $l_1 = \frac{1}{2}$, $l_2 = j_3 - \frac{1}{2}$, $l_3 = j_2 - \frac{1}{2}$ in (22) all symbols in which l_1 occurs can be expressed directly by (37e) and (39c), the one-j-symbols by (32a). Only two terms of the summation remain, corresponding to $\lambda_1 = \pm\frac{1}{2}$. One obtains in this way

$$\begin{pmatrix} j_1 & j_2 & j_3 \\ \iota_1 & \iota_2 & \iota_3 \end{pmatrix} = -\left[\frac{(j_2-\iota_2)(j_3+\iota_3)}{(J+1)(J-2j_1)}\right]^{1/2}\begin{pmatrix} j_1 & j_2-\frac{1}{2} & j_3-\frac{1}{2} \\ \iota_1 & \iota_2+\frac{1}{2} & \iota_3-\frac{1}{2} \end{pmatrix} + \left[\frac{(j_2+\iota_2)(j_3-\iota_3)}{(J+1)(J-2j_1)}\right]^{1/2}\begin{pmatrix} j_1 & j_2-\frac{1}{2} & j_3-\frac{1}{2} \\ \iota_1 & \iota_2-\frac{1}{2} & \iota_3+\frac{1}{2} \end{pmatrix}. \tag{45}$$

The three-j-symbols given in the next table[5] were calculated by this formula and (37e). Since the three-j-symbol vanishes unless the sum of the greek indices is zero, all symbols with the same value of the j can be given in one figure. The center of the figure corresponds to $\iota_1 = \iota_2 = \iota_3 = 0$—if all j are integers; ι_1 is constant in horizontal rows, ι_2 in rows of 60° to the right up, ι_3 in rows of 60° to the left up. The values of ι_1, ι_2, ι_3 corresponding to the different points of the diagrams in case of integer j is given in the first diagram, in case ι_1 and ι_2 are half integer (ι_3 therefore integer) in the second diagram. In this case, the center of the diagram is marked by $\otimes$. Each of the remaining diagrams contains the values of the three-j-symbol given at the left of the diagram. The notation in the remaining cases is analogous.

$$\begin{array}{ccccccccc}
 & -220 & & & -211 & & -202 & & \\
-12-1 & & & -110 & & -101 & & & -1-12 \\
 & 01-1 & & & 000 & & 0-11 & & \\
11-2 & & & 10-1 & & 1-10 & & & 1-21 \\
 & 20-2 & & & 2-1-1 & & 2-20 & &
\end{array}$$

$$\begin{array}{cccccc}
 & -\tfrac{3}{2}\ \tfrac{3}{2}\ 0 & & -\tfrac{3}{2}\ \tfrac{1}{2}\ 1 & & -\tfrac{3}{2}\ -\tfrac{1}{2}\ 2 \\
-\tfrac{1}{2}\ \tfrac{3}{2}-1 & & -\tfrac{1}{2}\ \tfrac{1}{2}\ 0 & & -\tfrac{1}{2}\ -\tfrac{1}{2}\ 1 & \\
 & & & \otimes & & \\
 & \tfrac{1}{2}\ \tfrac{1}{2}\ -1 & & \tfrac{1}{2}\ -\tfrac{1}{2}\ 0 & & \tfrac{1}{2}\ -\tfrac{3}{2}\ 1 \\
\tfrac{3}{2}\ \tfrac{1}{2}\ -2 & & \tfrac{3}{2}\ -\tfrac{1}{2}\ -1 & & \tfrac{3}{2}\ -\tfrac{3}{2}\ 0 &
\end{array}$$

$(0\ j\ j)$

$$\frac{(-1)^{2j}}{(2j+1)^{1/2}},\ \frac{(-1)^{2j-1}}{(2j+1)^{1/2}},\ \cdots,\ \frac{-1}{(2j+1)^{1/2}},\frac{1}{(2j+1)^{1/2}}$$

$(\tfrac{1}{2}\ \tfrac{1}{2}\ 1)$

$$\begin{array}{cccc}
 & 6^{-1/2} & & -3^{-1/2} \\
 & & \otimes & \\
-3^{-1/2} & & 6^{-1/2} &
\end{array}$$

$(\tfrac{1}{2}\ \tfrac{3}{2}\ 1)$

$$\begin{array}{cccccc}
-\tfrac{1}{2} & & 6^{-1/2} & & -12^{-1/2} & \\
 & & & \otimes & & \\
 & 12^{-1/2} & & -6^{-1/2} & & \tfrac{1}{2}
\end{array}$$

$(\tfrac{1}{2}\ \tfrac{3}{2}\ 2)$

$$\begin{array}{cccccccc}
 & 20^{-1/2} & & -10^{-1/2} & & (3/20)^{-1/2} & & -5^{-1/2} \\
 & & & & \otimes & & & \\
-5^{-1/2} & & (3/20)^{1/2} & & -10^{-1/2} & & 20^{-1/2} &
\end{array}$$

$(\tfrac{1}{2}\ \tfrac{5}{2}\ 2)$

$$\begin{array}{cccccccccc}
-6^{-1/2} & & (2/15)^{1/2} & & -10^{-1/2} & & 15^{-1/2} & & -30^{-1/2} & \\
 & & & & & \otimes & & & & \\
 & 30^{-1/2} & & -15^{-1/2} & & 10^{-1/2} & & -(2/15)^{-1/2} & & 6^{-1/2}
\end{array}$$

	$42^{-1/2}$ $\quad -21^{-1/2}$ $\quad 14^{-1/2}$ $\quad -(2/21)^{-1/2}$ $\quad (5/42)^{1/2}$ $\quad 7^{-1/2}$
$(\frac{1}{2}\ \frac{5}{2}\ 3)$	$\otimes$
	$-7^{-1/2}$ $\quad (5/42)^{1/2}$ $\quad -(2/21)^{-1/2}$ $\quad 14^{-1/2}$ $\quad -21^{-1/2}$ $\quad 42^{-1/2}$
	$-6^{-1/2}$ $\quad 6^{-1/2}$
$(1\ 1\ 1)$	$6^{-1/2}$ $\quad 0$ $\quad -6^{-1/2}$
	$-6^{-1/2}$ $\quad 6^{-1/2}$
	$30^{-1/2}$ $\quad -10^{-1/2}$ $\quad 5^{-1/2}$
$(1\ 1\ 2)$	$-10^{-1/2}$ $\quad (2/15)^{1/2}$ $\quad -10^{-1/2}$
	$5^{-1/2}$ $\quad -10^{-1/2}$ $\quad 30^{-1/2}$
	$-15^{-1/2}$ $\quad 10^{-1/2}$ $\quad -10^{-1/2}$ $\quad 15^{-1/2}$
$(1\ 2\ 2)$	$(2/15)^{1/2}$ $\quad -30^{-1/2}$ $\quad 0$ $\quad 30^{-1/2}$ $\quad -(2/15)^{-1/2}$
	$-15^{-1/2}$ $\quad 10^{-1/2}$ $\quad -10^{-1/2}$ $\quad 15^{-1/2}$
	$105^{-1/2}$ $\quad -35^{-1/2}$ $\quad (2/35)^{1/2}$ $\quad -(2/21)^{1/2}$ $\quad 7^{-1/2}$
$(1\ 2\ 3)$	$-21^{-1/2}$ $\quad (8/105)^{1/2}$ $\quad -(3/35)^{1/2}$ $\quad (8/105)^{1/2}$ $\quad -21^{-1/2}$
	$7^{-1/2}$ $\quad -(2/21)^{1/2}$ $\quad (2/35)^{1/2}$ $\quad -35^{-1/2}$ $\quad 105^{-1/2}$

12. *Calculation of the Representation Coefficients*

We go over, finally, to formulas concerning the representation coefficients themselves. On account of (30) these are essentially equivalent with formulas relating to the representatives of Y_β in $R = Z_\alpha Y_\beta Z_\gamma$.

Using (2a) with $j_3 = j_1 + j_2$, (43) gives a simple expression for the three-j-symbols occurring and we can set $\lambda_1 = j_1$ in order to have by (30b) and (33b) a simple form for the j_1 representation coefficient. Writing κ for κ_3, ν for κ_1 and $-j_1 - \lambda$ for λ_2 we obtain, using (33a), for $\lambda < j_2 - j_1$

$$\begin{bmatrix} j_1 + j_2 & Y_\beta \\ \kappa & \lambda \end{bmatrix} = \sum_\nu \frac{(2j_1)!}{(j_1+\nu)!(j_1-\nu)!} \left[\frac{(j_1+j_2-\kappa)!(j_1+j_2+\kappa)!(j_2-j_1-\lambda)!}{(j_2-\nu-\kappa)!(j_2+\nu+\kappa)!(j_2+j_1-\lambda)!}\right]^{1/2} \cdot \begin{bmatrix} j_2 & Y_\beta \\ \kappa+\nu & j_1+\lambda \end{bmatrix} \cos^{j_1+\nu}\tfrac{1}{2}\beta \sin^{j_1-\nu}\tfrac{1}{2}\beta. \tag{46}$$

The summation over ν must be extended, as in all similar formulas, over all values for which in the denominators only factorials of non-negative integers occur.

One can specialize this by assuming $j_1 = \frac{1}{2}$

$$\begin{bmatrix} j & Y_\beta \\ \kappa & \lambda \end{bmatrix} = \left(\frac{j+\kappa}{j-\lambda}\right)^{1/2} \begin{bmatrix} j-\frac{1}{2} & Y_\beta \\ \kappa-\frac{1}{2} & \lambda+\frac{1}{2} \end{bmatrix} \sin \tfrac{1}{2}\beta + \left(\frac{j-\kappa}{j-\lambda}\right)^{1/2} \begin{bmatrix} j-\frac{1}{2} & Y_\beta \\ \kappa+\frac{1}{2} & \lambda+\frac{1}{2} \end{bmatrix} \cos \tfrac{1}{2}\beta. \tag{46a}$$

This provides a useful formula for the calculation of the representation coefficients [except for $\lambda = +j$ which can be obtained from (30b), for example] while (46) allows one to take larger steps. For $j = 1$, (46a) gives easily

$$\begin{bmatrix} 1 & Y_\beta \\ \kappa & \lambda \end{bmatrix} = \begin{Vmatrix} \frac{1}{2}(1+\cos\beta) & -2^{-1/2}\sin\beta & \frac{1}{2}(1-\cos\beta) \\ 2^{-1/2}\sin\beta & \cos\beta & -2^{-1/2}\sin\beta \\ \frac{1}{2}(1-\cos\beta) & 2^{-1/2}\sin\beta & \frac{1}{2}(1+\cos\beta) \end{Vmatrix} \tag{46b}$$

(in all these matrices, the order of rows and columns is $-j, -j+1, \ldots, j-1, j$).

The disadvantage of Eqs. (46), (46a), and (30a) is that they express the representation coefficients by means of the powers of $\cos \frac{1}{2}\beta$ and $\sin \frac{1}{2}\beta$. The general behavior of these functions can be visualized much more readily if they are given in terms of cosines and sines of multiples of β. This is done, for the representatives of Z_α, in (30). For the Y_β one can start out from the remark that Y_β is in the class of Z_β and can be obtained from Z_β

$$Y_\beta = X Z_\beta X^{-1} \tag{47}$$

by transforming the latter with

$$X = X_{\pi/2} = \begin{Vmatrix} 2^{-1/2} & i2^{-1/2} \\ i2^{-1/2} & 2^{-1/2} \end{Vmatrix}. \tag{47a}$$

Hence

$$\begin{bmatrix} j & Y_\beta \\ \kappa & \lambda \end{bmatrix} = \begin{bmatrix} j & X_{\pi/2} \\ \kappa & \mu \end{bmatrix} \begin{bmatrix} j & Z_\beta \\ \mu & \mu' \end{bmatrix} \begin{bmatrix} j & X_{\pi/2} \\ \lambda & \mu' \end{bmatrix}^* = \begin{bmatrix} j & X_{\pi/2} \\ \kappa & \mu \end{bmatrix} \begin{bmatrix} j & X_{\pi/2} \\ \lambda & \mu \end{bmatrix}^* e^{i\mu\beta} \tag{47b}$$

where,

$$\begin{bmatrix} j & X_{\pi/2} \\ \kappa & \mu \end{bmatrix} = \begin{bmatrix} j & Z_{\pi/2} \\ \kappa & \nu \end{bmatrix} \begin{bmatrix} j & Y_{\pi/2} \\ \nu & \nu' \end{bmatrix} \begin{bmatrix} j & Z_{-\pi/2} \\ \nu' & \mu \end{bmatrix} = i^{\kappa-\mu} \begin{bmatrix} j & Y_{\pi/2} \\ \kappa & \mu \end{bmatrix}. \tag{47c}$$

Substituting this into (47b)

$$\begin{bmatrix} j & Y_\beta \\ \kappa & \lambda \end{bmatrix} = \sum_\mu \begin{bmatrix} j & Y \\ \mu & \kappa \end{bmatrix} \begin{bmatrix} j & Y \\ \mu & \lambda \end{bmatrix} i^{\lambda-\kappa} e^{i\mu\beta} \tag{48}$$

one sees that the knowledge of the representatives of $Y = Y_{\pi/2}$ suffices for obtaining the desired formula for the representatives of an arbitrary Y_β. It is possible, by means of (33d), to reduce the summation over μ to non-negative values of this quantity

$$\begin{bmatrix} j & Y_\beta \\ \kappa & \lambda \end{bmatrix} = 2 \sum{}' \begin{bmatrix} j & Y \\ \mu & \kappa \end{bmatrix} \begin{bmatrix} j & Y \\ \mu & \lambda \end{bmatrix} \operatorname{tr} \mu\beta \tag{48a}$$

where the summation over μ extends over all positive values of this quantity and over the value $\mu = 0$ with a factor $\frac{1}{2}$;

$$\operatorname{tr} \mu\beta = \begin{cases} \cos \mu\beta & \text{if} \quad \kappa - \lambda = 0 \pmod 4 \\ \sin \mu\beta & \text{if} \quad \kappa - \lambda = 1 \pmod 4 \\ -\cos \mu\beta & \text{if} \quad \kappa - \lambda = 2 \pmod 4 \\ -\sin \mu\beta & \text{if} \quad \kappa - \lambda = 3 \pmod 4). \end{cases} \tag{48b}$$

The calculation of the $\begin{bmatrix} j & Y \\ \mu & \kappa \end{bmatrix}$ can be easily carried out by means of (46a) setting $\sin \frac{1}{2}\beta = \cos \frac{1}{2}\beta = 2^{-1/2}$ and one is further aided in this calculation by the relations (33). The first few of them are

$$2^{-1/2} \begin{Vmatrix} 1 & -1 \\ 1 & 1 \end{Vmatrix}; \qquad 2^{-1} \begin{Vmatrix} 1 & -2^{1/2} & 1 \\ 2^{1/2} & 0 & -2^{1/2} \\ 1 & 2^{1/2} & 1 \end{Vmatrix};$$

$$2^{-3/2} \begin{Vmatrix} 1 & -3^{1/2} & 3^{1/2} & -1 \\ 3^{1/2} & -1 & -1 & 3^{1/2} \\ 3^{1/2} & 1 & -1 & -3^{1/2} \\ 1 & 3^{1/2} & 3^{1/2} & 1 \end{Vmatrix};$$

$$2^{-2} \begin{Vmatrix} 1 & -2 & 6^{1/2} & -2 & 1 \\ 2 & -2 & 0 & 2 & -2 \\ 6^{1/2} & 0 & -2 & 0 & 6^{1/2} \\ 2 & 2 & 0 & -2 & -2 \\ 1 & 2 & 6^{1/2} & 2 & 1 \end{Vmatrix}$$

III. Applications to Quantum Mechanics

13. *The Hilbert Space as a Union of Representation Spaces; Irreducible Tensors*

The application of the notions developed in the previous sections as it occurs in quantum mechanics shall be summarized in this section. Since it is not necessary for these applications to specify the group with which we are dealing (which is the symmetry group of the underlying problem), we shall not insert into our formulas the values of three-j-symbols etc., obtained in the previous sections for the rotation group. For the same reason, the real character of the three-j-symbols and one-j-symbols, although true in practically all applications, will not be assumed here.

In quantum mechanics, the different representation spaces are united into a single Hilbert space. In fact, the Hilbert space of quantum theory usually consists of the direct sum of an infinite number of each of the representation spaces. However, practically all the vectors in this Hilbert space which we shall have to deal with will be confined to one of the representation spaces, i.e., will have vanishing components in the direction of the axes of all but one representation space.

We introduce in each representation space l a set of orthogonal vectors ψ_ν^l. This is a slight change of notation from that used in Section 4: in that section ψ_ν was the ν component of a vector. Henceforth, ψ_ν^l will denote a unit vector, situated in the representation space l. The general vector of this representation space is

$$\phi^l = a^{l\nu}\psi_\nu^l \tag{49}$$

while the general vector of the Hilbert space is

$$\phi = \sum_l a^{l\nu}\psi_\nu^l . \tag{49a}$$

There are linear unitary operators P_R in the Hilbert space which permit one to express the invariance of this space with respect to the transformations R of the underlying group. The transformation of the unit vectors is given by

$$P_R\psi_\nu^l = \begin{bmatrix} l & R \\ \mu & \nu \end{bmatrix} \psi_\mu^l \tag{50}$$

i.e., the P_R leave the representation spaces within the Hilbert space invariant. Equation (50) conforms with the general covariant-contravariant formalism: it was noted after (14) that the first index of the representation coefficients is contravariant.

Because of its linearity, P_R applied to ϕ^l of (49) gives

$$P_R\phi_l = a^{l\nu}P_R\psi_\nu^l = a^{l\nu}\begin{bmatrix} R & l \\ \mu & \nu \end{bmatrix} \psi_\mu^l . \tag{51}$$

This equation permits P_R to be interpreted also as a transformation of the $a^{l\nu}$ to

$$a^{l\nu} \to \begin{bmatrix} l & R \\ \mu & \nu \end{bmatrix} a^{l\nu} . \tag{51a}$$

This equation expresses the fact that the contravariant components $a^{l\nu}$ of ϕ transform by the usual law of transformation for contravariant components.

If S is another group element the calculation of $P_S(P_R\psi^l_\nu)$ can be carried out using the linear nature of the P and (50)

$$\begin{aligned} P_S(P_R\psi^l_\nu) &= P_S\left(\begin{bmatrix} l & R \\ \mu & \nu \end{bmatrix} \psi^l_\mu\right) = \begin{bmatrix} l & R \\ \mu & \nu \end{bmatrix} P_S\psi^l_\mu \\ &= \begin{bmatrix} l & R \\ \mu & \nu \end{bmatrix}\begin{bmatrix} l & S \\ \eta & \mu \end{bmatrix}\psi^l_\eta = \begin{bmatrix} l & SR \\ \eta & \nu \end{bmatrix}\psi^l_\eta = P_{SR}\psi^l_\nu . \end{aligned} \tag{52}$$

A similar equation holds also for the general vector ϕ. It follows from this that

$$P_S P_R = P_{SR} \tag{52a}$$

The reader versed in quantum mechanics will recognize the Hilbert space as the space of all state vectors. The ψ^l_ν usually represent the characteristic functions of a definite eigenvalue of the energy operator; the ϕ^l of (49) is the general characteristic function of this eigenvalue. The energy operator H is assumed to be invariant under the transformations of the underlying group ($P_R H P_{R^{-1}} = H$) and a representation l of this group is associated with every eigenvalue of H. Usually, there is an infinity of eigenvalues with which the same representation l is associated and this holds for all l. This accounts for the remark made earlier in this section that the Hilbert space is, usually, a direct sum of infinitely many of each of the representation spaces.

The next important notions are "irreducible tensor operators of rank t" t_τ where τ runs from $-t$ to t. They satisfy the equations

$$P_R t_\tau P_{R^{-1}} = \begin{bmatrix} t & R \\ \sigma & \tau \end{bmatrix} t_\sigma \tag{53}$$

which is again consistent with (52a) inasmuch as

$$P_S P_R t_\tau P_{R^{-1}} P_{S^{-1}} = P_{SR} t_\tau P_{R^{-1}S^{-1}} .$$

It will be noted that, under this definition, the energy operator H is an irreducible tensor operator of rank 0, i.e., an invariant operator or a scalar, in the case which we have called above usual.

The operators of integer rank can be Hermitian. In fact, assuming the representation t to be real it is directly clear from (53) that the components $P_R t_\tau P_{R^{-1}}$ are hermitean if this holds for the original components t_τ. If one does not want to assume the representation in the real form, it is not useful to define the self adjoint character for the single operators t_τ, but only for their entity. This shall imply, at any rate, that the hermitean adjoints of the t_τ are linear combinations of the t_τ themselves

$$t^\dagger_\tau = a_{\sigma\tau} t_\sigma . \tag{53a}$$

Transforming this equation by P_R, yields, because of $P_R t_\tau^\dagger P_{R^{-1}} = (P_R t_\tau P_{R^{-1}})^\dagger$, the relation that $a_{\sigma\tau}$ transforms the representation t into its conjugate imaginary form. Hence, apart from a constant, $a_{\sigma\tau}$ must be the one-j-symbol corresponding to the representation t and we can assume the operators t_τ to be multiplied by a constant in such a way that

$$t_\tau^\dagger = \begin{pmatrix} & t & \\ \sigma & & \tau \end{pmatrix}^* t_\sigma = t^\tau \tag{54}$$

holds exactly. In our covariant-contravariant formalism (54) states that the contravariant components of a self-adjoint tensor operator are the hermitean adjoints of the covariant components.

It follows from (54) that tensor operators corresponding to a half integer rank cannot form a self-adjoint set, i.e., cannot satisfy (53a). This becomes evident if one takes the Hermitian adjoint of (54) and expresses on the right side the $t_\sigma^\dagger$ by the t_ρ themselves. Because the t_ρ are necessarily linearly independent, this gives (10) for the one-j-symbol of t, however, without the $(-1)^{2t}$ factor. It follows that $2t$ is even and t an integer. It should be remarked that the self-adjoint character of the set t_τ does not imply that all t_τ are Hermitian themselves. Thus, for instance, in the case of the rotation group only the t_0 are hermitean, and these only for even values of t. The t_0 for odd values of t are skew hermitean.

The simplest "tensor operator" is of rank 0, i.e., a scalar and the number 1 (or the energy operator of the above-mentioned example) is an example thereof. It is clear that linear combinations of the operators "multiplication with x, y, or z" will form the components of a tensor operator of rank 1, i.e., of a vector operator, and we shall proceed to determine these linear combinations explicitly. The following calculation will, at the same time, prove the vector operator character of (61).

Let us set

$$\begin{aligned} v_{-1} &= s_{x-1}x + s_{y-1}y + s_{z-1}z \\ v_0 &= s_{x0}x + s_{y0}y + s_{z0}z \\ v_1 &= s_{x1}x + s_{y1}y + s_{z1}z. \end{aligned} \tag{55}$$

Writing $x_1 = x$, $x_2 = y$, $x_3 = z$ one can abbreviate (55) as

$$v_\tau = \sum_a s_{a\tau} x_a. \tag{55a}$$

One easily convinces oneself that the definition

$$P_R f(x, y, z) = f(\sum R_{ax} x_a, \sum R_{ay} x_a, \sum R_{az} x_a) \tag{56}$$

or

$$P_R f(x) = f(R^{-1}x) \tag{56a}$$

satisfies (52a). It follows for the operators x_a (i.e., multiplication with x_a)

$$P_R x_a P_{R^{-1}} = \sum_b R_{ba} x_b \tag{57}$$

and for

$$P_R v_\tau P_{R^{-1}} = \sum_{ab} s_{a\tau} R_{ba} x_b. \tag{58}$$

Combined with (53), this gives

$$\sum_{ab} R_{ba} s_{a\tau} x_b = \begin{bmatrix} 1 & R \\ \sigma & \tau \end{bmatrix} v_\sigma = \sum_b \begin{bmatrix} 1 & R \\ \sigma & \tau \end{bmatrix} s_{b\sigma} x_b,$$

whence

$$Rs = s[1; R] \tag{59}$$

$[l; R]$ being the matrix which corresponds to R in the representation l. According to (59), s is the matrix which transforms R into the representation $l = 1$.

Naturally, (59) leaves a common multiplicative constant in all s free. This corresponds to the circumstances that a tensor operator does not lose its character if all its components are multiplied by the same constant. We shall choose this constant in such a way that v forms a self-adjoint set in the sense of (54); this will fix the argument of the indeterminate constant. The calculation of the s which satisfies (59) is quite straightforward. Thus if $R = Z_\alpha$ is a rotation about Z, the representation is given by (30) while Z_α itself is

$$Z_\alpha = \begin{Vmatrix} \cos\alpha & \sin\alpha & 0 \\ -\sin\alpha & \cos\alpha & 0 \\ 0 & 0 & 1 \end{Vmatrix}. \tag{60}$$

Hence (59) gives $s_{z-1} = s_{z1} = s_{x0} = s_{y0} = 0$; $s_{x-1} = is_{y-1}$; $s_{x1} = -is_{y1}$. The form of $[1; Y_\beta]$ is given in (46b) and the corresponding Y_β is obtained by a cyclic permutation of (60). Normalizing $s_{z0} = i$, one obtains in this way

$$\begin{aligned} v_{-1} &= 2^{-1/2} i(x - iy) \\ v_0 &= iz \\ v_1 &= -2^{-1/2} i(x + iy). \end{aligned} \tag{61}$$

It forms a self-adjoint set. Another, equally important set of vector operators is given by the infinitesimal rotations themselves. Let us define

$$M_a = \lim_{\varepsilon \to 0} \frac{1}{\varepsilon} (P_{a(\varepsilon)} - 1) \tag{62}$$

where a may be any of the three axes x, y, z and $a(\varepsilon)$ is the rotation by ε

about the axis a. Then

$$\begin{aligned} J_{-1} &= -\frac{i}{\sqrt{2}}(M_x - iM_y) \\ J_0 &= -iM_z \\ J_1 &= \frac{i}{\sqrt{2}}(M_x + iM_y) \end{aligned} \tag{62a}$$

form a set of (not self-adjoint) vector operators.

Since the effect of P_R on the wave functions is directly given by (50), the effect of those operators on the ψ^l_ν can be calculated directly. Let us calculate first $M_a\psi^l_\nu$. This is

$$M_a\psi^l_\nu = \lim \frac{1}{\varepsilon}(P_{a(\varepsilon)} - 1)\psi^l_\nu = \lim\left(\frac{1}{\varepsilon}\begin{bmatrix} l & a(\varepsilon) \\ \mu & \nu \end{bmatrix} - \frac{1}{\varepsilon}\delta_{\mu\nu}\right)\psi^l_\mu. \tag{63}$$

We need the $\begin{bmatrix} l & a(\varepsilon) \\ \mu & \nu \end{bmatrix}$ up to first powers of ε. For $a = Z$ and Y, these are directly given by (30) and (30a)

$$\lim \frac{1}{\varepsilon}\left(\begin{bmatrix} l & Z(\varepsilon) \\ \mu & \nu \end{bmatrix} - \delta_{\mu\nu}\right) = i\nu\delta_{\mu\nu} \tag{64}$$

and

$$\lim \frac{1}{\varepsilon}\left(\begin{bmatrix} l & Y(\varepsilon) \\ \mu & \nu \end{bmatrix} - \delta_{\mu\nu}\right)$$

$$= \tfrac{1}{2}\sqrt{(l+\nu+1)(l-\nu)}\delta_{\mu\nu+1} - \tfrac{1}{2}\sqrt{(l+\nu)(l-\nu+1)}\delta_{\mu\nu-1}. \tag{64a}$$

Both are, naturally, skew Hermitian. For M_y, one has from the analogue of (47) $Y_\beta = Z^{-1}X_\beta Z$

$$\lim \frac{1}{\varepsilon}\left(\begin{bmatrix} l & X(\varepsilon) \\ \mu & \nu \end{bmatrix} - \delta_{\mu\nu}\right)$$

$$= \frac{i}{2}\sqrt{(l+\nu+1)(l-\nu)}\delta_{\mu\nu+1} + \frac{i}{2}\sqrt{(l+\nu)(l-\nu+1)}\delta_{\mu\nu-1}. \tag{64b}$$

Hence, by (63)

$$\begin{aligned} M_x\psi^l_\nu &= \tfrac{1}{2}i\sqrt{(l+\nu+1)(l-\nu)}\psi^l_{\nu+1} + \tfrac{1}{2}i\sqrt{(l+\nu)(l-\nu+1)}\psi^l_{\nu-1} \\ M_y\psi^l_\nu &= \tfrac{1}{2}\sqrt{(l+\nu+1)(l-\nu)}\psi^l_{\nu+1} - \tfrac{1}{2}\sqrt{(l+\nu)(l-\nu+1)}\psi_{\nu-1} \\ M_z\psi^l_\nu &= i\nu\psi^l_\nu \end{aligned} \tag{65}$$

and

$$J_{-1}\psi^l_\nu = 2^{-1/2}\sqrt{(l+\nu)(l-\nu+1)}\,\psi^l_{\nu-1}$$
$$J_0\psi^l_\nu = \nu\psi^l_\nu \tag{65a}$$
$$J_1\psi^l_\nu = -\,2^{-1/2}\sqrt{(l+\nu+1)(l-\nu)}\,\psi^l_{\nu+1}.$$

Another, at least partial derivation of these formulas will be given at the end of this section.

For $t=2$, one obtains in a way similar to that which led to (61) that

$$\begin{aligned} q_{-2} &= -\tfrac{1}{2}(x-iy)^2 \qquad & q_2 &= -\tfrac{1}{2}(x+iy)^2 \\ q_{-1} &= -z(x-iy) \qquad & q_1 &= z(x+iy) \end{aligned} \tag{66}$$
$$q_0 = 6^{-1/2}(x^2+y^2-2z^2)$$

form the components of a self-adjoint set of irreducible tensor operators of rank 2. A simpler derivation of (66) will be given below.

Let us calculate now an expression of the form $(\psi^l_\nu, t_\tau\psi^l_\nu)$. Because of the unitary character of the P_R and the invariance of the sets ψ^l, ψ^l, t as expressed in (50) and (53), we have

$$\begin{aligned}(\psi^l_\kappa, t_\tau\,\psi^{l'}_\lambda) &= (P_R\psi^l_\kappa, P_R t_\tau P_{R^{-1}}P_R\psi^{l'}_\lambda) \\ &= \begin{bmatrix} l & R \\ \kappa' & \kappa\end{bmatrix}^* \begin{bmatrix} t & R \\ \tau' & \tau\end{bmatrix} \begin{bmatrix} l' & R \\ \lambda' & \lambda\end{bmatrix} (\psi^l_{\kappa'}, t_{\tau'}\psi^{l'}_{\lambda'}) \\ &= \begin{pmatrix} & l & \\ \mu' & & \kappa'\end{pmatrix} \begin{bmatrix} l & R \\ \mu' & \mu\end{bmatrix} \begin{pmatrix} & l & \\ \mu & & \kappa\end{pmatrix}^* \begin{bmatrix} t & R \\ \tau' & \tau\end{bmatrix} \begin{bmatrix} l' & R \\ \lambda' & \lambda\end{bmatrix} (\psi^l_{\kappa'}, t_{\tau'}\psi^{l'}_{\lambda'}).\end{aligned}$$

Integration of both sides over all elements of the group yields because of (5)

$$\begin{aligned}(\psi^l_\kappa, t_\tau\psi^{l'}_\lambda) &= \begin{pmatrix} l & t & l' \\ \mu & \tau & \lambda\end{pmatrix} \begin{pmatrix} & l & \\ \mu & & \kappa\end{pmatrix}^* t_{ll'} \\ &= \begin{pmatrix} \kappa & t & l' \\ l & \tau & \lambda\end{pmatrix} t_{ll'}\end{aligned} \tag{67}$$

where

$$t_{ll'} = \begin{pmatrix} \kappa' & t & l' \\ l & \tau' & \lambda'\end{pmatrix}^* (\psi^l_{\kappa'}, t_{\tau'}\psi^{l'}_{\lambda'}) \tag{67a}$$

is independent of κ, τ, λ. Equation (67) gives a complete analytic expression for the consequences of the invariance of the quantum mechanical equations, as long as one deals only with one Hilbert space. It shows, in particular, that the matrix element (67) vanishes unless the direct product of the representations l and t contains l', i.e., unless l, t, l' can form a vector triangle. In this case, it permits us to compare the matrix elements of the various components τ of t with each other and also gives the ratios of these matrix elements for the various κ, λ of the representation spaces. In the all-important case of a scalar operator, it shows that (67) vanishes unless $l = l'$ and, because of (7) and (32a), $\kappa = \lambda$. In this case, the value of (67) is independent of κ.

A simple calculation shows that if the set t_τ is self-adjoint

$$t_{ll'} = (-1)^{l-t-l'} t_{l'l}^*. \tag{67b}$$

Comparing (67) with (65a) one sees that the three-j-symbols the middle j of which is 1, are essentially equal to the matrix elements of the infinitesimal operators. This permits some simplification in the derivation of (65a).

14. *Transition to Contravariant Components as Time Inversion*

The form of (67) conforms with the covariant-contravariant formalism if we recollect that the covariant index κ on ψ^l_κ, the conjugate complex of which enters (67), is really a contravariant index. It may be worthwhile, however, to expound more fully at this point upon the connection between the covariant-contravariant formalism and the operation of time inversion than has been possible heretofore.

Let us denote the operation of time inversion by θ, an antiunitary operator. If the set of wave function ψ^j, i.e., the representation space j, is invariant under time inversion, $\theta\psi^j_\nu = \psi^{j\nu}$ (the appropriateness of this definition will become evident below) must be a linear combination of the ψ^j_ν. Since, furthermore, all operations of the underlying group commute with θ, application of θ to (50) gives for

$$P_R\psi^{j\nu} = P_R\theta\psi^j_\nu = \theta \begin{bmatrix} j & R \\ \mu & \nu \end{bmatrix} \psi^j_\mu = \begin{bmatrix} j & R \\ \mu & \nu \end{bmatrix}^* \psi^{j\mu} \tag{68}$$

(note that θ in antiunitary, i.e., $\theta a^\nu\psi_\nu = a^{\nu*}\theta\psi_\nu$). One sees that the ψ^j_ν transform by the conjugate complex representation. It follows then with the aid of (8) that

$$\begin{pmatrix} & j & \\ \nu & & \nu' \end{pmatrix} \psi^{j\nu'}$$

transforms under the P_R like ψ^j_ν. If it is not equal to ψ^j_ν, they can differ only by a factor of modulus 1 and a change of the base vectors ψ^j_ν by the square root of this factor will render

$$\psi^j_\nu = \begin{pmatrix} & j & \\ \nu & & \nu' \end{pmatrix} \psi^{j\nu'} = \begin{pmatrix} & j & \\ \nu & & \nu' \end{pmatrix} \theta\psi^j_\nu \tag{69}$$

or

$$\psi^{j\nu} = \theta\psi^j_\nu = \psi^j_{\nu'} \begin{pmatrix} & j & \\ \nu' & & \nu \end{pmatrix}^*. \tag{69a}$$

A base which satisfies (69a) may be called real and the preceding consideration, based on the existence of the operator θ, shows that it is always possible to choose the base real. This operator is simply complex conjugation in the simple Schrödinger theory. However, j is necessarily an integer in this case. If one includes the spin dependence of the wave function, the operation θ is somewhat more complicated but permits the definition of the contravariant vectors $\psi^{j\nu}$ and of a real base also if j is half integer.

If the ϕ^j of (49) is invariant with respect to time inversion, the vector with the components $a^{j\nu}$ will be called real. Since θ is an antiunitary operator, we have, for $\theta\phi$

$$\theta\phi^j = a^{j\nu *}\theta\psi^j_\nu = a^{j\nu *}\psi^j_{\nu'}\begin{pmatrix} & j & \\ \nu' & & \nu \end{pmatrix}^* \tag{70}$$

This will be equal to $a^{j\nu}\psi^j_\nu$ if

$$a^{j*}_\nu = \begin{pmatrix} & j & \\ \nu & & \nu' \end{pmatrix}^* a^{j\nu' *} = a^{j\nu} \tag{71}$$

i.e., if the covariant and the contravariant components of a^j are conjugate complex. It is easy to see that (71) can be satisfied only if j is integer, i.e., there are no real vectors in the spaces of half integer representations. This corresponds to the well known fact that no state of a system with an odd number of spinor particles can be invariant with respect to time inversion. The above consideration and Eq. (68) to (71) give a more precise meaning to the connection between the operation of time inversion in quantum mechanics and the covariant-contravariant transformations.

Whatever was said above about the reality of wave functions and their invariance with respect to time inversion applies *mutatis mutandis* also to operators t. It is more appropriate, however, to call an operator Q which satisfies the equation

$$Q\theta = \theta Q \tag{72}$$

"even" (with respect to time inversion) rather than "real" because the term "real" is used for operators synonymously with "self-adjoint." It remains true, of course, that the energy operator is even in the absence of magnetic fields.

15. *Union of Two Quantum Mechanical Systems; Nine-j-Symbols*

The states of the union of two quantum mechanical systems are best described in a Hilbert space which is the direct product of the Hilbert spaces of the two systems which are to be united. The axes of this Hilbert space can be characterized by a pair of the symbols which characterize the axes of the Hilbert spaces to be united. Thus if ψ^l and χ^k_κ are the axes of these latter

Hilbert spaces, $\psi_\lambda^l\chi_\kappa^k$ will be the axes of their direct product. We shall assume that the two Hilbert spaces undergo similar transformations when subjected to the operators P_R and that the transformation of the product space is given by

$$P_R\psi_\lambda^l\chi_\kappa^k = \begin{bmatrix} l & R \\ \lambda' & \lambda \end{bmatrix}\begin{bmatrix} k & R \\ \kappa' & \kappa \end{bmatrix}\psi_{\lambda'}^l\chi_{\kappa'}^k . \tag{73}$$

Quite apart from the aforementioned problem, one may be interested in the problem of the transformation properties of products such as those that occur in (73). We wish to form, in particular, linear combinations of these products which transform by irreducible representations. On the basis of our covariant-contravariant formalism, it seems reasonable to consider

$$\begin{aligned} \Psi_\iota^{\cdot j} &= [j]^{1/2}\begin{pmatrix} j & \lambda & \kappa \\ \iota & l & k \end{pmatrix}\psi_\lambda^l\chi_x^k \\ &= [j]^{1/2}\begin{pmatrix} & j & \\ \iota & & \iota' \end{pmatrix}\begin{pmatrix} j & l & k \\ \iota' & \lambda & \kappa \end{pmatrix}^* \psi_\lambda^l\chi_\kappa^k . \end{aligned} \tag{74}$$

The last line follows by (10b) and (10c). Applying (73) to this equation, one obtains in fact

$$P_R\Psi_\iota^{\cdot j} = \begin{bmatrix} j & R \\ \iota' & \iota \end{bmatrix}\Psi_{\iota'}^{\cdot j} . \tag{74a}$$

The factor $[j]^{1/2}$ was introduced in (74) in order to give $\Psi_\iota^{\cdot j}$ unit length. Conversely, one can express the $\psi_\lambda^l\chi_\kappa^k$ in terms of the $\Psi^{\cdot}$

$$\psi_\lambda^l\chi_\kappa^k = \sum_j [j]^{1/2}\begin{pmatrix} j & l & k \\ \iota & \lambda & \kappa \end{pmatrix}\Psi_{\iota'}^{\cdot j}\begin{pmatrix} & j & \\ \iota' & & \iota \end{pmatrix}^* . \tag{74b}$$

The expression following the three-j-symbol is the contravariant component $\Psi^{\cdot j\iota}$.

The above calculus describes the vector addition model. The ψ^l and χ^k can be the wave functions of two particles: $\Psi^{\cdot}$ is then the wave function of the system consisting of both. The possible j values of this range from $|l-k|$ to $l+k$. In another application $\psi^l=\psi^L$ is the positional wave function, and $\chi^k=\chi^S$ the spin function; $\Psi^{\cdot j}=\Psi^{\cdot J}$ is then the total wave function. J ranges from $|L-S|$ to $L+S$.

Concepts very similar to the above can be introduced also for operators. Let r_ρ be the components of an irreducible tensor operator of rank r which acts in the Hilbert space of the ψ, and s_σ have a similar significance in the Hilbert space of the χ. In other words,

$$r_\rho\psi_\lambda^l\chi_\kappa^k = (r_\rho\psi_\lambda^l)\chi_\kappa^k; \qquad s_\sigma\psi_\lambda^l\chi_\kappa^k = \psi_\lambda^l(s_\sigma\chi_\kappa^k). \tag{75}$$

One can define then, in analogy to (74), the operators

$$t_\tau = [t]^{1/2} \begin{pmatrix} & t & \\ \tau & & \tau' \end{pmatrix} \begin{pmatrix} t & r & s \\ \tau' & \rho & \sigma \end{pmatrix}^* r_\rho s_\sigma . \tag{76}$$

One easily convinces oneself that the t form a set of irreducible tensor operators of rank t. In fact, the proof thereof does not assume that the r and s commute—as follows from (75)—the t are components of a tensor operator even if they do not, as long as the order of the r and s is the same in each term of (76). We shall return to this point later.

The commutative nature of the r and s is necessary, however, to establish the fact that if those form self-adjoint sets, the t will be self-adjoint also.

$$\begin{aligned} t_\tau^\dagger &= [t]^{1/2} \begin{pmatrix} & t & \\ \tau & & \tau' \end{pmatrix}^* \begin{pmatrix} t & r & s \\ \tau' & \rho & \sigma \end{pmatrix} s_\sigma^\dagger r_\rho^\dagger \\ &= [t]^{1/2} \begin{pmatrix} & t & \\ \tau & & \tau' \end{pmatrix}^* \begin{pmatrix} t & r & s \\ \tau' & \rho' & \sigma' \end{pmatrix} \begin{pmatrix} & r & \\ \rho & & \rho' \end{pmatrix}^* \begin{pmatrix} & s & \\ \sigma & & \sigma' \end{pmatrix}^* s_\sigma r_\rho \qquad (76a) \\ &= [t]^{1/2} \begin{pmatrix} t & r & s \\ \tau & \rho & \sigma \end{pmatrix}^* s_\sigma r_\rho = \begin{pmatrix} & t & \\ \tau' & & \tau \end{pmatrix}^* t_{\tau'} . \end{aligned}$$

The second line follows by (54), the last one by the remark after (11a) and (76) together with the commutative nature of the r and s.

We shall now calculate expressions similar to (67), however with the Ψ^j instead of the ψ^j, and assume the form (76) for t. We shall encounter in the calculation expressions of the form

$$\begin{aligned} (\psi_\lambda^l \chi_\kappa^k, r_\rho s_\sigma \psi_{\lambda'}^{l'} \chi_{\kappa'}^{k'}) &= (\psi_\lambda^l, r_\rho \psi_{\lambda'}^{l'})(\chi_\kappa^k, s_\sigma \chi_{\kappa'}^{k'}) \\ &= \begin{pmatrix} l & r & l' \\ \lambda'' & \rho & \lambda' \end{pmatrix} \begin{pmatrix} k & s & k' \\ \kappa'' & \sigma & \kappa' \end{pmatrix} \begin{pmatrix} & l & \\ \lambda'' & & \lambda \end{pmatrix}^* \begin{pmatrix} & k & \\ \kappa'' & & \kappa \end{pmatrix}^* r_{ll'} s_{kk'} . \end{aligned} \tag{77}$$

The first line follows from (75) and the definition of the scalar product in the product of two Hilbert spaces; the second line is a consequence of (67). Using now (74) and (76), we can express the $t_{jj'}$ of

$$(\Psi_\iota^j, t_\tau \Psi_{\iota'}^{j'}) = \begin{pmatrix} j & t & j' \\ \iota'' & \tau & \iota' \end{pmatrix} \begin{pmatrix} & j & \\ \iota'' & & \iota \end{pmatrix}^* t_{jj'} \tag{67a}$$

in terms of $r_{ll'} s_{kk'}$

$$t_{jj'} = ([j][t][j'])^{1/2} \begin{Bmatrix} j & t & j' \\ l & r & l' \\ k & s & k' \end{Bmatrix} r_{ll'} s_{kk'} . \tag{78}$$

The expression obtained for the left side of (67a) by the direct substitution of (74) and of (76) for $\Psi^{j'}$, Ψ^j, and t, respectively, contains five three-j-symbols, neither of which is, however, the one which appears at the right side.

In order to give the expression directly obtained the form which appears on the right side, one has to apply, first, (17) to two of the three-j-symbols (it does not matter which) in order to exchange some of their columns. After this, (22) can be applied twice to give the form (67a). One obtains in this way for the nine-j-symbol of (78)

$$\begin{Bmatrix} j & t & j' \\ l & r & l' \\ k & s & k' \end{Bmatrix} = \sum_m (-1)^{2m} [m] \begin{Bmatrix} j & l & k \\ s & k' & m \end{Bmatrix} \begin{Bmatrix} t & r & s \\ l & m & l' \end{Bmatrix} \begin{Bmatrix} j' & l' & k' \\ m & j & t \end{Bmatrix}. \tag{78a}$$

There are, of course, numerous symmetry and orthogonality relations for (78a) which can be obtained either from (78a) through the relations of Section 3 or even more easily from the alternate expression for the nine-j-symbol

$$\begin{Bmatrix} j_{11} & j_{12} & j_{13} \\ j_{21} & j_{22} & j_{23} \\ j_{31} & j_{32} & j_{33} \end{Bmatrix} = \begin{pmatrix} j_{11} & j_{12} & j_{13} \\ \lambda_{11} & \lambda_{12} & \lambda_{13} \end{pmatrix} \begin{pmatrix} j_{21} & j_{22} & j_{23} \\ \lambda_{21} & \lambda_{22} & \lambda_{23} \end{pmatrix} \begin{pmatrix} j_{31} & j_{32} & j_{33} \\ \lambda_{31} & \lambda_{32} & \lambda_{33} \end{pmatrix}$$

$$\cdot \begin{pmatrix} j_{11} & j_{21} & j_{31} \\ \lambda_{11} & \lambda_{21} & \lambda_{31} \end{pmatrix}^* \begin{pmatrix} j_{12} & j_{22} & j_{32} \\ \lambda_{12} & \lambda_{22} & \lambda_{32} \end{pmatrix}^* \begin{pmatrix} j_{13} & j_{23} & j_{33} \\ \lambda_{13} & \lambda_{23} & \lambda_{33} \end{pmatrix}^*. \tag{78b}$$

The most direct derivation of (78b) starts from the equation, mentioned after (78), which expresses $(\Psi^{j}_{\iota}, t_{\tau}\Psi^{j}_{\iota'})$ directly in terms of (74) and (76) and which contains five six-j-symbols. One can calculate from this equation

$$\begin{pmatrix} & j & \\ \iota'' & & \iota \end{pmatrix} \begin{pmatrix} j & t & j' \\ \iota'' & \tau & \iota' \end{pmatrix}^* (\Psi^{j}_{\iota}, t_{\tau}\Psi^{j'}_{\iota'}) = t_{jj'}.$$

Comparison of this equation with (78) directly gives (78b). This equation shows that, in a sense, the nine-j-symbols are the most symmetric invariant expressions in the product space of the representation spaces. The fact that the invariant expression (78b) can be written in terms of the six-j-symbols, also follows from a theorem of Biedenharn.[10] We shall not discuss the symmetry and othogonality relation of the nine-j-symbols in detail.

There are a few special cases in which (78) assumes a much simpler form. In the most important of these t is a scalar operator, i.e., an invariant under the elements of the group. This is, of course, possible only if r and s have the same rank and the scalar is given in this case, on account of (76) and (7), (13)

$$S = (-)^{2r}[r]^{-1/2} \begin{pmatrix} & r & \\ \rho & & \sigma \end{pmatrix}^* r_{\rho} s_{\sigma}. \tag{79}$$

Inserting $t = 0$ into (78a) causes the last six-j-symbol to vanish unless $j = j'$ (which is natural, since t is a scalar) and unless $m = l'$. Hence, only one term of (78a) remains and the last two six-j-symbols are directly given by (18b). One obtains with (7)

$$(\Psi'^{j}_{\iota}, S\Psi^{j}_{\iota'}) = \delta_{\iota\iota'}[r]^{-1/2}(-1)^{j+k'+l+r}\begin{Bmatrix} j & k & l \\ r & l' & k' \end{Bmatrix} r_{ll'}s_{kk'}. \tag{79a}$$

The prime on the first Ψ should indicate that although this Ψ' has the same j as the second, it need not be identical with it: it is linear combination of the $\psi^l\chi^k$ while Ψ is a linear combination of the $\psi^{l'}\chi^{k'}$.

Another case in which the nine-j-symbol reduces to a six-j-symbol is in which the operator s (or the operator r) is a scalar. In this case

$$t_\tau = r_\tau s. \tag{80}$$

For $s = 0$, the first two six-j-symbols vanish unless $k = k'$, $t = r$ (which is natural), and unless $m = l$. Hence the summation in (78a) again reduces to a single term and the first two six-j-symbols are again given by (18b). We have for the nine-j-symbol

$$\begin{Bmatrix} j & t & j' \\ l & t & l' \\ k & 0 & k \end{Bmatrix} = \frac{(-1)^{j'+t+l'+k}}{([k][t])^{1/2}} \begin{Bmatrix} j & t & j' \\ l' & k & l \end{Bmatrix} \tag{80a}$$

and for

$$t_{jj'} = (-1)^{j'+t+l+k}\left(\frac{[j'][j]}{[k]}\right)^{1/2}\begin{Bmatrix} j & t & j' \\ l' & k & l \end{Bmatrix} r_{ll'}s. \tag{80b}$$

With $s = 1$, this formula gives the ratio $t_{j'j}/r_{l'l}$ of the matrix elements of an operator $r_\tau = t_\tau$ between the states Ψ of the compound system and the states ψ of one of the components. A well known problem covered by (80b) is the contribution of the orbital (or of the spin) magnetic moment to the magnetic moment of the atom. In this case the two systems which are united to a common system are the orbital motion and the spins of the particles. The former is described by a coordinate wave function ψ, its l is usually denoted by L; the latter is described by a spin wave function χ, its k is usually denoted by S. The operator of the orbital momentum is the vector operator J of (62a), its matrix elements with respect to the ψ can be obtained directly from (65a). The problem is to determine the matrix elements of this operator with respect to the wave functions Ψ of the united system, i.e., the total wave functions of the atom. This is given by (67) with the $t_{jj'}$, given by (80b). In fact for $t = 1, j = j' = J, k = S$, and $l = l' = L$, the six-j-symbol in (80b) is essentially Landé's g-factor.

16. *Combinations of Irreducible Tensors*

It has been mentioned before that the law of combination for irreducible tensor operators which is given in (76) has much wider applications than given in the preceding section. The expression given by (76) will be a tensor operator of rank t not only if r and s act in separate Hilbert spaces. It will be a tensor operator also if they are arbitrary irreducible tensor operators in the same Hilbert space; they do not even have to commute. This equation is therefore eminently suited to make tensor operators of a higher rank from lower rank tensor operators. Thus, for instance, the tensor operator of (66) can be obtained in a more simple way than the one used for its derivation by substituting in (76) the v of (61) for both r and s. The number t has to be set $t=2$, corresponding to the rank 2 of (66), both $r=s=1$ because of the vector operator character of the v of (61). A similar tensor operator can be constructed from the J_τ of (62a).

The tensor operator of rank 2 obtained from J_τ in this way, and the higher order tensor operators obtained by a repeated application of (76) to J_τ, permit a more natural and more symmetric formulation of several rather common questions. Thus, for instance, it is customary to define the quadrupole moment as the expectation value of the operator $-\sqrt{6}q_0$ for the wave function ψ_j^j (i.e., ψ_ν^j for $\nu=j$). This is then, by definition, the quadrupole moment of all ψ_τ^j. Instead of this, one may consider the scalar operator [cf. (79)]

$$Q = 5^{-1/2}\begin{pmatrix} 2 \\ \tau\rho \end{pmatrix}^* q_\rho J_\tau^2 \tag{81}$$

where J^2 is the aforementioned tensor operator

$$J_\tau^2 = 5^{1/2}\begin{pmatrix} & 2 & \\ \tau & & \tau' \end{pmatrix}\begin{pmatrix} 2 & 1 & 1 \\ \tau' & \rho' & \sigma' \end{pmatrix}^* J_{\rho'} J_{\sigma'} \tag{81a}$$

or

$$Q = \begin{pmatrix} 2 & 1 & 1 \\ \rho & \rho' & \sigma' \end{pmatrix}^* q_\rho J_{\rho'} J_{\sigma'}. \tag{81b}$$

The expectation value of this operator is the same for all states ψ_ν^j and, for a given j, its magnitude is proportional to the quadrupole moment as commonly defined.

References

1. G. Frobenius and I. Schur, *Sitzber. Deut. Akad. Wiss. Berlin, Kl. Math. Phys. Tech.*, p. 186 (1906).
2. Cf. F. D. Murnaghan, "Group Representations," Chapter 3 (Baltimore, Maryland, 1938).
3. The parametrization of the group and the form (29), (30a) of the representations is identical with that adopted in E. Wigner, "Gruppentheorie und ihre Anwendungen auf etc." (Braunschweig, 1931), Chapter XV. This chapter also gives the isomorphism of the three-dimensional rotation and two-dimensional unimodular unitary groups.

4. $(j' + \frac{1}{2}) \dotplus (j' - \frac{1}{2})$ denotes the sum of the representations $j' + \frac{1}{2}$ and $j' - \frac{1}{2}$ (i.e. the reducible representation containing $j' + \frac{1}{2}$ and $j' - \frac{1}{2}$). Cf. F. D. Murnaghan, Reference 2.

5. A collection of formulas for these quantities (which are essentially identical with the three-j-symbols) is given in "The Theory of Atomic Spectra" by E. U. Condon and G. H. Shortley (Cambridge Univ. Press, London and New York, 1935), Chapter III. Relations of this character are, of course, well known in quantum mechanics and many of the six-j-symbols are implicitly contained in the formulas for the relative intensities of spectral lines and the magnitude of the Zeeman effect (Hönl, Kronig, Rubinowicz). Cf. the monographs of Reference 3 and 5, particularly the latter.

6. E. P. Wigner, *Am. J. Math.*, **63**, 57, 1941. For the present article, the following definitions and theorems are of importance. An irreducible representation which can be transformed into real form is called an integer representation even if the form in which it is used is not real. An irreducible representation which is equivalent to its conjugate complex but cannot be transformed into real form is called a half integer representation. The representations $D^{(j)}$ of the three dimensional rotation group with integer j are examples for the former, those with half odd integer j, examples for the latter. The Kronecker product of two integer representations or of two half integer representations contains only integer representations; the Kronecker product of an integer and a half integer representation contains only half integer representations. A representation which occurs in the symmetrized part of the Kronecker product of an integer representation with itself, or in the antisymmetrized part of the Kronecker product of a half integer representation with itself, is called an even representation. Representations occurring in the antisymmetrized square of an integer representation, or the symmetrized square of a half integer representation are called odd representations. Both even and odd representations are integer; examples for the even and odd are again the $D^{(j)}$ with even and odd j. Some representations may not occur in the Kronecker product of any representation with itself; these are then neither even, nor odd. However, no representation can be both even and odd and every integer representation of the three-dimensional rotation group (or of the two-dimensional unitary group) is either even or odd. The identical representation's j is 0; it is an even representation.

7. G. Racah, *Phys. Rev.* **62**, 438 (1942); **78**, 622 (1950).

8. H. A. Jahn, *Proc. Roy. Soc.* **A205**, 192 (1951).

9. L. C. Biedenharn, Oak Ridge National Laboratory Report No. 1098 (1952).

10. L. C. Biedenharn, *J. Math. Phys.* **31**, 287 (1953).

FEBRUARY 1 AND 15, 1942 PHYSICAL REVIEW VOLUME 61

Theory of Complex Spectra. I

GIULIO RACAH
The Hebrew University, Jerusalem, Palestine
(Received November 14, 1941)

This paper gives a closed formula which entirely replaces for the two-electron spectra the previous lengthy calculations with the diagonal-sum method. Applications are also made to some configurations with three or more electrons and to the p^n configurations of the nuclei.

§1. INTRODUCTION

THE first-order perturbation energy for the terms of a given configuration was calculated at first by Slater.[1] In his classical paper he showed that the electrostatic interaction between two electrons depends on a very few integrals F^k and G^k, and he developed the diagonal-sum procedure for calculating the coefficients of these integrals; with this procedure he obtained numerical tables of coefficients for the two-electron configurations involving s, p, or d electrons. These tables were extended by several authors[2] to f electrons and to some configurations with three or more electrons.

But the diagonal-sum procedure has some deficiencies. Firstly, when two terms of a kind occur in a given configuration, this procedure will determine only the sum of their energies, and they can be separated only by other methods. Secondly, this method does not give general formulas, but only numerical tables; it is therefore impossible to make generalizations, and one must begin again for each new case with new and more complex calculations.[3]

It is the purpose of this paper to substitute for the numerical methods of Chapters VI and VII of TAS more general methods and more conformable to Chapter III of the same book.

---

[1] J. C. Slater, Phys. Rev. **34**, 1293 (1929).
[2] See E. U. Condon and G. H. Shortley, *Theory of Atomic Spectra* (Cambridge 1935), (which we shall denote by TAS), Chapters VI and VII, for definitions, notations and bibliographical indications.
[3] G. H. Shortley and B. Fried, Phys. Rev. **54**, 739 (1938).

§2. TWO-ELECTRON CONFIGURATIONS

If ω is the angle between the radii vectors of the two electrons, the coefficients f_k of F^k are[4] the eigenvalues of the matrix

$$(l_1l_2m_1m_2|P_k(\cos\omega)|l_1l_2m_1'm_2')\,; \tag{1}$$

here P_k is the Legendre polynomial of the order k. The transformation which diagonalizes this matrix is $(l_1l_2LM|l_1l_2m_1m_2)$, and therefore

$$f_k(l_1l_2L)=\sum_{m_1m_2m_1'm_2'}(l_1l_2LM|l_1l_2m_1m_2)(l_1l_2m_1m_2|P_k(\cos\omega)|l_1l_2m_1'm_2')(l_1l_2m_1'm_2'|l_1l_2LM), \tag{2}$$

or

$$f_k(l_1l_2L)=(l_1l_2LM|P_k(\cos\omega)|l_1l_2LM). \tag{3}$$

In the same way, if $\pm g_k$ are the coefficients of G^k for the singlet and for the triplet terms, we have

$$g_k(l_1l_2L)=\sum_{m_1m_2m'_1m'_2}(l_1l_2LM|l_1l_2m_1m_2)(l_1l_2m_1m_2|P_k(\cos\omega)|l_2l_1m'_2m'_1)(l_1l_2m'_1m'_2|l_1l_2LM), \tag{4}$$

and in view of[5]

$$(l_1l_2m'_1m'_2|l_1l_2LM)=(-1)^{l_1+l_2-L}(l_2l_1m'_2m'_1|l_2l_1LM), \tag{5}$$

this becomes

$$g_k(l_1l_2L)=(-1)^{l_1+l_2-L}(l_1l_2LM|P_k(\cos\omega)|l_2l_1LM). \tag{6}$$

Slater calculated the matrix elements of P_k (cos ω) in the $l_1l_2m_1m_2$ scheme, and then obtained the eigenvalues of this operator by means of the diagonal-sum procedure; we will calculate the matrix elements of cos ω directly in the l_1l_2LM scheme by the method of Güttinger and Pauli,[6] and then calculate f_k and g_k with the ordinary methods of matrix calculations.

If $\mathbf{u}_i$ is the unit vector in the direction from the origin to the electron i, by comparing TAS $4^3$21 with TAS $9^3$11, we have

$$(l_i\vdots u_i\vdots l_i)=0,\quad (l_i\vdots u_i\vdots l_i-1)=(l_i-1\vdots u_i\vdots l_i)=\frac{1}{[(2l_i-1)(2l_i+1)]^{\frac{1}{2}}}; \tag{7}$$

and since

$$\cos\omega=(\mathbf{u}_1\cdot\mathbf{u}_2), \tag{8}$$

introducing (7) in TAS $12^3$2 we find that the only non-vanishing elements of $(l_1l_2LM|\cos\omega|l_1'l_2'LM)$ are

$$\begin{aligned}
(l_1l_2LM|\cos\omega|l_1-1\;l_2-1\;LM)&=-\frac{[(l_1+l_2+L+1)(l_1+l_2+L)(l_1+l_2-L)(l_1+l_2-L-1)]^{\frac{1}{2}}}{2[(2l_1-1)(2l_1+1)(2l_2-1)(2l_2+1)]^{\frac{1}{2}}},\\
(l_1l_2LM|\cos\omega|l_1+1\;l_2-1\;LM)&=\frac{[(L+l_1-l_2+2)(L+l_1-l_2+1)(L+l_2-l_1)(L+l_2-l_1-1)]^{\frac{1}{2}}}{2[(2l_1+1)(2l_1+3)(2l_2-1)(2l_2+1)]^{\frac{1}{2}}},\\
(l_1l_2LM|\cos\omega|l_1-1\;l_2+1\;LM)&=\frac{[(L+l_1-l_2)(L+l_1-l_2-1)(L+l_2-l_1+2)(L+l_2-l_1+1)]^{\frac{1}{2}}}{2[(2l_1-1)(2l_1+1)(2l_2+1)(2l_2+3)]^{\frac{1}{2}}},\\
(l_1l_2LM|\cos\omega|l_1+1\;l_2+1\;LM)&=-\frac{[(l_1+l_2+L+3)(l_1+l_2+L+2)(l_1+l_2-L+2)(l_1+l_2-L+1)]^{\frac{1}{2}}}{2[(2l_1+1)(2l_1+3)(2l_2+1)(2l_2+3)]^{\frac{1}{2}}}.
\end{aligned} \tag{9}$$

From these formulas it is possible to calculate the matrix elements of P_k (cos ω) with the ordinary methods of matrix calculations; in order that these elements may have a value different from zero, k must satisfy the conditions

$$k+l_1+l'_1=2g_1,\quad k+l_2+l'_2=2g_2 \tag{10}$$

[4] TAS §8^6.
[5] TAS 14^3 7.
[6] Güttinger and Pauli, Zeits. f. Physik **67**, 743 (1931); TAS §10^3 *et seq.*

(g_1 and g_2 are integers), and the so-called triangular conditions

$$|l_1-l'_1|\leqslant k\leqslant l_1+l'_1,\quad |l_2-l'_2|\leqslant k\leqslant l_2+l'_2; \tag{11}$$

if these conditions are satisfied, the final result is

$$(l_1l_2LM|P_k(\cos\omega)|l'_1l'_2LM)=\frac{(-1)^{g_1+g_2-k}(2g_1-2l_1)!(2g_2-2l'_2)!g_1!g_2!}{(g_1-k)!(g_1-l_1)!(g_1-l'_1)!(g_2-k)!(g_2-l_2)!(g_2-l'_2)!(2g_1+1)!(2g_2+1)!}$$
$$\cdot\left[\frac{(2l_1+1)(2l'_1+1)(2l_2+1)(2l'_2+1)(l_1+l_2+L+1)!(l'_1+l'_2+L+1)!\times(l_1+l_2-L)!(l'_1+l'_2-L)!(L+l_1-l_2)!(L+l'_2-l'_1)!}{(L+l'_1-l'_2)!(L+l_2-l_1)!}\right]^{\frac{1}{2}}$$
$$\cdot\sum_u(-1)^u\frac{(u+l'_1-l'_2)!(u+l_2-l_1)!(k+l_1+l'_2-u)!}{(u+L+1)!(u-L)!(l_1+l_2-u)!(l'_1+l'_2-u)!(k-l_1-l'_2+u)!}, \tag{12}$$

where, in the summation, u takes on all integral values consistent with the factorial notation, the factorial of a negative number being meaningless.

In order to demonstrate this formula, it suffices to verify that (12) reduces to $\delta(l_1l'_1)\delta(l_2l'_2)$ for $k=0$ and to (9) for $k=1$ and that, introducing (12) for $k=n-1$ and $k=n-2$ in the formula[7]

$$P_n(\cos\omega)=\frac{2n-1}{n}\cos\omega P_{n-1}(\cos\omega)-\frac{n-1}{n}P_{n-2}(\cos\omega) \tag{13}$$

written in matrix form, we obtain again (12) for $k=n$. These verifications are somewhat long, but they are not difficult and will be omitted for brevity.

It is remarkable that (12) has an unsymmetrical aspect: it is however possible (as is shown in the appendix) to transform this formula by means of algebraic identities and to replace it with[8]

$$(l_1l_2LM|P_k(\cos\omega)|l'_1l'_2LM)$$
$$=\frac{(-1)^{g_1+g_2-k-L}(l_1+l'_1-k-1)!!(k+l_1-l'_1-1)!!(k+l'_1-l_1-1)!!\times(l_2+l'_2-k-1)!!(k+l_2-l'_2-1)!!(k+l'_2-l_2-1)!!}{(k+l_1+l'_1+1)!!(k+l_2+l'_2+1)!!}$$
$$\cdot\left[\frac{(2l_1+1)(2l_2+1)(2l'_1+1)(2l'_2+1)(l_1+l_2-L)!(l'_1+l'_2-L)!\times(L+l_1-l_2)!(L+l'_1-l'_2)!(L+l_2-l_1)!(L+l'_2-l'_1)!}{(l_1+l_2+L+1)!(l'_1+l'_2+L+1)!}\right]^{\frac{1}{2}}$$
$$\cdot\sum_v(-1)^v\frac{(l_1+l_2+l'_1+l'_2+1-v)!}{(l_1+l_2-L-v)!(l'_1+l'_2-L-v)!(l_1+l'_1-k-v)!(l_2+l'_2-k-v)!v!\times(k+L-l_1-l'_2+v)!(k+L-l'_1-l_2+v)!}. \tag{12'}$$

Introducing (12′) in (3) and (6), and putting

$$v=l_1+l_2-L-w,$$

[7] Courant and Hilbert, *Methoden der Mathematischen Physik* (Springer, 1931), p. 73, Eq. (19).

[8] With the symbol $n!!$ we indicate the *semifactorial* of n, that is the product 1.3.5 . . . n if n is odd, and the product 2.4.6 . . . n if n is even.

we obtain

$$f_k(l_1l_2L)=\frac{(k-1)!!^4(2l_1-k-1)!!(2l_2-k-1)!!(2l_1+1)(2l_2+1)}{(2l_1+k+1)!!(2l_2+k+1)!!}$$
$$\times\sum_w(-1)^w\binom{l_1+l_2+L+1+w}{w}\binom{l_1+l_2-L}{w}\binom{L+l_1-l_2}{k-w}\binom{L+l_2-l_1}{k-w} \quad (14)$$

and

$$g_k(l_1l_2L)=(-1)^{l_1+l_2-L}\frac{(k+l_1-l_2-1)!!^2(k+l_2-l_1-1)!!^2(l_1+l_2-k-1)!!^2(2l_1+1)(2l_2+1)}{(l_1+l_2+k+1)!!^2}$$
$$\times\sum_w(-1)^w\binom{l_1+l_2+L+1+w}{w}\binom{l_1+l_2-L}{w}\binom{L+l_1-l_2}{k+l_1-l_2-w}\binom{L+l_2-l_1}{k+l_2-l_1-w}. \quad (15)$$

The dependence on L of such formulas was already given by Kramers[9] for f_k and by Brinkman[10] for g_k by means of a group-theoretical procedure, but it was impossible, by such a general method, to give the first factor.

Putting

$$p=l_1(l_1+1)l_2(l_2+1),\quad s=l_1(l_1+1)+l_2(l_2+1),\quad \Lambda=L(L+1),\quad \lambda=\frac{\Lambda-s}{2}=(\mathrm{l}_1\cdot\mathrm{l}_2),\quad q=(l_1+l_2+1)^2, \quad (16)$$

we obtain, for the most common and important cases:

$$f_0=1,$$
$$f_2=\frac{6\lambda^2+3\lambda-2p}{(2l_1-1)(2l_1+3)(2l_2-1)(2l_2+3)}, \quad (17)$$
$$f_4=9\frac{70\lambda^4+350\lambda^3-10(6p-5s-39)\lambda^2-10(17p-6s-9)\lambda+3p(2p-4s-27)}{4(2l_1-3)(2l_1-1)(2l_1+3)(2l_1+5)(2l_2-3)(2l_2-1)(2l_2+3)(2l_2+5)},$$

and

$$g_{l_1-l_2}(l_1l_2L)=\frac{(-1)^{l_1+l_2-L}}{2^{2k}}\binom{2k}{k}\frac{\Lambda(\Lambda-2)(\Lambda-6)\cdots[\Lambda-(k-1)k]}{(2l_2+1)(2l_2+3)^2(2l_2+5)^2\cdots(2l_1-1)^2(2l_1+1)},$$
$$g_{l_1-l_2+2}(l_1l_2L)=\frac{(-1)^{l_1+l_2-L}}{2^{2k-1}}\binom{2k-2}{k-1}\Lambda(\Lambda-2)(\Lambda-6)\cdots[\Lambda-(k-3)(k-2)]. \quad (18)$$
$$\cdot\frac{k(2k-1)\Lambda^2-2(k-1)(2k-1)(q-k)\Lambda+(k-1)[(2k-3)q^2-(4k^2-6k-1)q+(k-1)k^2]}{(2l_2-1)^2(2l_2+1)(2l_2+3)^2(2l_2+5)^2\cdots(2l_1-1)^2(2l_1+1)(2l_1+3)^2}.$$

By means of these formulas all results of TAS and of Shortley and Fried[3] were checked, and a sole mistake was found: the coefficient of F_2 and of G_2 in the F terms of the configuration ff is not $+10$, as reported in TAS (p. 207), but -10, as given in the original paper of Condon and Shortley.[11]

§3. CONFIGURATIONS WITH THREE OR MORE ELECTRONS

The expression of the electrostatic interaction of two electrons as function of λ is not only important for a more rapid calculation of the two-electron terms: in the case of three or more electrons the methods of Chapter III of TAS give us the possibility of calculating the matrix elements of

[9] Kramers, Proc. Amst. Acad. **34**, 965 (1931).
[10] Brinkman, Zeits. f. Physik **79**, 753 (1932).
[11] Condon and Shortley, Phys. Rev. **37**, 1030 (1931).

$\lambda_{ij}=(\mathbf{l}_i\cdot\mathbf{l}_j)$ in every complex case of vector coupling; and it is therefore possible to calculate the terms of more complex configurations, even if two or more terms of the same kind occur.

We will at first pay attention to some particular applications to atomic and nuclear spectra, that can be treated without matrix calculations; then we shall treat in detail the p^2p configuration, and give the results for the p^2l configuration. For other important configurations and for the cases of (jj) coupling the calculations are more complex, and a new general procedure for this purpose will therefore be developed in a later paper.

The first application was already made to the p^n configurations by Van Vleck,[12] who found empirically the formula of f_2 for p electrons, and expressed λ^2_{ij} as a linear function of λ_{ij} and of $(\mathbf{s}_i\mathbf{s}_j)$ by means of Dirac's vector model; but, as Van Vleck himself pointed out, such a procedure is not generally sufficient for d^n configurations.

This procedure suffices however for the terms of d^n with higher multiplicity. In these states all spins are parallel, and it follows therefore from the principle of antisymmetry that the possible values of each λ are 0 or -5 (F or P resultant); from this we have that

$$\lambda^2+5\lambda=0; \tag{19}$$

introducing this relation into the expressions

$$f_2(dd)=\frac{2\lambda^2+\lambda-24}{147}, \quad f_4(dd)=\frac{35\lambda^4+175\lambda^3-585\lambda^2-2655\lambda-162}{7938}, \tag{20}$$

we obtain

$$f_2=-(3\lambda+8)/49, \quad f_4=(15\lambda-9)/441, \tag{21}$$

and therefore

$$E=\sum_{i<j}\left(F^0-\frac{3\lambda_{ij}+8}{49}F^2+\frac{15\lambda_{ij}-9}{441}F^4\right);$$

since for a d^n term

$$L(L+1)=6n+2\sum_{i<j}\lambda_{ij}, \tag{22}$$

we obtain that for all d^n configurations with $S=n/2$

$$E=\frac{n(n-1)}{2}F^0-\frac{\frac{3}{2}L(L+1)+n(4n-13)}{49}F^2+\frac{\frac{15}{2}L(L+1)-\frac{9}{2}n(n+9)}{441}F^4. \tag{23}$$

§4. THE NUCLEAR CONFIGURATIONS p^n

The calculation of the energy levels of a nuclear configuration p^n with symmetrical forces was made by Hund[13] with the diagonal-sum procedure. After very long calculations he obtained a numerical table for the energies of such configurations, and from this table he deduced empirical formulas for the interactions of Wigner and of Majorana. The direct calculation of such formulas is a remarkable application of the above developed methods.

Putting, as customary in order to avoid fractional coefficients,

$$F_0=F^0, \quad F_2=F^2/25, \tag{24}$$

we obtain from (17) that the normal (Wigner) interaction between two particles is

$$V_{ij}=F_0+(6\lambda_{ij}^2+3\lambda_{ij}-8)F_2; \tag{25}$$

[12] J. H. Van Vleck, Phys. Rev. **45**, 412 (1934).
[13] Hund, Zeits. f. Physik **105**, 202 (1937).

hence the Wigner interaction between all particles of the configuration is

$$V_W = \frac{n(n-1)}{2} F_0 + \sum_{i<j} (6\lambda_{ij}^2 + 3\lambda_{ij} - 8) F_2. \tag{26}$$

In order to calculate this sum, we cannot use Dirac's vector model, because the exclusion principle does not hold for a proton and a neutron; but we can observe that for p particles the operator

$$M_{ij} = \lambda_{ij}^2 + \lambda_{ij} - 1 \tag{27}$$

is Majorana's operator of position exchange, since it has the eigenvalue 1 for the symmetrical states S and D, and the eigenvalue -1 for the antisymmetrical state P: from (26) and (27) we obtain

$$V_W = \frac{n(n-1)}{2} F_0 + \sum_{i<j} (6M_{ij} - 3\lambda_{ij} - 2) F_2. \tag{28}$$

The sum

$$\mathfrak{M} = \sum_{i<j} M_{ij} \tag{29}$$

is Hund's $(\alpha-\beta)$ and depends only from the symmetry character of the positional eigenfunction of the level: it is the difference between the number of symmetrically connected couples and the number of antisymmetrically connected couples. From (29) and from

$$\sum_{i<j} \lambda_{ij} = \tfrac{1}{2} L(L+1) - n, \tag{30}$$

we obtain

$$V_W = \frac{n(n-1)}{2} F_0 + [6\mathfrak{M} - \tfrac{3}{2} L(L+1) - n(n-4)] F_2. \tag{31}$$

The Majorana interaction is

$$V_M = \sum_{i<j} M_{ij} V_{ij} = \mathfrak{M} F_0 + \sum_{i<j} (\lambda_{ij}^2 + \lambda_{ij} - 1)(6\lambda_{ij}^2 + 3\lambda_{ij} - 8) F_2 \tag{32}$$

and reduces to

$$V_M = \mathfrak{M} F_0 + \sum_{i<j} (M_{ij} - 3\lambda_{ij} + 3) F_2 \tag{33}$$

in virtue of

$$\lambda_{ij}^3 + 2\lambda_{ij}^2 - \lambda_{ij} - 2 = 0 \tag{34}$$

(which expresses the fact that λ_{ij} has only the eigenvalues 1, -1, and -2) and of (27). In view of (29) and (30), (33) becomes

$$V_M = \mathfrak{M} F_0 + [\mathfrak{M} - \tfrac{3}{2} L(L+1) + \tfrac{3}{2} n(n+1)] F_2. \tag{35}$$

Putting

$$F_0 = A - \tfrac{4}{3} B \quad \text{and} \quad F_2 = \tfrac{1}{3} B$$

in (31) and (35), we obtain Hund's formulas.

When two terms with the same L, R, S and $\mathfrak{M}$ occur, Hund could only calculate their sum, and said that his formulas hold for their mean values; since our method gives the energies of all terms separately, we can say that in the later case the first-order energies are the same, and that Hund's formulas hold for each term separately.

§5. THE CONFIGURATION p^2p

It follows from §2 that the interaction between two non-equivalent p electrons is

$$W_{pp} = F^0(np, n'p) + \frac{6\lambda^2 + 3\lambda - 8}{25} F^2(np, n'p) \pm (-1)^L \left[G^0(np, n'p) + \frac{6\lambda^2 + 3\lambda - 8}{25} G^2(np, n'p) \right], \tag{36}$$

where the upper sign holds when the spins are antiparallel, and the lower sign when the spins are parallel.

Following Dirac's vector model[14] we can substitute the operator

$$-\frac{1+4(\mathbf{s}\cdot\mathbf{s}')}{2}=-\frac{1+\mu}{2} \tag{37}$$

for the double sign. For $(-1)^L$ we can substitute the operator (27).

Considering that λ satisfies the Eq. (34), we have that

$$W_{pp}=F^0(np,n'p)+\frac{6\lambda^2+3\lambda-8}{25}F^2(np,n'p)$$
$$-\frac{1+\mu}{2}\left[(\lambda^2+\lambda-1)G^0(np,n'p)+\frac{\lambda^2-2\lambda+2}{25}G^2(np,n'p)\right]. \tag{38}$$

Putting

$$F_0=F^0(np,np)+2F^0(np,n'p),\quad F_2=\tfrac{1}{25}F^2(np,np),\quad F'_2=\tfrac{1}{25}F^2(np,n'p),$$
$$G_0=G^0(np,n'p),\quad G_2=\tfrac{1}{25}G^2(np,n'p),$$

and marking by 1 and 2 the two np electrons, and by 3 the $n'p$ electron, we obtain for the electrostatic interaction of the $np^2n'p$ configuration:

$$E=F_0+(6\lambda_{12}^2+3\lambda_{12}-8)F_2+\sum_1^2{}_i(6\lambda_{i3}^2+3\lambda_{i3}-8)F'_2$$
$$-\sum_1^2{}_i\frac{1+\mu_{i3}}{2}[(\lambda_{i3}^2+\lambda_{i3}-1)G_0+(\lambda_{i3}^2-2\lambda_{i3}+2)G_2]. \tag{39}$$

The most convenient scheme for calculating this energy operator is the LS scheme of the parent ion np^2. In this scheme λ_{12} is diagonal, and the coefficient of F_2 is therefore the diagonal matrix

$$\|(L|6\lambda_{12}^2+3\lambda_{12}-8|L')\|=\begin{array}{c|ccc|} & \mathrm{D} & \mathrm{P} & \mathrm{S}\\ \hline \mathrm{D} & 1 & 0 & 0\\ \mathrm{P} & 0 & -5 & 0\\ \mathrm{S} & 0 & 0 & 10\\ \hline \end{array}\,. \tag{40}$$

In order to calculate the coefficients of F'_2, G_0 and G_2 we must at first calculate the matrices of λ_{i3}. From TAS 10^32 we have

$$\|(L\vdots l_i\vdots L')\|=\begin{array}{c|ccc|} & \mathrm{D} & \mathrm{P} & \mathrm{S}\\ \hline \mathrm{D} & \frac{1}{2} & \pm\frac{1}{2\sqrt{3}} & 0\\ \mathrm{P} & \pm\frac{1}{2\sqrt{3}} & \frac{1}{2} & \pm\left(\frac{2}{3}\right)^{\frac{1}{2}}\\ \mathrm{S} & 0 & \pm\left(\frac{2}{3}\right)^{\frac{1}{2}} & \frac{1}{2}\\ \hline \end{array}\,, \tag{41}$$

[14] Dirac, Proc. Roy. Soc. **A123**, 714 (1929), and *Quantum Mechanics* (Oxford, 1930), Chapter 11.

the upper sign holding for $i=1$ and the lower for $i=2$; from this table and from TAS 12^32 we obtain

$$\|(\mathrm{L}L|\lambda_{i3}|\mathrm{L}'L)\| = \begin{array}{c|ccc|} & \mathrm{D} & \mathrm{P} & \mathrm{S} \\ \hline \mathrm{D} & \frac{\Lambda-8}{4} & \mp\frac{[\Lambda(12-\Lambda)]^{\frac{1}{2}}}{4\sqrt{3}} & 0 \\ \mathrm{P} & \mp\frac{[\Lambda(12-\Lambda)]^{\frac{1}{2}}}{4\sqrt{3}} & \frac{\Lambda-4}{4} & \mp\frac{[\Lambda(6-\Lambda)]^{\frac{1}{2}}}{\sqrt{6}} \\ \mathrm{S} & 0 & \mp\frac{[\Lambda(6-\Lambda)]^{\frac{1}{2}}}{\sqrt{6}} & \frac{\Lambda-2}{4} \\ \hline \end{array}, \tag{42}$$

Λ having the meaning of (16); and this matrix gives us easily the coefficient of F'_2:

$$\|(\mathrm{L}L|\sum_1^2{}_i\,(6\lambda_{i3}^2+3\lambda_{i3}-8)|\mathrm{L}'L)\| = \begin{array}{c|ccc|} & \mathrm{D} & \mathrm{P} & \mathrm{S} \\ \hline \mathrm{D} & \frac{\Lambda^2-15\Lambda+40}{2} & 0 & \frac{\Lambda[(12-\Lambda)(6-\Lambda)]^{\frac{1}{2}}}{\sqrt{2}} \\ \mathrm{P} & 0 & \frac{-3\Lambda^2+21\Lambda-20}{2} & 0 \\ \mathrm{S} & \frac{\Lambda[(12-\Lambda)(6-\Lambda)]^{\frac{1}{2}}}{\sqrt{2}} & 0 & \frac{-5\Lambda^2+42\Lambda-64}{4} \\ \hline \end{array}. \tag{43}$$

In order to calculate the coefficients of G_0 and G_2 we must also calculate the matrices of $\mu_{i3}=4(\mathbf{s}_i\cdot\mathbf{s}_3)$. From TAS 10^32 and 12^32 we obtain in the same way

$$\|(\mathrm{s}S|\mu_{i3}|\mathrm{s}'S)\| = \begin{array}{c|cc|} & \mathrm{s}=1 & \mathrm{s}=0 \\ \hline \mathrm{s}=1 & S(S+1)-1\tfrac{1}{4} & \mp[[S(S+1)+\tfrac{1}{4}][1\tfrac{3}{4}-S(S+1)]]^{\frac{1}{2}} \\ \mathrm{s}=0 & \mp[[S(S+1)+\tfrac{1}{4}][1\tfrac{3}{4}-S(S+1)]]^{\frac{1}{2}} & S(S+1)-\tfrac{3}{4} \\ \hline \end{array} \tag{44}$$

and therefore

$$\left\|\left({}^m\mathrm{L}\ {}^2L\left|\frac{1+\mu_{i3}}{2}\right|{}^{m'}\mathrm{L}\ {}^2L\right)\right\| = \begin{array}{c|cc|} & {}^3\mathrm{L} & {}^1\mathrm{L} \\ \hline {}^3\mathrm{L} & -\frac{1}{2} & \mp\frac{\sqrt{3}}{2} \\ {}^1\mathrm{L} & \mp\frac{\sqrt{3}}{2} & \frac{1}{2} \\ \hline \end{array} \tag{45}$$

and

$$\left({}^3\mathrm{L}\ {}^4L\left|\frac{1+\mu_{i3}}{2}\right|{}^3\mathrm{L}\ {}^4L\right)=1. \tag{45'}$$

TABLE I.

Atom	Configuration	$({}^4P-{}^4D)/({}^4D-{}^4S)$
Theory	$np^2n'p$	0.667
N I	$2p^23p$	0.521
N I	$2p^24p$	0.504
O II	$2p^23p$	0.520
S II	$3p^24p$	0.604

From (42) and (45) we get

$$\left\|\left({}^{m}L\ {}^2L\left|\sum_{1}^{2}{}_i\frac{1+\mu_{i3}}{2}(\lambda_{i3}{}^2+\lambda_{i3}-1)\right|{}^{m'}L'\ {}^2L\right)\right\|$$

$$=\begin{array}{c|ccc|} & {}^1D & {}^3P & {}^1S \\ \hline {}^1D & \dfrac{\Lambda^2-12\Lambda+24}{24} & \dfrac{(\Lambda-4)[\Lambda(12-\Lambda)]^{\frac{1}{2}}}{8} & \dfrac{\Lambda[(12-\Lambda)(6-\Lambda)]^{\frac{1}{2}}}{12\sqrt{2}} \\ {}^3P & \dfrac{(\Lambda-4)[\Lambda(12-\Lambda)]^{\frac{1}{2}}}{8} & \dfrac{\Lambda^2-8\Lambda+8}{8} & \dfrac{(\Lambda-1)[\Lambda(6-\Lambda)]^{\frac{1}{2}}}{2\sqrt{2}} \\ {}^1S & \dfrac{\Lambda[(12-\Lambda)(6-\Lambda)]^{\frac{1}{2}}}{12\sqrt{2}} & \dfrac{(\Lambda-1)[\Lambda(6-\Lambda)]^{\frac{1}{2}}}{2\sqrt{2}} & \dfrac{-5\Lambda^2+48\Lambda-60}{48} \\ \hline \end{array} \tag{46}$$

and

$$\left\|\left({}^{m}L\ {}^2L\left|\sum_{1}^{2}{}_i\frac{1+\mu_{i3}}{2}(\lambda_{i3}{}^2-2\lambda_{i3}+2)\right|{}^{m'}L'\ {}^2L\right)\right\|$$

$$=\begin{array}{c|ccc|} & {}^1D & {}^3P & {}^1S \\ \hline {}^1D & \dfrac{\Lambda^2-30\Lambda+240}{24} & \dfrac{(\Lambda-10)[\Lambda(12-\Lambda)]^{\frac{1}{2}}}{8} & \dfrac{\Lambda[(12-\Lambda)(6-\Lambda)]^{\frac{1}{2}}}{12\sqrt{2}} \\ {}^3P & \dfrac{(\Lambda-10)[\Lambda(12-\Lambda)]^{\frac{1}{2}}}{8} & \dfrac{\Lambda^2-2\Lambda-40}{8} & \dfrac{(\Lambda-7)[\Lambda(6-\Lambda)]^{\frac{1}{2}}}{2\sqrt{2}} \\ {}^1S & \dfrac{\Lambda[(12-\Lambda)(6-\Lambda)]^{\frac{1}{2}}}{12\sqrt{2}} & \dfrac{(\Lambda-7)[\Lambda(6-\Lambda)]^{\frac{1}{2}}}{2\sqrt{2}} & \dfrac{-5\Lambda^2+12\Lambda+156}{48} \\ \hline \end{array}. \tag{47}$$

From (42) and (45′) we get

$$\left({}^3P\ {}^4L\left|\sum_{1}^{2}{}_i\frac{1+\mu_{i3}}{2}(\lambda_{i3}{}^2+\lambda_{i3}-1)\right|{}^3P\ {}^4L\right)=-\frac{\Lambda^2-8\Lambda+8}{4} \tag{46′}$$

and

$$\left({}^3P\ {}^4L\left|\sum_{1}^{2}{}_i\frac{1+\mu_{i3}}{2}(\lambda_{i3}{}^2-2\lambda_{i3}+2)\right|{}^3P\ {}^4L\right)=-\frac{\Lambda^2-2\Lambda-40}{4}. \tag{47′}$$

Introducing our results into (39) we obtain·

$${}^4D=F_0-5F_2-F'_2-G_0-4G_2,$$

$${}^4P=F_0-5F_2+5F'_2-G_0-10G_2,$$

$${}^4S=F_0-5F_2-10F'_2+2G_0-10G_2,$$

$$^2S = F_0 - 5F_2 - 10F'_2 - G_0 + 5G_2,$$

$$^2F = F_0 + F_2 + 2F'_2 - G_0 - G_2.$$

The energy matrix for the ^2D terms is

	^{1}D	^{3}P
^{1}D	$F_0+F_2-7F'_2+\frac{1}{2}G_0-4G_2$	$-\frac{3}{2}G_0+3G_2$
^{3}P	$-\frac{3}{2}G_0+3G_2$	$F_0-5F_2-F'_2+\frac{1}{2}G_0+2G_2$

and has the eigenvalues

$$^2D = F_0 - 2F_2 - 4F'_2 + \tfrac{1}{2}G_0 - G_2 \pm 3[(F_2 - F'_2 - G_2)^2 + \tfrac{1}{4}(G_0 - 2G_2)^2]^{\frac{1}{2}}.$$

The energy matrix for the ^2P terms is

	^{1}D	^{3}P	^{1}S
^{1}D	$F_0+F_2+7F'_2-\frac{1}{6}G_0-\frac{23}{3}G_2$	$(-G_0+2G_2)\sqrt{5}$	$(4F'_2-\frac{1}{3}G_0-\frac{1}{3}G_2)\sqrt{5}$
^{3}P	$(-G_0+2G_2)\sqrt{5}$	$F_0-5F_2+5F'_2+\frac{1}{2}G_0+5G_2$	$-G_0+5G_2$
^{1}S	$(4F'_2-\frac{1}{3}G_0-\frac{1}{3}G_2)\sqrt{5}$	$-G_0+5G_2$	$F_0+10F_2-\frac{1}{3}G_0-\frac{10}{3}G_2$

It follows from our results that the ratio $(^4P-{}^4D)/(^4D-{}^2S)$ has the theoretical value of $\frac{2}{3}$. The comparison with the experimental ratios[15] is given in Table I.

The deviations are of the same order as those of the np^3 configurations of the same elements.[16]

§6. THE CONFIGURATION p^2l

The terms of the configuration $np^2n'l$ can be calculated in the same way as those of $np^2n'p$. The only difference is that the coefficients of G^{l-1} and G^{l+2} in W_{pl} are polynomials of higher degree in λ, and must be reduced to the second degree by means of the equation

$$\lambda^3 + 2\lambda^2 - (l^2 + l - 1)\lambda - l(l+1) = 0, \tag{48}$$

which corresponds to the Eq. (34) of the pp case. This reduction cannot be carried out without specifying the value of l; it is possible however to avoid such direct reduction, by calculating at first the single values of $g_{l-1}(l1L)$ and $g_{l+1}(l1L)$ for the three possible values of L by means of (15), and then determining the polynomials of the second degree which assume these values for the values l, -1, and $-(l+1)$ of the variable λ. By this procedure we find that the electrostatic interaction between a p and a l electron can be expressed by means of the formula

$$W_{pl} = F^0(np, n'l) + [6\lambda^2 + 3\lambda - 4l(l+1)]F'_2 - \frac{1+\mu}{2}[(\lambda^2 + l\lambda - l)G_{l-1} + (\lambda^2 - (l+1)\lambda + l + 1)G_{l+1}], \tag{49}$$

where[17]

$$F'_2 = \frac{F^2(np, n'l)}{5(2l-1)(2l+3)}, \quad G_{l-1} = \frac{3G^{l-1}(np, n'l)}{(2l+1)(2l-1)^2}, \quad G_{l+1} = \frac{3G^{l+1}(np, n'l)}{(2l+1)(2l+3)^2}. \tag{50}$$

From this point the calculations were carried out exactly in the same way as for p^2p, and give the following results

[15] See Bacher and Goudsmit, *Atomic Energy States* (McGraw-Hill, 1932).
[16] TAS, p. 198.
[17] Our definitions (50) differ in some cases by a factor 3 from the definitions of TAS, p. 177.

Quartets:

$$L=l+1:\quad F_0-5F_2-l(2l-1)F'_2-l(2l-1)G_{l-1}-2(l+1)G_{l+1},$$

$$L=l:\quad F_0-5F_2+(2l-1)(2l+3)F'_2-l(2l-1)G_{l-1}-(l+1)(2l+3)G_{l+1},$$

$$L=l-1:\quad F_0-5F_2-(l+1)(2l+3)F'_2+2lG_{l-1}-(l+1)(2l+3)G_{l+1}.$$

Doublets:

$$L=l+2:\quad F_0+F_2+2l(2l-1)F'_2-l(2l-1)G_{l-1}-G_{l+1},$$

$$L=l-2:\quad F_0+F_2+2(l+1)(2l+3)F'_2-G_{l-1}-(l+1)(2l+3)G_{l+1},$$

$$L=l+1:\quad F_0-2F_2-(l+3)(2l-1)F'_2+\tfrac{1}{2}(2l-1)G_{l-1}-G_{l+1}$$
$$\pm[[3F_2-3(2l-1)F'_2-\tfrac{1}{2}(l-1)(2l-1)G_{l-1}-(l+2)G_{l+1}]^2+3l(l+2)[\tfrac{1}{2}(2l-1)G_{l-1}-G_{l+1}]^2]^{\frac{1}{2}},$$

$$L=l-1:\quad F_0-2F_2-(l-2)(2l+3)F'_2-G_{l-1}-\tfrac{1}{2}(2l+3)G_{l+1}$$
$$\pm[[3F_2+3(2l+3)F'_2+(l-1)G_{l-1}-\tfrac{1}{2}(l+2)(2l+3)G_{l+1}]^2+3(l^2-1)[G_{l-1}+\tfrac{1}{2}(2l+3)G_{l+1}]^2]^{\frac{1}{2}}.$$

The energy matrix for the doublets with $L=l$ is of the third order and has the elements:

$$({}^1\mathrm{D}|E|{}^1\mathrm{D})=F_0+F_2-(2l-3)(2l+5)F'_2-\frac{2l^2-13l+12}{6}G_{l-1}-\frac{2l^2+17l+27}{6}G_{l+1},$$

$$({}^3\mathrm{P}|E|{}^3\mathrm{P})=F_0-5F_2+(2l-1)(2l+3)F'_2+\tfrac{1}{2}l(2l-1)G_{l-1}+\tfrac{1}{2}(l+1)(2l+3)G_{l+1},$$

$$({}^1\mathrm{S}|E|{}^1\mathrm{S})=F_0+10F_2-\tfrac{1}{3}l(2l-1)G_{l-1}-\tfrac{1}{3}(l+1)(2l+3)G_{l+1},$$

$$({}^1\mathrm{D}|E|{}^3\mathrm{P})=\tfrac{1}{2}[(2-l)G_{l-1}+(l+3)G_{l+1}][(2l-1)(2l+3)]^{\frac{1}{2}},$$

$$({}^3\mathrm{P}|E|{}^1\mathrm{S})=\tfrac{1}{2}[-(2l-1)G_{l-1}+(2l+3)G_{l+1}][2l(l+1)]^{\frac{1}{2}},$$

$$({}^1\mathrm{D}|E|{}^1\mathrm{S})=(2F'_2-\tfrac{1}{6}G_{l-1}-\tfrac{1}{6}G_{l+1})[2l(l+1)(2l-1)(2l+3)]^{\frac{1}{2}}.$$

APPENDIX

From the addition theorem for binomial coefficients

$$\sum_s \binom{x}{s}\binom{y}{z-s}=\binom{x+y}{z}, \tag{51}$$

putting $x=a-b$, $y=b$, $z=a-c$, we have

$$\frac{a!}{b!c!}=\sum_s \frac{(a-b)!(a-c)!}{(a-b-s)!(a-c-s)!(b+c-a+s)!s!}. \tag{52}$$

If y is negative, we can transform (51) by means of

$$\binom{y}{z-s}=(-1)^{z-s}\binom{z-s-y-1}{z-s} \tag{53}$$

and obtain

$$\sum_s (-1)^s\binom{x}{s}\binom{z-s-y-1}{z-s}=(-1)^z\binom{x+y}{z} \tag{54}$$

or

$$\sum_s (-1)^s\binom{x}{s}\binom{z-s-y-1}{z-s}=\binom{z-x-y-1}{z}; \tag{54'}$$

putting $y=z-t-1$ we have from (54)

$$\sum_s (-1)^s \frac{(t-s)!}{s!(x-s)!(z-s)!} = (-1)^z \frac{(t-z)!(x+z-t-1)!}{x!z!(x-t-1)!} \quad \text{if} \quad x>t\geqslant z\geqslant 0, \tag{55}$$

and from (54′)

$$\sum_s (-1)^s \frac{(t-s)!}{s!(x-s)!(z-s)!} = \frac{(t-x)!(t-z)!}{x!z!(t-x-z)!} \quad \text{if} \quad t\geqslant x\geqslant 0,\ t\geqslant z\geqslant 0. \tag{55′}$$

Using repeatedly (52), and also (55) and (55′), we can transform the sum in (12) as follows:

$$\sum_u (-1)^u \frac{(u+l'_1-l'_2)!}{(u+L+1)!} \cdot \frac{(u+l_2-l_1)!}{(u-L)!(k-l_1-l'_2+u)!} \cdot \frac{(k+l_1+l'_2-u)!}{(l_1+l_2-u)!(l_1'+l'_2-u)!}$$

$$= \sum_{\alpha\beta u} (-1)^u \frac{(u+l'_1-l'_2)!}{(u+L+1)!} \cdot \frac{(L+l_2-l_1)!(l_2+l'_2-k)!}{(L+l_2-l_1-\alpha)!(l_2+l_2'-k-\alpha)!(k-l_2-l'_2-L+u+\alpha)!\alpha!}$$

$$\cdot \frac{(k+l'_2-l_2)!(k+l_1-l'_1)!}{(k+l'_2-l_2-\beta)!(k+l_1-l'_1-\beta)!(l'_1+l_2-k-u+\beta)!\beta!}$$

$$= \sum_{\alpha\beta} (-1)^{k-l_2-l'_2-L+\alpha} \frac{(L+l_2-l_1)!(l_2+l'_2-k)!(k+l'_2-l_2)!(k+l_1-l'_1)!}{(L+l_2-l_1-\alpha)!(l_2+l'_2-k-\alpha)!(k+l'_2-l_2-\beta)!(k+l_1-l'_1-\beta)!}$$

$$\cdot \frac{(l'_1+l_2+L-k-\alpha)!}{(l'_1+l_2+L+1-k+\beta)!} \cdot \frac{(\alpha+\beta)!}{(l'_1-l'_2-L+\alpha+\beta)!(L+l'_2-l'_1)!\alpha!\beta!}$$

$$= \sum_{\alpha\beta\gamma} (-1)^{k-l_2-l'_2-L+\alpha} \frac{(L+l_2-l_1)!(l_2+l'_2-k)!(k+l'_2-l_2)!(k+l_1-l'_1)!(l'_1+l_2+L-k-\alpha)!}{(L+l_2-l_1-\alpha)!(l_2+l'_2-k-\alpha)!(k+l'_2-l_2-\beta)! \times (k+l_1-l'_1-\beta)!(l'_1+l_2+L+1-k+\beta)!}$$

$$\cdot \frac{1}{(L+l'_2-l'_1-\gamma)!(\alpha-\gamma)!(l'_1-l'_2-L+\beta+\gamma)!\gamma!}$$

$$= \sum_\gamma (-1)^{k-l_2-l'_2-L+\gamma} \frac{(L+l_2-l_1)!(l_2+l'_2-k)!(k+l'_2-l_2)!(k+l_1-l'_1)!}{(L+l'_2-l'_1-\gamma)!\gamma!}$$

$$\cdot \frac{(l_1+l'_1-k)!(L+l'_1-l'_2)!}{(L+l_2-l_1-\gamma)!(l_2+l'_2-k-\gamma)!(l_1+l'_1-l_2-l'_2+\gamma)!}$$

$$\cdot \frac{(k+l_1+l'_1+1+\gamma)!}{(l'_1+l'_2+L+1)!(l_1+l_2+L+1)!(k+l_1-l'_2-L+\gamma)!(k+l'_1-l_2-L+\gamma)!}.$$

Putting $\gamma=l_2+l'_2-k-v$, this expression becomes

$$\sum_v (-1)^{v-L} \frac{(L+l_2-l_1)!(l_2+l'_2-k)!(k+l'_2-l_2)!(k+l_1-l'_1)! \times (l_1+l'_1-k)!(L+l'_1-l'_2)!(l_1+l_2+l'_1+l'_2+1-v)!}{(k+L-l'_1-l_2+v)!(l_2+l'_2-k-v)!(k+L-l_1-l'_2+v)!v!(l_1+l'_1-k-v)! \times (l_1+l_2-L-v)!(l'_1+l'_2-L-v)!(l'_1+l'_2+L+1)!(l_1+l_2+L+1)!};$$

introducing this result into (12) we obtain (12′).

NOVEMBER 1 AND 15, 1942 PHYSICAL REVIEW VOLUME 62

Theory of Complex Spectra. II

GIULIO RACAH
The Hebrew University, Jerusalem, Palestine
(Received August 5, 1942)

The spectra of two-electron configurations in (jj) and (jl) coupling and of the configurations d^n, f^3, d^2p, and d^8p in (LS) coupling are calculated with tensor operators. The agreement with the odd terms of Ti II and Ni II is satisfactory. It is also proved that $G^k/(2k+1)$ is a positive decreasing function of k.

---

§1. INTRODUCTION

IN a first paper on this matter[1] a general formula was given for the coefficients of Slater's integrals for the two-electron configurations in (LS) coupling. It was also shown that this formula gives the possibility of calculating the terms of more complex configurations, and some simple examples were given. But the method used for the configurations p^2l still necessitates for d^3 very long calculations, because we must calculate the fourth power of the matrix of the scalar product of two angular momenta. It appeared therefore more convenient to develop a new method, based on tensor operators.

The algebra of tensor operators is developed in analogy to the treatment of Chapter III of TAS[2] for the vector operators. With this method some group-theoretical results of Wigner and of Kramers are obtained by a direct algebraical way, and in some cases also in a more simple and general form.

Expressing the coefficients of Slater's integrals as scalar products of tensors, we give a direct demonstration of Eq. (12′) I, and obtain also its extension to (jj) and (jl) coupling. This new method is more suitable for calculations of many-electron spectra, and applications are made to the configurations d^n, f^3, d^2p, and d^8p.

Since the whole method is based on Wigner's[3] transformation formula for vector addition (TAS 14^35), we shall begin with a direct algebraical derivation of this formula, without the use of the theory of groups.

§2. THE ALGEBRAIC CALCULATION OF $(j_1j_2m_1m_2|j_1j_2jm)$

It is shown in $§14^3$ of TAS that the transformation coefficients $(m_1m_2|jm)$ for the addition of two angular momenta are defined by the relation

$$\psi(\gamma j_1j_2jm)=\sum_{m_1m_2}\phi(\gamma j_1j_2m_1m_2)(m_1m_2|jm) \tag{1}$$

and are completely determined by the initial condition

$$(j_1j_2|j_1+j_2j_1+j_2)=1 \tag{2}$$

and by the two recursion formulas

$$[(j+m)(j-m+1)]^{\frac{1}{2}}(m_1m_2|jm-1)=[(j_1+m_1+1)(j_1-m_1)]^{\frac{1}{2}}(m_1+1m_2|jm) \\ +[(j_2+m_2+1)(j_2-m_2)]^{\frac{1}{2}}(m_1m_2+1|jm) \tag{3}$$

and

$$[(j-m)(j+m)]^{\frac{1}{2}}(j-1\vdots J_1\vdots j)(m_1m_2|j-1m) \\ =[m_1-m(j\vdots J_1\vdots j)](m_1m_2|jm)-[(j-m+1)(j+m+1)]^{\frac{1}{2}}(j+1\vdots J_1\vdots j)(m_1m_2|j+1m). \tag{4}$$

---

[1] G. Racah, Phys. Rev. **61**, 186 (1942), which will be referred to as I.
[2] E. U. Condon and G. H. Shortley, *Theory of Atomic Spectra* (Cambridge, 1935). We refer to this book (TAS) and to I for definitions, notations, and bibliographical indications.
[3] E. Wigner, *Gruppentheorie* (Vieweg, 1931), Chapter 17, Eq. (27).

But it is also pointed out there that a general formula for such coefficients is very difficult to obtain from these relations.

The calculation is however much simplified if we add a third recursion formula, which follows also, as (3), from TAS $3^3 3$ if we take the upper sign instead of the lower:

$$[(j-m)(j+m+1)]^{\frac{1}{2}}(m_1m_2|jm+1)$$
$$=[(j_1-m_1+1)(j_1+m_1)]^{\frac{1}{2}}(m_1-1m_2|jm)+[(j_2-m_2+1)(j_2+m_2)]^{\frac{1}{2}}(m_1m_2-1|jm). \quad (5)$$

In order to avoid the irrational factors we put

$$(m_1m_2|jm)=(-1)^{j_1-m_1}f(m_1m_2;\,jm)[(j_1+m_1)!(j_2+m_2)!(j+m)!]^{\frac{1}{2}}/[(j_1-m_1)!(j_2-m_2)!(j-m)!]^{\frac{1}{2}}, \quad (6)$$

and obtain from (3) and (5)

$$f(m_1m_2;\,jm-1)=(j_2+m_2+1)(j_2-m_2)f(m_1m_2+1;\,jm)-(j_1+m_1+1)(j_1-m_1)f(m_1+1m_2;\,jm) \quad (3')$$

and

$$(j-m)(j+m+1)f(m_1m_2;\,jm+1)=f(m_1m_2-1;\,jm)-f(m_1-1m_2;\,jm). \quad (5')$$

Putting $m=j$ in (5′), we see that $f(m_1m_2;\,jj)$ is independent of m_1 and m_2 and we may write

$$f(m_1m_2;\,jj)=A_j. \quad (7)$$

From (7) and (3′) we get

$$f(m_1m_2;\,jj-1)=[(j_2+m_2+1)(j_2-m_2)-(j_1+m_1+1)(j_1-m_1)]A_j; \quad (7')$$

from (7′) and (3′) we get

$$f(m_1m_2;\,jj-2)=[(j_2+m_2+1)(j_2+m_2+2)(j_2-m_2)(j_2-m_2-1)$$
$$-2(j_2+m_2+1)(j_2-m_2)(j_1+m_1+1)(j_1-m_1)+(j_1+m_1+1)(j_1+m_1+2)(j_1-m_1)(j_1-m_1-1)]A_j \quad (7'')$$

and we see that the general formula will be

$$f(m_1m_2;\,jj-u)=A_j\sum_t(-1)^t\binom{u}{t}\frac{(j_1+m_1+t)!(j_1-m_1)!(j_2+m_2+u-t)!(j_2-m_2)!}{(j_1+m_1)!(j_1-m_1-t)!(j_2+m_2)!(j_2-m_2-u+t)!}, \quad (8)$$

where, as in all formulas of this paper, the summation parameter takes on all integral values consistent with the factorial notation, the factorial of a negative number being meaningless. To demonstrate (8) it suffices to verify that it satisfies (3′); this verification is very simple and will be omitted for brevity.

Introducing (8) in (6) and remembering that

$$m=m_1+m_2, \quad (9)$$

we obtain the dependence of $(m_1m_2|jm)$ on m_1, m_2, and m:

$$(m_1m_2|jm)=\delta(m_1+m_2,\,m)A_j\left[\frac{(j_1-m_1)!(j_2-m_2)!(j-m)!(j+m)!}{(j_1+m_1)!(j_2+m_2)!}\right]^{\frac{1}{2}}\sum_t(-1)^{j_1-m_1+t}$$
$$\times\frac{(j_1+m_1+t)!(j+j_2-m_1-t)!}{t!(j-m-t)!(j_1-m_1-t)(j_2-j+m_1+t)!}. \quad (10)$$

In order to obtain from (4) the dependence of A_j on j, we calculate at first from (10) the expression of $(m_1m_2|j+1j)$: owing to the δ factor and to the expression of $(j\vdots J_1\vdots j)$ (TAS $10^3 2a$), we have

$$(m_1m_2|j+1j)=\delta(m_1+m_2,\,j)(-1)^{j_1-m_1}A_{j+1}\left[\frac{(j_1+m_1)!(j_2+m_2)!(2j+1)!}{(j_1-m_1)!(j_2-m_2)!}\right]^{\frac{1}{2}}2(j+1)[m_1-j(j\vdots J_1\vdots j)]. \quad (11)$$

The left side of (4) vanishes for $m=j$; introducing (7) and (11) and eliminating the common factors, we get

$$0=A_j-2(j+1)(2j+1)(j+1\vdots J_1\vdots j)A_{j+1},$$

and owing to the expression of $(j+1\vdots J_1\vdots j)$ (TAS 10^3 2b), this becomes

$$A_j=[(j_1+j_2+j+2)(j_1+j_2-j)(j+j_1-j_2+1)(j+j_2-j_1+1)(2j+1)/(2j+3)]^{\frac{1}{2}}A_{j+1} \tag{12}$$

and is satisfied by

$$A_j=B[(2j+1)(j_1+j_2-j)!]^{\frac{1}{2}}/[(j_1+j_2+j+1)!(j+j_1-j_2)!(j+j_2-j_1)!]^{\frac{1}{2}}, \tag{13}$$

where B is also independent of j.

It follows from (2) that

$$B=1; \tag{14}$$

and collecting (10), (13), and (14) we have at last

$$(m_1m_2|jm)=\delta(m_1+m_2,m)\left[\frac{(2j+1)(j_1+j_2-j)!(j_1-m_1)!(j_2-m_2)!(j-m)!(j+m)!}{(j_1+j_2+j+1)!(j+j_1-j_2)!(j+j_2-j_1)!(j_1+m_1)!(j_2+m_2)!}\right]^{\frac{1}{2}}$$

$$\times\sum_t(-1)^{j_1-m_1+t}\frac{(j_1+m_1+t)!(j+j_2-m_1-t)!}{t!(j-m-t)!(j_1-m_1-t)!(j_2-j+m_1+t)!}. \tag{15}$$

This formula is similar to Wigner's formula (TAS $14^3$5), and is, also, unsymmetrical and unpractical for the use; it is, however, possible to obtain a more symmetrical and useful form, by transforming it with the methods shown in the appendix of I: using (52) I and (55′) I we have

$$\sum_t(-1)^{j_1-m_1+t}\frac{(j_1+m_1+t)!(j+j_2-m_1-t)!}{t!(j-m-t)!(j_1-m_1-t)!(j_2-j+m_1+t)!}$$

$$=\sum_{tu}(-1)^{j_1-m_1+t}\frac{(j_1+m_1+t)!}{t!(j_2-j+m_1+t)!}\cdot\frac{(j_2+m_2)!(j+j_2-j_1)!}{(j_2+m_2-u)!(j+j_2-j_1-u)!(j_1-j_2-m-t+u)!u!}$$

$$=\sum_u(-1)^{j_2+m_2-u}\frac{(j_1+m_1)!(j+j_1-j_2)!(j_2+m_2)!(j+j_2-j_1)!}{(j_1-j_2-m+u)!(j_1-j-m_2+u)!(j+m-u)!(j_2+m_2-u)!(j+j_2-j_1-u)!u!},$$

and putting $z=j_2+m_2-u$ we get

$$(j_1j_2m_1m_2|j_1j_2jm)=\delta(m_1+m_2,m)[(2j+1)(j_1+j_2-j)!(j+j_1-j_2)!(j+j_2-j_1)!/(j_1+j_2+j+1)!]^{\frac{1}{2}}$$

$$\cdot\sum_z(-1)^z\frac{[(j_1+m_1)!(j_1-m_1)!(j_2+m_2)!(j_2-m_2)!(j+m)!(j-m)!]^{\frac{1}{2}}}{z!(j_1+j_2-j-z)!(j_1-m_1-z)!(j_2+m_2-z)!(j-j_2+m_1+z)!(j-j_1-m_2+z)!}. \tag{16}$$

We might also transform TAS $14^3$5 in the same way, and the result would of course be the same.

For further use it is convenient to introduce the abbreviations

$$v(abc;\alpha\beta\gamma)=\delta(\alpha+\beta+\gamma,0)\sum_z(-1)^{c-\gamma+z}$$

$$\times\frac{[(a+\alpha)!(a-\alpha)!(b+\beta)!(b-\beta)!(c+\gamma)!(c-\gamma)!]^{\frac{1}{2}}}{z!(a+b-c-z)!(a-\alpha-z)!(b+\beta-z)!(c-b+\alpha+z)!(c-a-\beta+z)!} \tag{17}$$

and

$$V(abc;\alpha\beta\gamma)=[(a+b-c)!(a+c-b)!(b+c-a)!/(a+b+c+1)!]^{\frac{1}{2}}v(abc;\alpha\beta\gamma) \tag{17'}$$

and to write

$$(j_1j_2m_1m_2|j_1j_2jm)=(-1)^{j+m}(2j+1)^{\frac{1}{2}}V(j_1j_2j;m_1m_2-m). \tag{16'}$$

The functions v and V are defined for integral and half-integral values of the arguments, with the limitation that $a-\alpha$, $b-\beta$, $c-\gamma$ must be integers; it follows from this limitation and from the factor $\delta(\alpha+\beta+\gamma, 0)$ that all the nine numbers

$$a+\alpha, \quad a-\alpha, \quad b+\beta, \quad b-\beta, \quad c+\gamma, \quad c-\gamma, \quad a+b-c, \quad a+c-b, \quad b+c-a \tag{18}$$

must be integers.

Since in (17) z takes on only such integral values for which the argument of every factorial is not negative, the number of terms in this sum is one more than the *smallest* of the nine numbers (18) (and not only of four of them, as in (15) or in TAS 14[3]5); therefore V vanishes if one of the numbers (18) is negative, and the summation reduces to one term if one of these numbers vanishes.

Assuming the argument of one of the five other factorials instead of z as summation parameter in (17), we obtain some symmetry properties for v and V:

$$V(abc;\alpha\beta\gamma)=(-1)^{a+b-c}V(bac;\beta\alpha\gamma)=(-1)^{a+b+c}V(acb;\alpha\gamma\beta)=(-1)^{a-b+c}V(cba;\gamma\beta\alpha)$$
$$=(-1)^{2b}V(cab;\gamma\alpha\beta)=(-1)^{2c}V(bca;\beta\gamma\alpha). \tag{19a}$$

Interchanging in (17) a with b and α with $-\beta$, we have

$$V(abc;\alpha\beta\gamma)=(-1)^{2\gamma}V(bac;\,-\beta-\alpha-\gamma);$$

and owing to the first of (19a) and to the fact that $2(c-\gamma)$ is even, we get also

$$V(abc;\alpha\beta\gamma)=(-1)^{a+b+c}V(abc;\,-\alpha-\beta-\gamma). \tag{19b}$$

Since the transformation matrix $(j_1j_2m_1m_2|j_1j_2jm)$ is a unitary one, it follows from (16′) that the real function V must satisfy the orthogonality relations

$$\sum_{\alpha\beta} V(abc;\alpha\beta\gamma)\,V(abc';\alpha\beta\gamma')=\delta(c,c')\delta(\gamma,\gamma')/(2c+1) \quad (a+b\geqslant c\geqslant |a-b|,\ c\geqslant|\gamma|), \tag{20a}$$

and

$$\sum_{c\gamma} (2c+1)\,V(abc;\alpha\beta\gamma)\,V(abc;\alpha'\beta'\gamma)=\delta(\alpha,\alpha')\delta(\beta,\beta') \quad (a\geqslant|\alpha|,\ b\geqslant|\beta|); \tag{20b}$$

if the inequalities in parentheses are not satisfied, the left side vanishes.

The sum in (17) cannot generally be transformed into a closed form; it is, however, possible to do so for the particular case $\alpha=\beta=\gamma=0$ (a, b, c integers!): if $a+b+c$ is odd, it follows from (19b) that $V(abc;000)$ vanishes; if

$$a+b+c=2g \tag{21}$$

with g integer, it is shown in Appendix A that

$$v(abc;000)=(-1)^g g!/[(g-a)!(g-b)!(g-c)!] \tag{22}$$

and therefore

$$\begin{cases} V(abc;000)=(-1)^g\left[\dfrac{(a+b-c)!(a+c-b)!(b+c-a)!}{(a+b+c+1)!}\right]^{\frac{1}{2}}\dfrac{g!}{(g-a)!(g-b)!(g-c)!}, & (a+b+c \text{ even}) \\ V(abc;000)=0, & (a+b+c \text{ odd}). \end{cases} \tag{22′}$$

§3. THE ALGEBRA OF TENSOR OPERATORS

(1) Definition of Tensor Operator

It is shown in §8[6] of TAS that the matrix components of the electrostatic interaction between two electrons depend on the matrix elements of the spherical harmonics $\Theta(km)\Phi(m)$; in this case the spherical harmonics play the role of operators and not of eigenfunctions, and it appears convenient to consider in a general way the algebra of such operators.

In Chapter III of TAS the algebra of vector operators was developed from the sole assumption that their components satisfy the commutation rule 8^32 with respect to **J**; this assumption is indeed equivalent to the definition of a vector, because the operators J_x, J_y, and J_z are proportional to the rotation operators,[4] and therefore the commutation law with respect to **J** determines completely the transformation law of each quantity considered for a rotation of the axes; since 8^32 holds for x, y, and z, each group of three quantities which satisfies 8^32 has the same transformation law as x, y, and z, and is therefore a vector.

It is shown in the theory of tensors that by means of symmetrizations and contractions each tensor may be decomposed in parts which transform themselves independently for a rotation of the axes, the transformation law of each irreducible part being the same as that of the spherical harmonics of a determinate degree. We may therefore define as "irreducible tensor operator of the degree k" each operator $\mathbf{T}^{(k)}$ whose $2k+1$ components $T_q^{(k)}$ $(q=-k,\ -k+1,\ \cdots,\ k-1,\ k)$ satisfy the same commutation rule with respect to **J** as the spherical-harmonic operators $\Theta(kq)\Phi(q)$; this commutation rule is easily derived from §§3^3 and 4^3 of TAS, and is

$$[(J_x \pm iJ_y),\ T_q^{(k)}] = [(k\mp q)(k\pm q+1)]^{\frac{1}{2}} T_{q\pm1}^{(k)}, \tag{23a}$$

$$[J_z,\ T_q^{(k)}] = qT_q^{(k)}. \tag{23b}$$

It is easily seen that for $k=1$ (23) reduce to TAS 8^32, if we put

$$T_1^{(1)} = -(T_x+iT_y)/(2)^{\frac{1}{2}}, \quad T_0^{(1)} = T_z, \quad T_{-1}^{(1)} = (T_x-iT_y)/(2)^{\frac{1}{2}}. \tag{24}$$

In view of TAS 4^318 we shall say that an irreducible tensor operator is Hermitian, if

$$T_q^{(k)\dagger} = (-1)^q T_{-q}^{(k)}. \tag{25}$$

(2) Dependence of the Matrix of $T^{(k)}$ on m

The dependence on m of the matrix elements of $T_q^{(k)}$ in the jm scheme will readily be derived from (23). The relation (23b) gives us in the usual manner the selection rule: the only non-vanishing elements of $(\alpha jm|T_q^{(k)}|\alpha' j'm')$ are those for which

$$m' = m-q. \tag{26}$$

The two relations (23a), written for a general non-vanishing element, give

$$\begin{aligned} &[(j+m)(j-m+1)]^{\frac{1}{2}}(\alpha jm-1|T_q^{(k)}|\alpha' j'm-q-1) = [(j'+m-q)(j'-m+q+1)]^{\frac{1}{2}}(\alpha jm|T_q^{(k)}|\alpha' j'm-q) \\ &\qquad\qquad + [(k-q)(k+q+1)]^{\frac{1}{2}}(\alpha jm|T_{q+1}^{(k)}|\alpha' j'm-q-1), \\ &[(j-m)(j+m+1)]^{\frac{1}{2}}(\alpha jm+1|T_q^{(k)}|\alpha' j'm-q+1) = [(j'-m+q)(j'+m-q+1)]^{\frac{1}{2}} \\ &\qquad\qquad \times(\alpha jm|T_q^{(k)}|\alpha' j'm-q) + [(k+q)(k-q+1)]^{\frac{1}{2}}(\alpha jm|T_{q-1}^{(k)}|\alpha' j'm-q+1). \end{aligned} \tag{27}$$

We observe now that if we replace $(\alpha jm|T_q^{(k)}|\alpha' j'm')$ by $(j'km'q|j'kjm)$, (27), we reduce exactly to (3) and (5); since we saw that these equations were sufficient to determine the dependence of $(j_1j_2m_1m_2|j_1j_2jm)$ on m_1, m_2, and m, it follows that

$$(\alpha jm|T_q^{(k)}|\alpha' j'm') = A(j'km'q|j'kjm), \tag{28}$$

where A is independent of m, m', and q.[5] Owing to (16′) and (19), we write

$$(\alpha jm|T_q^{(k)}|\alpha' j'm') = (-1)^{j+m}(\alpha j\|T^{(k)}\|\alpha' j')\,V(jj'k\,;\ -mm'q). \tag{29}$$

[4] P. A. M. Dirac, *Quantum Mechanics* (Oxford, 1935), §29.

[5] This relation was already given by Wigner, Reference 3, Chapter 21, Eq. (19), with group-theoretical methods.

This formula is the tensorial extension of TAS 9³ II; in order to avoid mistakes we wrote the quantities which are independent of m, m', and q with $\|$ instead of $\vdots$, since for $k=1$ these quantities differ from the analogous quantities defined in §9³ of TAS; it is easy to see that they are related by the following relations:

$$\begin{aligned}(\alpha j\|T^{(1)}\|\alpha' j)&=[j(j+1)(2j+1)]^{\frac{1}{2}}(\alpha j\vdots T\vdots\alpha' j),\\ (\alpha j\|T^{(1)}\|\alpha' j-1)&=[j(2j-1)(2j+1)]^{\frac{1}{2}}(\alpha j\vdots T\vdots\alpha' j-1),\\ (\alpha j\|T^{(1)}\|\alpha' j+1)&=-[(j+1)(2j+1)(2j+3)]^{\frac{1}{2}}(\alpha j\vdots T\vdots\alpha' j+1).\end{aligned}\tag{30}$$

It must also be observed that for a Hermitian tensor $\mathbf{T}^{(k)}$ the matrix $(\alpha j\|T^{(k)}\|\alpha' j')$ is not Hermitian, but satisfies the relation

$$(\alpha j\|T^{(k)}\|\alpha' j')=(-1)^{j-j'}\overline{(\alpha' j'\|T^{(k)}\|\alpha j)}\,;\tag{31}$$

the general relation for any tensor is

$$(\alpha j\|T^{(k)}\|\alpha' j')=(-1)^{j-j'}\overline{(\alpha' j'\|T^{(k)\dagger}\|\alpha j)}.\tag{31'}$$

The reasons which brought us to this choice of phases are similar to those which fixed the phases in TAS 4³17.

(3) Scalar Product of Tensors

If two irreducible tensors of the same degree are given, we consider the quantity

$$Q=\textstyle\sum_q(-1)^q T_q^{(k)}U_{-q}^{(k)}\,;\tag{32}$$

owing to (29), (19), and (20a), the matrix elements of Q are

$$(\alpha jm|Q|\alpha' j'm')=\sum_{\alpha'' j''}(-1)^{j-j''}(\alpha j\|T^{(k)}\|\alpha'' j'')(\alpha'' j''\|U^{(k)}\|\alpha' j')\delta(j,j')\delta(m,m')/(2j+1).\tag{33}$$

The matrix of Q is then diagonal with respect to j and m, and entirely independent of m; it follows that Q commutes with $\mathbf{J}$, and is therefore a scalar. Since, owing to (24), Q is for $k=1$ the scalar product of the two vectors, we shall in general name Q the scalar product of $\mathbf{T}^{(k)}$ and $\mathbf{U}^{(k)}$, and write

$$Q=(\mathbf{T}^{(k)}\cdot\mathbf{U}^{(k)})=\textstyle\sum_q(-1)^q T_q^{(k)}U_{-q}^{(k)}.\tag{32'}$$

The most important example of such scalar products is given by the spherical-harmonic addition theorem (TAS 4³22).

The tensorial extension of TAS 12³2 may be obtained by a direct use of (16′) and (29). If $\mathbf{T}^{(k)}$ and $\mathbf{U}^{(k)}$ are of such a character that, when a resolution of the type TAS 6³5 is made of the states in question, $\mathbf{T}^{(k)}$ operates only on ϕ_1 and $\mathbf{U}^{(k)}$ only on ϕ_2, the matrix elements of Q will be

$$\begin{aligned}(\gamma j_1j_2jm|(\mathbf{T}^{(k)}\cdot\mathbf{U}^{(k)})|\gamma' j'_1j'_2j'm')&=\sum_{\gamma'' qm_1m_2m'_1m'_2}(-1)^q(j_1j_2jm|j_1j_2m_1m_2)(\gamma j_1m_1|T_q^{(k)}|\gamma'' j_1'm_1')\\ &\times(\gamma'' j_2m_2|U_{-q}^{(k)}|\gamma' j'_2m'_2)(j'_1j'_2m'_1m'_2|j'_1j'_2j'm')=(-1)^{j_1+j_2-j-j'-m}[(2j+1)(2j'+1)]^{\frac{1}{2}}\\ &\times\textstyle\sum_{\gamma''}(\gamma j_1\|T^{(k)}\|\gamma'' j'_1)(\gamma'' j_2\|U^{(k)}\|\gamma' j'_2)\sum(-1)^q V(j_1j_2j\,;m_1m_2-m)\,V(j_1j'_1k\,;-m_1m'_1q)\\ &\times V(j_2j'_2k\,;-m_2m'_2-q)\,V(j'_1j'_2j'\,;m'_1m'_2-m').\end{aligned}\tag{34}$$

The last sum is very difficult to evaluate for general values of the parameters; but, owing to (33), it suffices to calculate it for the particular case $j=j'=m=m'$. It is shown in Appendix B that

$$\begin{aligned}\sum_{\alpha\beta\gamma\delta\varphi}&(-1)^{f+\varphi}v(abe\,;\alpha\beta-e)v(acf\,;-\alpha\gamma\varphi)v(bdf\,;-\beta\delta-\varphi)v(cde\,;\gamma\delta-e)\\ &=(-1)^{2e+f+d-b}w(abcd\,;ef)/(2e+1),\end{aligned}\tag{35}$$

where

$$w(abcd\,;ef)=\sum_z(-1)^z\frac{(a+b+c+d+1-z)!}{(a+b-e-z)!(c+d-e-z)!(a+c-f-z)!\times(b+d-f-z)!z!(e+f-a-d+z)!(e+f-b-c+z)!}\tag{36}$$

Putting also

$$W(abcd\,;ef)=\left[\frac{(a+b-e)!(a+e-b)!(b+e-a)!(c+d-e)!(c+e-d)!(d+e-c)!\cdot(a+c-f)!(a+f-c)!(c+f-a)!(b+d-f)!(b+f-d)!(d+f-b)!}{(a+b+e+1)!(c+d+e+1)!(a+c+f+1)!(b+d+f+1)!}\right]^{\frac{1}{2}}w(abcd\,;ef),\tag{36'}$$

we have

$$\sum_{\alpha\beta\gamma\delta\varphi}(-1)^{f+\varphi}V(abe\,;\alpha\beta-e)\,V(acf\,;-\alpha\gamma\varphi)\,V(bdf\,;-\beta\delta-\varphi)\,V(cde\,;\gamma\delta-e)$$
$$=(-1)^{2e+f+d-b}W(abcd\,;ef)/(2e+1)\,;\tag{35'}$$

and owing to (34) and (33) we get

$$\sum_{\alpha\beta\gamma\delta\varphi}(-1)^{f+\varphi}V(abe\,;\alpha\beta-\epsilon)\,V(acf\,;-\alpha\gamma\varphi)\,V(bdf\,;-\beta\delta-\varphi)\,V(cdg\,;\gamma\delta-\eta)$$
$$=(-1)^{e+\epsilon+f+d-b}W(abcd\,;ef)\delta(e,g)\delta(\epsilon,\eta)/(2e+1)\tag{37}$$

and

$$(\gamma j_1j_2jm|(\mathbf{T}^{(k)}\cdot\mathbf{U}^{(k)})|\gamma'j'_1j'_2jm)$$
$$=(-1)^{j_1+j'_2-j}\sum_{\gamma''}(\gamma j_1\|T^{(k)}\|\gamma''j'_1)(\gamma''j_2\|U^{(k)}\|\gamma'j'_2)W(j_1j_2j'_1j'_2\,;jk),\tag{38}$$

which is the tensorial extension of TAS 12^32.[6]

(4) Properties of W

The functions w and W are defined for integral and half-integral values of the parameters, with the limitation that each of the four triads

$$(a,b,e),\quad(c,d,e),\quad(a,c,f),\quad(b,d,f)\tag{39}$$

has an integral sum. Since in (36) z takes on only such integral values for which the argument of every factorial is not negative, W vanishes unless the elements of each triad (39) satisfy the triangular inequalities; if one of these triangles reduces to a segment, the summation reduces to one term.

It follows from the symmetry of (36) and (36′) that

$$W(abcd;ef)=W(badc;ef)=W(cdab;ef)=W(acbd;fe);\tag{40a}$$

assuming the argument of one of the two last factorials instead of z as summation parameter in (36), we obtain other symmetry properties of W:

$$W(abcd\,;ef)=(-1)^{e+f-a-d}W(ebcf\,;ad)=(-1)^{e+f-b-c}W(aefd\,;bc)\,;\tag{40b}$$

combining (40a) and (40b) we obtain 24 different permutations of the parameters of W, which correspond to all possible permutations between the four triads (39).

Since (33) and (34) have a meaning only for integral values of k, our demonstration of (37) holds only for integral values of f; but it follows from the symmetry properties of V and W that (37) holds also for half-integral values of f.

[6] For the diagonal elements of this matrix an equivalent formula was given by H. A. Kramers, Proc. Amst. Akad. Sci. **34**, 965 (1931).

If we multiply the two sides of (37) by $(2g+1)V(cdg;\gamma'\delta'-\eta)$ and extend a summation over all possible values of g and η, we obtain, owing to (20b),

$$\sum_{\alpha\beta\varphi}(-1)^{f+\varphi}V(abe;\alpha\beta-\epsilon)V(acf;-\alpha\gamma'\varphi)V(bdf;-\beta\delta'-\varphi)=(-1)^{e+\epsilon+f+d-b}W(abcd;ef)V(cde;\gamma'\delta'-\epsilon),$$

or, owing to (19) and (40) and omitting the dashes,

$$\sum_{\alpha\beta\varphi}(-1)^{f+\varphi}V(abe;\alpha\beta-\epsilon)V(afc;-\alpha\varphi\gamma)V(fbd;\varphi\beta-\delta)$$
$$=(-1)^{b+c-a-d+e+\epsilon}W(aefd;bc)V(edc;-\epsilon\delta\gamma). \quad (41)$$

We rewrite (41) with slightly different parameters,

$$\sum_{\alpha'\beta'\eta}(-1)^{g+\eta}V(abe;\alpha'\beta'-\epsilon)V(agc;-\alpha'\eta\gamma)V(gbd;\eta\beta'-\delta)$$
$$=(-1)^{b+c-a-d+e+\epsilon}W(aegd;bc)V(edc;-\epsilon\delta\gamma), \quad (41')$$

and multiply the two sides by the two sides of (41) and by $(2d+1)(2e+1)$; extending a summation over all possible values of γ, ϵ, and e, and owing to (20), we obtain an orthogonality relation between the W:

$$\sum_e(2e+1)W(aefd;bc)W(aegd;bc)=\delta(f,g)/(2f+1). \quad (42)$$

Interchanging a with b and α' with β' in (41') and operating as before, we obtain another useful relation between the W:

$$\sum_e(-1)^{a+b+c+d+e+f+g}(2e+1)W(acbd;fe)W(abdc;eg)=W(acdb;fg). \quad (43)$$

(5) Matrix of a Tensor $T^{(k)}$ Which Commutes with J_2 or with J_1

From (16'), (29), and (41) we have

$$(\gamma j_1j_2jm|T_q^{(k)}|\gamma'j'_1j'_2j'm')=\sum_{m_1m'_1m_2}(j_1j_2jm|j_1j_2m_1m_2)(\gamma j_1m_1|T_q^{(k)}|\gamma'j'_1m'_1)(j'_1j_2m'_1m_2|j'_1j_2j'm')$$
$$=(-1)^{j_2+k-j'_1+m}[(2j+1)(2j'+1)]^{\frac{1}{2}}(\gamma j_1\|T^{(k)}\|\gamma'j'_1)W(j_1jj'_1j';j_2k)V(jj'k;-mm'q),$$

and owing again to (29) we get

$$(\gamma j_1j_2j\|T^{(k)}\|\gamma'j'_1j_2j')=(-1)^{j_2+k-j'_1-j}(\gamma j_1\|T^{(k)}\|\gamma'j'_1)[(2j+1)(2j'+1)]^{\frac{1}{2}}W(j_1jj'_1j';j_2k), \quad (44a)$$

which is the tensorial extension of TAS 11^38. In the same way we obtain also

$$(\gamma j_1j_2j\|U^{(k)}\|\gamma'j_1j'_2j')=(-1)^{j_1+k-j_2-j'}(\gamma j_2\|U^{(k)}\|\gamma'j'_2)[(2j+1)(2j'+1)]^{\frac{1}{2}}W(j_2jj'_2j';j_1k). \quad (44b)$$

§4. THE ELECTROSTATIC INTERACTION BETWEEN TWO ELECTRONS

It was shown in I that the coefficients of Slater's integrals F^k and G^k in the two-electron configurations are the matrix elements of $P_\kappa(\cos\omega)$, where ω is the angle between the radii vectors of the two electrons. It follows from (32') and from the spherical-harmonic addition theorem (TAS 4^322) that

$$P_k(\cos\omega)=(\mathbf{C}_1^{(k)}\cdot\mathbf{C}_2^{(k)}), \quad (45)$$

where $\mathbf{C}^{(k)}$ is the tensor operator defined by

$$C_q^{(k)}=[4\pi/(2k+1)]^{\frac{1}{2}}\Theta(kq)\Phi(q); \quad (46)$$

we see from TAS 8^66 that the non-vanishing matrix elements of $C_q^{(k)}$ are the c^k of Condon and Shortley.

Confronting the expressions TAS 9^68 and (29) of $(l0\|C_0^{(k)}\|l'0)$, we obtain

$$(-1)^l(l\|C^{(k)}\|l')V(ll'k;000)=\tfrac{1}{2}[(2l+1)(2l'+1)]^{\frac{1}{2}}C_{lkl'}, \quad (47)$$

where (TAS 9^67)

$$C_{lkl'}=\int_0^\pi P_l(\cos\omega)P_k(\cos\omega)P_{l'}(\cos\omega)\sin\omega d\omega. \tag{48}$$

In order to calculate algebraically $C_{lkl'}$ we express the Legendre polynomials by means of TAS 4^322 and 4^318,

$$P_l(\cos\omega)=\bar{P}_l(\cos\omega)=[4\pi/(2l+1)]\sum_m(-1)^m\bar{\Theta}(lm)\bar{\Phi}(m)\bar{\Theta}'(l-m)\bar{\Phi}'(-m),$$
$$P_k(\cos\omega)=[4\pi/(2k+1)]\sum_q(-1)^q\Theta(kq)\Phi(q)\Theta'(k-q)\Phi'(-q),$$
$$P_{l'}(\cos\omega)=[4\pi/(2l'+1)]\sum_{m'}(-1)^{m'}\Theta(l'm')\Phi(m')\Theta'(l'-m')\Phi'(-m'),$$

and integrate their product over the spheres $(\theta\varphi)$ and $(\theta'\varphi')$; we obtain

$$8\pi^2\int_0^\pi P_l(\cos\omega)P_k(\cos\omega)P_{l'}(\cos\omega)\sin\omega d\omega=64\pi^3[(2l+1)(2k+1)(2l'+1)]^{-1}$$
$$\sum_{mqm'}\int_0^\pi\int_0^{2\pi}\bar{\Theta}(lm)\bar{\Phi}(m)\Theta(kq)\Phi(q)\Theta(l'm')\Phi(m')\sin\theta d\theta d\varphi$$
$$\cdot\int_0^\pi\int_0^{2\pi}\bar{\Theta}'(l-m)\bar{\Phi}'(-m)\Theta'(k-q)\Phi'(-q)\Theta'(l'-m')\Phi'(-m')\sin\theta' d\theta' d\varphi',$$

or, if we use successively (46), (29), (19b), and (20a),

$$(2l+1)(2l'+1)C_{lkl'}=2\sum_{mqm'}(lm|C_q^{(k)}|l'm')(l-m|C_{-q}^{(k)}|l'-m')$$
$$=2(l\|C^{(k)}\|l')^2\sum_{mqm'}V(ll'k;-mm'q)V(ll'k;m-m'-q)=2(-1)^{l+l'+k}(l\|C^{(k)}\|l')^2. \tag{49}$$

It follows from the comparison of (47) and (49) that

$$C_{lkl'}=2(-1)^{l+l'+k}[V(ll'k;000)]^2, \tag{50}$$

and owing to (22′) we obtain

$$\begin{cases} C_{lkl'}=0 & (l+l'+k \text{ odd}) \\ C_{lkl'}=\dfrac{2(l+l'-k)!(l+k-l')!(l'+k-l)!g!^2}{(l+l'+k+1)!(g-l)!^2(g-l')!^2(g-k)!^2} & (l+l'+k=2g \text{ even}), \end{cases} \tag{50′}$$

which agrees with TAS 9^69. The numerical values of $C_{lkl'}$ are tabulated by Shortley and Fried.[7]

It follows also from (47), (49), and (22′) that

$$(l\|C^{(k)}\|l')=(-1)^{g-l}[\tfrac{1}{2}(2l+1)(2l'+1)C_{lkl'}]^{\frac{1}{2}}, \tag{51}$$

and therefore

$$c^k(lm,l'm')=(lm|C_{m-m'}^{(k)}|l'm')=(-1)^{g+m}[\tfrac{1}{2}(2l+1)(2l'+1)C_{lkl'}]^{\frac{1}{2}}V(ll'k;-mm'm-m'). \tag{52}$$

We do not know to what extent this derivation is different from Gaunt's derivation, because we had no opportunity of consulting his paper;[8] in every case (52) has the same advantages in comparison to Gaunt's formula (TAS 8^611) as (16) in comparison to Wigner's formula.

Introducing (45) and (51) in (38) we have a direct demonstration of Eq. (12′) I, from which we obtained the coefficients of F^k and G^k for two-electron configurations in (LS) coupling.

[7] G. H. Shortley and B. Fried, Phys. Rev. **54**, 739 (1938), Table III.
[8] J. A. Gaunt, Phil. Trans. Roy. Soc. **A228**, 195 (1929).

In order to obtain from (45) and (38) the matrix elements of the electrostatic interaction in (jj) coupling, it suffices to calculate the elements $(\frac{1}{2}lj\|C^{(k)}\|\frac{1}{2}l'j')$; it follows from (44b) and (51) that

$$(\tfrac{1}{2}ll\pm\tfrac{1}{2}\|C^{(k)}\|\tfrac{1}{2}l'l'\pm\tfrac{1}{2})=(-1)^{(j'+k-j)/2}\left[\frac{(j+j'-k)!(j+k-j')!(j'+k-j)!}{(j+j'+k+1)!}\right]^{\frac{1}{2}}$$
$$\times\frac{(j+j'+k+1)!!}{(j+j'-k-1)!!(j+k-j')!!(j'+k-j)!!},\qquad(53a)$$

$$(\tfrac{1}{2}ll\pm\tfrac{1}{2}\|C^{(k)}\|\tfrac{1}{2}l'l'\mp\tfrac{1}{2})=(-1)^{(j'+k-j-1)/2}\left[\frac{(j+j'-k)!(j+k-j')!(j'+k-j)!}{(j+j'+k+1)!}\right]^{\frac{1}{2}}$$
$$\times\frac{(j+j'+k)!!}{(j+j'-k)!!(j+k-j'-1)!!(j'+k-j-1)!!},\qquad(53b)$$

and then

$$f_k(l_1j_1l_2j_2J)=\frac{(k-1)!!^4(2j_1-k)!!(2j_2-k)!!}{(2j_1+k)!!(2j_2+k)!!}$$
$$\cdot\sum_w(-1)^w\binom{j_1+j_2+J+1+w}{w}\binom{j_1+j_2-J}{w}\binom{J+j_1-j_2}{k-w}\binom{J+j_2-j_1}{k-w}.\qquad(54)$$

It is remarkable that (54) depends only on the j's and not on the l's of the electrons, and therefore the coefficients of F^k for the interaction between two $p_{1\frac{1}{2}}$ electrons are the same as between two $d_{1\frac{1}{2}}$ or between a $p_{1\frac{1}{2}}$ and a $d_{1\frac{1}{2}}$.

For the coefficients of the G^k the situation is somewhat different, because, although g_k does not depend explicitly on l, in the case that both the electrons have their spins parallel or antiparallel to their orbital momenta the formula is not the same as in the case that one spin is parallel to its orbital momentum and the other is antiparallel. In the first case

$$g_k(l_1l_1\pm\tfrac{1}{2}l_2l_2\pm\tfrac{1}{2}J)=(-1)^{j_1+j_2-J}\frac{(j_1+j_2-k)!!^2(k+j_1-j_2-1)!!^2(k+j_2-j_1-1)!!^2}{(j_1+j_2+k)!!^2}$$
$$\cdot\sum_w(-1)^w\binom{j_1+j_2+J+1+w}{w}\binom{j_1+j_2-J}{w}\binom{J+j_1-j_2}{k+j_1-j_2-w}\binom{J+j_2-j_1}{k+j_2-j_1-w}\qquad(55a)$$

and in the second case

$$g_k(l_1l_1\pm\tfrac{1}{2}l_2l_2\mp\tfrac{1}{2}J)=(-1)^{j_1+j_2-J}\frac{(j_1+j_2-k-1)!!^2(k+j_1-j_2)!!^2(k+j_2-j_1)!!^2}{(j_1+j_2+k+1)!!^2}$$
$$\cdot\sum_w(-1)^w\binom{j_1+j_2+J+1+w}{w}\binom{j_1+j_2-J}{w}\binom{J+j_1-j_2}{k+j_1-j_2-w}\binom{J+j_2-j_1}{k+j_2-j_1-w}.\qquad(55b)$$

It must also be noted that in (LS) coupling g_k is preceded by different signs, corresponding to the singlet and triplet states; but in (jj) coupling g_k is always preceded by a minus sign, since in (jj) coupling only antisymmetrical eigenfunctions are possible.

By means of these formulas the results of Inglis[9] were checked, and the following mistakes were found: the coefficient of $-G^1$ 225 for the level $p_{\frac{1}{2}}d_{1\frac{1}{2}}$ with $J=2$ is not 25, but 75; the denominators of F^2 and F^4 in the configuration dd' are not 25 and 49, but 1225 and 441.

[9] D. R. Inglis, Phys. Rev. **38**, 862 (1931), Table II.

The non-diagonal elements of the electrostatic interaction in (jj) coupling may also easily be calculated in this way.

From our results follows also that the coefficients of F^k in (jl) coupling[10] are

$$f_k(jlK)=\frac{(k-1)!!^4(2j-k)!!(2l-k-1)!!(2l+1)}{(2j+k)!!(2l+k+1)!!}$$
$$\cdot\sum_w(-1)^w\binom{j+l+K+1+w}{w}\binom{j+l-K}{w}\binom{K+j-l}{k-w}\binom{K+l-j}{k-w}. \quad (56)$$

The particular expression for $k=2$ which we gave already for the rare-gas spectra differs from (56) in the sign, owing to the fact that in the rare-gas configurations there is an almost closed shell (see §6).

§5. MANY-ELECTRON SPECTRA. GENERAL PART

In §4 we expressed the coefficients of the F^k in the interaction between two electrons as scalar products of tensors, each of which operates on a definite electron (type (34)); thus by the general methods of §3 we may calculate for every configuration that part of the energy matrix which depends on the F^k, in a schema in which each state is defined by its "genealogical characterizations" (TAS §2[8]).

The problem is more complex for the coefficients of the G^k, since the g_k are not scalar products of tensors of the type (34); in the particular cases considered in §§5 and 6 of I we gave to g_k the form of a polynomial in λ, but in more general cases it appears convenient to develop g_k in a sum of scalar products of tensors of the type (34).[11] It follows from (6) I and (45) that

$$g_k(l_1l_2L)=(-1)^{l_1+l_2-L}(l_1l_2LM|(\mathbf{C}_1^{(k)}\cdot\mathbf{C}_2^{(k)})|l_2l_1LM), \quad (57)$$

and owing to (38), (31), and (43) we obtain

$$g_k(l_1l_2L)=(l_1\|C^{(k)}\|l_2)^2\sum_r(-1)^{L+k+r}(2r+1)W(l_1l_2l_1l_2;Lr)W(l_1l_1l_2l_2;rk);$$

if we define the tensor $\mathbf{u}^{(r)}$ by

$$(l\|u^{(r)}\|l')=\delta(l,l')$$

and take into account (38) and the fact that l_1+l_2+k is even, we may also write

$$g_k(l_1l_2L)=(l_1\|C^{(k)}\|l_2)^2\sum(-1)^r(2r+1)W(l_1l_1l_2l_2;rk)(l_1l_2LM|(\mathbf{u}_1^{(r)}\cdot\mathbf{u}_2^{(r)})|l_1l_2LM). \quad (59)$$

Following Dirac's vector model we can also substitute the operator (37) I for the double sign which precedes g_k and assume as coefficient of G^k in the exchange interaction between two electrons the expression

$$-[\tfrac{1}{2}+2(\mathbf{s}_1\cdot\mathbf{s}_2)](l_1\|C^{(k)}\|l_2)^2\sum_r(-1)^r(2r+1)W(l_1l_1l_2l_2;rk)(\mathbf{u}_1^{(r)}\cdot\mathbf{u}_2^{(r)}). \quad (60)$$

It will also be convenient to consider the quantities $(\mathbf{s}_1\cdot\mathbf{s}_2)(\mathbf{u}_1\cdot\mathbf{u}_2)$ as scalar products of "double tensors";[12] a double tensor of the degree (κ, k) is defined as a quantity which behaves as an irreducible tensor of the degree κ with respect to $\mathbf{S}$ and as an irreducible tensor of the degree k with respect to $\mathbf{L}$. The algebra of these double tensors is a trivial extension of the tensor algebra developed in §3; it must be noted only that such a double tensor does not satisfy the commutation rule (23) with respect to $\mathbf{J}$, because with respect to $\mathbf{J}$ it is reducible and may be decomposed in a sum of irreducible tensors, the degrees of which lie between $|k-\kappa|$ and $k+\kappa$. From this point of view the scalar product $(\mathbf{l}\cdot\mathbf{s})$ is the scalar part of the decomposition of the double vector $\mathbf{ls}$.

---

[10] G. Racah, Phys. Rev. **61**, 537 (1942).

[11] These two different possibilities correspond to the developments of a function in series of powers or of Legendre polynomials.

[12] Reference 3, p. 295.

If x electrons of a configuration are equivalent, it has no sense to speak of the tensor (or double tensor) $\mathbf{t}_L^{(\kappa k)}$ which operates on the electron i (TAS §I[8]), but we must be content to consider tensors of the type

$$\mathbf{T}^{(\kappa k)} = \sum_1^x {}_i \, \mathbf{t}_i^{(\kappa k)}, \tag{61}$$

which operate on the whole group of equivalent electrons; it is easily seen that every symmetrical operator may be built up with such tensors. The matrix elements of a tensor (61) must be calculated in the scheme of the allowed states of the group: for $x \geqslant 3$ this calculation is not very simple, as a general method of vector coupling for equivalent electrons is not yet known; however, it is possible at first to couple the vectors and then to antisymmetrize the obtained states, but in this paper we shall not deal with such cases.

§6. CONFIGURATIONS CONTAINING ALMOST CLOSED SHELLS

Let us consider a shell "$\mathfrak{S}$" which is complete except for ϵ missing electrons; if the number of places in the shell is m, this configuration "$\mathfrak{R}$" will contain $m-\epsilon$ electrons. We shall in this paragraph determine a simple relation between the matrix of a tensor of the type (61) which operates on $\mathfrak{R}$ and the matrix of the corresponding tensor for the simpler group "$\mathfrak{L}$" of ϵ equivalent electrons.

Let us denote by $\Phi_{\mathfrak{L}}(aSLM_SM_L)$ the eigenfunctions of the allowed states of $\mathfrak{L}$ and by $\Phi_{\mathfrak{R}}(bSLM_SM_L)$ those of $\mathfrak{R}$, the parameters a and b being introduced in order to distinguish the different multiplets of the same type which may occur in the given configurations. We consider now a fictive configuration $\mathfrak{L}+\mathfrak{R}$ in which the exclusion principle does not hold between the electrons of $\mathfrak{L}$ and those of $\mathfrak{R}$: a complete set of eigenfunctions of such configuration is given by the functions

$$\Phi_{\mathfrak{L}}(aS'L'M'_SM'_L)\Phi_{\mathfrak{R}}(bS''L''M''_SM''_L)$$

or by the functions

$$\Psi(abS'S''L'L''SLM_SM_L) = \sum_{M'_SM''_SM'_LM''_L} \Phi_{\mathfrak{L}}(aS'L'M'_SM'_L)\Phi_{\mathfrak{R}}(bS''L''M''_SM''_L) \times (S'S''M'_SM''_S\|S'S''SM_S)(L'L''M'_LM''_L\|L'L''LM_L). \tag{62}$$

If the exclusion principle holds also between the two groups of electrons, only one state will be allowed, a particular ${}^1S$ state which we shall denote by ${}^1S^*$; its eigenfunction will be

$$\Psi({}^1S^*) = \sum_{abSL} q(abSL)\Psi(abSSLL0000). \tag{63}$$

In order to establish a correlation between the states of $\mathfrak{L}$ and those of $\mathfrak{R}$ and between the phases of their eigenfunctions, we consider, for S and L given, the Hermitian matrix

$$A_{ac} = \sum_b q(ab)\bar{q}(cb) \tag{64}$$

and the unitary transformation $u_{c\gamma}$ which diagonalizes it: in view of the special form of (64) it will be

$$\sum_{ac} \bar{u}_{a\alpha}u_{c\gamma}A_{ac} = Q^2(\alpha)\delta(\alpha\gamma), \tag{65}$$

where the $Q(\alpha)$ are real numbers, which we may assume are not negative. If we put for each value of β for which $Q(\beta)$ does not vanish

$$v_{b\beta} = \sum_c \bar{u}_{c\beta}q(cb)/Q(\beta) \quad (Q(\beta)\neq 0), \tag{66}$$

it follows from (64) and (65) that

$$\sum_b \bar{v}_{b\beta}v_{b\alpha} = \delta(\alpha\beta). \tag{67}$$

If $Q(\beta)$ vanishes for some values of β, we complete the matrix $v_{b\beta}$ so that it be unitary: owing to (64) and (65) it will be anyway

$$Q(\beta)v_{b\beta}=\sum_c \bar{u}_{c\beta}q(cb). \tag{66'}$$

We change now the schemes of the states of $\mathfrak{L}$ and $\mathfrak{R}$, putting

$$\begin{aligned}\Phi_{\mathfrak{L}}(\alpha SLM_SM_L)&=\sum_a \Phi_{\mathfrak{L}}(aSLM_SM_L)u_{a\alpha}(SL),\\ \Phi_{\mathfrak{R}}(\alpha SLM_SM_L)&=\sum_b \Phi_{\mathfrak{R}}(bSLM_SM_L)v_{b\alpha}(SL)\end{aligned} \tag{68}$$

and shall consider two terms of $\mathfrak{L}$ and $\mathfrak{R}$ as correlated, if they have the same values of α, S, and L. It follows from (62) that

$$\Psi(\alpha'\alpha''S'S''L'L''SLM_SM_L)=\sum_{ab}\Psi(abS'S''L'L''SLM_SM_L)u_{a\alpha'}(S'L')v_{b\alpha''}(S''L''), \tag{68'}$$

and from the unitarity of $u_{a\alpha}$ and $v_{b\beta}$ that

$$\Psi(abS'S''L'L''SLM_SM_L)=\sum_{\alpha'\alpha''}\bar{u}_{a\alpha'}(S'L')\bar{v}_{b\alpha''}(S''L'')\Psi(\alpha'\alpha''S'S''L'L''SLM_SM_L); \tag{68''}$$

introducing this result in (63) and owing to (66') we obtain

$$\Psi({}^1S^*)=\sum_{\alpha SL}Q(\alpha SL)\Psi(\alpha\alpha SSLL0000). \tag{63'}$$

Since $\Psi({}^1S^*)$ is the only antisymmetrical eigenfunction of the configuration $\mathfrak{L}+\mathfrak{R}$, the matrix elements connecting ${}^1S^*$ with every state (68') will vanish for every symmetrical operator, unless $S=L=0$. If $\mathbf{T}_{\mathfrak{L}}^{(\kappa k)}$ and $\mathbf{T}_{\mathfrak{R}}^{(\kappa k)}$ are two tensors which operate on the groups $\mathfrak{L}$ and $\mathfrak{R}$ according to (61), $\mathbf{T}_{\mathfrak{L}}^{(\kappa k)}+\mathbf{T}_{\mathfrak{R}}^{(\kappa k)}$ is symmetrical in all the electrons, and then

$$(\alpha'\alpha''S'S''L'L''SL\|T_{\mathfrak{L}}^{(\kappa k)}\|{}^1S^*)+(\alpha'\alpha''S'S''L'L''SL\|T_{\mathfrak{R}}^{(\kappa k)}\|{}^1S^*) \tag{69}$$

vanishes, unless $S=L=0$. Owing to the triangular conditions, each term vanishes alone unless $S=\kappa$ and $L=\kappa$; we shall therefore consider only the remaining equations

$$(\alpha'\alpha''S'S''L'L''\kappa k\|T_{\mathfrak{L}}^{(\kappa k)}\|{}^1S^*)+(\alpha'\alpha''S'S''L'L''\kappa k\|T_{\mathfrak{R}}^{(\kappa k)}\|{}^1S^*)=0 \tag{70}$$

which hold for every double tensor, excepting the double scalars.

Owing to (63') and to the fact that $\mathbf{T}_{\mathfrak{L}}^{(\kappa k)}$ is diagonal with respect to $\alpha''S''L''$ and $\mathbf{T}_{\mathfrak{R}}^{(\kappa k)}$ with respect to $\alpha'S'L'$, we get

$$\begin{aligned}&(\alpha'\alpha''S'S''L'L''\kappa k\|T_{\mathfrak{L}}^{(\kappa k)}\|\alpha''\alpha''S''S''L''L''00)Q(\alpha''S''L'')\\ &\qquad+(\alpha'\alpha''S'S''L'L''\kappa k\|T_{\mathfrak{R}}^{(\kappa k)}\|\alpha'\alpha'S'S'L'L'00)Q(\alpha'S'L')=0,\end{aligned}$$

and introducing (44) and (36') this becomes

$$\begin{aligned}&(\alpha'S'L'\|T_{\mathfrak{L}}^{(\kappa k)}\|\alpha''S''L'')Q(\alpha''S''L'')/[(2S''+1)(2L''+1)]^{\frac{1}{2}}+(-1)^{S'+L'-S''-L''+\kappa+k}\\ &\qquad\times(\alpha''S''L''\|T_{\mathfrak{R}}^{(\kappa k)}\|\alpha'S'L')Q(\alpha'S'L')/[(2S'+1)(2L'+1)]^{\frac{1}{2}}=0.\end{aligned} \tag{71a}$$

Also, since $\mathbf{T}_{\mathfrak{L}}^{(\kappa k)\dagger}+\mathbf{T}_{\mathfrak{R}}^{(\kappa k)\dagger}$ is a symmetrical operator, and since the $Q(\alpha SL)$ are real, it follows from (71a) that

$$\begin{aligned}&\langle(\alpha''S''L''\|T_{\mathfrak{L}}^{(\kappa k)}\dagger\|\alpha'S'L')\rangle Q(\alpha'S'L')/[(2S'+1)(2L'+1)]^{\frac{1}{2}}+(-1)^{S''+L''-S'-L'+\kappa+k}\\ &\qquad\times\langle(\alpha'S'L'\|T_{\mathfrak{R}}^{(\kappa k)\dagger}\|\alpha''S''L'')\rangle Q(\alpha''S''L'')/[(2S''+1)(2L''+1)]^{\frac{1}{2}}=0,\end{aligned}$$

and owing to (31′) we get

$$(\alpha'S'L'\|T_{\mathfrak{L}}^{(\kappa k)}\|\alpha''S''L'')Q(\alpha'S'L')/[(2S'+1)(2L'+1)]^{\frac{1}{2}}+(-1)^{S'+L'-S''-L''+\kappa+k}$$
$$\times(\alpha''S''L''\|T_{\mathfrak{R}}^{(\kappa k)}\|\alpha'S'L')Q(\alpha''S''L'')/[(2S''+1)(2L''+1)]^{\frac{1}{2}}=0. \quad (71b)$$

It follows from the homogeneous equation system (71) that the matrix elements connecting two terms $\alpha'S'L'$ and $\alpha''S''L''$ vanish for every tensor (61) operating on $\mathfrak{L}$ or on $\mathfrak{R}$, unless

$$\frac{Q(\alpha'S'L')}{[(2S'+1)(2L'+1)]^{\frac{1}{2}}}=\frac{Q(\alpha''S''L'')}{[(2S''+1)(2L''+1)]^{\frac{1}{2}}}; \quad (72)$$

since every symmetrical operator may be expressed as a function of tensors (61), and since there exists always at least one symmetrical operator connecting two allowed terms, it follows that (72) must hold for every couple of allowed terms, or that

$$Q(\alpha SL)=C[(2S+1)(2L+1)]^{\frac{1}{2}}.$$

Since $(2S+1)(2L+1)$ is the number of states of the term αSL, it follows from the normalization of $\Psi(^1S^*)$ that $1/C^2$ equals the number of states of the configuration $\mathfrak{L}$, or that

$$C=\binom{m}{\epsilon}^{-\frac{1}{2}}. \quad (73)$$

It follows from (65) and (72′) that the matrix (64) is a multiple of the unit matrix: the unitary matrix $u_{c\gamma}$ is then entirely arbitrary and to every αSL scheme of $\mathfrak{L}$ a scheme of $\mathfrak{R}$ may be correlated.

From (71) and (72) we have also

$$(\alpha'S'L'\|T_{\mathfrak{L}}^{(\kappa k)}\|\alpha''S''L'')=-(-1)^{S'+L'-S''-L''+\kappa+k}(\alpha''S''L''\|T_{\mathfrak{R}}^{(\kappa k)}\|\alpha'S'L'),$$

and owing to (31′)

$$(\alpha''S''L''\|T_{\mathfrak{R}}^{(\kappa k)}\|\alpha'S'L')=-(-1)^{\kappa+k}\langle(\alpha''S''L''\|T_{\mathfrak{L}}^{(\kappa k)\dagger}\|\alpha'S'L')\rangle, \quad (74)$$

which is the requested relation between the matrices of $T_{\mathfrak{R}}^{(\kappa k)}$ and $T_{\mathfrak{L}}^{(\kappa k)}$.

This demonstration, however, does not hold for scalars. In this case $(lm\|\mathbf{t}^{(00)}\|lm')$ is a multiple of the unit matrix; if its value is a, the matrix of $\mathbf{T}_{\mathfrak{L}}^{(00)}$ has the value ϵa and the matrix of $\mathbf{T}_{\mathfrak{R}}^{(00)}$ has the value $ma-\epsilon a$; we may, therefore, say that apart from a constant diagonal term (74) holds also for scalars, and if only differences of energies are considered we may use (74) even in this case.

The coefficients of F^k for the terms of a configuration l^x are, apart from a constant term, the squares of the tensor $\sum_i C_i^{(k)}$; the relative electrostatic energies of correlated terms are then the same.

Since the expression for the coefficients of F^k contains only tensors with $\kappa=0$ and k even, we obtain immediately the known rule that the coefficients of F^k in the interaction between the group $l^{m-\epsilon}$ and an electron l' are the negatives of those for l^ϵ and l'.

For the coefficients of G^k the situation is more complex, since in (60) there are tensors of even and of odd degrees. For two-electron-like configurations $l^{m-1}l'$ the result is, however, very simple, since it follows from (60), (74), and (58) that the coefficient of G^k is

$$[\tfrac{1}{2}-2(\mathbf{s}\cdot\mathbf{s}')](l\|C^{(k)}\|l')^2(-1)^{L+k}\textstyle\sum_r(2r+1)W(ll'll';Lr)W(lll'l';rk),$$

and reduces, owing to (40), (42), and (51), to

$$\frac{1-4(s\cdot s')}{4}\cdot\frac{(2l+1)(2l'+1)}{(2k+1)}C_{ll'k}\delta(L,k); \quad (75)$$

this formula agrees with the result of Shortley and Fried,[7] since the first coefficient vanishes for triplets and equals unity for singlets.

Some interesting results may be obtained for the configurations with $\epsilon = m/2$, which may be considered as "self-corresponding." Since the electrostatic energies of two corresponding terms are the same, it follows that in (LS) coupling each term is self-corresponding;[13] it is therefore

$$\Phi_{\mathfrak{L}}(\alpha SLM_SM_L) = c\Phi_{\mathfrak{R}}(\alpha SLM_SM_L)$$

with $|c| = 1$, and since with our choice of phases all transformation coefficients are real,

$$\Phi_{\mathfrak{L}}(\alpha SLM_SM_L) = \pm\Phi_{\mathfrak{R}}(\alpha SLM_SM_L). \tag{76}$$

According to these two possibilities the terms of a self-corresponding configuration split in two classes, and a remarkable selection rule follows from (74): the elements of $\mathbf{T}^{(\kappa k)}$ connecting two terms of the same class vanish if $k+\kappa$ is even, the elements connecting two terms of different classes vanish if $k+\kappa$ is odd. A particular case of it is the vanishing of the diagonal elements of the double vector $\sum_i \mathbf{s}_i \mathbf{l}_i$, which causes the vanishing of the spin-orbit interaction constants for all terms of $l^{m/2}$.

This splitting in two classes is also the cause of the unexpected number of rational roots found by Laporte[14] in the electrostatic-energy matrix of d^5.

§7. THE CONFIGURATIONS d^n

d^2

The formulas for this configuration are well known (TAS p. 202); we wish only to point out that putting

$$A = F_0 - 49F_4 = F^0 - F^4/9, \quad B = F_2 - 5F_4 = (9F^2 - 5F^4)/441, \quad C = 35F_4 = 5F^4/63, \tag{77}$$

they get the simpler form

$${}^1S = A + 14B + 7C, \quad {}^3P = A + 7B, \quad {}^1D = A - 3B + 2C, \quad {}^3F = A - 8B, \quad {}^1G = A + 4B + 2C. \tag{78}$$

d^3

Condon and Shortley calculated the formulas for this configuration, but they could not separate the energies of the two ${}^2D$ terms; this separation was performed by Ufford and Shortley by calculating the eigenfunctions of these terms. We shall calculate in detail this configuration with the tensor method, since the same method will be used without greater complication in the cases d^4, d^5, and f^3.

The term energies of d^3 are the eigenvalues of that part of

$$3F^0 + [(\mathbf{C}_1^{(2)} \cdot \mathbf{C}_2^{(2)}) + (\mathbf{C}_1^{(2)} \cdot \mathbf{C}_3^{(2)}) + (\mathbf{C}_2^{(2)} \cdot \mathbf{C}_3^{(2)})]F^2$$
$$+ [(\mathbf{C}_1^{(4)} \cdot \mathbf{C}_2^{(4)}) + (\mathbf{C}_1^{(4)} \cdot \mathbf{C}_3^{(4)}) + (\mathbf{C}_2^{(4)} \cdot \mathbf{C}_3^{(4)})]F^4 \tag{79}$$

which operates in the space of the antisymmetrical states with $l_1 = l_2 = l_3 = 2$. Since this operator does not affect the spins, we shall calculate its matrix in a m_SL scheme, and it will be sufficient to consider the elements corresponding to $m_{S1} = m_{S2} = \frac{1}{2}$ and $m_{S3} = -\frac{1}{2}$; in this scheme only the two first electrons are to be considered as equivalent, and we can thus avoid the difficulties arising from the coupling of three equivalent electrons.

It follows from (51) that

$$(2\|C^{(2)}\|2) = -(10/7)^{\frac{1}{2}}, \quad (2\|C^{(4)}\|2) = (10/7)^{\frac{1}{2}}, \tag{80}$$

[13] This holds only in (LS) coupling: it is, for instance, evident that in (jj) coupling of the three levels with $J = \frac{3}{2}$ in the p^3 configuration, only the level $p_{1\frac{1}{2}}{}^2p_{\frac{1}{2}}$ is self-corresponding.

[14] O. Laporte, Phys. Rev. **61**, 302 (1942).

and hence and from (44) that

$$(22L\|C_1^{(2)}+C_2^{(2)}\|22L')=\begin{array}{c|cc|} & P & F \\ \hline P & (6/5)^{\frac{1}{2}} & -(48/35)^{\frac{1}{2}} \\ F & -(48/35)^{\frac{1}{2}} & -(12/35)^{\frac{1}{2}} \\ \hline \end{array}$$

$$(22L\|C_1^{(4)}+C_2^{(4)}\|22L')=\begin{array}{c|cc|} & P & F \\ \hline P & 0 & -(4/7)^{\frac{1}{2}} \\ F & -(4/7)^{\frac{1}{2}} & -(22/7)^{\frac{1}{2}} \\ \hline \end{array}. \tag{81}$$

From (38) we obtain now easily the interaction matrix between the d^2 group and the third d electron:

$$W(d^2,d)=2F_0+\left\|\begin{matrix} (-1)^L\frac{7\Lambda}{30}\binom{7}{3-L} & \frac{2\Lambda}{15}\left[\binom{6+L}{2}\binom{5-L}{2}\right]^{\frac{1}{2}} \\ \frac{2\Lambda}{15}\left[\binom{6+L}{2}\binom{5-L}{2}\right]^{\frac{1}{2}} & \frac{\Lambda^2-35\Lambda+210}{30} \end{matrix}\right\|F_2$$

$$+(-1)^L\left\|\begin{matrix} 0 & \left[\binom{9}{5-L}\binom{9}{3-L}\right]^{\frac{1}{2}} \\ \left[\binom{9}{5-L}\binom{9}{3-L}\right]^{\frac{1}{2}} & \frac{\Lambda}{10}\binom{11}{5-L} \end{matrix}\right\|F_4, \tag{82}$$

where

$$\Lambda=L(L+1). \tag{83}$$

It suffices now to add to (82) the diagonal matrix of the internal energy of d^2, and to diagonalize this sum for each possible value of L; with this method we cannot specify to which value of S each eigenvalue belongs, but the quartet terms may be recognized, according to (23) I, as those for which the coefficient of C vanishes. The results are

$$\begin{aligned} &{}^2P=3A-6B+3C, && {}^4F=3A-15B, \\ &{}^4P=3A, && {}^2G=3A-11B+3C, \\ &{}^2D=3A+5B+5C\pm(193B^2+8BC+4C^2)^{\frac{1}{2}}, && {}^2H=3A-6B+3C, \\ &{}^2F=3A+9B+3C, && \end{aligned} \tag{84}$$

and agree with those of the above-mentioned authors (TAS pp. 206 and 233).

d^4

This configuration was calculated by Ostrofsky,[15] but Laporte and Platt[16] found some mistakes in his results.

If we assume $ms_1=ms_2=\frac{1}{2}$ and $ms_3=ms_4=-\frac{1}{2}$, the scheme will be an L_1L_2L one, where $\mathbf{L}_1=\mathbf{l}_1+\mathbf{l}_2$ and $\mathbf{L}_2=\mathbf{l}_3+\mathbf{l}_4$; the interaction matrix $W(d^2, d^2)$ between the two d^2 groups is of the fourth degree and was calculated in the same way as for d^3. Adding to it the diagonal matrix of the internal energies of the two d^2 groups, we obtained the complete energy matrix, the eigenvalues of which are the requested energies of the configuration d^4; the singlet and the triplet terms were distinguished by the

[15] M. Ostrofsky, Phys. Rev. **46**, 604 (1934).
[16] O. Laporte and J. R. Platt, Phys. Rev. **62**, 305 (1942).

irrationality of the results in the cases of two expected terms of the same kind. The results are

$$\begin{aligned}
{}^1S &= 6A+10B+10C\pm 2(193B^2+8BC+4C^2)^{\frac{1}{2}},\\
{}^3P &= 6A-5B+(11/2)C\pm \tfrac{1}{2}(912B^2-24BC+9C^2)^{\frac{1}{2}},\\
{}^1D &= 6A+9B+(15/2)C\pm \tfrac{3}{2}(144B^2+8BC+C^2)^{\frac{1}{2}},\\
{}^3D &= 6A-5B+4C,\\
{}^5D &= 6A-21B,\\
{}^1F &= 6A+6C,\\
{}^3F &= 6A-5B+(11/2)C\pm \tfrac{3}{2}(68B^2+4BC+C^2)^{\frac{1}{2}},\\
{}^1G &= 6A-5B+(15/2)C\pm \tfrac{1}{2}(708B^2-12BC+9C^2)^{\frac{1}{2}},\\
{}^3G &= 6A-12B+4C,\\
{}^3H &= 6A-17B+4C,\\
{}^1I &= 6A-15B+6C,
\end{aligned} \tag{85}$$

and agree with those of Laporte and Platt, with the exception of a misprint for the term ^1F, which must be $6F_0-84F_4$.

d^5

Catalan and Antunes[17] and also Bowman[18] calculated the formulas for this configuration, but they could not separate the energies of the terms of the same kind.

Assuming $ms_1=ms_2=ms_3=\frac{1}{2}$ and $ms_4=ms_5=-\frac{1}{2}$, we need at first the elements of $(222L\|C_1^{(k)}+C_2^{(k)}+C_3^{(k)}\|222L')$ for the antisymmetrical states; but if we consider only electrons with $m_S=\frac{1}{2}$, d^3 is the almost closed shell corresponding to d^2, and according to (74) the needed matrix elements are the same as (81) with inverted signs; it follows that the interaction between the d^3 and the d^2 group is

$$W(d^3, d^2)=10F^0-W(d^2, d^2) \tag{86}$$

and we can proceed as for d^4. It must, however, be noted that the irrationality criterion is not sufficient in this case for the distinction between quartets and doublets, since almost all the eigenvalues are rational; this distinction may be based on the property that the relative positions of the quartets and of the sextet are exactly opposed to those of the terms of d^2 with the same L; this property follows immediately from the possibility of calculating these terms in a scheme in which $ms_1=ms_2=ms_3=ms_4=\frac{1}{2}$ and $ms_5=-\frac{1}{2}$. The results are

$$\begin{aligned}
{}^2S &= 10A-3B+8C, & {}^2F' &= 10A-25B+10C,\\
{}^6S &= 10A-35B, & {}^4F &= 10A-13B+7C,\\
{}^2P &= 10A+20B+10C, & {}^2G &= 10A-13B+8C,\\
{}^4P &= 10A-28B+7C, & {}^2G' &= 10A+3B+10C,\\
{}^2D &= 10A-3B+11C\pm 3(57B^2+2BC+C^2)^{\frac{1}{2}}, & {}^4G &= 10A-25B+5C,\\
{}^2D' &= 10A-4B+10C, & {}^2H &= 10A-22B+10C,\\
{}^4D &= 10A-18B+5C, & {}^2I &= 10A-24B+8C,\\
{}^2F &= 10A-9B+8C, & &
\end{aligned} \tag{87}$$

and agree with the recent results of Laporte.[14]

---

[17] M. A. Catalan and M. T. Antunes, Zeits. f. Physik **102**, 432 (1936).
[18] D. S. Bowman, Phys. Rev. **59**, 386 (1941).

The Coefficients of C

It is pointed out by Laporte and Platt[16] that if B vanishes all the energies of the d^n terms are rational and show high degree degeneracies; this fact may be explained by general considerations similar to those of §3 of I.

It follows from (20)I and (77) that if B vanishes the interaction between two electrons is

$$W_{ij}=A+\frac{\lambda_{ij}^4+5\lambda_{ij}^3-15\lambda_{ij}^2-75\lambda_{ij}}{18}C; \tag{88}$$

owing to the relation

$$(\mathbf{s}_i\cdot\mathbf{s}_j)=-\frac{\lambda_{ij}^4+6\lambda_{ij}^3-13\lambda_{ij}^2-90\lambda_{ij}-18}{72}, \tag{89}$$

which corresponds for equivalent d electrons to Van Vleck's relation for p electrons,[19] we may also write

$$W_{ij}=A+[\tfrac{1}{2}-2(\mathbf{s}_i\cdot\mathbf{s}_j)+q_{ij}]C, \tag{88'}$$

where

$$q_{ij}=\frac{(\lambda_{ij}-4)\lambda_{ij}(\lambda_{ij}+3)(\lambda_{ij}+5)}{36} \tag{90}$$

is an operator which has the eigenvalue 0 in all cases, except if the resultant of the two electrons is a ^1S state. Owing to the relation

$$S(S+1)=\tfrac{3}{4}n+2\sum_{i<j}(\mathbf{s}_i\cdot\mathbf{s}_j), \tag{91}$$

we find that

$$W=\sum_{i<j}W_{ij}=\frac{n(n-1)}{2}A+\left[\frac{n(n+2)}{4}-S(S+1)+Q\right]C, \tag{92}$$

where

$$Q=\sum_{i<j}q_{ij}. \tag{93}$$

If in the configuration d^n a term occurs which does not occur in d^{n-2}, it is impossible that in this state two electrons are connected so as to have a ^1S resultant, and therefore Q vanishes in all such terms.

For the terms which occur also in d^{n-2} the calculation of Q may be made as follows. Each m_sm_l state of d^{n-2} gives rise in d^n to a "family" of states in which the first $n-2$ electrons have the quantum numbers of the "parent" state and the last two electrons have the quantum numbers m^+ and $-m^-$. The number of states of each family equals five minus the number of possibilities forbidden by the exclusion principle: if in the parent state there are no other couples of electrons of the type m'^+ and $-m'^-$ (and this is always the case when the values of M_S and M_L under consideration are not allowed for d^{n-4}), each electron of the parent state excludes a possibility, and the residual number is then $7-n$.

Since the matrix elements of q_{ij} are

$$(m_im_j|q_{ij}|m_i'm_j')=(-1)^{m_i-m_i'}\delta(m_i,\,-m_j)\delta(m_i',\,-m_j'), \tag{94}$$

and the exchange term compensates the direct term for electrons of like spins, the only non-vanishing q_{ij} is q_{n-1n} and the only non-vanishing elements of q_{n-1n} are those connecting states of the same family, the values of these elements being $(-1)^{m-m'}$. The eigenvalues of these submatrices, owing to this particular expression of their elements, are all 0 except one, which equals the trace of the submatrix, i.e., $7-n$. We can then say that each term of d^{n-2} gives rise in d^n to a term with $Q=7-n$, and that the other eigenvalues of Q vanish also in d^n.

[19] J. H. Van Vleck, Phys. Rev. **45**, 412 (1934), Eq. (33).

If more than a couple of the type m^+ and $-m^-$ may be simultaneously present in a $m_s m_l$ state, the calculation is more complex, but the result is almost the same: each term which occurs in d^{n-2} has in d^n its Q value increased by $7-n$; each further term has $Q=0$.

Since $F_2 \geqslant 9F_4$ (TAS p. 177) and it is therefore impossible that B could vanish, we omitted in §3 of I this application; but the recent paper of Laporte and Platt induced us to put here these considerations. The high degree degeneracies remarked by these authors for this hypothetical case and the relation between their Table II and an old table of Hund[20] are really based on the simple form of (92) and on the properties of the operator Q.

We take this opportunity for pointing out that Eq. (6) of Laporte and Platt is right only for certain m, since its exact form is

$$\sum_k \frac{2k+1}{2}\frac{c^k(lmlm')c^k(lm_1lm'_1)}{c^k(l0l0)} = \delta(mm_1)\delta(m'm'_1)+(-1)^{m-m_1}\delta(m,\ -m'_1)\delta(m_1,\ -m')\frac{2l+1}{4}; \quad (95)$$

its demonstration follows almost immediately from (52), (20b), and (19a).

The ratios between the distances of the ^1S terms in f^2 and g^2 from the other singlets and the distances between these singlets and the triplets are also incorrect by a factor 2.

§8. THE CONFIGURATION f^3

For the configurations f^n it is also convenient to introduce new parameters; if we put

$$A=F_0-21F_4-468F_6,\quad B=(5F_2+6F_4-91F_6)/5,\quad C=7(F_4-6F_6)/5,\quad D=462F_6, \quad (96)$$

the formulas for f^2 assume the simpler form

$$\begin{array}{ll} ^1S=A+60B+105C+9D, & ^1G=A-30B+110C+2D, \\ ^3P=A+45B, & ^3H=A-25B, \\ ^1D=A+19B-72C+2D, & ^1I=A+25B+2D. \\ ^3F=A-10B, & \end{array} \quad (97)$$

The calculation for f^3 was made exactly in the same way as for d^3, and gave the following results:

$$\begin{aligned}
&^4S=3A-30B,\\
&^2P=3A-25B+35C+3D,\\
&^2D=3A-7B+(57/2)C+3D\pm\tfrac{1}{2}(2176B^2-18096BC+74529C^2)^{\frac{1}{2}},\\
&^4D=3A+25B,\\
&^2F=3A+55B+(75/2)C+6D\pm\tfrac{1}{2}[9700B^2-180B(45C+2D)+9(45C+2D)^2]^{\frac{1}{2}},\\
&^4F=3A-30B,\\
&^2G=3A+7B+(113/2)C+3D\pm\tfrac{1}{2}(12676B^2-31676BC+36169C^2)^{\frac{1}{2}},\\
&^4G=3A-10B,\\
&^2H=3A-23B+(63/2)C+3D\pm\tfrac{1}{2}(5056B^2-18816BC+29169C^2)^{\frac{1}{2}},\\
&^2I=3A-5B+45C+3D,\\
&^4I=3A-65B,\\
&^2K=3A-40B+80C+3D,\\
&^2L=3A+3D.
\end{aligned} \quad (98)$$

[20] F. Hund, Zeits. f. Physik **33**, 345 (1925), Table IV.

§9. THE CONFIGURATIONS d^2p AND d^3p

The energy matrix of the configuration d^2p will be calculated in a scheme in which each term is characterized by a definite state of the core d^2. In this scheme the interaction between the two d electrons is diagonal and has the values given by (78); the interaction between each d electron and the p electron follows from (60) and is

$$W(d,p)=F_0+10(21)^{\frac{1}{2}}(\mathbf{u}_d^{(2)}\cdot\mathbf{u}_p^{(2)})F_2-[\tfrac{1}{2}+2(\mathbf{s}_d\cdot\mathbf{s}_p)]\{[2+9\sqrt{5}(\mathbf{u}_d^{(1)}\cdot\mathbf{u}_p^{(1)})+5(21)^{\frac{1}{2}}(\mathbf{u}_d^{(2)}\cdot\mathbf{u}_p^{(2)})]G_1+[7-21\sqrt{5}(\mathbf{u}_d^{(1)}\cdot\mathbf{u}_p^{(1)})+5(21)^{\frac{1}{2}}(\mathbf{u}_d^{(2)}\cdot\mathbf{u}_p^{(2)})]G_3\}, \tag{99}$$

where[21]

$$F_0=F^0(nd,n'p),\quad F_2=F^2(nd,n'p)/35,\quad G_1=G^1(nd,n'p)/15,\quad G_3=3G^3(nd,n'p)/245. \tag{100}$$

In order to obtain the matrix of

$$W(d^2,p)=W(d_1,p)+W(d_2,p), \tag{101}$$

we calculate at first, by means of (44), the matrices of the tensors

$$\mathbf{U}^{(k)}=\mathbf{u}_1^{(k)}+\mathbf{u}_2^{(k)} \tag{102a}$$

and of the double tensors

$$\mathbf{V}^{(1k)}=\mathbf{s}_1\mathbf{u}_1^{(k)}+\mathbf{s}_2\mathbf{u}_2^{(k)} \tag{102b}$$

for two equivalent d electrons; the results are

$(SL\|U^{(1)}\|S'L')=$ (103a)

	^1S	^3P	^1D	^3F	^1G
^1S	0	0	0	0	0
^3P	0	$(1/5)^{\frac{1}{2}}$	0	0	0
^1D	0	0	1	0	0
^3F	0	0	0	$(14/5)^{\frac{1}{2}}$	0
^1G	0	0	0	0	$(6)^{\frac{1}{2}}$

$(SL\|U^{(2)}\|S'L')=$ (103b)

	^1S	^3P	^1D	^3F	^1G
^1S	0	0	$(4/5)^{\frac{1}{2}}$	0	0
^3P	0	$-(21/25)^{\frac{1}{2}}$	0	$(24/25)^{\frac{1}{2}}$	0
^1D	$(4/5)^{\frac{1}{2}}$	0	$-3/7$	0	$(144/245)^{\frac{1}{2}}$
^3F	0	$(24/25)^{\frac{1}{2}}$	0	$(6/25)^{\frac{1}{2}}$	0
^1G	0	0	$(144/245)^{\frac{1}{2}}$	0	$(198/49)^{\frac{1}{2}}$

$(SL\|V^{(11)}\|S'L')=$ (103c)

	^1S	^3P	^1D	^3F	^1G
^1S	0	$(2/5)^{\frac{1}{2}}$	0	0	0
^3P	$(2/5)^{\frac{1}{2}}$	$(1/5)^{\frac{1}{2}}$	$-(7/10)^{\frac{1}{2}}$	0	0
^1D	0	$-(7/10)^{\frac{1}{2}}$	0	$(4/5)^{\frac{1}{2}}$	0
F^3	0	0	$(4/5)^{\frac{1}{2}}$	$(14/5)^{\frac{1}{2}}$	$-(3/5)^{\frac{1}{2}}$
^1G	0	0	0	$-(3/5)^{\frac{1}{2}}$	0

[21] This definition of G_3 differs by a factor 3 from that of TAS, but agrees with (50) I.

$$(SL\|V^{(12)}\|S'L') = \begin{array}{c|ccccc} & {}^1S & {}^3P & {}^1D & {}^3F & {}^1G \\ \hline {}^1S & 0 & 0 & 0 & 0 & 0 \\ {}^3P & 0 & -(21/25)^{\frac{1}{2}} & -(3/10)^{\frac{1}{2}} & (24/25)^{\frac{1}{2}} & 0 \\ {}^1D & 0 & -(3/10)^{\frac{1}{2}} & 0 & (32/35)^{\frac{1}{2}} & 0 \\ {}^3F & 0 & (24/25)^{\frac{1}{2}} & (32/35)^{\frac{1}{2}} & (6/25)^{\frac{1}{2}} & -(9/7)^{\frac{1}{2}} \\ {}^1G & 0 & 0 & 0 & -(9/7)^{\frac{1}{2}} & 0 \end{array} \quad (103d)$$

The matrix of (101) follows from (99), (103), and (38); adding to it the energy of the core we obtain the following results for the terms of d^2p [22]

$$^2S = A+7B-14F_2-3G_1+7G_3$$
$$^4S = A+7B-14F_2+6G_1-14G_3$$
$$^4P = A+7B+7F_2-6G_1-21G_3$$
$$^4F = A-8B-3F_2-G_1-16G_3$$
$$^4G = A-8B+F_2-9G_1-4G_3$$
$$^2H = A+4B+2C+4F_2-6G_1-G_3$$

$$^4D = \left\| \begin{array}{cc} A+7B-(7F_2+24G_1+49G_3)/5 & (12F_2-6G_1-6G_3)(14/25)^{\frac{1}{2}} \\ (12F_2-6G_1-6G_3)(14/25)^{\frac{1}{2}} & A-8B+(12F_2+4G_1-146G_3)/5 \end{array} \right\|$$

$$^2G = \left\| \begin{array}{cc} A-8B+F_2+(9/2)G_1+2G_3 & (-\tfrac{3}{2}G_1+G_3)(15)^{\frac{1}{2}} \\ (-\tfrac{3}{2}G_1+G_3)(15)^{\frac{1}{2}} & A+4B+2C-11F_2+\tfrac{3}{2}G_1-6G_3 \end{array} \right\|$$

$$^2P = \left\| \begin{array}{ccc} A+14B+7C-2G_1-7G_3 & (-3G_1+7G_3)(3)^{\frac{1}{2}} & (4F_2-G_1-G_3)(7)^{\frac{1}{2}} \\ (-3G_1+7G_3)(3)^{\frac{1}{2}} & A+7B+7F_2+3G_1+(21/2)G_3 & (5/2)G_3(21)^{\frac{1}{2}} \\ (4F_2-G_1-G_3)(7)^{\frac{1}{2}} & (5/2)G_3(21)^{\frac{1}{2}} & A-3B+2C-3F_2+G_1-(23/2)G_3 \end{array} \right\|$$

$$^2D = \left\| \begin{array}{ccc} A+7B+(-7F_2+12G_1+(49/2)G_3)/5 & (-6G_1+9G_3)(21/20)^{\frac{1}{2}} & (12F_2+3G_1+3G_3)(14/25)^{\frac{1}{2}} \\ (-6G_1+9G_3)(21/20)^{\frac{1}{2}} & A-3B+2C+3F_2-2G_1-(19/2)G_3 & (G_1+11G_3)(6/5)^{\frac{1}{2}} \\ (12F_2+3G_1+3G)(14/25)^{\frac{1}{2}} & (G_1+11G_3)(6/5)^{\frac{1}{2}} & A-8B+(12F_2-2G_1+73G_3)/5 \end{array} \right\|$$

$$^2F = \left\| \begin{array}{ccc} A-3B+2C-(6F_2+23G_1+23G_3)/7 & (-10G_1+10G_3)(3/7)^{\frac{1}{2}} & (24F_2-6G_1-6G_3)(3/49)^{\frac{1}{2}} \\ (-10G_1+10G_3)(3/7)^{\frac{1}{2}} & A-8B-3F_2+\tfrac{1}{2}G_1+8G_3 & (9G_1+54G_3)/(28)^{\frac{1}{2}} \\ (24F_2-6G_1-6G_3)(3/49)^{\frac{1}{2}} & (9G_1+54G_3)/(28)^{\frac{1}{2}} & A+4B+2C+(55F_2-\tfrac{3}{2}G_1-124G_3)/7 \end{array} \right\|$$

For calculating the terms of d^8p it suffices to take into account the fact that the constant matrix and the matrices of $\mathbf{U}^{(2)}$ and $\mathbf{V}^{(11)}$ change their signs at passing from the core d^2 to d^8, according to (74). The results are:

$$^2S = {}^4S = A+7B+14F_2$$
$$^4P = A+7B-7F_2$$
$$^4F = A-8B+3F_2$$
$$^4G = A-8B-F_2$$
$$^2H = A+4B+2C-4F_2+15G_3$$

[22] The parameters have the meaning given in (77) and (100), with the exception of A which contains also $2F^0(nd, n'p)$.

TABLE I. The configuration $3d^24p$ of Ti II.

Term	Obs.	Calc. A	Calc. B
$(^3F)\ ^4G$	29936	29748	29823
$(^3F)\ ^4F$	31108	31028	31125
$(^3F)\ ^2F$	31369	31375	31471
$(^3F)\ ^2D$	31918	32132	32251
$(^3F)\ ^4D$	32690	32813	32890
$(^3F)\ ^2G$	34657	33951	34109
$(^3P)\ ^2S$	37431	37623	37628
$(^1D)\ ^2D$	39380	40461	40000
$(^1D)\ ^2P$	39627	40035	39496
$(^1D)\ ^2F$	40011	39983	39507
$(^3P)\ ^4S$	40027	40233	40237
$(^3P)\ ^4D$	40612	40518	40441
$(^3P)\ ^4P$	42127	42213	42100
$(^1G)\ ^2G$	43763	43259	43675
$(^3P)\ ^2D$	44907	44807	44737
$(^3P)\ ^2P$	45524	45672	45673
$(^1G)\ ^2H$	45802	44822	45184
$(^1G)\ ^2F$	47535	47751	48078
$(^1S)\ ^2P$	=	64692	(64465)
Parameters:			
B		669	=
C		2563	=
ϵ		=	0.9484
F_2		290	288
G_1		332	337
G_3		18	20.2

TABLE II. The configuration $3d^84p$ of Ni II.

Term	Obs.	Calc.
$(^3F)\ ^4D$	52588	52581
$(^3F)\ ^4G$	53883	53707
$(^3F)\ ^4F$	55394	55163
$(^3F)\ ^2G$	55775	55735
$(^3F)\ ^2F$	57685	57492
$(^3F)\ ^2D$	57933	57892
$(^3P)\ ^4P$	66649	66853
$(^1D)\ ^2F$	67943	68729
$(^1D)\ ^2D$	68442	68929
$(^1D)\ ^2P$	68636	69013
$(^3P)\ ^4D$	70716	70527
$(^3P)\ ^2D$	72011	72159
$(^3P)\ ^2P$	73256	73245
$(^3P)\ ^2S$	74282	74497
$(^3P)\ ^4S$	74299	74497
$(^1G)\ ^2H$	75460	74713
$(^1G)\ ^2F$	75904	75494
$(^1G)\ ^2G$	79878	79505
$(^1S)\ ^2P$	=	109339
Parameters:		
B		1022
C		4509
F_2		364
G_1		303
G_3		54.4

$$^4D=\left\|\begin{matrix} A+7B+7F_2/5 & -12F_2(14/25)^{\frac{1}{2}} \\ -12F_2(14/25)^{\frac{1}{2}} & A-8B-12F_2/5 \end{matrix}\right\|$$

$$^2G=\left\|\begin{matrix} A-8B-F_2+75G_3/2 & -5G_3(15/4)^{\frac{1}{2}} \\ -5G_3(15/4)^{\frac{1}{2}} & A+4B+2C+11F_2+5G_3/2 \end{matrix}\right\|$$

$$^2P=\left\|\begin{matrix} A+14B+7C+2G_1+7G_3 & (3G_1-7G_3)(3)^{\frac{1}{2}} & (-4F_2+G_1+G_3)(7)^{\frac{1}{2}} \\ (3G_1-7G_3)(3)^{\frac{1}{2}} & A+7B-7F_2+(27/2)G_1+21G_3 & (3G_1-2G_3)(21/4)^{\frac{1}{2}} \\ (-4F_2+G_1+G_3)(7)^{\frac{1}{2}} & (3G_1-2G_3)(21/4)^{\frac{1}{2}} & A-3B+2C+3F_2+(7/2)G_1+G_3 \end{matrix}\right\|$$

$$^2D=\left\|\begin{matrix} A+7B+(7F_2+(27/2)G_1+126G_3)/5 & (3G_1-12G_3)(21/20)^{\frac{1}{2}} & (-12F_2+9G_1+9G_3)(14/25)^{\frac{1}{2}} \\ (3G_1-12G_3)(21/20)^{\frac{1}{2}} & A-3B+2C-3F_2+(7/2)G_1+6G_3 & (7G_1-3G_3)(6/5)^{\frac{1}{2}} \\ (-12F_2+9G_1+9G_3)(14/25)^{\frac{1}{2}} & (7G_1-3G_3)(6/5)^{\frac{1}{2}} & A-8B+(-12F_2+84G_1+9G_3)/5 \end{matrix}\right\|$$

$$^2F=\left\|\begin{matrix} A-3B+2C+(6F_2+2G_1+72G_3)/7 & (2G_1-18G_3)(3/7)^{\frac{1}{2}} & (-24F_2+6G_1+6G_3)(3/49)^{\frac{1}{2}} \\ (2G_1-18G_3)(3/7)^{\frac{1}{2}} & A-8B+3F_2+6G_1+(27/2)G_3 & (18G_1-(9/2)G_3)/(7)^{\frac{1}{2}} \\ (-24F_2+6G_1+6G_3)(3/49)^{\frac{1}{2}} & (18G_1-(9/2)G_3)/(7)^{\frac{1}{2}} & A+4B+2C+(-55F_2+54G_1+\frac{3}{2}G_3)/7 \end{matrix}\right\|$$

We fitted the experimental values[23] of Ti II and Ni II to the theoretical formulas by least squares, and obtained the results given in Table I (column A) and Table II. The mean deviations are ± 430 for Ti II and ± 347 for Ni II, and confirm the known fact that the agreement is better for the elements on the right side of the periodic table.

It must also be observed that these deviations show a certain regularity: in almost all the cases the differences between observed and calculated values are positive for the terms grounded on the

[23] R. F. Bacher and S. Goudsmit, *Atomic Energy States* (McGraw-Hill, 1932).

^3F and ^1G terms of the core, and negative for those grounded on ^3P and ^1D. Since the configurations d^2, d^2s, d^8, d^8s of the whole iron group show the same regularities in the deviations,[24] the main part of these deviations may be attributed to second-order effects in the coupling of the core electrons.

It seems therefore reasonable to substitute in the formulas of d^2p and d^8p for the part depending on the energy of the core (terms with B and C) the experimental values of the corresponding ions d^2 and d^8; these values must, however, be multiplied with a convenient reduction factor ϵ, since with increasing ionization all coupling parameters increase slightly.

The result of fitting the terms of Ti II to such semi-theoretical formulas by least squares is shown in Table I, column B; the mean deviation reduces to ± 332, although the number of free parameters is also reduced (ϵ instead of B and C).

For writing down the matrix of the ^2P terms we needed also the value of the term $3d^2\ {}^1S$ of Ti III: since this is still unknown, we calculated it from the other $3d^2$ terms by least squares; this approximated value suffices for calculating the perturbation of $(^1S)\ ^2P$ of Ti II on the other ^2P terms, but cannot be of use for predicting with the same approximation the position of $(^1S)\ ^2P$ himself.

We could not apply this method of calculation to Ni II, since the terms of Ni III are still unknown, as far as known to us.

§10. THE INTEGRALS G^k

It is noted in TAS (p. 177) that F^k is essentially positive and a decreasing function of k, and it is stated as an empirical fact that also G^k shares the same properties, although they do not follow from its definition. Since this fact was sometimes questioned,[25] it seemed worth while to us to seek for a mathematical demonstration: we found that the first property may be proved, but only $G^k/(2k+1)$ is necessarily a decreasing function of k.

According to its definition, G^k has the form

$$G^k=\int_0^\infty\int_0^\infty (r_<^k/r_>^{k+1})f(r_1)f(r_2)dr_1dr_2, \tag{104}$$

where $r_<$ is the lesser and $r_>$ the greater of r_1 and r_2; we may also write

$$G^k=\int_0^\infty f(x)\,\varphi(x)dx \tag{105}$$

with

$$\varphi(x)=x^{-k-1}\int_0^x y^kf(y)dy+x^k\int_x^\infty y^{-k-1}f(y)dy. \tag{106}$$

It follows from (106) that

$$f(x)=-x^{k+1}[x^{-2k}(x^{k+1}\varphi)']'/(2k+1)\,; \tag{106'}$$

introducing this expression in (105) and integrating by parts we get

$$G^k=\int_0^\infty x^{-2k}[(x^{k+1}\varphi)']^2dx/(2k+1)\geqslant 0. \tag{107}$$

In the same way, putting

$$G^k/(2k+1)-G^{k+1}/(2k+3)=\int_0^\infty f(x)\psi(x)dx \tag{108}$$

[24] TAS, Figs. 3[7] and 4[7]; G. Racah, Phys. Rev. **61**, 538 (1942), Table I.
[25] TAS, p. 366.

with

$$\psi(x)=\left[x^{-k-1}\int_0^x y^k f(y)dy+x^k\int_x^\infty y^{-k-1}f(y)dy\right]/(2k+1)$$

$$+\left[x^{-k-2}\int_0^x y^{k+1}f(y)dy+x^{k+1}\int_x^\infty y^{-k-2}f(y)dy\right]/(2k+3), \qquad (109)$$

we have

$$f(x)=x^{k+2}[x^{-2k}(x^{k+2}\psi)'']''/(2k+2), \qquad (109')$$

and from (108) we get with a double integration by parts

$$G^k/(2k+1)-G^{k+1}/(2k+3)=\int_0^\infty x^{-2k}[(x^{k+2}\psi)'']^2dx/(2k+2)\geqslant 0. \qquad (110)$$

It is also possible to find particular functions $f(x)$ for which the ratio G^{k+1}/G^k tends to the theoretical limit $(2k+3)/(2k+1)$; an example is given by

$$f(x)=\delta(x-a)-\delta(x-a-b) \qquad (111)$$

with $b\ll a$.

APPENDIX A

From the definition (17) we have

$$v(abc\,;000)=\frac{a!b!c!}{a+b-c}\Sigma_z(-1)^{c+z}\frac{(a+b-c-z)+z}{z!(a+b-c-z)!(a-z)!(b-z)!(c-a+z)!(c-b+z)!}$$

$$=\frac{a!b!c!}{a+b-c}\Sigma_z(-1)^{c+z}\{[z!(a+b-c-z-1)!(a-z)!(b-z)!(c-a+z)!(c-b+z)!]^{-1}$$
$$+[(z-1)!(a+b-c-z)!(a-z)!(b-z)!(c-a+z)!(c-b+z)!]^{-1}\},$$

and changing the summation parameter in the second term of the brackets,

$$v(abc\,;000)=\frac{a!b!c!}{a+b-c}\Sigma_z(-1)^{c+z}\{[z!(a+b-c-z-1)!(a-z)!(b-z)!(c-a+z)!(c-b+z)!]^{-1}$$
$$-[z!(a+b-c-z-1)!(a-z-1)!(b-z-1)!(c-a+z+1)!(c-b+z+1)!]^{-1}\}$$

$$=\frac{a!b!(c+1)!}{a+b-c}\Sigma_z(-1)^{c+z+1}\frac{a+b-c-2z-1}{z!(a+b-c-z-1)!(a-z)!(b-z)!(c-a+z+1)!(c-b+z+1)!};$$

owing to the identity

$$a(a+b-c-2z-1)=(a-z)(a+b-c-z-1)-z(c-b+z+1),$$

we have also

$$v(abc\,;000)=\frac{(a-1)!b!(c+1)!}{a+b-c}\Sigma_z(-1)^{c+z+1}\{[z!(a+b-c-z-2)!(a-z-1)!(b-z)!(c-a+z+1)!(c-b+z+1)!]^{-1}$$
$$-[(z-1)!(a+b-c-z-1)!(a-z)!(b-z)!(c-a+z+1)!(c-b+z)!]^{-1}\},$$

and changing again the summation parameter in the second term of the brackets,

$$v(abc\,;000)=\frac{(a-1)!b!(c+1)!}{a+b-c}\Sigma_z(-1)^{c+z+1}\{[z!(a+b-c-z-2)!(a-z-1)!(b-z)!(c-a+z+1)!(c-b+z+1)!]^{-1}$$
$$+[z!(a+b-c-z-2)!(a-z-1)!(b-z-1)!(c-a+z+2)!(c-b+z+1)!]^{-1}\}$$

$$=\frac{b+c-a+2}{a+b-c}\Sigma_z(-1)^{c+z+1}\frac{(a-1)!b!(c+1)!}{z!(a+b-c-z-2)!(a-z-1)!(b-z)!(c-a+z+2)!(c-b+z+1)!}$$
$$=[(g-a+1)/(g-c)]v(a-1bc+1\,;000).$$

From this recursion formula we have

$$v(abc\,;000)=(g-a+x)!(g-c-x)!v(a-xbc+x\,;000)/[(g-a)!(g-c)!],$$

and for $x=g-c$,

$$v(abc\,;000)=b!v(g-bbg\,;000)/[(g-a)!(g-c)!];$$

since

$$v(g-bbg\,;000)=(-1)^g g!/[b!(g-b)!],$$

we obtain at last (22).

APPENDIX B

Using repeatedly Eqs. (52), and (55′) of I, and also the relation

$$\Sigma_s \frac{(a+s)!(b-s)!}{(c+s)!(d-s)!} = \frac{(a+b+1)!(a-c)!(b-d)!}{(c+d)!(a+b-c-d+1)!}, \tag{112}$$

which follows from (51) I for x and y negative, we obtain from the definition (17):

$$\begin{aligned}
&\sum_{\alpha\beta\gamma\delta\varphi} (-1)^{f+\varphi} v(abe;\alpha\beta-e)v(acf;-\alpha\gamma\varphi)v(bdf;-\beta\delta-\varphi)v(cde;\gamma\delta-e) \\
&= \sum_{\alpha\gamma} (-1)^{f+\alpha-\gamma} v(abe;\alpha e-\alpha-e)v(acf;-\alpha\gamma\ \alpha-\gamma)v(bdf;\alpha-e\ e-\gamma\ \gamma-\alpha)v(cde;\gamma e-\gamma-e) \\
&= \sum_{\alpha\gamma tu} \frac{(-1)^{a-c-f+t+u} 2e!(a+\alpha)!(b+e-\alpha)!(c+\gamma)!(d+e-\gamma)!(f+\alpha-\gamma)!(f-\alpha+\gamma)!}{(e+a-b)!(e+b-a)!(e+c-d)!(e+d-c)!t!(a+c-f-t)!u!(b+d-f-u)!(a+\alpha-t)!(f-a-\gamma+t)! \cdot (d+e-\gamma-u)!(f-d+\alpha-e+u)!(c+\gamma-t)!(f-c-\alpha+t)!(b+e-\alpha-u)!(f-b+\gamma-e+u)!} \\
&= \sum_{\alpha\gamma tuvw} \frac{(-1)^{a-c-f+t+u} 2e!(a+\alpha)!(b+e-\alpha)!(c+\gamma)!(d+e-y)!}{(e+a-b)!(e+b-a)!(e+c-d)!(e+d-c)!t!(a+c-f-t)!u!(b+d-f-u)! \cdot (f-a+c+t-v)!(c+d+e-u-v)!(a-c-d-e+\alpha-t+u+v)!(v-c-\gamma)! \cdot (v-t-w)!(f-b-c-e+u+v-w)!(b+e-\alpha+t-u-v+w)!(c+\gamma-v+w)!} \\
&= \sum_{tuvw} \frac{(-1)^{a-c-f+t+u} 2e!(a+b+e+1)!(c+d+e+1)!(c+d+e-v)!(v-w)!(u+v-t-w)!(c+d+e+t-u-v)!}{(e+a-b)!(e+b-a)!(e+c-d)!(e+d-c)!w!(a+b-c-d+w)!(c+d+e+1-w)!^2(f-b-c-e+u+v-w)! \cdot u!(v-t-w)!(a+c-f-t)!(f-a+c+t-v)!t!(c+d+e-u-v)!(b+d-f-u)!} \\
&= \sum_{tuvwxz} \frac{(-1)^{a-c-f+t+u} 2e!(a+b+e+1)!(c+d+e+1)!(c+d+e-v)!(v-w)!(f-a-c+u+v-w)!(c+e+f-b+t-v)!}{(e+a-b)!(e+b-a)!(e+c-d)!(e+d-c)!w!(a+b-c-d+w)!(c+d+e+1-w)!^2 \cdot (f-b-c-e+u+v-w)!(u-x)!(a+c-f-t-x)!(f-a-c+v-w+x)!x! \cdot (f-a+c+t-v)!(t-z+x)!(b+d-f-u-z+x)!(c+e+f-b-v+z-x)!(z-x)!} \\
&= \sum_{vwxz} \frac{(-1)^{2e+f+d-b+z} 2e!(a+b+e+1)!(c+d+e+1)!(c+d+e-v)!(v-w)!}{(e+c-d)!(e+d-c)!w!(a+b-c-d+w)!(c+d+e+1-w)!^2(2c-v-x)!(d-c-e+v-w+x-z)! \cdot x!(z-x)!(a+c-f-z)!(b+d-f-z)!(e+f-a-d+z)!(e+f-b-c+z)!} \\
&= \sum_{wz} \frac{(-1)^{2e+f+d-b+z} 2e!(a+b+e+1)!(c+d+e+1)!(d+e-c+x)!(c+e-d+z-x)!}{(e+c-d)!(e+d-c)!w!(a+b-c-d+w)!(c+d+e+1-w)!(c+d-e-w-z)!(2e+z+1)! \cdot x!(z-x)!(a+c-f-z)!(b+d-f-z)!(e+f-a-d+z)!(e+f-b-c+z)!} \\
&= \frac{(-1)^{2e+f+d-b}}{2e+1} \Sigma_z \frac{(-1)^z (a+b+c+d+1-z)!}{(a+b-e-z)!(c+d-e-z)!z!(a+c-f-z)!(b+d-f-z)!(e+f-a-d+z)!(e+f-b-c+z)!}.
\end{aligned}$$

ERRATA OF PART I

Last line of the summary: read p^n for p''; element $({}^1D|E|{}^3P)$ of the ${}^2P$ matrix on p. 195: read $(\frac{1}{2}G_0+2G_2)\sqrt{5}$ for $(-G_0+2G_2)\sqrt{5}$; second line of §6: read G^{l+1} for G^{l+2}.

PHYSICAL REVIEW VOLUME 63, NUMBERS 9 AND 10 MAY 1 AND 15, 1943

Theory of Complex Spectra. III

GIULIO RACAH
The Hebrew University, Jerusalem, Palestine
(Received February 8, 1943)

The consideration of the phases of the fractional-parentage coefficients allows the extension of the matrix methods to configurations with more than two equivalent electrons. Tables are given for the parentages of the terms of p^n and d^n. Applications are made to the spin-orbit interaction of the d^n terms and to the electrostatic interaction between the configurations d^n, $d^{n-1}s$, and $d^{n-2}s^2$. Errata in Part II are indicated.

§1. INTRODUCTION

THIS paper deals chiefly with the application of matrix methods to calculations within configurations with more than two equivalent electrons.

It is known that the eigenfunctions built up with the usual vector-coupling formulas[1] are not antisymmetrical as required from the exclusion principle and they must be antisymmetrized afterwards. But if certain of the electrons are equivalent, these antisymmetrized states are no longer normalized and some of them are linearly dependent, so that the calculations become very complicated.

An escape from these difficulties was proposed by Gray and Wills[2] who started from the nlm_sm_l scheme with antisymmetrized eigenfunctions and computed the SL eigenfunctions using angular-momentum operators and orthogonality considerations. This method leads to an orthonormal system of eigenfunctions, but since it gives up the vector-coupling formulas, the matrix of each operator must at first be calculated in the nlm_lm_l scheme and then transformed to the SL scheme, and no use may be made of the powerful matrix methods developed in Chapter III of TAS[1] and also extended in a previous paper of the author.[3]

In order to make full use of the above-mentioned methods, we shall calculate the eigenfunctions of the configuration l^n as linear combinations of the eigenfunctions obtained by the addition of a further electron l to the configuration l^{n-1}. This possibility was already indicated by Goudsmit and Bacher,[4] who introduced the concept of fractional parentage; but they were interested only in the squares of the coefficients of these linear combinations and calculated them with a procedure which, being based on a diagonal-sum method, did not permit them to separate the fractional parentages of duplicated terms.[5] The consideration of the phases of the coefficients of fractional parentage will enable us to calculate them separately also for terms of the same kind occurring in a given configuration and to calculate the matrix elements of every symmetrical operator between configurations containing equivalent electrons.

The fractional parentages of the configurations p^n and d^n are calculated in §3 and §4, whilst §2 contains a lemma on which these calculations are based and §5 deals with the matrices of symmetrical operators. In §6 an analysis is made of the structure of the configurations l^n in connection with the appearance of more terms of the same kind, and §7 contains an application to configuration interactions.

§2. TRANSFORMATIONS BETWEEN THE DIFFERENT COUPLING SCHEMES OF THREE ANGULAR MOMENTA

If we add two angular momenta j_1 and j_2, the magnitude J of the resulting vector and its z-component M characterize completely the states of the system; but if we add three angular momenta, several states may occur with the

[1] E. U. Condon and G. H. Shortley, *Theory of Atomic Spectra* (Cambridge, 1935) (which we shall denote by TAS), §14[3].
[2] N. M. Gray and L. A. Wills, Phys. Rev. **38**, 248 (1931); TAS §5[8].
[3] G. Racah, Phys. Rev. **62**, 438 (1942) (which we shall denote by II). We refer to this paper and to TAS for definitions, notations, and bibliographical indications.
[4] S. Goudsmit and R. F. Bacher, Phys. Rev. **46**, 948 (1934).
[5] D. H. Menzel and L. Goldberg, Astrophys. J. **84**, 1 (1936).

TABLE I. $(p^3SL\{\!\!\{p^2(S'L')pSL)$. The different rows are normalized separately, and N is the normalization factor of each linear combination.

p^3	N	p^2 ${}^1S$	${}^3P$	${}^1D$
${}^4S$	1	0	1	0
${}^2P$	$18^{-\frac{1}{2}}$	2	-3	$-5^{\frac{1}{2}}$
${}^2D$	$2^{-\frac{1}{2}}$	0	1	-1

same J and M and a complete characterization of the states needs the specification of the type of coupling of the vectors.

We may for instance couple at first j_1 and j_2 and then add j_3 to their resultant J': In this case the eigenfunctions are

$$\psi({}_1jj_2(J')j_3JM)=\sum_{m_3M'}\psi(j_1j_2J'M')\phi(j_3m_3)\cdot(J'j_3M'm_3|J'j_3JM)$$
$$=\sum_{m_1m_2m_3M'}\phi(j_1m_1)\phi(j_2m_2)\phi(j_3m_3)\cdot(j_1j_2m_1m_2|j_1j_2J'M')\cdot(J'j_3M'm_3|J'j_3JM);\quad(1)$$

but we may also couple at first j_2 and j_3 and then add their resultant J'' to j_1, and in this case the eigenfunctions are

$$\psi(j_1,\,j_2j_3(J''),\,JM)=\sum_{m_1m_2m_3M''}\phi(j_1m_1)\phi(j_2m_2)\phi(j_3m_3)\cdot(j_2j_3m_2m_3|j_2j_3J''M'')\cdot(j_1J''m_1M''|j_1J''JM).\quad(2)$$

The unitary transformation which connects these two representations of the same system is

$$(j_1j_2(J')j_3J|j_1,\,j_2j_3(J''),\,J)=\sum_{m_1m_2m_3M'M''}(J'j_3JM|J'j_3M'm_3)\cdot(j_1j_2J'M'|j_1j_2m_1m_2)\cdot(j_2j_3m_2m_3|j_2j_3J''M'')\cdot(j_1J''m_1M''|j_1J''JM);\quad(3)$$

introducing the expression (16')II for the transformation coefficients for the addition of two angular momenta and using Eqs. (19)II and (37)II we obtain

$$(j_1j_2(J')j_3J|j_1,\,j_2j_3(J''),\,J)=[(2J'+1)(2J''+1)]^{\frac{1}{2}}W(j_1j_2Jj_3;J'J''),\quad(4)$$

where W is the function defined by (36')II.

It is sometimes useful to consider the changing of the coupling together with a change in the order of the vectors; the same way as before yields

$$(j_1j_2(J')j_3J|j_1j_3(J'')j_2J)=[(2J'+1)(2J''+1)]^{\frac{1}{2}}W(J'j_3j_2J'';Jj_1).\quad(5)$$

If we have three electrons or groups of electrons, the transformations between the different parentages in SL coupling are obvious extensions of (4) and (5): For instance,

$$(s_1l_1s_2l_2(S'L')s_3l_3SL|s_1l_1,\,s_2l_2s_3l_3(S''L''),\,SL)=[(2S'+1)(2S''+1)(2L'+1)(2L''+1)]^{\frac{1}{2}}\cdot W(s_1s_2Ss_3;S'S'')W(l_1l_2Ll_3;L'L'');\quad(6)$$

a particular case of this transformation was considered in TAS 6[8] 14.

§3. THE EIGENFUNCTIONS OF GROUPS OF EQUIVALENT ELECTRONS

If we couple two equivalent electrons with the usual vector-coupling formulas, we obtain antisymmetric or symmetric eigenfunctions according to whether $S+L$ is even or odd (TAS, p. 231); the eigenfunctions of the states with $S+L$ even are therefore the normalized eigenfunctions of the allowed states of l^2.

If we add in the same way to the allowed states of l^2 a third l electron, the obtained eigenfunctions are in general antisymmetric only with respect to the first two electrons, but not with

TABLE II. $(d^3vSL\{\!\!\{d^2(v'S'L)dSL)$.

d^3	N	d^2 ${}_0{}^1S$	${}_2{}^3P$	${}_2{}^1D$	${}_2{}^3F$	${}_2{}^1G$
${}_3{}^2P$	$30^{-\frac{1}{2}}$	0	$7^{\frac{1}{2}}$	$15^{\frac{1}{2}}$	$-8^{\frac{1}{2}}$	0
${}_3{}^4P$	$15^{-\frac{1}{2}}$	0	$-8^{\frac{1}{2}}$	0	$-7^{\frac{1}{2}}$	0
${}_1{}^2D$	$60^{-\frac{1}{2}}$	4	-3	$-5^{\frac{1}{2}}$	$-21^{\frac{1}{2}}$	-3
${}_3{}^2D$	$140^{-\frac{1}{2}}$	0	-7	$45^{\frac{1}{2}}$	$21^{\frac{1}{2}}$	-5
${}_3{}^2F$	$70^{-\frac{1}{2}}$	0	$28^{\frac{1}{2}}$	$-10^{\frac{1}{2}}$	$7^{\frac{1}{2}}$	-5
${}_3{}^4F$	$5^{-\frac{1}{2}}$	0	-1	0	2	0
${}_3{}^2G$	$42^{-\frac{1}{2}}$	0	0	$-10^{\frac{1}{2}}$	$21^{\frac{1}{2}}$	$11^{\frac{1}{2}}$
${}_3{}^2H$	$2^{-\frac{1}{2}}$	0	0	0	-1	1

respect to the third. If we apply in effect to $\psi(l^2(S'L')lSL)$ the transformation[6]

$$\psi(l^2(S'L')lSL) = \sum_{S''L''} \psi(l, ll(S''L''), SL) \cdot (l, ll(S''L''), SL | l^2(S'L')lSL), \quad (7)$$

where the transformation matrix is given by (6), we obtain in general in the sum (7) allowed and forbidden values of $S''L''$ and, therefore, $\psi(l^2(S'L')lSL)$ cannot be an eigenfunction of l^3.

Only such a linear combination

$$\Psi(l^3\alpha SL) = \sum_{S'L'} \psi(l^2(S'L')lSL) \cdot (l^2(S'L')lSL]\!]l^3\alpha SL) \quad (8)$$

may be the eigenfunction of l^3 for which the coefficients of $\psi(l, ll(S''L''), SL)$ vanish for every forbidden value of $S''L''$ after the application of the transformation (7); the "coefficients of fractional parentage" $(l^2(S'L')lSL]\!]l^3\alpha SL)$ must therefore satisfy the equation system

$$\sum_{S'L'} (l, ll(S''L''), SL | l^2(S'L')lSL) \cdot (l^2(S'L')lSL]\!]l^3\alpha SL) = 0 \quad (S''+L'' \text{ odd}). \quad (9)$$

Since a function antisymmetric with respect to the electrons 1 and 2 and also with respect to the electrons 2 and 3 is antisymmetric with respect to all three electrons, the condition (9) is necessary and sufficient for the determination of the coefficients of fractional parentage of the terms of l^3, and the number of independent non-vanishing solutions of (9) for a given SL equals the number of allowed terms of this kind in l^3; if this number is greater than one, the different terms may be distinguished by a parameter α.

As an illustration of this method let us calculate the eigenfunction of the term $p^3\,{}^2D$. It follows from (6) that

$$\psi(p^2(^3P)p\,{}^2D) = (3/16)^{1/2}\psi(p, pp(^3D), {}^2D) - (3/16)^{1/2}\psi(p, pp(^1P), {}^2D) + (3/4)\psi(p, pp(^1D), {}^2D) - (1/4)\psi(p, pp(^3P), {}^2D),$$

TABLE III. $(d^4vSL]\!]d^3(v'S'L')dSL)$.

d^4	N	d^2 ${}_3{}^2P$	${}_3{}^4P$	${}_1{}^2D$	${}_3{}^2D$	${}_3{}^2F$	${}_3{}^4F$	${}_3{}^2G$	${}_3{}^2H$
${}_0{}^1S$	1	0	0	1	0	0	0	0	0
${}_4{}^1S$	1	0	0	0	1	0	0	0	0
${}_2{}^3P$	$360^{-1/2}$	$-14^{1/2}$	-8	$135^{1/2}$	$-35^{1/2}$	$-56^{1/2}$	$-56^{1/2}$	0	0
${}_4{}^3P$	$90^{-1/2}$	5	$-14^{1/2}$	0	$10^{1/2}$	-5	4	0	0
${}_2{}^1D$	$280^{-1/2}$	$-42^{1/2}$	0	$105^{1/2}$	$45^{1/2}$	$28^{1/2}$	0	$-60^{1/2}$	0
${}_4{}^1D$	$140^{-1/2}$	$42^{1/2}$	0	0	$20^{1/2}$	$63^{1/2}$	0	$15^{1/2}$	0
${}_4{}^3D$	$210^{-1/2}$	$-14^{1/2}$	7	0	$60^{1/2}$	$-21^{1/2}$	$-21^{1/2}$	$45^{1/2}$	0
${}_4{}^5D$	$10^{-1/2}$	0	$3^{1/2}$	0	0	0	$7^{1/2}$	0	0
${}_4{}^1F$	$560^{-1/2}$	$120^{1/2}$	0	0	$200^{1/2}$	$-105^{1/2}$	0	$-3^{1/2}$	$-132^{1/2}$
${}_2{}^3F$	$840^{-1/2}$	4	$-56^{1/2}$	$315^{1/2}$	$15^{1/2}$	$-14^{1/2}$	$224^{1/2}$	$90^{1/2}$	$110^{1/2}$
${}_4{}^3F$	$1680^{-1/2}$	$-200^{1/2}$	$-448^{1/2}$	0	$120^{1/2}$	$-175^{1/2}$	$-112^{1/2}$	$-405^{1/2}$	$220^{1/2}$
${}_2{}^1G$	$504^{-1/2}$	0	0	$189^{1/2}$	-5	$70^{1/2}$	0	$66^{1/2}$	$-154^{1/2}$
${}_4{}^1G$	$1008^{-1/2}$	0	0	0	$88^{1/2}$	$385^{1/2}$	0	$-507^{1/2}$	$-28^{1/2}$
${}_4{}^3G$	$1680^{-1/2}$	0	0	0	$200^{1/2}$	$315^{1/2}$	$-560^{1/2}$	$297^{1/2}$	$308^{1/2}$
${}_4{}^3H$	$60^{-1/2}$	0	0	0	0	$5^{1/2}$	$20^{1/2}$	-3	$26^{1/2}$
${}_4{}^1I$	$10^{-1/2}$	0	0	0	0	0	0	$3^{1/2}$	$7^{1/2}$

$$\psi(p^2(^1D)p\,{}^2D) = (3/16)^{1/2}\psi(p, pp(^3D), {}^2D) - (3/16)^{1/2}\psi(p, pp(^1P), {}^2D) - (1/4)\psi(p, pp(^1D), {}^2D) + (3/4)\psi(p, pp(^3P), {}^2D);$$

since in the development of

$$\Psi(p^3\,{}^2D) = x\psi(p^2(^3P)p\,{}^2D) + y\psi(p^2(^1D)p\,{}^2D)$$

the coefficients of

$$\psi(p, pp(^3D), {}^2D) \quad \text{and} \quad \psi(p, pp(^1P), {}^2D)$$

must vanish, the only possibility, apart from a phase factor, is

$$\Psi(p^3\,{}^2D) = (1/2)^{1/2}\psi(p^2(^3P)p\,{}^2D) - (1/2)^{1/2}\psi(p^2(^1D)p\,{}^2D).$$

The same method may also be extended to the configurations l^n, if the fractional parentages of l^{n-1} are known. In this case

$$\Psi(l^n\alpha SL) = \sum_{\alpha'S'L'} \psi(l^{n-1}(\alpha'S'L')lSL) \cdot (l^{n-1}(\alpha'S'L')lSL]\!]l^n\alpha SL)$$
$$= \sum_{\alpha'S'L'\alpha''S''L''} \psi(l^{n-2}(\alpha''S''L'')l(S'L')lSL) \cdot (l^{n-2}(\alpha''S''L'')lS'L']\!]l^{n-1}\alpha'S'L') \cdot (l^{n-1}(\alpha'S'L')lSL]\!]l^n\alpha SL), \quad (10)$$

and the coefficients of fractional parentage $(l^{n-1}(\alpha'S'L')lSL]\!]l^n\alpha SL)$ must satisfy the equa-

[6] Since we shall mostly consider transformations and operators which are diagonal with respect to M_S and M_L and independent of them, we shall in general neglect these quantum numbers.

TABLE IV. $(d^5vSL[\![d^4(v'S'L')dSL)$.

d^5	N	d^4: ${}_0{}^1S$	${}_4{}^1S$	${}_2{}^3P$	${}_4{}^3P$	${}_2{}^1D$	${}_4{}^1D$	${}_4{}^3D$	${}_4{}^5D$	${}_4{}^1F$	${}_2{}^3F$	${}_4{}^3F$	${}_2{}^1G$	${}_4{}^1G$	${}_4{}^3G$	${}_4{}^3H$	${}_4{}^1I$
${}_5{}^2S$	$5^{-1/2}$	0	0	0	0	0	$-2^{1/2}$	$3^{1/2}$	0	0	0	0	0	0	0	0	0
${}_5{}^6S$	1	0	0	0	0	0	0	0	1	0	0	0	0	0	0	0	0
${}_3{}^2P$	$150^{-1/2}$	0	0	$14^{1/2}$	5	$30^{1/2}$	$15^{1/2}$	$10^{1/2}$	0	$-15^{1/2}$	-4	-5	0	0	0	0	0
${}_3{}^4P$	$300^{-1/2}$	0	0	-8	$14^{1/2}$	0	0	$35^{1/2}$	$-75^{1/2}$	0	$-56^{1/2}$	$56^{1/2}$	0	0	0	0	0
${}_1{}^2D$	$50^{-1/2}$	$6^{1/2}$	0	-3	0	$-5^{1/2}$	0	0	0	0	$-21^{1/2}$	0	-3	0	0	0	0
${}_3{}^2D$	$350^{-1/2}$	0	$-14^{1/2}$	-7	$-14^{1/2}$	$45^{1/2}$	$-10^{1/2}$	$60^{1/2}$	0	$35^{1/2}$	$21^{1/2}$	$-21^{1/2}$	-5	$-11^{1/2}$	$45^{1/2}$	0	0
${}_5{}^2D$	$700^{-1/2}$	0	$-56^{1/2}$	0	$126^{1/2}$	0	$90^{1/2}$	$60^{1/2}$	0	$35^{1/2}$	0	$189^{1/2}$	0	$99^{1/2}$	$45^{1/2}$	0	0
${}_5{}^4D$	$700^{-1/2}$	0	0	0	$126^{1/2}$	0	0	$-135^{1/2}$	$-175^{1/2}$	0	0	$-84^{1/2}$	0	0	$180^{1/2}$	0	0
${}_3{}^2F$	$2800^{-1/2}$	0	0	$448^{1/2}$	$-200^{1/2}$	$-160^{1/2}$	$180^{1/2}$	$120^{1/2}$	0	$105^{1/2}$	$112^{1/2}$	$-175^{1/2}$	-20	$275^{1/2}$	$-405^{1/2}$	$220^{1/2}$	0
${}_5{}^2F$	$2800^{-1/2}$	0	0	0	$360^{1/2}$	0	-10	$600^{1/2}$	0	$-525^{1/2}$	0	$-315^{1/2}$	0	$495^{1/2}$	-3	$-396^{1/2}$	0
${}_3{}^4F$	$700^{-1/2}$	0	0	$-56^{1/2}$	-4	0	0	$-15^{1/2}$	$-175^{1/2}$	0	$224^{1/2}$	$14^{1/2}$	0	0	$-90^{1/2}$	$-110^{1/2}$	0
${}_3{}^2G$	$8400^{-1/2}$	0	0	0	0	$-800^{1/2}$	-10	$600^{1/2}$	0	$-7^{1/2}$	$1680^{1/2}$	$945^{1/2}$	$880^{1/2}$	$845^{1/2}$	$891^{1/2}$	$924^{1/2}$	$-728^{1/2}$
${}_5{}^2G$	$18480^{-1/2}$	0	0	0	0	0	$1452^{1/2}$	$968^{1/2}$	0	$-2541^{1/2}$	0	$4235^{1/2}$	0	$-1215^{1/2}$	$-5577^{1/2}$	$-308^{1/2}$	$-2184^{1/2}$
${}_5{}^4G$	$420^{-1/2}$	0	0	0	0	0	0	5	$-105^{1/2}$	0	0	$-70^{1/2}$	0	0	$-66^{1/2}$	$154^{1/2}$	0
${}_3{}^2H$	$1100^{-1/2}$	0	0	0	0	0	0	0	0	$33^{1/2}$	$-220^{1/2}$	$55^{1/2}$	$220^{1/2}$	$-5^{1/2}$	$-99^{1/2}$	$286^{1/2}$	$182^{1/2}$
${}_5{}^2I$	$550^{-1/2}$	0	0	0	0	0	0	0	0	0	0	0	0	$-45^{1/2}$	$99^{1/2}$	$231^{1/2}$	$-175^{1/2}$

tion system

$$\sum_{\alpha'S'L'} (S''L'', ll(S'''L'''), SL\,|\,S''L''l(S'L')lSL) \cdot (l^{n-2}(\alpha''S''L'')lS'L'[\![l^{n-1}\alpha'S'L') \cdot (l^{n-1}(\alpha'S'L')lSL[\![l^n\alpha SL) = 0 \qquad (S'''+L''' \text{ odd}). \quad (11)$$

The systems (9) and (11) do not fix the phases of the eigenfunctions of the different terms, nor the scheme in the case of more terms of the same kind, but give the fractional parentages in any arbitrary orthonormal scheme; the convenience of a particular choice of the scheme will be considered in §6.

The fractional parentages of the terms of p^3, d^3, d^4, and d^5 calculated with this method are given in Tables I–IV. The phases of the eigenfunctions of p^3 and d^3 are in agreement with those of TAS 4^8 $6j$ and 5^8 6 with the exception of $p^3\ {}^2P$; it must however be remarked that these phases differ from those of Ufford[7] for the terms ${}^4P$, ${}^2F$, ${}^4F$, and ${}^2G$ of d^3.

The coefficients of fractional parentage considered by Goudsmit and Bacher and by Menzel and Goldberg are n times the squares of our coefficients.

It must be pointed out that the matrix $(l^{n-1}(\alpha'S'L')lSL[\![l^n\alpha SL)$ is not an ordinary unitary matrix, but only a rectangular matrix which is a part of a unitary one, since its columns do not exhaust all states of $l^{n-1}l$, but only those which are allowed in l^n; the hermitian conjugate

$$(l^n\alpha SL[\![l^{n-1}(\alpha'S'L')lSL) = [l^{n-1}(\alpha'S'L')lSL[\![l^n\alpha SL]^* \quad (12)$$

does therefore satisfy the relation

$$\sum_{\alpha'S'L'} (l^n\alpha SL[\![l^{n-1}(\alpha'S'L')lSL) \cdot (l^{n-1}(\alpha'S'L')lSL[\![l^n\alpha''SL) = \delta(\alpha\alpha''); \quad (13)$$

but a matrix multiplication in the opposite order has no sense, if the sum is limited to the antisymmetric states of l^n. In calculations with only symmetrical operators we may however, treat formally the matrix $(l^{n-1}(\alpha'S'L')lSL[\![l^n\alpha SL)$ as a common unitary matrix without weakening the general laws of matrix calculations, since symmetrical operators do not connect states of different symmetry and, therefore, the sum over the neglected states vanishes.

§4. FRACTIONAL PARENTAGES IN ALMOST CLOSED SHELLS

We shall determine in this section a relation between the fractional parentages of the terms of an almost closed shell l^{4l+2-n} and those of the terms of l^{n+1}. This relation will not only avoid long numerical calculations, but will also give us the eigenfunctions of the terms of l^{4l+2-n} with the phases fixed by the convention of §6 of II.

[7] C. W. Ufford, Phys. Rev. **44**, 732 (1933).

According to §6 of II two terms of l^n and of l^{4l+2-n} will be called conjugated[8] if their eigenfunctions appear multiplied with each other in the relation

$$\Psi(l^{4l+2}\,{}^1S)=\binom{4l+2}{n}^{-\frac{1}{2}}\sum_{\alpha SLM_SM_L}[(2S+1)\cdot(2L+1)]^{\frac{1}{2}}\Psi_{\mathfrak{L}}(l^n\alpha SLM_SM_L)\cdot\Psi_{\mathfrak{R}}(l^{4l+2-n}\alpha SL-M_S-M_L)\cdot(SSM_S-M_S|SS00)\cdot(LLM_L-M_L|LL00), \quad (14)$$

where $\mathfrak{L}$ denotes the group of the first n electrons of the shell and $\mathfrak{R}$ the group of the remaining $4l+2-n$; owing to (16′)II we have

$$\Psi(l^{4l+2}\,{}^1S)=\binom{4l+2}{n}^{-\frac{1}{2}}\sum_{\alpha SLM_SM_L}(-1)^{S+L-M_S-M_L}\Psi_{\mathfrak{L}}(l^n\alpha SLM_SM_L)\cdot\Psi_{\mathfrak{R}}(l^{4l+2-n}\alpha SL-M_S-M_L). \quad (15a)$$

In the same way, if we consider the group $\mathfrak{L}'$ of the first $n+1$ electrons of the shell and the group $\mathfrak{R}'$ of the remaining $4l+1-n$, we may also write

$$\Psi(l^{4l+2}\,{}^1S)=\binom{4l+2}{n+1}^{-\frac{1}{2}}\sum_{\alpha'S'L'M_S'M_L'}(-1)^{S'+L'-M_S'-M_L}\cdot\Psi_{\mathfrak{L}'}(l^{n+1}\alpha'S'L'M_S'M_L')\cdot\Psi_{\mathfrak{R}'}(l^{4l+1-n}\alpha'S'L'-M_S'-M_L'). \quad (15b)$$

It follows from (10) that

$$\Psi_{\mathfrak{L}'}(l^{n+1}\alpha'S'L'M_S'M_L')=\sum_{\alpha SLM_SM_Lm_sm_l}\Psi_{\mathfrak{L}}(l^n\alpha SLM_SM_L)\cdot\phi_{n+1}(m_sm_l)(S\tfrac{1}{2}M_Sm_s|S\tfrac{1}{2}S'M_S')\cdot(LlM_Lm_l|LlL'M_L')\cdot(l^n(\alpha SL)lS'L'|\!\}l^{n+1}\alpha'S'L'), \quad (16)$$

and

$$\Psi_{\mathfrak{R}}(l^{4l+2-n}\alpha SL-M_S-M_L)=\sum_{\alpha'S'L'M_S'M_L'm_sm_l}\cdot\Psi_{\mathfrak{R}''}(l^{4l+1-n}\alpha'S'L'-M_S'-M_L')\phi_{4l+2}(m_sm_l)\cdot(S'\tfrac{1}{2}-M_S'm_s|S'\tfrac{1}{2}S-M_S)\cdot(L'l-M_L'm_l|L'lL-M_L)\cdot(l^{4l+1-n}(\alpha'S'L')lSL|\!\}l^{4l+2-n}\alpha SL);$$

here $\mathfrak{R}''$ is the group of the electrons $n+1$, $n+2,\cdots 4l+1$. Since $\Psi_{\mathfrak{R}}$ is antisymmetric, the substitution of the electron $4l+2$ by the electron $n+1$ and of the group $\mathfrak{R}''$ by the group $\mathfrak{R}'$ multiplies $\Psi_{\mathfrak{R}}$ by $(-1)^{n+1}$, and then

$$\Psi_{\mathfrak{R}}(l^{4l+2-n}\alpha SL-M_S-M_L)=\sum_{\alpha'S'L'M_S'M_L'm_sm_l}(-1)^{n+1}\phi_{n+1}(m_sm_l)\cdot\Psi_{\mathfrak{R}'}(l^{4l+1-n}\alpha'S'L'-M_S'-M_L')\cdot(S'\tfrac{1}{2}-M_S'm_s|S'\tfrac{1}{2}S-M_S)\cdot(L'l-M_L'm_l|L'lL-M_L)\cdot(l^{4l+1-n}(\alpha'S'L')lSL|\!\}l^{4l+2-n}\alpha SL). \quad (17)$$

If we introduce (17) in (15a) and (16) in (15b), we may equate the coefficients of each product $\Psi_{\mathfrak{L}}\phi_{n+1}\Psi_{\mathfrak{R}'}$ separately, since for different quantum numbers these products are orthogonal, and obtain

$$\binom{4l+2}{n}^{-\frac{1}{2}}(-1)^{S+L-M_S-M_L+n+1}\cdot(S'\tfrac{1}{2}-M_S'm_s|S'\tfrac{1}{2}S-M_S)\cdot(L'l-M_L'm_l|L'lL-M_L)\cdot(l^{4l+1-n}(\alpha'S'L')lSL|\!\}l^{4l+2-n}\alpha SL)$$
$$=\binom{4l+2}{n+1}^{-\frac{1}{2}}(-1)^{S'+L'-M_S'-M_L'}\cdot(S\tfrac{1}{2}M_Sm_s|S\tfrac{1}{2}S'M_S')(LlM_Lm_l|LlL'M_L')\cdot(l^n(\alpha SL)lS'L'|\!\}l^{n+1}\alpha'S'L'). \quad (18)$$

Owing to (16′)II and (19a)II, and to the fact that $n+1$ has the same parity as $2(S'+L')$, we get

$$(l^{4l+1-n}(\alpha'S'L')lSL|\!\}l^{4l+2-n}\alpha SL)=(-1)^{S+S'+L+L'-l-\frac{1}{2}}\cdot\left[\frac{(n+1)(2S'+1)(2L'+1)}{(4l+2-n)(2S+1)(2L+1)}\right]^{\frac{1}{2}}\cdot(l^n(\alpha SL)lS'L'|\!\}l^{n+1}\alpha'S'L'), \quad (19)$$

which is the requested relation.

[8] To the correlation defined in §6 of II we shall reserve the word "conjugation," since other types of correlations between terms of different configurations will be found in §6 of this paper.

TABLE V. $(p^2SL\| U^{(2)} \| p^2S'L')$.

	^1S	^3P	^1D
^1S	0	0	$\frac{2}{3}(3)^{1/2}$
^3P	0	-1	0
^1D	$\frac{2}{3}(3)^{1/2}$	0	$\frac{1}{3}(21)^{1/2}$

TABLE VI. $(p^3SL\| U^{(2)} \| p^3SL')$.

	^4S	^2P	^2D
^4S	0	0	0
^2P	0	0	$-(3)^{1/2}$
^2D	0	$(3)^{1/2}$	0

From (19) and (13) we obtain also

$$\sum_{\alpha SL} (2S+1)(2L+1)(l^{n-1}(\alpha' S'L')lSL]\!]l^n\alpha SL) \cdot (l^n\alpha SL[\![l^{n-1}(\alpha'' S'L')lSL) = [(4l+3-n)/n](2S'+1)(2L'+1)\delta(\alpha'\alpha''). \quad (20)$$

For the determination of the fractional parentages of the terms of l^{2l+2} it must, however, be observed that (19) gives the parentages of $\Psi_{\mathfrak{N}}(l^{2l+2})$ with respect to $\Psi_{\mathfrak{N}}(l^{2l+1})$, and that the eigenfunctions of l^{2l+1} determined by the methods of the preceding section are $\Psi_{\mathfrak{L}}(l^{2l+1})$; since it was shown in §6 of II that the terms of l^{2l+1} split in two classes, according to the two possibilities of (76)II, we must change the sign in the relation (19) if $\Psi_{\mathfrak{N}}(l^{2l+1}\alpha' S'L')$ belongs to the class for which the minus sign holds in (76)II. The classification of the terms from this point of view will be considered in subsection (5) of §6.

§5. MATRIX COMPONENTS OF SYMMETRIC OPERATORS

We are at first interested in the matrix components $(\lambda^{\mathrm{I}}|F|\lambda^{\mathrm{II}})$ of the quantity

$$F = \sum_1^n {}_i f_i, \quad (21)$$

where f_i is an operator which operates on the

TABLE VII. $(d^3vSL\|35U^{(2)}\|d^3v'S'L')$.

	$_3{}^2P$	$_3{}^4P$	$_1{}^2D$	$_3{}^2D$	$_3{}^2F$	$_3{}^4F$	$_3{}^2G$	$_3{}^2H$
$_3{}^2P$	$-2(21)^{1/2}$	0	$-\frac{21}{2}(10)^{1/2}$	$\frac{1}{2}(210)^{1/2}$	$-4(21)^{1/2}$	0	0	0
$_3{}^4P$	0	$7(21)^{1/2}$	0	0	0	$-14(6)^{1/2}$	0	0
$_1{}^2D$	$\frac{21}{2}(10)^{1/2}$	0	$\frac{35}{2}$	$\frac{15}{2}(21)^{1/2}$	$-7(15)^{1/2}$	0	$-15(7)^{1/2}$	0
$_3{}^2D$	$-\frac{1}{2}(210)^{1/2}$	0	$\frac{15}{2}(21)^{1/2}$	$\frac{15}{2}$	$-9(35)^{1/2}$	0	$-5(3)^{1/2}$	0
$_3{}^2F$	$-4(21)^{1/2}$	0	$7(15)^{1/2}$	$9(35)^{1/2}$	$7(6)^{1/2}$	0	$2(210)^{1/2}$	$-(2310)^{1/2}$
$_3{}^4F$	0	$-14(6)^{1/2}$	0	0	0	$-7(6)^{1/2}$	0	0
$_3{}^2G$	0	0	$-15(7)^{1/2}$	$-5(3)^{1/2}$	$-2(210)^{1/2}$	0	$3(22)^{1/2}$	$-(462)^{1/2}$
$_3{}^2H$	0	0	0	0	$-(2310)^{1/2}$	0	$(462)^{1/2}$	$(3003)^{1/2}$

TABLE VIIIa. $(d^4\ {}_v{}^1L\|35U^{(2)}\|d^4\ {}_{v'}{}^1L')$.

	$_0{}^1S$	$_4{}^1S$	$_2{}^1D$	$_4{}^1D$	$_4{}^1F$	$_2{}^1G$	$_4{}^1G$	$_4{}^1I$
$_0{}^1S$	0	0	$7(30)^{1/2}$	0	0	0	0	0
$_4{}^1S$	0	0	$3(70)^{1/2}$	$4(35)^{1/2}$	0	0	0	0
$_2{}^1D$	$7(30)^{1/2}$	$3(70)^{1/2}$	-5	$-30(2)^{1/2}$	0	$4(5)^{1/2}$	$8(55)^{1/2}$	0
$_4{}^1D$	0	$4(35)^{1/2}$	$-30(2)^{1/2}$	-15	$10(14)^{1/2}$	$10(10)^{1/2}$	$2(110)^{1/2}$	0
$_4{}^1F$	0	0	0	$-10(14)^{1/2}$	$\frac{35}{2}(6)^{1/2}$	$-7(70)^{1/2}$	$-\frac{1}{2}(770)^{1/2}$	0
$_2{}^1G$	0	0	$4(5)^{1/2}$	$10(10)^{1/2}$	$7(70)^{1/2}$	$5(22)^{1/2}$	$5(2)^{1/2}$	$-2(455)^{1/2}$
$_4{}^1G$	0	0	$8(55)^{1/2}$	$2(110)^{1/2}$	$\frac{1}{2}(770)^{1/2}$	$5(2)^{1/2}$	$-\frac{125}{22}(22)^{1/2}$	$-\frac{8}{11}(5005)^{1/2}$
$_4{}^1I$	0	0	0	0	0	$-2(455)^{1/2}$	$-\frac{8}{11}(5005)^{1/2}$	$\frac{35}{11}(143)^{1/2}$

TABLE VIIIb. $(d^4vSL\|35U^{(2)}\|d^4v'S'L')$ for $S=1, 2$.

	$_2{}^3P$	$_4{}^3P$	$_4{}^3D$	$_4{}^5D$	$_2{}^3F$	$_4{}^3F$	$_4{}^3G$	$_4{}^3H$
$_2{}^3P$	$-\frac{7}{3}(21)^{1/2}$	$\frac{14}{3}(6)^{1/2}$	$-\frac{28}{3}(15)^{1/2}$	0	$\frac{14}{3}(6)^{1/2}$	$\frac{28}{3}(6)^{1/2}$	0	0
$_4{}^3P$	$\frac{14}{3}(6)^{1/2}$	$\frac{19}{3}(21)^{1/2}$	$-\frac{4}{3}(210)^{1/2}$	0	$\frac{22}{3}(21)^{1/2}$	$\frac{8}{3}(21)^{1/2}$	0	0
$_4{}^3D$	$\frac{28}{3}(15)^{1/2}$	$\frac{4}{3}(210)^{1/2}$	5	0	$-4(35)^{1/2}$	$4(35)^{1/2}$	$20(3)^{1/2}$	0
$_4{}^5D$	0	0	0	-35	0	0	0	0
$_2{}^3F$	$\frac{14}{3}(6)^{1/2}$	$\frac{22}{3}(21)^{1/2}$	$4(35)^{1/2}$	0	$\frac{7}{3}(6)^{1/2}$	$\frac{49}{3}(6)^{1/2}$	$-3(210)^{1/2}$	$\frac{2}{3}(2310)^{1/2}$
$_4{}^3F$	$\frac{28}{3}(6)^{1/2}$	$\frac{8}{3}(21)^{1/2}$	$-4(35)^{1/2}$	0	$\frac{49}{3}(6)^{1/2}$	$\frac{77}{6}(6)^{1/2}$	$-\frac{1}{2}(210)^{1/2}$	$\frac{2}{3}(2310)^{1/2}$
$_4{}^3G$	0	0	$20(3)^{1/2}$	0	$3(210)^{1/2}$	$\frac{1}{2}(210)^{1/2}$	$-\frac{3}{2}(22)^{1/2}$	$-2(462)^{1/2}$
$_4{}^3H$	0	0	0	0	$\frac{2}{3}(2310)^{1/2}$	$\frac{2}{3}(2310)^{1/2}$	$2(462)^{1/2}$	$\frac{1}{3}(3003)^{1/2}$

TABLE IXa. $(d^5\ {}_v{}^2L\|35\,U^{(2)}\|d^5\ {}_{v'}{}^2L')$.

	${}_5{}^2S$	${}_3{}^2P$	${}_1{}^2D$	${}_3{}^2D$	${}_5{}^2D$	${}_3{}^2F$	${}_5{}^2F$	${}_3{}^2G$	${}_5{}^2G$	${}_3{}^2H$	${}_5{}^2I$
${}_5{}^2S$	0	0	0	$4(70)^{\frac{1}{2}}$	0	0	0	0	0	0	0
${}_3{}^2P$	0	0	$-7(30)^{\frac{1}{2}}$	0	$5(105)^{\frac{1}{2}}$	0	$4(105)^{\frac{1}{2}}$	0	0	0	0
${}_1{}^2D$	0	$7(30)^{\frac{1}{2}}$	0	$15(7)^{\frac{1}{2}}$	0	$-14(5)^{\frac{1}{2}}$	0	$-10(21)^{\frac{1}{2}}$	0	0	0
${}_3{}^2D$	$4(70)^{\frac{1}{2}}$	0	$15(7)^{\frac{1}{2}}$	0	$5(2)^{\frac{1}{2}}$	0	$10(7)^{\frac{1}{2}}$	0	$-6(55)^{\frac{1}{2}}$	0	0
${}_5{}^2D$	0	$-5(105)^{\frac{1}{2}}$	0	$5(2)^{\frac{1}{2}}$	0	0	0	$20(6)^{\frac{1}{2}}$	0	0	0
${}_3{}^2F$	0	0	$14(5)^{\frac{1}{2}}$	0	0	0	$-7(30)^{\frac{1}{2}}$	0	0	0	0
${}_5{}^2F$	0	$4(105)^{\frac{1}{2}}$	0	$-10(7)^{\frac{1}{2}}$	0	$-7(30)^{\frac{1}{2}}$	0	$4(42)^{\frac{1}{2}}$	0	$-2(462)^{\frac{1}{2}}$	0
${}_3{}^2G$	0	0	$-10(21)^{\frac{1}{2}}$	0	$20(6)^{\frac{1}{2}}$	0	$-4(42)^{\frac{1}{2}}$	0	$9(30)^{\frac{1}{2}}$	0	$4(273)^{\frac{1}{2}}$
${}_5{}^2G$	0	0	0	$-6(55)^{\frac{1}{2}}$	0	0	0	$9(30)^{\frac{1}{2}}$	0	$6(70)^{\frac{1}{2}}$	0
${}_3{}^2H$	0	0	0	0	0	0	$-2(462)^{\frac{1}{2}}$	0	$-6(70)^{\frac{1}{2}}$	0	$7(13)^{\frac{1}{2}}$
${}_5{}^2I$	0	0	0	0	0	0	0	$4(273)^{\frac{1}{2}}$	0	$-7(13)^{\frac{1}{2}}$	0

TABLE IXb. $(d^5\ {}_v{}^4L\|35\,U^{(2)}\|d^5\ {}_{v'}{}^4L')$.

	${}_3{}^4P$	${}_5{}^4D$	${}_3{}^4F$	${}_5{}^4G$
${}_3{}^4P$	0	$7(15)^{\frac{1}{2}}$	0	0
${}_5{}^4D$	$-7(15)^{\frac{1}{2}}$	0	$8(35)^{\frac{1}{2}}$	0
${}_3{}^4F$	0	$-8(35)^{\frac{1}{2}}$	0	$15(14)^{\frac{1}{2}}$
${}_5{}^4G$	0	0	$-15(14)^{\frac{1}{2}}$	0

TABLE X. $(p^2SL\|6^{\frac{1}{2}}V^{(11)}\|p^2S'L')$.

	${}^1S$	${}^3P$	${}^1D$
${}^1S$	0	$-\frac{1}{2}(30)^{\frac{1}{2}}$	0
${}^3P$	$-\frac{1}{2}(30)^{\frac{1}{2}}$	3	$(6)^{\frac{1}{2}}$
${}^1D$	0	$(6)^{\frac{1}{2}}$	0

TABLE XI. $(p^3SL\|6^{\frac{1}{2}}V^{(11)}\|p^3S'L')$.

	${}^4S$	${}^2P$	${}^2D$
${}^4S$	0	$2(3)^{\frac{1}{2}}$	0
${}^2P$	$2(3)^{\frac{1}{2}}$	0	$(15)^{\frac{1}{2}}$
${}^2D$	0	$-(15)^{\frac{1}{2}}$	0

electron i, and λ^{I} and λ^{II} are states of the configurations I and II; owing to the antisymmetry of $\Psi(\lambda^{\mathrm{I}})$ and $\Psi(\lambda^{\mathrm{II}})$ we have

$$(\lambda^{\mathrm{I}}|F|\lambda^{\mathrm{II}})=\sum_1^n{}_i\int\bar{\Psi}(\lambda^{\mathrm{I}})f_i\Psi(\lambda^{\mathrm{II}})d\tau$$
$$=n\int\bar{\Psi}(\lambda^{\mathrm{I}})f_i\Psi(\lambda^{\mathrm{II}})d\tau. \quad (22)$$

If $\mathrm{I}=\mathrm{II}=l^n$, putting $i=n$ and assuming for Ψ the expression (10), we obtain

$$(l^n\alpha SLM_SM_L|F|l^n\alpha'S'L'M_S'M_L')$$
$$=n\sum_{\alpha_1S_1L_1}(l^n\alpha SL\{|l^{n-1}(\alpha_1S_1L_1)lSL)$$
$$\cdot(S_1L_1l_nSLM_SM_L|f_n|S_1L_1l_nS'L'M_S'M_L')$$
$$\cdot(l^{n-1}(\alpha_1S_1L_1)lS'L'|\}l^n\alpha'S'L'), \quad (23)$$

where $(S_1L_1l_nSLM_SM_L|f_n|S_1L_1l_nS'L'M_S'M_L')$ may now be calculated with the ordinary matrix methods of Chapter III of TAS and of II.

As application of this formula we calculated the matrix components of the tensors $U^{(2)}$ and $V^{(11)}$, defined by (102)II, for the configurations p^2, p^3, d^3, d^4, and d^5; the results are given in Tables V–XIV. For d^2 the matrices were already given by (103)II; it must, however, be noted that an error occurred in the final form of the manuscript, and all the elements of (103c)II and (103d)II must be multiplied by $(3/2)^{\frac{1}{2}}$.

From the elements of $V^{(11)}$ the matrix components of the spin-orbit interaction may easily be obtained: it follows in effect from the relation

$$\mathbf{l}=[l(l+1)(2l+1)]^{\frac{1}{2}}\mathbf{u}^{(1)} \quad (24)$$

and from (38)II and (102)II that

$$(l^n\alpha SLJM|\textstyle\sum_i(\mathbf{s}_i\cdot\mathbf{l}_i)|l^n\alpha'S'L'JM)$$
$$=(-1)^{S+L'-J}[l(l+1)(2l+1)]^{\frac{1}{2}}$$
$$\cdot(l^n\alpha SL\|V^{(11)}\|l^n\alpha'S'L')W(SLS'L';J1). \quad (25)$$

If in (22) $\mathrm{I}=l^n$ and $\mathrm{II}=l^{n-1}l'$, the terms of II are characterized by S and L and by the quantum numbers of the parent ion l^{n-1}; $\Psi(\lambda^{\mathrm{II}})$ has in this case the expression

$$\Psi(l^{n-1}(\alpha_1S_1L_1)l'S'L'M_S'M_L')$$
$$=(1/n)^{\frac{1}{2}}\sum_1^n{}_i(-1)^{P_i}$$
$$\cdot\psi(l^{n-1}(\alpha_1S_1L_1)l_i'S'L'M_S'M_L'), \quad (26)$$

where in the right side we consider the group l^{n-1} as composed by the electrons $1, 2, \cdots, i-1, i+1, \cdots, n$, and P_i is the parity of the permutation which exchanges i with n. Introducing

TABLE XII. $(d^3vSL\|30^{1/2}V^{(11)}\|d^3v'S'L')$.

	$_3{}^2P$	$_3{}^4P$	$_1{}^2D$	$_3{}^2D$	$_3{}^2F$	$_3{}^4F$	$_3{}^2G$	$_3{}^2H$
$_3{}^2P$	2	$-2(14)^{1/2}$	$-\frac{1}{2}(42)^{1/2}$	$\frac{9}{2}(2)^{1/2}$	0	0	0	0
$_3{}^4P$	$2(14)^{1/2}$	$(10)^{1/2}$	$-4(3)^{1/2}$	0	0	0	0	0
$_1{}^2D$	$\frac{1}{2}(42)^{1/2}$	$-4(3)^{1/2}$	$\frac{3}{2}(5)^{1/2}$	$-\frac{1}{2}(105)^{1/2}$	$(42)^{1/2}$	$-(42)^{1/2}$	0	0
$_3{}^2D$	$-\frac{9}{2}(2)^{1/2}$	0	$-\frac{1}{2}(105)^{1/2}$	$-\frac{1}{2}(5)^{1/2}$	$(2)^{1/2}$	$5(2)^{1/2}$	0	0
$_3{}^2F$	0	0	$-(42)^{1/2}$	$-(2)^{1/2}$	$-\frac{1}{2}(14)^{1/2}$	$-(14)^{1/2}$	$\frac{3}{2}-(10)^{1/2}$	0
$_3{}^4F$	0	0	$-(42)^{1/2}$	$5(2)^{1/2}$	$(14)^{1/2}$	$2(35)^{1/2}$	$-3(10)^{1/2}$	0
$_3{}^2G$	0	0	0	0	$\frac{3}{2}(10)^{1/2}$	$-3(10)^{1/2}$	$\frac{9}{10}(30)^{1/2}$	$\frac{6}{5}(55)^{1/2}$
$_3{}^2H$	0	0	0	0	0	0	$-\frac{6}{5}(55)^{1/2}$	$\frac{3}{5}(55)^{1/2}$

TABLE XIII. $(d^4vSL\|30^{1/2}V^{(11)}\|d^4v'S'L')$.

	$_0{}^1S$	$_4{}^1S$	$_2{}^3P$	$_4{}^3P$	$_2{}^1D$	$_4{}^1D$	$_4{}^3D$	$_4{}^5D$	$_4{}^1F$	$_2{}^3F$	$_4{}^3F$	$_2{}^1G$	$_4{}^1G$	$_4{}^3G$	$_4{}^3H$	$_4{}^1I$
$_0{}^1S$	0	0	$3(3)^{1/2}$	0	0	0	0	0	0	0	0	0	0	0	0	0
$_4{}^1S$	0	0	$-(7)^{1/2}$	$2(2)^{1/2}$	0	0	0	0	0	0	0	0	0	0	0	0
$_2{}^3P$	$3(3)^{1/2}$	$-(7)^{1/2}$	1	$-2(14)^{1/2}$	$-\frac{1}{2}(14)^{1/2}$	$2(7)^{1/2}$	0	$-4(5)^{1/2}$	0	0	0	0	0	0	0	0
$_4{}^3P$	0	$2(2)^{1/2}$	$-2(14)^{1/2}$	2	2	$\frac{1}{2}(2)^{1/2}$	$\frac{9}{2}(2)^{1/2}$	$\frac{1}{2}(70)^{1/2}$	0	0	0	0	0	0	0	0
$_2{}^1D$	0	0	$-\frac{1}{2}(14)^{1/2}$	2	0	0	$2(10)^{1/2}$	0	0	2	4	0	0	0	0	0
$_4{}^1D$	0	0	$2(7)^{1/2}$	$\frac{1}{2}(2)^{1/2}$	0	0	$-(5)^{1/2}$	0	0	$(2)^{1/2}$	$-4(2)^{1/2}$	0	0	0	0	0
$_4{}^3D$	0	0	0	$-\frac{9}{2}(2)^{1/2}$	$-2(10)^{1/2}$	$(5)^{1/2}$	$-\frac{1}{2}(5)^{1/2}$	$\frac{5}{2}(7)^{1/2}$	$-2(5)^{1/2}$	$5(2)^{1/2}$	$(2)^{1/2}$	0	0	0	0	0
$_4{}^5D$	0	0	$-4(5)^{1/2}$	$\frac{1}{2}(70)^{1/2}$	0	0	$-\frac{5}{2}(7)^{1/2}$	$\frac{15}{2}$	0	$-(70)^{1/2}$	$(70)^{1/2}$	0	0	0	0	0
$_4{}^1F$	0	0	0	0	0	0	$-2(5)^{1/2}$	0	0	$(35)^{1/2}$	$\frac{1}{2}(35)^{1/2}$	0	0	$-\frac{9}{2}$	0	0
$_2{}^3F$	0	0	0	0	2	$(2)^{1/2}$	$-5(2)^{1/2}$	$-(70)^{1/2}$	$-(35)^{1/2}$	$(14)^{1/2}$	$-(14)^{1/2}$	$-(3)^{1/2}$	$(33)^{1/2}$	$3(10)^{1/2}$	0	0
$_4{}^3F$	0	0	0	0	4	$-4(2)^{1/2}$	$-(2)^{1/2}$	$(70)^{1/2}$	$-\frac{1}{2}(35)^{1/2}$	$-(14)^{1/2}$	$-\frac{1}{2}(14)^{1/2}$	$5(3)^{1/2}$	$-\frac{1}{2}(33)^{1/2}$	$-\frac{3}{2}(10)^{1/2}$	0	0
$_2{}^1G$	0	0	0	0	0	0	0	0	0	$-(3)^{1/2}$	$5(3)^{1/2}$	0	0	3	$-(66)^{1/2}$	0
$_4{}^1G$	0	0	0	0	0	0	0	0	0	$(33)^{1/2}$	$-\frac{1}{2}(33)^{1/2}$	0	0	$\frac{3}{2}(11)^{1/2}$	$2(6)^{1/2}$	0
$_4{}^3G$	0	0	0	0	0	0	0	0	$-\frac{9}{2}$	$-3(10)^{1/2}$	$\frac{3}{2}(10)^{1/2}$	-3	$-\frac{3}{2}(11)^{1/2}$	$\frac{9}{10}(30)^{1/2}$	$\frac{6}{5}(55)^{1/2}$	0
$_4{}^3H$	0	0	0	0	0	0	0	0	0	0	0	$-(66)^{1/2}$	$2(6)^{1/2}$	$-\frac{6}{5}(55)^{1/2}$	$\frac{3}{5}(55)^{1/2}$	$-\frac{3}{2}(26)$
$_4{}^1I$	0	0	0	0	0	0	0	0	0	0	0	0	0	0	$-\frac{3}{2}(26)^{1/2}$	0

(10) and (26) in (22) and putting $i=n$, we have

$$(l^n\alpha SLM_SM_L|F|l^{n-1}(\alpha_1S_1L_1)l'S'L'M_S'M_L')$$
$$=n^{1/2}(l^n\alpha SL\{\!|l^{n-1}(\alpha_1S_1L_1)lSL)$$
$$\cdot(S_1L_1l_nSLM_SM_L|f_n|S_1L_1l_n'S'L'M_S'M_L'). \quad (27)$$

This formula, which is the extension of TAS 6^817, gives a rigorous demonstration to the method of Menzel and Goldberg[9] and also fixes for such transitions the phases of the matrix components, which are necessary for transformations to other types of vector coupling (TAS, p. 252).

If in (22) $\text{I}=l^{n-p}l'^p$ and $\text{II}=l^{n-p-1}l'^{p+1}$, the terms of each configuration are characterized by S and L and by the quantum numbers of the groups of equivalent electrons; in the same way as for the precedent case we obtain

$$(l^{n-p}(\alpha_1S_1L_1),\, l'^p(\alpha_2S_2L_2),$$
$$SLM_SM_L|F|l^{n-p-1}(\alpha_1'S_1'L_1'),$$
$$l'^{p+1}(\alpha_2'S_2'L_2'),\, S'L'M_S'M_L')$$
$$=[(n-p)(p+1)]^{1/2}\sum_{S_3L_3}$$
$$(l^{n-p}\alpha_1S_1L_1\cdot\{\!|l^{n-p-1}(\alpha_1'S_1'L_1')lS_1L_1)$$
$$\cdot(S_1'L_1'l_{n-p}(S_1L_1),\, S_2L_2,$$
$$SLM_SM_L|f_{n-p}|S_1'L_1'l_{n-p}'(S_3L_3),$$
$$S_2L_2,\, S'L'M_S'M_L')$$
$$\cdot(S_1'L_1'l'(S_3L_3),$$
$$S_2L_2,\, S'L'|S_1'L_1',\, l'S_2L_2(S_2'L_2'),\, S'L')$$
$$\cdot(l',\, l'^p(\alpha_2S_2I_2),\, S_2'L_2'|\!\}l'^{p+1}\alpha_2'S_2'L_2'). \quad (28)$$

In connection with this result it must be observed that $(l,\, l^{n-1}(\alpha'S'L'),\, SL|\!\}l^n\alpha SL)$ is not $(l^{n-1}(\alpha'S'L')lSL|\!\}l^n\alpha SL)$, but it is easy to see that the two coefficients are connected by the relation

$$(l,\, l^{n-1}(\alpha'S'L'),\, SL|\!\}l^n\alpha SL)$$
$$=(-)^{S+L+S'+L'-l-\frac{1}{2}}$$
$$\cdot(l^{n-1}(\alpha'S'L')lSL|\!\}l^n\alpha SL). \quad (29)$$

It is unnecessary to consider the matrix elements of F for transitions between more com-

[9] D. H. Menzel and L. Goldberg, Phys. Rev. **47**, 424 (1935) and reference 5.

TABLE XIV. $(d^5vSL\|30^{\frac{1}{2}}V^{(11)}\|d^5v'S'L')$.

	${}_5^2S$	${}_5^6S$	${}_3^2P$	${}_3^4P$	${}_1^2D$	${}_3^2D$	${}_5^2D$	${}_5^4D$	${}_3^2F$	${}_5^2F$	${}_3^4F$	${}_3^2G$	${}_5^2G$	${}_5^4G$	${}_3^2H$	${}_5^2I$
${}_5^2S$	0	0	-4	$(14)^{\frac{1}{2}}$	0	0	0	0	0	0	0	0	0	0	0	0
${}_5^6S$	0	0	0	$3(10)^{\frac{1}{2}}$	0	0	0	0	0	0	0	0	0	0	0	0
${}_3^2P$	4	0	0	0	$-(14)^{\frac{1}{2}}$	0	1	$2(2)^{\frac{1}{2}}$	0	0	0	0	0	0	0	0
${}_3^4P$	$(14)^{\frac{1}{2}}$	$3(10)^{\frac{1}{2}}$	0	0	-8	0	$-2(14)^{\frac{1}{2}}$	$-(70)^{\frac{1}{2}}$	0	0	0	0	0	0	0	0
${}_1^2D$	0	0	$(14)^{\frac{1}{2}}$	-8	0	$-(35)^{\frac{1}{2}}$	0	0	$2(14)^{\frac{1}{2}}$	0	$-2(14)^{\frac{1}{2}}$	0	0	0	0	0
${}_3^2D$	0	0	0	0	$-(35)^{\frac{1}{2}}$	0	$(10)^{\frac{1}{2}}$	$-4(5)^{\frac{1}{2}}$	0	$-2(10)^{\frac{1}{2}}$	0	0	0	0	0	0
${}_5^2D$	0	0	-1	$-2(14)^{\frac{1}{2}}$	0	$(10)^{\frac{1}{2}}$	0	0	8	0	-2	0	0	0	0	0
${}_5^4D$	0	0	$2(2)^{\frac{1}{2}}$	$(70)^{\frac{1}{2}}$	0	$4(5)^{\frac{1}{2}}$	0	0	$4(2)^{\frac{1}{2}}$	0	$-4(5)^{\frac{1}{2}}$	0	0	0	0	0
${}_3^2F$	0	0	0	0	$-2(14)^{\frac{1}{2}}$	0	-8	$4(2)^{\frac{1}{2}}$	0	$-\frac{1}{2}(70)^{\frac{1}{2}}$	0	0	$-\frac{1}{2}(66)^{\frac{1}{2}}$	$5(6)^{\frac{1}{2}}$	0	0
${}_5^2F$	0	0	0	0	0	$2(10)^{\frac{1}{2}}$	0	0	$-\frac{1}{2}(70)^{\frac{1}{2}}$	0	$-(70)^{\frac{1}{2}}$	$\frac{9}{2}(2)^{\frac{1}{2}}$	0	0	0	0
${}_3^4F$	0	0	0	0	$-2(14)^{\frac{1}{2}}$	0	-2	$4(5)^{\frac{1}{2}}$	0	$(70)^{\frac{1}{2}}$	0	0	$-(66)^{\frac{1}{2}}$	$(15)^{\frac{1}{2}}$	0	0
${}_3^2G$	0	0	0	0	0	0	0	0	0	$-\frac{3}{2}(2)^{\frac{1}{2}}$	0	0	$-\frac{3}{2}(22)^{\frac{1}{2}}$	$-3(2)^{\frac{1}{2}}$	0	0
${}_5^2G$	0	0	0	0	0	0	0	0	$\frac{1}{2}(66)^{\frac{1}{2}}$	0	$-(66)^{\frac{1}{2}}$	$-\frac{3}{2}(22)^{\frac{1}{2}}$	0	0	$-4(3)^{\frac{1}{2}}$	0
${}_5^4G$	0	0	0	0	0	0	0	0	$5(6)^{\frac{1}{2}}$	0	$2(15)^{\frac{1}{2}}$	$3(2)^{\frac{1}{2}}$	0	0	$-2(33)^{\frac{1}{2}}$	0
${}_3^2H$	0	0	0	0	0	0	0	0	0	0	0	0	$4(3)^{\frac{1}{2}}$	$-2(33)^{\frac{1}{2}}$	0	$-3(13)^{\frac{1}{2}}$
${}_5^2I$	0	0	0	0	0	0	0	0	0	0	0	0	0	0	$3(13)^{\frac{1}{2}}$	0

plicated configurations, since all other cases may be reduced to these three by means of TAS I[8]16.

The calculation by the same method of the matrix components of the scalar operator

$$G=\sum_{i<j} g_{ij} \tag{30}$$

needs in some cases the knowledge of $\Psi(l^n\alpha SL)$ as linear combination of the eigenfunctions of $l^{n-2}l^2$:

$$\Psi(l^n\alpha SL)=\sum_{\alpha_1S_1L_1S_2L_2}\psi(l^{n-2}(\alpha_1S_1L_1),\, l^2(S_2L_2),\, SL)(l^{n-2}(\alpha_1S_1L_1),\, l^2(S_2L_2),\, SL]\!]l^n\alpha SL)\,; \tag{31}$$

the coefficients of this expression are given by the formula

$$\begin{aligned}(l^{n-2}(\alpha_1S_1L_1),\, l^2(S_2L_2),\, SL]\!]l^nSL) &= \sum_{\alpha'S'L'}(S_1L_1,\, l^2(S_2L_2),\, SL]\!]S_1L_1l(S'L')lSL)\\ &\quad\cdot(l^{n-2}(\alpha_1S_1L_1)lS'L']\!]l^{n-1}\alpha'S'L')\\ &\quad\cdot(l^{n-1}(\alpha'S'L')lSL]\!]l^n\alpha SL).\end{aligned} \tag{32}$$

The following results are easily derived:

$$\begin{aligned}(l^n\alpha SL|G|l^n\alpha'SL) &= \tfrac{1}{2}n(n-1)\sum_{\alpha_1S_1L_1S_2L_2}(l^n\alpha SL[\![l^{n-2}(\alpha_1S_1L_1),\\ &\quad\cdot l^2(S_2L_2),\, SL)(l^2S_2L_2|g|l^2S_2L_2)\\ &\quad\cdot(l^{n-2}(\alpha_1S_1L_1),\, l^2(S_2L_2),\, SL]\!]l^n\alpha'SL),\end{aligned} \tag{33a}$$

$$\begin{aligned}(l^n\alpha SL|G|l^{n-1}(\alpha'S'L')l'SL) &= (n-1)n^{\frac{1}{2}}\sum_{\alpha_1S_1L_1S_2L_2}(l^n\alpha SL[\![l^{n-2}(\alpha_1S_1L_1),\\ &\quad l^2(S_2L_2),\, SL)(l_il_jS_2L_2|g_{ij}|l_il_j'S_2L_2)\\ &\quad\cdot(S_1L_1,\, ll'(S_2L_2),\, SL|S_1L_1l(S'L')l'SL)\\ &\quad\cdot(l^{n-2}(\alpha_1S_1L_1)lS'L']\!]l^{n-1}\alpha'S'L'),\end{aligned} \tag{33b}$$

$$\begin{aligned}(l^n\alpha SL|G|l^{n-2}(\alpha_1S_1L_1),\, l'^2(S_2L_2),\, SL) &= [n(n-1)/2]^{\frac{1}{2}}(l^n\alpha SL[\![l^{n-2}(\alpha_1S_1L_1),\\ &\quad l^2(S_2L_2),\, SL)(l^2S_2L_2|g|l'^2S_2L_2).\end{aligned} \tag{33c}$$

Since the actual application of (32) needs generally very long calculations, the formulas (33) are of practical use only in a few particular cases; in other cases it is simpler to express g_{ij} as a sum of scalar products of tensors and to reduce the problem to the calculation of tensors of the type F. Applications of both methods will be shown in the next sections.

§6. THE STRUCTURE OF THE CONFIGUATIONS l^n

In this section we shall classify the terms of the configuration l^n according to the eigenvalues of

$$Q=\sum_{i<j} q_{ij}, \tag{34}$$

where q_{ij} is a scalar operator which operates on the two equivalent electrons i and j and is defined by the relation

$$(l^2LM|q_{ij}|l^2LM)=(2l+1)\delta(L,\,0). \tag{35}$$

It will be shown that to every term of l^n with non-vanishing Q a term of the same kind corresponds in l^{n-2}, and this fact will allow us to assign to each term a "seniority number" according to the value of n for which the term appeared for the first time. Some useful relation between the fractional parentages of corresponding terms will be obtained and it will also be shown that the classification of the terms of l^{2l+1} according to the two possibilities of (76)II depends only on the seniority of the term.

(1) The Eigenvalues of Q

It follows from (42)II and (40a)II that

$$\sum_0^{2l}{}_r (2r+1) W(llll;0r) W(llll;Lr) = \delta(L,0); \quad (36)$$

expressing $W(llll;0r)$ by (36′)II and using also (38)II and (58)II we get for q_{ij} the expression

$$q_{ij} = \sum_0^{2l}{}_r (-1)^r (2r+1)(\mathbf{u}_i^{(r)} \cdot \mathbf{u}_j^{(r)}). \quad (37)$$

Since $\mathbf{u}^{(0)}$ is a scalar and, owing to (33)II,

$$\mathbf{u}^{(r)2} = (2l+1)^{-1}, \quad (38)$$

we have also

$$q_{ij} = (2l+1)^{-1} + \sum_1^{2l}{}_r (-1)^r (2r+1)(\mathbf{u}_i^{(r)} \cdot \mathbf{u}_j^{(r)}) \quad (37')$$

and

$$Q = \tfrac{1}{2} n(n-2l-1)(2l+1)^{-1} + \tfrac{1}{2} \sum_1^{2l}{}_r (-1)^r (2r+1) U^{(r)2}. \quad (39)$$

We shall henceforth consider only schemes for which also Q is diagonal, i.e., schemes for which

$$(l^n\alpha SL|Q|l^n\alpha' SL) = Q(l^n\alpha SL)\delta(\alpha\alpha'). \quad (40)$$

It follows from (39) and (74)II that if Q is diagonal in a given scheme of l^n, it is also diagonal in the conjugate scheme of l^{4l+2-n}, and that

$$Q(l^{4l+2-n}\alpha SL) = Q(l^n\alpha SL) + 2l + 1 - n. \quad (41)$$

In order to calculate the possible values of $Q(l^n\alpha SL)$, we express it by means of (33a) and (35):

$$Q(l^n\alpha SL)\delta(\alpha\alpha') = \tfrac{1}{2} n(n-1)(2l+1) \cdot \sum_{\beta'} (l^n\alpha SL\{|l^{n-2}(\beta' SL), l^2(^1S), SL) \cdot (l^{n-2}(\beta' SL), l^2(^1S), SL|\}l^n\alpha' SL). \quad (42)$$

Multiplying the two sides by

$$(l^n\alpha' SL\{|l^{n-2}(\beta SL), l^2(^1S), SL)$$

and adding with respect to α' we have

$$Q(l^n\alpha SL)(l^n\alpha SL\{|l^{n-2}(\beta SL), l^2(^1S), SL) = \sum_{\alpha'\beta'} (l^n\alpha SL\{|l^{n-2}(\beta' SL), l^2(^1S), SL) \cdot (l^{n-2}(\beta' SL), l^2(^1S), SL|\}l^n\alpha' SL) \cdot (l^n\alpha' SL\{|l^{n-2}(\beta SL), l^2(^1S), SL);$$

the summation with respect to α' may be made by means of (42) after transforming the last two factors by the relation

$$(l^{n-2}(\beta SL), l^2(^1S), SL|\}l^n\alpha SL) = \left[\frac{(4l+3-n)(4l+4-n)}{n(n-1)}\right]^{\frac{1}{2}} \cdot (l^{4l+2-n}(\alpha SL), l^2(^1S), SL|\}l^{4l+4-n}\beta SL), \quad (43)$$

which is analogous to (19) and may be obtained in the same way; we get

$$Q(l^n\alpha SL)(l^n\alpha SL\{|L^{n-2}(\beta SL), l^2(^1S), SL) = (l^n\alpha SL\{|l^{n-2}(\beta SL), l^2(^1S), SL) Q(l^{4l+4-n}\beta SL). \quad (44)$$

Owing to (41) $(l^n\alpha SL\{|l^{n-2}(\beta SL), l^2(^1S), SL)$ may be different from zero only if

$$Q(l^n\alpha SL) = Q(l^{n-2}\beta SL) + 2l + 3 - n, \quad (45)$$

and according to (42) the only non-vanishing values of $Q(l^n\alpha SL)$ are those which are connected to a $Q(l^{n-2}\beta SL)$ by (45).

(2) The "Seniority Number"

Putting for $Q \neq 0$

$$v_{\alpha\beta}(QSL) = (l^n\alpha SL\{|l^{n-2}(\beta SL), l^2(^1S), SL), \quad (46)$$

where α may assume all values for which $Q(l^n\alpha SL) = Q$, and β all values which satisfy (45), we have from (42) that

$$\tfrac{1}{2} n(n-1)(2l+1) \cdot \sum_\beta v_{\alpha\beta}(QSL) v^\dagger_{\beta\alpha'}(QSL) = Q\delta(\alpha\alpha'), \quad (47)$$

and also from (43) and again (42) that

$$\tfrac{1}{2} n(n-1)(2l+1) \cdot \sum_\alpha v^\dagger_{\beta\alpha}(QSL) v_{\alpha\beta'}(QSL) = Q\delta(\beta\beta'); \quad (47')$$

it follows that for two given values of Q which are connected by (45) the number of independent states of given S and L is the same in l^n and l^{n-2}, and that the matrix

$$u_{\alpha\beta}(QSL)=[\tfrac{1}{2}n(n-1)(2l+1)/Q]^{\frac{1}{2}}v_{\alpha\beta}(QSL) \quad (48)$$

is a unitary one. If we apply to the eigenfunctions of l^n the transformation u, and consider the states with eigenfunctions

$$\Psi(l^n\beta SL)=\sum_\alpha \Psi(l^n\alpha SL)u_{\alpha\beta},$$

we obtain

$$(l^{n-2}(\beta' SL), l^2(^1S), SL]\!]l^n\beta SL)$$
$$=[Q(l^n\beta SL)]^{\frac{1}{2}}[\tfrac{1}{2}n(n-1)(2l+1)]^{-\frac{1}{2}}\delta(\beta\beta'); \quad (49)$$

i.e., it is possible to find a scheme of l^n in which not only Q is diagonal, but also each term of l^n with $Q\neq 0$ corresponds to a well-defined term of l^{n-2} whose Q is connected to $Q(l^n\beta SL)$ by (45). If also $Q(l^{n-2}\beta SL)\neq 0$, this term corresponds to a term of l^{n-4} and so forth; each chain of corresponding terms begins with a term $l^v\beta SL$ which has $Q=0$.

We may thus assign to each term in the QSL scheme a "seniority number" v, which indicates the number of electrons of the first member of its chain; it follows immediately from (45) that Q depends only on n and v and that its values are given by

$$Q(n, v)=\tfrac{1}{4}(n-v)(4l+4-n-v). \quad (50)$$

Confronting (41) and (50) we see that conjugate terms have the same seniority.

The seniority number suffices for distinguishing the different terms of the same kind in the configurations d^n but not in f^n, since there are in f^n terms of the same kind which have also the same seniority. For such configurations an unspecified parameter α must be maintained besides v; terms corresponding according to (49) will have the same values of v and of α.

With this convention Eq. (49) which defines the correspondence between terms of the same chain becomes

$$(l^{n-2}(\alpha' v' SL), l^2(^1S), SL]\!]l^n\alpha v SL)$$
$$=[Q(n, v)]^{\frac{1}{2}}[\tfrac{1}{2}n(n-1)$$
$$\cdot(2l+1)]^{-\frac{1}{2}}\delta(vv')\delta(\alpha\alpha'). \quad (49')$$

In this paper all tables of matrix elements are given in the QSL scheme and the seniority number is indicated by a prefix under the multiplicity number of each term: for instance the two ^2D terms of d^3, which were indicated in TAS (p. 228) by a^2D and b^2D, are therefore, respectively, denoted by ${}_1^2D$ and ${}_3^2D$.

(3) The High Degeneracies

Majorana's operator of position exchange may be defined for equivalent electrons by the relation

$$(l^2LM|M_{ij}|l^2LM)=(-1)^L \quad (51)$$

and may also be expressed, according to Dirac's vector model, by

$$M_{ij}=-[\tfrac{1}{2}+2(\mathbf{s}_i\cdot\mathbf{s}_j)]. \quad (52)$$

From (43)II we have

$$\sum_{r=0}^{2l}(-1)^r(2r+1)W(llll;0r)W(llll;Lr)$$
$$=(-1)^L W(llll;L0); \quad (53)$$

expressing $W(llll;0r)$ and $W(llll;L0)$ by (36)II and using also (38)II and (58)II we get for M_{ij} the expression

$$M_{ij}=\sum_{r=0}^{2l}(2r+1)(\mathbf{u}_i^{(r)}\cdot\mathbf{u}_j^{(r)}). \quad (54)$$

Adding this equation to (37) and introducing (52) we have

$$2\sum_{t=0}^{l}(4t+1)(\mathbf{u}_i^{(2t)}\cdot\mathbf{u}_j^{(2t)})$$
$$=q_{ij}-[\tfrac{1}{2}+2(\mathbf{s}_i\cdot\mathbf{s}_j)]. \quad (55)$$

It follows therefore from §4 of II and particularly from (51)II that if Slater's integrals F^k are proportional to $(2k+1)/C_{llk}$, the electrostatic interaction between two equivalent electrons is proportional to (55) and then the electrostatic-energy matrix is diagonal in the QSL scheme and its eigenvalues are only functions of n, v, and S. This fact explains the high degeneracies observed by Laporte and Platt[10] for these particular ratios of the parameters; unfortunately these ratios are only hypothetical, since they are excluded by the property of F^k of being a decreasing function of k (TAS, p. 177).

[10] O. Laporte and J. R. Platt, Phys. Rev. 61, 305 (1942).

(4) Relations Between Parentages of Corresponding Terms

If we express $\Psi(l^n)$ as linear combination of $\psi(l^{n-1}l)$ with $\Psi(l^{n-1})$ as linear combination of $\psi(l^{n-3}l^2)$, express on the other hand $\Psi(l^n)$ as combination of $\psi(l^{n-2}l^2)$ with $\Psi(l^{n-2})$ as combination of $\psi(l^{n-3}l)$, and compare the two developments, we obtain

$$\sum_{\alpha''v''} (l^{n-3}(\alpha'v'S'L'),\, l^2(^1S),\, S'L'\rbrace l^{n-1}\alpha''v''S'L') \cdot (l^{n-1}(\alpha''v''S'L')lSL\rbrace l^n\alpha vSL) = \sum_{\alpha'''v'''} (l^{n-3}(\alpha'v'S'L')lSL\rbrace l^{n-2}\alpha'''v'''SL) \cdot (l^{n-2}(\alpha'''v'''SL),\, l^2(^1S),\, SL\rbrace l^n\alpha vSL), \quad (56)$$

and owing to (49′) and (50)

$$[(n-v'-1)(4l+5-n-v')/(n-2)]^{\frac{1}{2}} \cdot (l^{n-1}(\alpha'v'S'L')lSL\rbrace l^n\alpha vSL) = (l^{n-3}(\alpha'v'S'L')lSL\rbrace l^{n-2}\alpha vSL) \cdot [(n-v)(4l+4-n-v)/n]^{\frac{1}{2}}. \quad (57)$$

It is easy to deduce from this recursion formula that

$$(l^{n-1}(\alpha'v'S'L')lSL\rbrace l^n\alpha vSL)=0 \qquad (v'\neq v\pm 1), \quad (58a)$$

$$(l^{n-1}(\alpha'v-1S'L')lSL\rbrace l^n\alpha vSL) = [(4l+4-n-v)v/2n(2l+2-v)]^{\frac{1}{2}} \cdot (l^{v-1}(\alpha'v-1S'L')lSL\rbrace l^v\alpha vSL), \quad (58b)$$

$$(l^{n-1}(\alpha'v+1S'L')lSL\rbrace l^n\alpha vSL) = [(n-v)(v+2)/2n]^{\frac{1}{2}} \cdot (l^{v+1}(\alpha'v+1S'L')lSL\rbrace l^{v+2}\alpha vSL). \quad (58c)$$

Comparing (58b) with (13) we obtain the more accurate orthogonality relations

$$\sum_{\alpha'S'L'} (l^n\alpha vSL\lbrace l^{n-1}(\alpha'v-1S'L')lSL) \cdot (l^{n-1}(\alpha'v-1S'L')lSL\rbrace l^n\alpha''vSL) = [(4l+4-n-v)v/2n(2l+2-v)]\delta(\alpha\alpha'') \quad (59a)$$

and

$$\sum_{\alpha'S'L'} (l^n\alpha vSL\lbrace l^{n-1}(\alpha'v+1S'L')lSL) \cdot (l^{n-1}(\alpha'v+1S'L')lSL\rbrace l^n\alpha''vSL) = [(n-v)(4l+4-v)/2n(2l+2-v)]\delta(\alpha\alpha''); \quad (59b)$$

comparing (58c) with (20) we obtain also

$$\sum_{\alpha SL} (2S+1)(2L+1) \cdot (l^{n-1}(\alpha'v+1S'L')lSL\rbrace l^n\alpha vSL) \cdot (l^n\alpha vSL\lbrace l^{n-1}(\alpha''v+1S'L')lSL) = (2S'+1)(2L'+1) \cdot [(n-v)(v+1)/2n(2l+1-v)]\delta(\alpha'\alpha'') \quad (60a)$$

and

$$\sum_{\alpha SL} (2S+1)(2L+1) \cdot (l^{n-1}(\alpha'v-1S'L')lSL\rbrace l^n\alpha vSL) \cdot (l^n\alpha vSL\lbrace l^{n-1}(\alpha''v-1S'L')lSL) = (2S'+1)(2L'+1)[(4l+4-n-v) \cdot (4l+5-v)/2n(2l+3-v)]\delta(\alpha'\alpha''). \quad (60b)$$

Another useful relation is the following:

$$(l^{v+1}(\alpha'v+1S'L')lSL\rbrace l^{v+2}\alpha vSL) = (-1)^{S+L+l+\frac{1}{2}-S'-L'} \cdot \left[\frac{(2S'+1)(2L'+1)(v+1)}{(2S+1)(2L+1)(v+2)(2l+1-v)}\right]^{\frac{1}{2}} \cdot (l^v(\alpha vSL)lS'L'\rbrace l^{v+1}\alpha'v+1S'L'); \quad (61)$$

since this relation is verified for $v=0$ ($S=L=0$, $S'=\frac{1}{2}$, $L'=l$), it suffices to prove that if it holds for $v=v'-1$ it holds also for $v=v'$.

We use for this purpose the expressions (32) and (49′) of $(l^{v'}(\alpha v'SL),\, l^2(^1S),\, SL\rbrace l^{v'+2}\alpha v'SL)$: owing to (6), (58), and (50) we have

$$\sum_{\alpha_1S_1L_1} (-1)^{S+L+l+\frac{1}{2}-S_1-L_1} \cdot \left[\frac{(2S_1+1)(2L_1+1)}{2(2l+1)(2S+1)(2L+1)}\right]^{\frac{1}{2}} \cdot (l^{v'}(\alpha v'SL)lS_1L_1\rbrace l^{v'+1}\alpha_1v'-1S_1L_1) \cdot \left[\frac{(2l+1-v')v'}{(v'+2)(2l+2-v')}\right]^{\frac{1}{2}} \cdot (l^{v'-1}(\alpha_1v'-1S_1L_1)lSL\rbrace l^{v'}\alpha v'SL)$$
$$+\sum_{\alpha_2S_2L_2} (-1)^{S+L+l+\frac{1}{2}-S_2-L_2} \cdot \left[\frac{(2S_2+1)(2L_2+1)}{2(2l+1)(2S+1)(2L+1)}\right]^{\frac{1}{2}} \cdot (l^{v'}(\alpha v'SL)lS_2L_2\rbrace l^{v'+1}\alpha_2v'+1S_2L_2) \cdot (l^{v'+1}(\alpha_2v'+1S_2L_2)lSL\rbrace l^{v'+2}\alpha v'SL) = \left[\frac{2(2l+1-v')}{(v'+1)(v'+2)(2l+1)}\right]^{\frac{1}{2}}. \quad (62)$$

If (61) holds for $v=v'-1$ we may calculate the first sum with the aid of (59a) and obtain

$$-\frac{v'}{2l+2-v'}\left[\frac{2l+1-v'}{2(v'+1)(v'+2)(2l+1)}\right]^{\frac{1}{2}};$$

the second sum must then have the value

$$\frac{4l+4-v'}{2l+2-v'}\left[\frac{2l+1-v'}{2(v'+1)(v'+2)(2l+1)}\right]^{\frac{1}{2}},$$

and owing to (59b), to (60b) and to the well-known corollary of Schwarz's inequality, this fact is possible only if (61) holds also for $v=v'$.

By the use of the formulas (58) and (61) the calculation of the fractional parentages is considerably simplified: Only the parentages of the "new" terms ($v=n$) must really be calculated by the methods of §3; all others may be quickly deduced from them.

(5) Relations Between Correspondence and Conjugation

It follows from (43) that if two eigenfunctions $\Psi_{\mathfrak{L}}(l^n\alpha vSL)$ and $\Psi_{\mathfrak{L}}(l^{n+2}\alpha vSL)$ correspond according to (49′), also the eigenfunctions of their conjugate states $\Psi_{\mathfrak{R}}(l^{4l+2-n}\alpha vSL)$ and $\Psi_{\mathfrak{R}}(l^{4l-n}\alpha vSL)$ correspond in the same way. But if, in order to make full use of (74)II, we assume as standard scheme the scheme of the $\Psi_{\mathfrak{L}}$ for $n\leqslant 2l+1$ and that of the $\Psi_{\mathfrak{R}}$ for $n\geqslant 2l+2$, we cannot use (49′) for the determination of $(l^{2l}(\alpha'vSL),\ l^2(^1S),\ SL]\!]l^{2l+2}\alpha vSL)$, nor can we use (19) for $n=2l$ without knowing which of the two possibilities of (76)II holds for each term of l^{2l+1}.

In order to solve these questions we consider provisorily the system of functions $\Psi_{\mathfrak{L}}(l^{2l+2}\alpha vSL)$ defined by means of (49′) and the system of functions $\Psi_{\mathfrak{R}}(l^{2l+1}\alpha'v'S'L')$ defined by means of (14), and seek the relation between the parentages of $\Psi_{\mathfrak{L}}(l^{2l+2}\alpha vSL)$ with respect to $\Psi_{\mathfrak{L}}(l^{2l+1}\alpha'v'S'L')$ and the parentages of $\Psi_{\mathfrak{R}}(l^{2l+2}\alpha vSL)$ with respect to $\Psi_{\mathfrak{R}}(l^{2l+1}\alpha'v'S'L')$. Using (19), (58), and (61), and owing to the fact that $2S$ is even and $2S'$ is odd, we get

$$(l^{2l+1}(\alpha'v+1S'L')lSL]\!]l^{2l+2}\alpha vSL)_{\mathfrak{R}} = (l^{2l+1}(\alpha'v+1S'L')lSL]\!]l^{2l+2}\alpha vSL)_{\mathfrak{L}} \quad (63a)$$

and

$$(l^{2l+1}(\alpha'v-1S'L')lSL]\!]l^{2l+2}\alpha vSL)_{\mathfrak{R}} = -(l^{2l+1}(\alpha'v-1S'L')lSL]\!]l^{2l+2}\alpha vSL). \quad (63b)$$

If we assume

$$\Psi_{\mathfrak{R}}(l^{2l+1}\ {}_1{}^2L)=\Psi_{\mathfrak{L}}(l^{2l+1}\ {}_1{}^2L), \quad (64)$$

it follows by the alternate use of (63a) and (63b) that

$$\begin{aligned}\Psi_{\mathfrak{R}}(l^{2l+1}\alpha vSL)&=(-1)^{v-1/2}\Psi_{\mathfrak{L}}(l^{2l+1}\alpha vSL),\\ \Psi_{\mathfrak{R}}(l^{2l+2}\alpha vSL)&=(-1)^{v/2}\Psi_{\mathfrak{L}}(l^{2l+2}\alpha vSL).\end{aligned} \quad (65)$$

If we had assumed a minus sign in (64), the relations (65) would have also the opposite sign; the choice between these two possibilities depends on the phase of $\Psi(l^{4l+2}\ {}^1{}_0S)$, and it may be shown that (64) is in agreement with the convention of §5 of TAS for the eigenfunctions of closed shells.

The relations (65) must be taken in account if we use (19) for $n=21$ or (49′) for $n=2l+2$ and $n=2l+3$, and also if we calculate the coefficients of fractional parentage for $n\geqslant 2l+2$ by means of (58) instead of (19).

(6) Relations Between Matrix Components of Tensors

It follows immediately from (23) and (58) that the matrix components of every operator F between two states of l^n may be different from zero only if

$$\Delta v=0,\quad \pm 2, \quad (66)$$

and that

$$\begin{aligned}&(l^n\alpha vSLM_SM_L|F|l^n\alpha'v-2S'L'M_S'M_L')\\ &=\tfrac{1}{2}[(n+2-v)(4l+4-n-v)/(2l+2-v)]^{\frac{1}{2}}\\ &\cdot(l^v\alpha vSLM_SM_L|F|l^v\alpha'v-2S'L'M_S'M_L');\end{aligned} \quad (67)$$

owing to (65) a minus sign, however, must be introduced in this formula for $n\geqslant 2l+2$.

If $\Delta v=0$ the sum (23) splits in two sums according to the two possibilities $v-1$ and $v+1$ of the seniority numbers of l^{n-1}, but only the first sum may be immediately expressed as in the preceding case by means of the matrix components for l^v; the second sum is to be expressed by means of the matrix components for l^v and those for another arbitrary configuration $l^{n'}$. If we assume $n'=4l+2-v$, the final result is found to be

$$(l^n\alpha vSLM_SM_L|F|l^n\alpha'vS'L'M_S'M_L')$$
$$=[(4l+2-n-v)/2(2l+1-v)]$$
$$\cdot(l^v\alpha vSLM_SM_L|F|l^v\alpha'vS'L'M_S'M_L')$$
$$+[(n-v)/2(2l+1-v)]$$
$$\cdot(l^{4l+2-v}\alpha vSLM_SM_L|F|l^{4l+2-v}$$
$$\cdot\alpha'vS'L'M_S'M_L'). \quad (68)$$

If F is an irreducible tensor, it follows from (68) and (74)II that

$$(l^n\alpha vSL\|T^{(\kappa k)}\|l^n\alpha'vS'L')$$
$$=(l^v\alpha vSL\|T^{(\kappa k)}\|l^v\alpha'vS'L') \quad (\kappa+k \text{ odd}), \quad (69a)$$
$$(l^n\alpha vSL\|T^{(\kappa k)}\|l^n\alpha'vS'L')$$
$$=\frac{2l+1-n}{2l+1-v}(l^v\alpha vSL\|T^{(\kappa k)}\|l^v\alpha'vS'L')$$
$$(\kappa+k \text{ even}). \quad (69b)$$

From (67), (65), and (74)II we get also

$$(l^n\alpha vSL\|T^{(\kappa k)}\|l^n\alpha'v-2S'L')=0$$
$$(\kappa+k \text{ odd}). \quad (70)$$

The remarkable result that a tensor of odd degree is diagonal with respect to v and that its submatrices are independent of n may be obtained also in a more direct way. It follows from the triangular conditions and from the fact that in l^2 only states with even $S+L$ are allowed, that for $\kappa+k$ odd

$$(l^2\,{}^1S|t_1^{(\kappa k)}+t_2^{(\kappa k)}|l^2SLM_SM_L)$$
$$=(l^2SLM_SM_L|t_1^{(\kappa k)}+t_2^{(\kappa k)}|l^2\,{}^1S)=0$$

TABLE XV. $(p^2SL\|2V^{(12)}\|p^2S'L')$.

	^1S	^3P	^1D
^1S	0	0	0
^3P	0	$-(6)^{\frac{1}{2}}$	-3
^1D	0	-3	0

TABLE XVI. $(p^3SL\|2V^{(12)}\|p^3S'L')$.

	^4S	^2P	^2D
^4S	0	0	$-2(2)^{\frac{1}{2}}$
^2P	0	$(6)^{\frac{1}{2}}$	0
^2D	$2(2)^{\frac{1}{2}}$	0	$-(14)^{\frac{1}{2}}$

and, therefore,

$$q_{ij}(t_i^{(\kappa k)}+t_j^{(\kappa k)})=(t_i^{(\kappa k)}+t_j^{(\kappa k)})q_{ij}=0$$
$$(\kappa+k \text{ odd}); \quad (71)$$

since all other $t_h^{(\kappa k)}$ commute with q_{ij}, $T^{(\kappa k)}$ commutes with q_{ij} and also with Q, and is therefore diagonal with respect to v. From (71) we have also

$$QT=TQ=\sum_{\substack{i<j\\ i\neq h\neq j}} t_hq_{ij};$$

calculating the matrix of this operator with the methods of §5 and owing to (49′) we obtain

$$(l^n\alpha vSL\|T^{(\kappa k)}\|l^n\alpha'vS'L')$$
$$=(l^{n-2}\alpha vSL\|T^{(\kappa k)}\|l^{n-2}\alpha'vS'L'),$$

which is equivalent to (69a).

The matrices of the tensor $V^{(12)}$ defined by (102)II were calculated for the configurations p^2, p^3, d^3, d^4 and d^5 using also (69a); the results are given in Tables XV–XIX. The matrices given in Tables V and XIX are sufficient for the calculation of the spectra of the configurations p^nl and d^np with the methods of §8 of II.

§7. THE ELECTROSTATIC INTERACTION BETWEEN d^n, $d^{n-1}s$ AND $d^{n-2}s^2$

The electrostatic interaction between $d^2\,{}^1S$ and $s^2\,{}^1S$ is given by

$$(d^2\,{}^1S|e^2/r|s^2\,{}^1S)$$
$$=R^2(dd,ss)(d^2\,{}^1S|P_2(\cos\omega)|s^2\,{}^1S), \quad (72)$$

where R^2 is defined by TAS 8^68 and ω is the angle between the radii vectors of the two electrons. From (51)II we have

$$(2\|C^{(2)}\|0)=1, \quad (73)$$

and hence from (45)II and (38)II we get

$$(d^2\,{}^1S|P_2(\cos\omega)|s^2\,{}^1S)=1/5^{\frac{1}{2}};$$

since

$$R^2(dd,ss)=R^2(ds,sd)=G^2(ds)=5G_2(ds),$$

(72) becomes

$$(d^2\,{}^1S|e^2/r|s^2\,{}^1S)=5^{\frac{1}{2}}G_2. \quad (74)$$

Introducing this result in (33c) and owing to

TABLE XVII. $(d^3vSL\|70V^{(12)}\|d^3v'S'L')$.

	$_3{}^2P$	$_3{}^4P$	$_1{}^2D$	$_3{}^2D$	$_3{}^2F$	$_3{}^4F$	$_3{}^2G$	$_3{}^2H$
$_3{}^2P$	$-19(14)^{\frac12}$	-28	0	$8(35)^{\frac12}$	$-8(14)^{\frac12}$	$-22(14)^{\frac12}$	0	0
$_3{}^4P$	28	$14(35)^{\frac12}$	0	$-28(10)^{\frac12}$	56	$-28(10)^{\frac12}$	0	0
$_1{}^2D$	0	0	$35(6)^{\frac12}$	0	0	0	0	0
$_3{}^2D$	$-8(35)^{\frac12}$	$-28(10)^{\frac12}$	0	$-5(6)^{\frac12}$	$-4(210)^{\frac12}$	$4(210)^{\frac12}$	$-60(2)^{\frac12}$	0
$_3{}^2F$	$-8(14)^{\frac12}$	-56	0	$4(210)^{\frac12}$	-77	-98	$3(35)^{\frac12}$	$-4(385)^{\frac12}$
$_3{}^4F$	$22(14)^{\frac12}$	$-28(10)^{\frac12}$	0	$4(210)^{\frac12}$	98	$-14(10)^{\frac12}$	$-18(35)^{\frac12}$	$4(385)^{\frac12}$
$_3{}^2G$	0	0	0	$-60(2)^{\frac12}$	$-3(35)^{\frac12}$	$-18(35)^{\frac12}$	$3(33)^{\frac12}$	$12(77)^{\frac12}$
$_3{}^2H$	0	0	0	0	$-4(385)^{\frac12}$	$-4(385)^{\frac12}$	$-12(77)^{\frac12}$	$-(2002)^{\frac12}$

TABLE XVIII. $(d^4 4SL\|70V^{(12)}\|d^4 4S'L')$.

	$_4{}^1S$	$_4{}^3P$	$_4{}^1D$	$_4{}^3D$	$_4{}^5D$	$_4{}^1F$	$_4{}^3F$	$_4{}^1G$	$_4{}^3G$	$_4{}^3H$	$_4{}^1I$
$_4{}^1S$	0	0	0	$4(210)^{\frac12}$	0	0	0	0	0	0	0
$_4{}^3P$	0	$6(14)^{\frac12}$	$-15(35)^{\frac12}$	$-3(35)^{\frac12}$	-105	$-12(35)^{\frac12}$	$12(14)^{\frac12}$	0	0	0	0
$_4{}^1D$	0	$-15(35)^{\frac12}$	0	$5(6)^{\frac12}$	0	0	0	0	$60(2)^{\frac12}$	0	0
$_4{}^3D$	$-4(210)^{\frac12}$	$3(35)^{\frac12}$	$-5(6)^{\frac12}$	$-\frac{15}{2}(6)^{\frac12}$	$\frac{15}{2}(210)^{\frac12}$	$-10(21)^{\frac12}$	$9(210)^{\frac12}$	$6(165)^{\frac12}$	$15(2)^{\frac12}$	0	0
$_4{}^5D$	0	-105	0	$-\frac{15}{2}(210)^{\frac12}$	$-\frac{35}{2}(30)^{\frac12}$	0	$35(6)^{\frac12}$	0	$15(70)^{\frac12}$	0	0
$_4{}^1F$	0	$12(35)^{\frac12}$	0	$-10(21)^{\frac12}$	0	0	$-21(10)^{\frac12}$	0	$12(14)^{\frac12}$	$-6(154)^{\frac12}$	0
$_4{}^3F$	0	$12(14)^{\frac12}$	0	$-9(210)^{\frac12}$	$35(6)^{\frac12}$	$21(10)^{\frac12}$	-42	0	$-12(35)^{\frac12}$	$6(385)^{\frac12}$	0
$_4{}^1G$	0	0	0	$-6(165)^{\frac12}$	0	0	0	0	$27(10)^{\frac12}$	$6(210)^{\frac12}$	0
$_4{}^3G$	0	0	$-60(2)^{\frac12}$	$15(2)^{\frac12}$	$-15(70)^{\frac12}$	$12(14)^{\frac12}$	$12(35)^{\frac12}$	$-27(10)^{\frac12}$	$-6(33)^{\frac12}$	$-6(77)^{\frac12}$	$-12(91)^{\frac12}$
$_4{}^3H$	0	0	0	0	0	$6(154)^{\frac12}$	$6(385)^{\frac12}$	$6(210)^{\frac12}$	$6(77)^{\frac12}$	$-3(2002)^{\frac12}$	$-7(39)^{\frac12}$
$_4{}^1I$	0	0	0	0	0	0	0	0	$12(91)^{\frac12}$	$-7(39)^{\frac12}$	0

TABLE XIX. $(d^5 5SL\|70V^{(12)}\|d^5 5S'L')$.

	$_5{}^2S$	$_5{}^6S$	$_5{}^2D$	$_5{}^4D$	$_5{}^2F$	$_5{}^2G$	$_5{}^4G$	$_5{}^2I$
$_5{}^2S$	0	0	$-4(210)^{\frac12}$	$-6(105)^{\frac12}$	0	0	0	0
$_5{}^6S$	0	0	0	$70(3)^{\frac12}$	0	0	0	0
$_5{}^2D$	$-4(210)^{\frac12}$	0	$15(6)^{\frac12}$	$60(3)^{\frac12}$	$-20(21)^{\frac12}$	-4	$-20(15)^{\frac12}$	0
$_5{}^4D$	$6(105)^{\frac12}$	$-70(3)^{\frac12}$	$-60(3)^{\frac12}$	$10(15)^{\frac12}$	0	$8(330)^{\frac12}$	$-40(3)^{\frac12}$	0
$_5{}^2F$	0	0	$20(21)^{\frac12}$	0	-105	$(1155)^{\frac12}$	$14(105)^{\frac12}$	0
$_5{}^2G$	0	0	$-4(165)^{\frac12}$	$-8(330)^{\frac12}$	$-(1155)^{\frac12}$	$\frac{125}{11}(33)^{\frac12}$	$-10(3)^{\frac12}$	$\frac{8}{11}(30030)^{\frac12}$
$_5{}^4G$	0	0	$20(15)^{\frac12}$	$-40(3)^{\frac12}$	$14(105)^{\frac12}$	$10(3)^{\frac12}$	$-10(330)^{\frac12}$	$-2(2730)^{\frac12}$
$_5{}^2I$	0	0	0	0	0	$\frac{8}{11}(30030)^{\frac12}$	$2(2730)^{\frac12}$	$-\frac{35}{11}(858)^{\frac12}$

(49′) we obtain

$$(d^n vSL|\sum e^2/r_{ij}|d^{n-2}s^2v'SL) = [Q(n,v)]^{\frac12}\delta(vv')G_2; \quad (75)$$

owing to the conventions of subsection (5) of §6, a minus sign must be introduced for $n=6$, $v=2$ and for $n=7$, $v=3$.

The calculation of the interaction between the configurations d^n and $d^{n-1}s$ by means of (33b) is easy only for $n=3$, since in this case

$$(d, d^2(^1D), SL]d^3vSL)$$

may be obtained from Table II by means of (29); but, as it was already mentioned at the end of §5, the calculations for $n \geqslant 4$ become very long, and it appears more convenient to use the following method.

The interaction in question is given by

$$(d^n vSL|\sum e^2/r_{ij}|d^{n-1}(v'S'L)sSL) = R^2(dd, ds) \cdot (d^n vSL|\sum P_2(\cos\omega_{ij})|d^{n-1}(v'S'L)sSL); \quad (76)$$

owing to (45)II and to the fact that $\sum_i \mathbf{C}_i^{(2)2}$ is a scalar and its non-diagonal matrix components vanish, we have

$$(d^n vSL|\sum_{i<j} P_2(\cos\omega_{ij})|d^{n-1}(v'S'L)sSL) = \tfrac12(d^n vSL|[\sum_i \mathbf{C}_i^{(2)}]^2|d^{n-1}(v'S'L)sSL), \quad (77)$$

TABLE XX. $(d^3vSL|\Sigma e^2/r_{ij}|d^2(v'S'L')sSL)$.

d^3	d^2s	H_2
${}_3{}^2P$	$({}_2{}^3P){}^2P$	$3(35)^{\frac{1}{2}}$
${}_3{}^4P$	$({}_2{}^3P){}^4P$	0
${}_1{}^2D$	$({}_2{}^1D){}^2D$	$-\frac{1}{2}(70)^{\frac{1}{2}}$
${}_3{}^2D$	$({}_2{}^1D){}^2D$	$\frac{3}{2}(30)^{\frac{1}{2}}$
${}_3{}^2F$	$({}_2{}^3F){}^2F$	$-3(10)^{\frac{1}{2}}$
${}_3{}^4F$	$({}_2{}^3F){}^4F$	0
${}_3{}^2G$	$({}_2{}^1G){}^2G$	$-5(2)^{\frac{1}{2}}$

and using (33)II we obtain

$$2(2L+1)(d^nvSL|\sum_{i<j} P_2(\cos\omega_{ij})|d^{n-1}(v'S'L)sSL)$$
$$= \sum_{v''L''} (-1)^{L-L''}$$
$$\cdot(d^nvSL\|\textstyle\sum_i C_i^{(2)}\|d^nv''SL'')$$
$$\cdot(d^nv''SL''\|\textstyle\sum_i C_i^{(2)}\|d^{n-1}(v'S'L)sSL)$$
$$+\sum_{v''L''} (-1)^{L-L''}$$
$$\cdot(d^nvSL\|\textstyle\sum_i C_i^{(2)}\|d^{n-1}(v''S'L'')sSL'')$$
$$\cdot(d^{n-1}(v''S'L'')sSL''\|\textstyle\sum_i C_i^{(2)}\|d^{n-1}(v'S'L)sSL). \quad (78)$$

From (27), (44)II, (80)II, and (73) we obtain finally

$$(d^nvSL|\sum e^2/r_{ij}|d^{n-1}(v'S'L)sSL)$$
$$=(n/14)^{\frac{1}{2}}[\sum_{v''L''}(-1)^{L-L''}$$
$$\cdot(d^nvSL\|U^{(2)}\|d^nv''SL'')$$
$$\cdot(d^nv''SL''\{\!|d^{n-1}(v'S'L)dSL)$$
$$\cdot(2L''+1)^{\frac{1}{2}}/(2L+1)+\sum_{v''L''}(-1)^{L-L''}$$
$$\cdot(d^nvSL\{\!|d^{n-1}(v''S'L'')dSL)$$
$$\cdot(d^{n-1}v''S'L''\|U^{(2)}\|d^{n-1}v'S'L)/$$
$$(2L+1)^{\frac{1}{2}}]R^2(dd,ds). \quad (79)$$

By means of this formula the interaction between the configurations d^n and $d^{n-1}s$ was calculated for $n=3$, 4, 5; the results are different from zero only if $v'=v\pm 1$ and are given in Tables XX–XXII, where the quantity

$$H_2=R^2(dd,ds)/35 \quad (80)$$

was assumed as parameter. For $n=3$ our results agree with those given by Marvin.[11]

TABLE XXI. $(d^4vSL|\Sigma e^2/r_{ij}|d^3(v'S'L')sSL)$.

d^4	d^3s	H_2	d^4	d^3s	H_2
${}_2{}^3P$	$({}_3{}^2P){}^3P$	$(70)^{\frac{1}{2}}$	${}_4{}^1F$	$({}_3{}^2F){}^1F$	-15
${}_2{}^3P$	$({}_3{}^4P){}^3P$	0	${}_2{}^3F$	$({}_3{}^2F){}^3F$	$-2(5)^{\frac{1}{2}}$
${}_4{}^3P$	$({}_3{}^2P){}^3P$	$-2(5)^{\frac{1}{2}}$	${}_2{}^3F$	$({}_3{}^4F){}^3F$	0
${}_4{}^3P$	$({}_3{}^4P){}^3P$	$2(70)^{\frac{1}{2}}$	${}_4{}^3F$	$({}_3{}^2F){}^3F$	$3(5)^{\frac{1}{2}}$
${}_2{}^1D$	$({}_1{}^2D){}^1D$	$-(105)^{\frac{1}{2}}$	${}_4{}^3F$	$({}_3{}^4F){}^3F$	$-4(5)^{\frac{1}{2}}$
${}_2{}^1D$	$({}_3{}^2D){}^1D$	$-3(5)^{\frac{1}{2}}$	${}_2{}^1G$	$({}_3{}^2G){}^1G$	$\frac{10}{3}(3)^{\frac{1}{2}}$
${}_4{}^1D$	$({}_3{}^2D){}^1D$	$6(10)^{\frac{1}{2}}$	${}_4{}^1G$	$({}_3{}^2G){}^1G$	$\frac{5}{3}(33)^{\frac{1}{2}}$
${}_4{}^3D$	$({}_3{}^2D){}^3D$	$-4(5)^{\frac{1}{2}}$	${}_4{}^3G$	$({}_3{}^2G){}^3G$	$3(5)^{\frac{1}{2}}$
			${}_4{}^3H$	$({}_3{}^2H){}^3H$	$-2(5)^{\frac{1}{2}}$

TABLE XXII. $(d^5vSL|\Sigma e^2/r_{ij}|d^4(v'S'L')sSL)$.

d^5	d^4s	H_2	d^5	d^4s	H_2
${}_5{}^2S$	$({}_4{}^1S){}^2S$	$8(5)^{\frac{1}{2}}$	${}_3{}^2F$	$({}_2{}^3F){}^2F$	$-2(15)^{\frac{1}{2}}$
${}_3{}^2P$	$({}_2{}^3P){}^2P$	$(210)^{\frac{1}{2}}$	${}_3{}^2F$	$({}_4{}^3F){}^2F$	$-\frac{3}{2}(15)^{\frac{1}{2}}$
${}_3{}^2P$	$({}_4{}^3P){}^2P$	$(15)^{\frac{1}{2}}$	${}_5{}^2F$	$({}_4{}^1F){}^2F$	$\frac{7}{2}(5)^{\frac{1}{2}}$
${}_3{}^4P$	$({}_2{}^3P){}^4P$	0	${}_5{}^2F$	$({}_4{}^3F){}^2F$	$\frac{15}{2}(3)^{\frac{1}{2}}$
${}_3{}^4P$	$({}_4{}^3P){}^4P$	$(105)^{\frac{1}{2}}$	${}_3{}^4F$	$({}_2{}^3F){}^4F$	0
${}_1{}^2D$	$({}_2{}^1D){}^2D$	$-(35)^{\frac{1}{2}}$	${}_3{}^4F$	$({}_4{}^3F){}^4F$	$-(30)^{\frac{1}{2}}$
${}_3{}^2D$	$({}_2{}^1D){}^2D$	$3(5)^{\frac{1}{2}}$	${}_3{}^2G$	$({}_2{}^1G){}^2G$	$-\frac{10}{3}(3)^{\frac{1}{2}}$
${}_3{}^2D$	$({}_4{}^1D){}^2D$	$3(10)^{\frac{1}{2}}$	${}_3{}^2G$	$({}_4{}^1G){}^2G$	$\frac{5}{6}(33)^{\frac{1}{2}}$
${}_3{}^2D$	$({}_4{}^3D){}^2D$	$2(15)^{\frac{1}{2}}$	${}_3{}^2G$	$({}_4{}^3G){}^2G$	$-\frac{3}{2}(15)^{\frac{1}{2}}$
${}_5{}^2D$	$({}_4{}^1D){}^2D$	$-(5)^{\frac{1}{2}}$	${}_5{}^2G$	$({}_4{}^1G){}^2G$	$-\frac{9}{2}(5)^{\frac{1}{2}}$
${}_5{}^2D$	$({}_4{}^3D){}^2D$	$-3(30)^{\frac{1}{2}}$	${}_5{}^2G$	$({}_4{}^3G){}^2G$	$-\frac{5}{2}(11)^{\frac{1}{2}}$
${}_5{}^4D$	$({}_4{}^3D){}^4D$	$-\frac{3}{2}(30)^{\frac{1}{2}}$	${}_5{}^4G$	$({}_4{}^3G){}^4G$	$5(2)^{\frac{1}{2}}$
${}_5{}^4D$	$({}_4{}^5D){}^4D$	$-\frac{5}{2}(14)^{\frac{1}{2}}$	${}_3{}^2H$	$({}_4{}^3H){}^2H$	$(15)^{\frac{1}{2}}$
${}_3{}^2F$	$({}_4{}^1F){}^2F$	$-\frac{15}{2}$	${}_5{}^2I$	$({}_4{}^1I){}^2I$	$(5)^{\frac{1}{2}}$

The interaction between the configurations $d^{n-1}s$ and $d^{n-2}s^2$ may be calculated in the same way; the result is

$$(d^{n-1}(v'S'L)sSL|\sum e^2/r_{ij}|d^{n-2}s^2vSL)$$
$$=(-1)^{S+\frac{1}{2}-S'}\left[\frac{2S'+1}{2S+1}\right]^{\frac{1}{2}}\cdot$$
$$\cdot(d^{n-1}v'S'L|\sum e^2/r_{ij}|d^{n-2}(vSL)sS'L). \quad (81)$$

[11] H. H. Marvin, Phys. Rev. **47**, 521 (1935).

PHYSICAL REVIEW VOLUME 76, NUMBER 9 NOVEMBER 1, 1949

Theory of Complex Spectra. IV

GIULIO RACAH
The Hebrew University, Jerusalem, Israel
(Received February 7, 1949)

The calculation of the coefficients of fractional parentage and of the energy matrices for the configurations f^n is simplified very much by the use of the theory of groups. Tables of results are given.

1. INTRODUCTION

IT was shown in two previous papers[1] that the calculations on complex spectra may be simplified by the introduction of tensor operators and coefficients of fractional parentage. These coefficients may be calculated by Eqs. (9) of III and (11) of III, but it appears that for the configurations f^n Eqs. (11) of III are too cumbersome for practical use.

By considering the meaning and the properties of the coefficients of fractional parentage from the standpoint of the theory of groups, we shall see that these calculations may be somewhat simplified and that a very fortunate and important simplification takes place exactly for the configurations f^n.

In Section 4 we shall classify the states of f^n as the basis of some group representations and in Section 5 we shall find some properties of the coefficients of fractional parentage which will avoid the use of Eqs. (11) of III; the results of the calculations will be given in Tables III and IV. The energy matrices will be calculated in Section 6, and also these calculations will be simplified by group-theoretical considerations.

Before treating the very argument of this paper, we shall give in Section 2 a formula which should have its natural place in Section 5 of III, but was unfortunately obtained only after the publication of that paper, and we shall prove in Section 3 a corollary of Schur's lemma, which will be very useful in the following calculations.

2. THE MATRIX OF SYMMETRIC SCALAR OPERATORS

The matrix components between two states of l^n of the scalar operator (30) of III were calculated in (33a) of III by taking only the last term of the summation and then multiplying the result by $\frac{1}{2}n(n-1)$. It appears, on the contrary, more convenient to limit the sum of (30) of III to the first $n-1$ electrons and then to multiply by $n/(n-2)$. Thus, we obtain easily

$$\begin{aligned}(l^n\alpha SL|G|l^n\alpha' SL) &= [n/(n-2)]\sum_{\alpha_1\alpha_1'S_1L_1}(l^n\alpha SL\{|l^{n-1}(\alpha_1S_1L_1)lSL)\\ &\times(l^{n-1}\alpha_1S_1L_1|G|l^{n-1}\alpha_1'S_1L_1)\\ &\times(l^{n-1}(\alpha_1'S_1L_1)lSL|\}l^n\alpha' SL).\quad(1)\end{aligned}$$

[1] G. Racah, Phys. Rev. 62, 438 (1942) and 63, 367 (1943) (which will be referred to as II and III. We refer to these papers for definitions and notations.

This formula advantageously replaces (33a) of III, since it does not need the use of (32) of III. Moreover, if it is used for operators the eigenvalues of which are already known, it gives a great number of equations between the coefficients of fractional parentage. Operators of this type are not only the operator Q defined by (34) of III and the operator R which will be defined by (23), but, first of all, the operators

$$\sum_{i<j}(\mathbf{l}_i\cdot\mathbf{l}_j)=[L(L+1)-nl(l+1)]/2 \tag{2}$$

and

$$\sum_{i<j}(\mathbf{s}_i\cdot\mathbf{s}_j)=\tfrac{1}{2}S(S+1)-3n/8. \tag{3}$$

3. A LEMMA

The irreducible representations $(a|U_A(s)|a')$ of a group $\mathfrak{g}$ are generally reducible as representations of a subgroup $\mathfrak{h}$ of $\mathfrak{g}$; i.e., a constant matrix R_A exists, so that for every element t of $\mathfrak{h}$,

$$\sum_{aa'}(\beta Bb|R_A^{-1}|a)(a|U_A(t)|a')(a'|R_A|\beta'B'b')=(b|V_B(t)|b')\delta(BB')\delta(\beta\beta'), \tag{4}$$

where the $(b|V_B(t)|b')$ are the irreducible representations of $\mathfrak{h}$. Instead of the representations $U_A(s)$, we shall always consider the equivalent representations ("reduced with respect to $\mathfrak{h}$"),

$$(\beta Bb|W_A(s)|\beta'B'b')=(\beta Bb|R_A^{-1}U_A(s)R_A|\beta'B'b'); \tag{5}$$

it follows from (4) and (5) that for every element t of $\mathfrak{h}$,

$$(\beta Bb|W_A(t)|\beta'B'b')=(b|V_B(t)|b')\delta(BB')\delta(\beta\beta'). \tag{6}$$

The external (Kronecker's) product of two irreducible representations of $\mathfrak{g}$ is generally reducible, i.e., a constant matrix S exists, so that

$$\sum_{\beta_1B_1b_1\beta_2B_2b_2\beta_1'B_1'b_1'\beta_2'B_2'b_2'}(A_1A_2\alpha A\beta Bb|S^{-1}|A_1\beta_1B_1b_1;A_2\beta_2B_2b_2)(\beta_1B_1b_1|W_{A_1}(s)|\beta_1'B_1'b_1')$$
$$\times(\beta_2B_2b_2|W_{A_2}(s)|\beta_2'B_2'b_2')(A_1\beta_1'B_1'b_1';A_2\beta_2'B_2'b_2'|S|A_1A_2\alpha'A'\beta'B'b')=(\beta Bb|W_A(s)|\beta'B'b')\delta(AA')\delta(\alpha\alpha'). \tag{7}$$

Also the external product of two irreducible representations of $\mathfrak{h}$ is generally reducible and a constant matrix T exists, so that

$$\sum_{b_1b_2b_1'b_2'}(B_1B_2\gamma Bb|T^{-1}|B_1b_1B_2b_2)(b_1|V_{B_1}(t)|b_1')\times(b_2|V_{B_2}(t)|b_2')(B_1b_1'B_2b_2'|T|B_1B_2\gamma'B'b')=(b|V_B(t)|b')\delta(BB')\delta(\gamma\gamma'). \tag{8}$$

From (8), (6), and (7) we obtain

$$\sum_{b_1b_2b''}(b|V_B(t)|b'')(B_1B_2\gamma Bb''|T^{-1}|B_1b_1B_2b_2)\times(A_1\beta_1B_1b_1;A_2\beta_2B_2b_2|S|A_1A_2\alpha A\beta B'b')$$
$$=\sum_{b_1b_2b''}(B_1B_2\gamma Bb|T^{-1}|B_1b_1B_2b_2)\times(A_1\beta_1B_1b_1;A_2\beta_2B_2b_2|S|A_1A_2\alpha A\beta B'b'')\times(b''|V_{B'}(t)|b'), \tag{9}$$

and it follows from the lemma of Schur that the constant matrix $T^{-1}S$ is diagonal with respect to B and b and is independent of b:

$$\sum_{b_1b_2}(B_1B_2\gamma Bb|T^{-1}|B_1b_1B_2b_2)\times(A_1\beta_1B_1b_1;A_2\beta_2B_2b_2|S|A_1A_2\alpha A\beta B'b')=(A_1\beta_1B_1+A_2\beta_2B_2|X_\gamma|\alpha A\beta B)\delta(BB')\delta(bb'). \tag{10}$$

Multiplying from the left by T we have, finally,

$$(A_1\beta_1B_1b_1;A_2\beta_2B_2b_2|S|A_1A_2\alpha A\beta Bb)=\sum_\gamma(B_1b_1B_2b_2|T|B_1B_2\gamma Bb)\times(A_1\beta_1B_1+A_2\beta_2B_2|X_\gamma|\alpha A\beta B); \tag{11}$$

this formula will be very useful for practical calculations, since it expresses the dependence of the matrix S on b_1, b_2, and b by means of the simpler matrices T.

4. THE GROUP-THEORETICAL CLASSIFICATION OF THE TERMS OF l^n

1. The Spin and the Orbital Momentum

The configuration l^n has $\binom{4l+2}{n}$ independent states which may be characterized by a set of quantum numbers Γ; if the $4l+2$ eigenfunctions $\phi(m_sm_l)$ of the individual electrons undergo a linear unimodular transformation,

$$\phi'(m_s'm_l')=\sum_{m_sm_l}\phi(m_sm_l)c(m_sm_l;m_s'm_l'), \tag{12}$$

the eigenfunctions $\psi(l^n,\Gamma)$ undergo the linear transformation which is induced by (12) on the ansitymmetrical tensors of degree n in the $(4l+2)$-dimensional space; i.e., the $\Psi(l^n,\Gamma)$ are the basis of the antisymmetrical representation $\{\mathfrak{c}_{4l+2}\}^n$ of the linear unimodular group $\mathfrak{c}_{4l+2}$. The rows and columns of this representation are characterized by the quantum numbers Γ.

If we limit $\mathfrak{c}_{4l+2}$ to the subgroup $\mathfrak{c}_2\times\mathfrak{c}_{2l+1}$ defined by

$$c(m_sm_l;m_s'm_l')=\gamma(m_sm_s')c(m_lm_l'), \tag{13}$$

where γ and c are two independent linear unimodular transformations, the representation $\{\mathfrak{c}_{4l+2}\}^n$ breaks up into irreducible representations of $\mathfrak{c}_2\times\mathfrak{c}_{2l+1}$, each of which is the external product of a representation $\mathfrak{D}_S$ of $\mathfrak{c}_2$ and a representation $\mathfrak{H}_{n,S}$ of $\mathfrak{c}_{2l+1}$. It is well known that the symmetry schemes of $\mathfrak{D}_S$ and $\mathfrak{H}_{n,S}$ must be

TABLE I. Reduction of $\mathfrak{B}_W$ as representation of G_2.

Values of S and v I: S	I: v	II: S	II: v	W	U						
0	0	7/2	7	(000)	(00)						
1/2	1	3	6	(100)	(10)						
1	2	5/2	5	(110)	(10)	(11)					
0	2	5/2	7	(200)	(20)						
3/2	3	2	4	(111)	(00)	(10)	(20)				
1/2	3	2	6	(210)	(11)	(20)	(21)				
1	4	3/2	5	(211)	(10)	(11)	(20)	(21)	(30)		
0	4	3/2	7	(220)	(20)	(21)	(22)				
1/2	5	1	6	(221)	(10)	(11)	(20)	(21)	(30)	(31)	
0	6	1/2	7	(222)	(00)	(10)	(20)	(30)	(40)		

Table II. Reduction of $\mathfrak{C}_U$ as representation of $\mathfrak{d}_3$.

U	$12g(U)$	L
(00)	0	S
(10)	6	F
(11)	12	PH
(20)	14	DGI
(21)	21	$DFGHKL$
(30)	24	$PFGHIKM$
(22)	30	$SDGHILN$
(31)	32	$PDFFGHHIIKKLMNO$
(40)	36	$SDFGGHIIKLLMNQ$

dual; since the scheme of $\mathfrak{D}_S$ has two lines, the lengths of which are, respectively, $(n/2)+S$ and $(n/2)-S$, the scheme of $\mathfrak{H}_{n,S}$ will have two columns of these lengths. The basis of these representations of $\mathfrak{c}_2\times\mathfrak{c}_{2l+1}$ are the functions $\Psi(l^nSM_S\Delta)$; the quantum number M_S characterizes the rows and columns of $\mathfrak{D}_S$, the quantum numbers Δ those of $\mathfrak{H}_{n,S}$.

If we limit $\mathfrak{c}_{2l+1}$ to its subgroup composed by the elements of the representation $\mathfrak{D}_l$ of order $2l+1$ of the three-dimensional rotation group $\mathfrak{d}_3$, the representation $\mathfrak{H}_{n,S}$ breaks up into representations $\mathfrak{D}_L$ of $\mathfrak{d}_3$, the basis of which are the functions $\Psi(l^n\alpha SLM_SM_L)$. The quantum number (or set of quantum numbers) α must be introduced in order to distinguish the different equivalent representations of $\mathfrak{d}_3$ which may appear in the reduction of $\mathfrak{H}_{n,S}$, i.e., the different terms of the same kind which are allowed in l^n.

In order to classify in a suitable way these different terms, it is convenient to perform the passage from $\mathfrak{c}_{2l+1}$ to $\mathfrak{d}_3$ by successive steps.

2. The Seniority Number

As a first step, we limit $\mathfrak{c}_{2l+1}$ to the orthogonal subgroup $\mathfrak{d}_{2l+1}$ which leaves invariant the quadratic form

$$\sum_{-l}^{l}{}_m(-1)^m\phi(m)\phi(-m), \tag{14}$$

and the representations $\mathfrak{H}_{n,S}$ then break up into irreducible representations $\mathfrak{B}_W$ of $\mathfrak{d}_{2l+1}$; since the group $\mathfrak{d}_{2l+1}$ is of rank l, each $\mathfrak{B}_W$ is characterized by a set W of l integral numbers

$$w_1\geqslant w_2\geqslant\cdots\geqslant w_l\geqslant 0, \tag{15}$$

and since in the symmetry scheme of $\mathfrak{H}_{n,S}$ no row has a length greater than 2, also the w_i will not be greater than 2 and it will be

$$w_1=\cdots=w_a=2,\quad w_{a+1}=\cdots=w_{a+b}=1,\quad w_{a+b+1}=\cdots=w_l=0. \tag{16}$$

It is known from the theory of tensors that the passage from the linear to the metric space (or from $\mathfrak{c}_r$ to $\mathfrak{d}_r$) allows the decomposition of tensors by trace operation or contraction, i.e., some linear combinations of the components of tensors of degree n transform themselves as components of tensors of degree $n-2$; the classification of the terms of l^n according to the representations of $\mathfrak{d}_{2l+1}$ will therefore introduce a correspondence between some of them and the terms of l^{n-2}. It is easy to see that this correspondence is the same which was introduced in Section 6, Subsection 2 of III i.e., that the separation of the terms with $Q\neq 0$ from those with $Q=0$ is equivalent to the decomposition of a tensor by trace operation.

If we subtract (54), III, from (37), III, and add (52), III, we get

$$q_{ij}=-\tfrac{1}{2}-2(\mathbf{s}_i\cdot\mathbf{s}_j)-2\sum_{1}^{l}{}_t(4t-1)(\mathbf{u}_i^{(2t-1)}\cdot\mathbf{u}_j^{(2t-1)}); \tag{17}$$

owing to (38) of III and to (3) we have, for Q, the expression

$$Q=\tfrac{1}{4}n(4l+4-n)-S(S+1)-\sum_{1}^{l}{}_t(4t-1)\mathbf{U}^{(2t-1)^2}, \tag{18}$$

and it may be shown that

$$\sum_{1}^{l}{}_t(4t-1)\mathbf{U}^{(2t-1)^2}=(2l-1)G(\mathfrak{d}_{2l+1}), \tag{19}$$

where $G(\mathfrak{d}_{2l+1})$ is Casimir's[2] operator G for the group $\mathfrak{d}_{2l+1}$.

It may also be shown that the numbers a and b, which characterize the representations $\mathfrak{B}_W$ according to (16), are connected to the spin and the seniority number by the relations

$$a=(v/2)-S,\quad b=\min(2S,\,2l+1-v). \tag{20}$$

The basis of the representations $\mathfrak{B}_W$ are the functions $\Psi(l^n\alpha vSLM_SM_L)$.

The seniority number could also be introduced before the spin number, by limiting $\mathfrak{c}_{4l+2}$ to its symplectic subgroup which leaves invariant the bilinear antisymmetric form

$$\sum_{-\frac{1}{2}}^{\frac{1}{2}}{}_{m_s}\sum_{-l}^{l}{}_{m_l}(-1)^{m_s+m_l-\frac{1}{2}}\phi_1(m_sm_l)\phi_2(-m_s,-m_l). \tag{21}$$

[2] H. Casimir, Proc. Roy. Acad. Amsterdam 34, 844 (1931).

TABLE IIIa. $(WU|W'U'+f)$ for W'=(000), (100), (110), (200), (111), (210).

		$W'U'$										
		(000)	(100)	(110)		(200)		(111)			(210)	
W	U	(00)	(10)	(10)	(11)	(20)	(00)	(10)	(20)	(11)	(20)	(21)
(000)	(00)	0	1	0	0	0						
(100)	(10)	1	0	$(1/3)^{\frac{1}{2}}$	$(2/3)^{\frac{1}{2}}$	1						
(110)	(10)	0	1	0	0	0	$(3/35)^{\frac{1}{2}}$	$(2/5)^{\frac{1}{2}}$	$(18/35)^{\frac{1}{2}}$	$(2/5)^{\frac{1}{2}}$	$(3/5)^{\frac{1}{2}}$	0
	(11)	0	1	0	0	0	0	$-(1/10)^{\frac{1}{2}}$	$-(9/10)^{\frac{1}{2}}$	0	$(3/35)^{\frac{1}{2}}$	$(32/35)^{\frac{1}{2}}$
(200)	(20)	0	1	0	0	0	0	0	0	$(2/15)^{\frac{1}{2}}$	$(9/35)^{\frac{1}{2}}$	$(64/105)^{\frac{1}{2}}$
(111)	(00)			1	0	0	0	-1	0	0	0	0
	(10)			$(2/3)^{\frac{1}{2}}$	$-(1/3)^{\frac{1}{2}}$	0	$-(1/7)^{\frac{1}{2}}$	$-(3/8)^{\frac{1}{2}}$	$(27/56)^{\frac{1}{2}}$	0	0	0
	(20)			$(2/9)^{\frac{1}{2}}$	$-(7/9)^{\frac{1}{2}}$	0	0	$(1/8)^{\frac{1}{2}}$	$-(7/8)^{\frac{1}{2}}$	0	0	0
(210)	(11)			1	0	1	0	0	0	0	0	0
	(20)			$(7/9)^{\frac{1}{2}}$	$(2/9)^{\frac{1}{2}}$	1	0	0	0	0	0	0
	(21)			0	1	1	0	0	0	0	0	0
(211)	(10)						$(27/35)^{\frac{1}{2}}$	$-(9/40)^{\frac{1}{2}}$	$(1/280)^{\frac{1}{2}}$	$-(3/5)^{\frac{1}{2}}$	$(2/5)^{\frac{1}{2}}$	0
	(11)						0	$(9/10)^{\frac{1}{2}}$	$-(1/10)^{\frac{1}{2}}$	0	$(32/35)^{\frac{1}{2}}$	$-(3/35)^{\frac{1}{2}}$
	(20)						0	$(7/8)^{\frac{1}{2}}$	$(1/8)^{\frac{1}{2}}$	$(1/3)^{\frac{1}{2}}$	$(2/7)^{\frac{1}{2}}$	$-(8/21)^{\frac{1}{2}}$
	(21)						0	0	1	$(3/16)^{\frac{1}{2}}$	$-(25/112)^{\frac{1}{2}}$	$(33/56)^{\frac{1}{2}}$
	(30)						0	0	1	0	$(1/7)^{\frac{1}{2}}$	$(6/7)^{\frac{1}{2}}$
(220)	(20)									$(8/15)^{\frac{1}{2}}$	$-(16/35)^{\frac{1}{2}}$	$(1/105)^{\frac{1}{2}}$
	(21)									$-(1/8)^{\frac{1}{2}}$	$(27/56)^{\frac{1}{2}}$	$(11/28)^{\frac{1}{2}}$
	(22)									0	0	1

3. The Special Case of f^n

It was remarked in Section 6 of III that the seniority number suffices to distinguish the different terms of the same kind in d^n, but not for greater l; for $l \geqslant 3$ we must therefore seek for a subgroup of $\mathfrak{d}_{2l+1}$ which contains $\mathfrak{D}_l$, and it is a very fortunate chance that such a subgroup exists exactly for $l=3$: it is the subgroup of $\mathfrak{d}_7$ which leaves invariant the trilinear antisymmetric form

$$\sum_{mm'm''} V(333;mm'm'')\phi_1(m)\phi_2(m')\phi_3(m''), \quad (22)$$

where $V(abc;\alpha\beta\gamma)$ is defined by (17′) of II. This group is the first of the five simple groups which exist besides the four great classes of simple groups, and is usually denoted as G_2.

If we limit $\mathfrak{d}_7$ to its subgroup G_2, the representations $\mathfrak{B}_W$ break up into irreducible representations $\mathfrak{C}_U$ of G_2; since G_2 is of rank 2, the $\mathfrak{C}_U$ are characterized by a set $U\equiv(u_1u_2)$ of two integral numbers. If we limit G_2 to its subgroup composed by the elements of the representation $\mathfrak{D}_3$ of $\mathfrak{d}_3$, the $\mathfrak{C}_U$ also break up into representations $\mathfrak{D}_L$ of $\mathfrak{d}_3$, the basis of which are the functions $\Psi(f^n\alpha UvSLM_SM_L)$, and these functions will form our definitive system of eigenfunctions of f^n.

The law of reduction of $\mathfrak{B}_W$ as representation of G_2 is given in Table I, that of $\mathfrak{C}_U$ as representation of $\mathfrak{d}_3$ is given in Table II; we see from this last table that the quantum number α must be maintained only for $U\equiv(31)$ and $U\equiv(40)$.

Also the quantum numbers U could be introduced in a similar way as the seniority number in Section 6

TABLE IIIb. $(WU|(211)U'+f)$.

			U'				
W	U	N	(10)	(11)	(20)	(21)	(30)
(111)	(00)	1	1	0	0	0	0
	(10)	$24^{-\frac{1}{2}}$	-1	$8^{\frac{1}{2}}$	$15^{\frac{1}{2}}$	0	0
	(20)	$5832^{-\frac{1}{2}}$	1	$-56^{\frac{1}{2}}$	$135^{\frac{1}{2}}$	$2560^{\frac{1}{2}}$	$3080^{\frac{1}{2}}$
(210)	(11)	$42^{-\frac{1}{2}}$	$-7^{\frac{1}{2}}$	0	$15^{\frac{1}{2}}$	$20^{\frac{1}{2}}$	0
	(20)	$1701^{-\frac{1}{2}}$	$98^{\frac{1}{2}}$	$448^{\frac{1}{2}}$	$270^{\frac{1}{2}}$	$-500^{\frac{1}{2}}$	$385^{\frac{1}{2}}$
	(21)	$672^{-\frac{1}{2}}$	0	$-7^{\frac{1}{2}}$	$-60^{\frac{1}{2}}$	$220^{\frac{1}{2}}$	$385^{\frac{1}{2}}$
(211)	(10)	$72^{-\frac{1}{2}}$	$-5^{\frac{1}{2}}$	$40^{\frac{1}{2}}$	$-27^{\frac{1}{2}}$	0	0
	(11)	$126^{-\frac{1}{2}}$	$35^{\frac{1}{2}}$	0	$-27^{\frac{1}{2}}$	$64^{\frac{1}{2}}$	0
	(20)	$2520^{-\frac{1}{2}}$	$-245^{\frac{1}{2}}$	$-280^{\frac{1}{2}}$	$-867^{\frac{1}{2}}$	$-512^{\frac{1}{2}}$	$616^{\frac{1}{2}}$
	(21)	$315^{-\frac{1}{2}}$	0	$35^{\frac{1}{2}}$	$-27^{\frac{1}{2}}$	$-176^{\frac{1}{2}}$	$77^{\frac{1}{2}}$
	(30)	$315^{-\frac{1}{2}}$	0	0	$27^{\frac{1}{2}}$	$64^{\frac{1}{2}}$	$-224^{\frac{1}{2}}$

of III, by classifying the terms of f^n according to the eigenvalues of

$$R=\sum_{i<j} r_{ij}, \quad (23)$$

where the scalar operator r_{ij} is defined by the relation

$$(f^2LM|r_{ij}|f^2LM)=6\delta(L,3); \quad (24)$$

the equations which correspond to (17), (18), and (19) are

$$r_{ij}=\tfrac{1}{2}-2(\mathbf{s}_i\cdot\mathbf{s}_j)-2q_{ij}-18(\mathbf{u}_i^{(1)}\cdot\mathbf{u}_j^{(1)})-66(\mathbf{u}_i^{(5)}\cdot\mathbf{u}_j^{(5)}), \quad (25)$$

$$R=\tfrac{1}{4}n(n+26)-S(S+1)-2Q-9\mathbf{U}^{(1)2}-33\mathbf{U}^{(5)2}, \quad (26)$$

and

$$9\mathbf{U}^{(1)2}+33\mathbf{U}^{(5)2}=12G(G_2); \quad (27)$$

TABLE IVa. $(UL|U'L'+f)$ for $U'=(00), (10), (11), (20)$.

U	L	(00) S	(10) F	(11) P	(11) H	(20) D	(20) G	(20) I
(00)	S	0	-1	0	0	0	0	0
(10)	F	1	1	$(3/14)^{1/2}$	$(11/14)^{1/2}$	$-(5/27)^{1/2}$	$-(1/3)^{1/2}$	$-(13/27)^{1/2}$
(11)	P	0	1	0	0	$(10/21)^{1/2}$	$-(11/21)^{1/2}$	0
	H	0	1	0	0	$(20/189)^{1/2}$	$(65/231)^{1/2}$	$-(182/297)^{1/2}$
(20)	D	0	1	$-(27/49)^{1/2}$	$-(22/49)^{1/2}$	$-4/7$	$(33/49)^{1/2}$	0
	G	0	1	$(33/98)^{1/2}$	$-(65/98)^{1/2}$	$(55/147)^{1/2}$	$-(125/539)^{1/2}$	$(13/33)^{1/2}$
	I	0	1	0	1	0	$(3/11)^{1/2}$	$(8/11)^{1/2}$
(21)	D			$-(22/49)^{1/2}$	$(27/49)^{1/2}$	$(33/49)^{1/2}$	$4/7$	0
	F			$-(11/14)^{1/2}$	$(3/14)^{1/2}$	$-(55/126)^{1/2}$	$-(8/77)^{1/2}$	$(91/198)^{1/2}$
	G			$(65/98)^{1/2}$	$(33/98)^{1/2}$	$(13/882)^{1/2}$	$(104/147)^{1/2}$	$(5/18)^{1/2}$
	H			0	1	$(13/27)^{1/2}$	$-(16/33)^{1/2}$	$-(10/297)^{1/2}$
	K			0	1	0	$(16/33)^{1/2}$	$-(17/33)^{1/2}$
	L			0	1	0	0	1
(30)	P					$(11/21)^{1/2}$	$(10/21)^{1/2}$	0
	F					$(143/378)^{1/2}$	$-(130/231)^{1/2}$	$(35/594)^{1/2}$
	G					$(11/18)^{1/2}$	$(2/33)^{1/2}$	$-(65/198)^{1/2}$
	H					$(26/63)^{1/2}$	$(18/77)^{1/2}$	$(35/99)^{1/2}$
	I					0	$(8/11)^{1/2}$	$-(3/11)^{1/2}$
	K					0	$(17/33)^{1/2}$	$(16/33)^{1/2}$
	M					0	0	1

TABLE IVb. $(UL|(21)L'+f)$.

U	L	N	L': D	F	G	H	K	L
(11)	P	$1344^{-1/2}$	$220^{1/2}$	$-539^{1/2}$	$-585^{1/2}$	0	0	0
	H	$4928^{-1/2}$	$-270^{1/2}$	$147^{1/2}$	$-297^{1/2}$	$1078^{1/2}$	$1470^{1/2}$	$-1666^{1/2}$
(20)	D	$31360^{-1/2}$	$8910^{1/2}$	$8085^{1/2}$	$351^{1/2}$	$-14014^{1/2}$	0	0
	G	$4312^{-1/2}$	$330^{1/2}$	$147^{1/2}$	$1287^{1/2}$	$1078^{1/2}$	$-1470^{1/2}$	0
	I	$18304^{-1/2}$	0	$-1911^{1/2}$	$1485^{1/2}$	$220^{1/2}$	$4590^{1/2}$	$10098^{1/2}$
(21)	D	$5390^{-1/2}$	$375^{1/2}$	$1960^{1/2}$	$-1144^{1/2}$	$1911^{1/2}$	0	0
	F	$154^{-1/2}$	$-40^{1/2}$	7	$-65^{1/2}$	0	0	0
	G	$630630^{-1/2}$	$-74360^{1/2}$	455	$226941^{1/2}$	$-51744^{1/2}$	$70560^{1/2}$	0
	H	$15730^{-1/2}$	$-2535^{1/2}$	0	$1056^{1/2}$	$-3179^{1/2}$	$7260^{1/2}$	$1700^{1/2}$
	K	$2860^{-1/2}$	0	0	$-192^{1/2}$	$968^{1/2}$	$-85^{1/2}$	$1615^{1/2}$
	L	$572^{-1/2}$	0	0	0	$-40^{1/2}$	$-285^{1/2}$	$247^{1/2}$
(30)	P	$2688^{-1/2}$	34	$-245^{1/2}$	$1287^{1/2}$	0	0	0
	F	$112^{-1/2}$	$-39^{1/2}$	0	$24^{1/2}$	7	0	0
	G	$7920^{-1/2}$	$1375^{1/2}$	$2450^{1/2}$	$-858^{1/2}$	$1617^{1/2}$	$1620^{1/2}$	0
	H	$640640^{-1/2}$	$-42250^{1/2}$	-455	$3971^{1/2}$	$261954^{1/2}$	$490^{1/2}$	$124950^{1/2}$
	I	$9152^{-1/2}$	0	$1274^{1/2}$	$2750^{1/2}$	$1320^{1/2}$	$2125^{1/2}$	$-1683^{1/2}$
	K	$1040^{-1/2}$	0	0	$-204^{1/2}$	$-136^{1/2}$	$605^{1/2}$	$95^{1/2}$
	M	$64^{-1/2}$	0	0	0	0	$-15^{1/2}$	-7
(22)	S	1	0	1	0	0	0	0
	D	$640^{-1/2}$	$-130^{1/2}$	$195^{1/2}$	$297^{1/2}$	$18^{1/2}$	0	0
	G	$51480^{-1/2}$	$18590^{1/2}$	65	$-429^{1/2}$	$-23826^{1/2}$	$-4410^{1/2}$	0
	H	$6160^{-1/2}$	$980^{1/2}$	$-2450^{1/2}$	$1078^{1/2}$	$-132^{1/2}$	$245^{1/2}$	$-1275^{1/2}$
	I	$18304^{-1/2}$	0	$1105^{1/2}$	$9163^{1/2}$	$-2244^{1/2}$	$-4802^{1/2}$	$990^{1/2}$
	L	$364^{-1/2}$	0	0	0	$152^{1/2}$	$-147^{1/2}$	$-65^{1/2}$
	N	$16^{-1/2}$	0	0	0	0	$-5^{1/2}$	$11^{1/2}$

it may also be shown[3] that the eigenvalues of $G(G_2)$ are

$$g(U)=g(u_1u_2)=(u_1^2+u_1u_2+u_2^2+5u_1+4u_2)/12. \quad (28)$$

Although this method of introducing the quantum numbers U avoids the explicit use of the theory of groups, the group-theoretical definition appeared this time more convenient, since the properties of the coefficients of fractional parentage and of the energy matrices, which are connected with this classification and will be obtained in the next sections, could be demonstrated only with the use of the theory of groups.

[3] The general expression of the eigenvalues of Casimir's operator G for every semisimple group will be published elsewhere.

TABLE IVc. $(UL|(30)L'+f)$.

U	L	N	L': P	F	G	H	I	K	M
(20)	D	$490^{-\frac{1}{2}}$	$-54^{\frac{1}{2}}$	$-91^{\frac{1}{2}}$	$189^{\frac{1}{2}}$	$-156^{\frac{1}{2}}$	0	0	0
	G	$5929^{-\frac{1}{2}}$	$-330^{\frac{1}{2}}$	$910^{\frac{1}{2}}$	$126^{\frac{1}{2}}$	$-594^{\frac{1}{2}}$	$2184^{\frac{1}{2}}$	$-1785^{\frac{1}{2}}$	0
	I	$22022^{-\frac{1}{2}}$	0	$-245^{\frac{1}{2}}$	$-1755^{\frac{1}{2}}$	$-2310^{\frac{1}{2}}$	$-2106^{\frac{1}{2}}$	$-4320^{\frac{1}{2}}$	$-11286^{\frac{1}{2}}$
(21)	D	$2695^{-\frac{1}{2}}$	$-578^{\frac{1}{2}}$	$1092^{\frac{1}{2}}$	$700^{\frac{1}{2}}$	$325^{\frac{1}{2}}$	0	0	0
	F	$1694^{-\frac{1}{2}}$	$-55^{\frac{1}{2}}$	0	$-560^{\frac{1}{2}}$	$-715^{\frac{1}{2}}$	$-364^{\frac{1}{2}}$	0	0
	G	$630630^{-\frac{1}{2}}$	$-83655^{\frac{1}{2}}$	$-87360^{\frac{1}{2}}$	$-56784^{\frac{1}{2}}$	$-3971^{\frac{1}{2}}$	$227500^{\frac{1}{2}}$	$171360^{\frac{1}{2}}$	0
	H	$55055^{-\frac{1}{2}}$	0	$12740^{\frac{1}{2}}$	$-7644^{\frac{1}{2}}$	$18711^{\frac{1}{2}}$	$-7800^{\frac{1}{2}}$	$-8160^{\frac{1}{2}}$	0
	K	$165165^{-\frac{1}{2}}$	0	0	$-16848^{\frac{1}{2}}$	$77^{\frac{1}{2}}$	$-27625^{\frac{1}{2}}$	$79860^{\frac{1}{2}}$	$-40755^{\frac{1}{2}}$
	L	$17017^{-\frac{1}{2}}$	0	0	0	$-1785^{\frac{1}{2}}$	$-1989^{\frac{1}{2}}$	$-1140^{\frac{1}{2}}$	$12103^{\frac{1}{2}}$
(30)	P	$16^{-\frac{1}{2}}$	0	$13^{\frac{1}{2}}$	$3^{\frac{1}{2}}$	0	0	0	0
	F	$1232^{-\frac{1}{2}}$	$429^{\frac{1}{2}}$	$-77^{\frac{1}{2}}$	$273^{\frac{1}{2}}$	$-33^{\frac{1}{2}}$	$-420^{\frac{1}{2}}$	0	0
	G	$55440^{-\frac{1}{2}}$	$-3465^{\frac{1}{2}}$	$-9555^{\frac{1}{2}}$	$7203^{\frac{1}{2}}$	$27797^{\frac{1}{2}}$	$1300^{\frac{1}{2}}$	$-6120^{\frac{1}{2}}$	0
	H	$11440^{-\frac{1}{2}}$	0	$-195^{\frac{1}{2}}$	$-4693^{\frac{1}{2}}$	$1232^{\frac{1}{2}}$	$2600^{\frac{1}{2}}$	$2720^{\frac{1}{2}}$	0
	I	$32032^{-\frac{1}{2}}$	0	$5880^{\frac{1}{2}}$	$520^{\frac{1}{2}}$	$-6160^{\frac{1}{2}}$	$-1911^{\frac{1}{2}}$	$15680^{\frac{1}{2}}$	$-1881^{\frac{1}{2}}$
	K	$510510^{-\frac{1}{2}}$	0	0	$33813^{\frac{1}{2}}$	$89012^{\frac{1}{2}}$	$-216580^{\frac{1}{2}}$	$8085^{\frac{1}{2}}$	$163020^{\frac{1}{2}}$
	M	$3808^{-\frac{1}{2}}$	0	0	0	0	$153^{\frac{1}{2}}$	$960^{\frac{1}{2}}$	$2695^{\frac{1}{2}}$

5. THE CALCULATION OF THE FRACTIONAL PARENTAGES OF f^n

1. General Properties

The eigenfunctions of l^n, which are the basis of $\{\mathfrak{c}_{4l+2}\}^n$, may be obtained by reduction of the representation $\{\mathfrak{c}_{4l+2}\}^{n-1}\times(\mathfrak{c}_{4l+2})$:

$$\begin{aligned}\Psi(l^n\alpha vSLM_SM_L) &= \sum \Psi(l^{n-1}\alpha'v'S'L'M_S'M_L')\phi(m_sm_l)\\ &\times(l^{n-1}\alpha'v'S'L'M_S'M_L', lm_sm_l|l^n\alpha vSLM_SM_L); \quad (29)\end{aligned}$$

owing to the particular choice of the scheme which was made in the preceding section, it follows from the lemma (11) that the coefficients of this transformation break up in a product of different factors, each of which depends only from a smaller number of variables:

$$\begin{aligned}&(l^{n-1}\alpha'v'S'L'M_S'M_L', lm_sm_l|l^n\alpha vSLM_SM_L)\\ &= (S'\tfrac{1}{2}M_S'm_s|S'\tfrac{1}{2}SM_S)(L'lM_L'm_l|L'lLM_L)\\ &\times(W'\alpha'L'+l|W\alpha L)(l^{n-1}v'S'+l|\}l^nvS). \quad (30)\end{aligned}$$

Confronting this expression with (10) of III, we see that the coefficients of fractional parentage are the product of two factors:

$$\begin{aligned}&(l^{n-1}(\alpha'v'S'L')lSL|\}l^n\alpha vSL)\\ &= (W'\alpha'L'+l|W\alpha L)(l^{n-1}v'S'+l|\}l^nvS); \quad (31)\end{aligned}$$

the relations (58) of III are particular cases of this result. Owing to the unitary of all our transformations, the factors of (31) satisfy the orthogonality relations

$$\sum_{v'S'}(l^nvS\{|l^{n-1}v'S'+l)(l^{n-1}v'S'+l|\}l^nvS)=1 \quad (32)$$

and

$$\sum_{\alpha'L'}(W\alpha L|W'\alpha'L'+l)(W'\alpha'L'+l|W''\alpha''L) = \delta(WW'')\delta(\alpha\alpha''). \quad (33)$$

For the particular case $l=3$, owing to the existence of the intermediate group G_2, the coefficients of fractional parentage are the product of three factors:

$$\begin{aligned}&(f^{n-1}(\alpha'U'v'S'L')fSL|\}f^n\alpha UvSL)\\ &= (U'\alpha'L'+f|U\alpha L)(W'U'+f|WU)\\ &\times(f^{n-1}v'S'+f|\}f^nvS), \quad (34)\end{aligned}$$

and the orthogonality relations (33) break up into

$$\sum_{U'}(WU|W'U'+f)(W'U'+f|W''U)=\delta(WW'') \quad (35)$$

and

$$\sum_{\alpha'L'}(U\alpha L|U'\alpha'L'+f)(U'\alpha'L'+f|U''\alpha''L) = \delta(UU'')\delta(\alpha\alpha''). \quad (36)$$

In order to find also relations of the type (61) of III, we consider now the identical representations of $\mathfrak{d}_{2l+1}$ which appears in the reduction of $\mathfrak{B}_{W_1}\times\mathfrak{B}_{W_2}$. Such representations may only appear if $W_1=W_2$, and since the tensors of odd degree are diagonal with respect to W (see (70) of III), we obtain in the same way as in Section 6 of II that

$$\begin{aligned}&(W_1\alpha_1L_1+W_2\alpha_2L_2|(00\cdots0)0)\\ &= [(2L_1+1)/g_{W_1}]^{\frac{1}{2}}\delta(W_1W_2)\delta(\alpha_1\alpha_2)\delta(L_1L_2), \quad (37)\end{aligned}$$

where g_{W_1} is the order of the representation $\mathfrak{B}_{W_1}$. Owing to the value (16′) of II of $(LLMM'|LL00)$, we get also

$$\begin{aligned}&(W_1\alpha_1L_1M_1, W_2\alpha_2L_2M_2|W_1W_2, (00\cdots0)00)\\ &= g_{W_1}^{-\frac{1}{2}}(-1)^{L_1-M_1}\delta(W_1W_2)\delta(\alpha_1\alpha_2)\\ &\times\delta(L_1L_2)\delta(M_1, -M_2). \quad (38)\end{aligned}$$

TABLE V. $c(UU'(40))$.

	(00)	(10)	(11)	(20)	(21)	(30)	(22)	(31)	(40)
(00)	0	0	0	0	0	0	0	0	1
(10)	0	0	0	0	0	1	0	1	1
(11)	0	0	0	0	1	1	0	1	1
(20)	0	0	0	1	1	1	1	2	2
(21)	0	0	1	1	2	2	2	3	2
(30)	0	1	1	1	2	2	1	3	2
(22)	0	0	0	1	2	1	1	2	2
(31)	0	1	1	2	3	3	2	5	3
(40)	1	1	1	2	2	2	2	3	3

If, in the general formula,[4]

$$\int (W_1\alpha_1L_1M_1|R|W_1\alpha_1'L_1'M_1')\times(W_2\alpha_2L_2M_2|R|W_2\alpha_2'L_2'M_2')\times(W\alpha LM|R|W\alpha'L'M')^*dR$$
$$=g_W^{-1}(W_1\alpha_1L_1M_1,\ W_2\alpha_2L_2M_2|W_1W_2,\ W\alpha LM)\times(W_1W_2,\ W\alpha'L'M'|W_1\alpha_1'L_1'M_1',\ W_2\alpha_2'L_2'M_2')\times\int dR, \quad (39)$$

we consider the special case $W\equiv(00\cdots0)$ and introduce (38), we get

$$\int (W_1\alpha_1L_1M_1|R|W_1\alpha_1'L_1'M_1')\times(W_2\alpha_2L_2M_2|R|W_2\alpha_2'L_2'M_2')dR$$
$$=g_W^{-1}(-1)^{L_1+L_1'-M_1-M_1'}\delta(W_1W_2)\times\delta(\alpha_1\alpha_2)\delta(L_1L_2)\delta(M_1,-M_2)\times\delta(\alpha_1'\alpha_2')\delta(L_1'L_2')\delta(M_1',-M_2')\int dR, \quad (40)$$

and confronting this result with the orthogonality relation[5]

$$\int (W_1\alpha_1L_1M_1|R|W_1\alpha_1'L_1'M_1')^*\times(W_2\alpha_2L_2M_2|R|W_2\alpha_2'L_2'M_2')dR$$
$$=g_{W_1}^{-1}\delta(W_1W_2)\delta(\alpha_1\alpha_2)\delta(L_1L_2)\delta(M_1M_2)\times\delta(\alpha_1'\alpha_2')\delta(L_1'L_2')\delta(M_1'M_2')\int dR, \quad (41)$$

we obtain from the well-known corollary of Schwarz's inequality that

$$(W\alpha LM|R|W\alpha'L'M')^*=(-1)^{L+L'-M-M'}(W\alpha L-M|R|W\alpha'L'-M'). \quad (42)$$

Applying this result to the first and third factor in the left side of (39), we have

$$(W_1\alpha_1L_1M_1,\ W_2\alpha_2L_2M_2|W_1W_2,\ W\alpha LM(W_1W_2,\ W\alpha'L'M'|W_1\alpha_1'L_1'M_1',\ W_2\alpha_2'L_2'M_2')$$
$$=(-1)^{L+L'-M-M'-L_1-L_1'+M_1+M_1'}(g_W/g_{W_1})(W\alpha L-M,\ W_2\alpha_2L_2M_2|WW_2,\ W_1\alpha_1L_1-M_1)$$
$$\times(WW_2,\ W_1\alpha_1'L_1'-M_1'|W\alpha'L'-M',\ W_2\alpha_2'L_2'M_2'), \quad (43)$$

and since, owing to (16′) of II and (19a) of II,

$$(L_1L_2M_1M_2|L_1L_2LM)=(-1)^{L_2+M-M_1}[(2L+1)/(2L_1+1)]^{\frac{1}{2}}\times(LL_2-MM_2|LL_2L_1-M_1), \quad (44)$$

we have also

$$(W_1\alpha_1L_1+W_2\alpha_2L_2|W\alpha L)(W\alpha'L'|W_1\alpha_1'L_1'+W_2\alpha_2'L_2')$$
$$=(-1)^{L+L_2-L_1+L'+L_2'-L_1'}[(2L_1+1)(2L_1'+1)/(2L+1)(2L'+1)]^{\frac{1}{2}}(g_W/g_{W_1})$$
$$\times(W\alpha L+W_2\alpha_2L_2|W_1\alpha_1L_1)\times(W_1\alpha_1'L_1'|W\alpha'L'+W_2\alpha_2'L_2'). \quad (45)$$

This equation may be satisfied only if

$$(W\alpha L+W_2\alpha_2L_2|W_1\alpha_1L_1)=(-1)^{L_1-L_2-L+x}[(2L+1)g_{W_1}/(2L_1+1)g_W]^{\frac{1}{2}}\times(W_1\alpha_1L_1+W_2\alpha_2L_2|W\alpha L), \quad (46)$$

where x is independent of the L and depends only on the W. The value of x is to a some extent arbitrary, since it depends from our choice of phases. For the particular case $W_2\equiv(10\cdots0)\equiv l$, which is important for us, we put $x=l=L_2$, and therefore,

$$(W\alpha L+l|W'\alpha'L')=(-1)^{L-L'}\times[(2L+1)g_{W'}/(2L'+1)g_W]^{\frac{1}{2}}(W'\alpha'L'+l|W\alpha L); \quad (47)$$

the relation (61) of III is a particular case of this result.

It is easy to see that for $l=3$ the relation (47) breaks up into

$$(WU+f|W'U')=(g_Ug_{W'}/g_{U'}g_W)^{\frac{1}{2}}(W'U'+f|WU) \quad (48)$$

and

$$(U\alpha L+f|U'\alpha'L')=(-1)^{L-L'}\times[(2L+1)g_{U'}/(2L'+1)g_U]^{\frac{1}{2}}(U'\alpha'L'+f|U\alpha L). \quad (49)$$

2. The Calculation of $(l^{n-1}v'S'+l|\}l^nvS)$

Applying (1) to (3) and owing to (31) and (33), we have

[4] See E. Wigner, *Gruppentheorie* (Friedrich Vieweg and Sohn, Braunschweig, 1931), p. 204, Eq. (22).

[5] Reference 4, p. 110, Eq. (11).

TABLE VIa. $(U|\chi(L)|U')$ for U, $U' \neq (31)$, (40).

$(U\|\chi\|U')$	S	P	D	F	G	H	I	K	L	M	N
$(20\|\chi\|20)$	0	0	143	0	-130	0	35	0	0	0	0
$(11\|\chi\|21)$	0	0	0	0	0	1	0	0	0	0	0
$(20\|\chi\|21)$	0	0	$-39\sqrt{2}$	0	$4(65)^{\frac{1}{2}}$	0	0	0	0	0	0
$(21\|\chi_1\|21)$	0	0	377	455	-561	49	0	-315	245	0	0
$(21\|\chi_2\|21)$	0	0	13	-65	55	-75	0	133	-75	0	0
$(10\|\chi\|30)$	0	0	0	1	0	0	0	0	0	0	0
$(11\|\chi\|30)$	0	$-13(11)^{\frac{1}{2}}$	0	0	0	$(39)^{\frac{1}{2}}$	0	0	0	0	0
$(20\|\chi\|30)$	0	0	0	0	$-13(5)^{\frac{1}{2}}$	0	30	0	0	0	0
$(21\|\chi\|30)$	0	0	0	$12(195)^{\frac{1}{2}}$	$8(143)^{\frac{1}{2}}$	$11(42)^{\frac{1}{2}}$	0	$-4(17)^{\frac{1}{2}}$	0	0	0
$(30\|\chi\|30)$	0	-52	0	38	-52	88	25	-94	0	25	0
$(20\|\chi\|22)$	0	0	$3(429)^{\frac{1}{2}}$	0	$-38(65)^{\frac{1}{2}}$	0	$21(85)^{\frac{1}{2}}$	0	0	0	0
$(21\|\chi\|22)$	0	0	$45(78)^{\frac{1}{2}}$	0	$12(11)^{\frac{1}{2}}$	$-12(546)^{\frac{1}{2}}$	0	0	$-8(665)^{\frac{1}{2}}$	0	0
$(22\|\chi\|22)$	260	0	-25	0	94	104	-181	0	-36	0	40

TABLE VIb. $(U|\chi(L)|31)$.

L	$(10\|\chi\|31)$	$(11\|\chi\|31)$	$(20\|\chi\|31)$	$(21\|\chi\|31)$	$(30\|\chi\|31)$	$(31\|\chi\|31)$
P	0	$11(330)^{\frac{1}{2}}$	0	0	$76(143)^{\frac{1}{2}}$	-6644
D	0	0	$-8(78)^{\frac{1}{2}}$	$-60(39/7)^{\frac{1}{2}}$	0	4792
F	0	0	0	$-312(5)^{\frac{1}{2}}$	$-48(39)^{\frac{1}{2}}$	‖ 4420 $336(143)^{\frac{1}{2}}$ ‖
F'	1	0	0	$12(715)^{\frac{1}{2}}$	$-98(33)^{\frac{1}{2}}$	‖ $336(143)^{\frac{1}{2}}$ -902 ‖
G	0	0	$5(65)^{\frac{1}{2}}$	$2024/(7)^{\frac{1}{2}}$	$20(1001)^{\frac{1}{2}}$	-2684
H	0	$11(85)^{\frac{1}{2}}$	0	$31(1309/3)^{\frac{1}{2}}$	$-20(374)^{\frac{1}{2}}$	‖ -2024 $-48(6545)^{\frac{1}{2}}$ ‖
H'	0	$-25(77)^{\frac{1}{2}}$	0	$103(5/3)^{\frac{1}{2}}$	$-44(70)^{\frac{1}{2}}$	‖ $-48(6545)^{\frac{1}{2}}$ 2680 ‖
I	0	0	$10(21)^{\frac{1}{2}}$	0	$-57(33)^{\frac{1}{2}}$	‖ $-12661/5$ $-3366(34)^{\frac{1}{2}}/5$ ‖
I'	0	0	0	0	$18(1122)^{\frac{1}{2}}$	‖ $-3366(34)^{\frac{1}{2}}/5$ $17336/5$ ‖
K	0	0	0	$-52(323/23)^{\frac{1}{2}}$	$-494(19/23)^{\frac{1}{2}}$	‖ $123506/23$ $144(21318)^{\frac{1}{2}}/23$ ‖
K'	0	0	0	$-336(66/23)^{\frac{1}{2}}$	$73(1122/23)^{\frac{1}{2}}$	‖ $144(21318)^{\frac{1}{2}}/23$ $-85096/23$ ‖
L	0	0	0	$-24(190)^{\frac{1}{2}}$	0	-4712
M	0	0	0	0	$-21(385)^{\frac{1}{2}}$	-473
N	0	0	0	0	0	1672
O	0	0	0	0	0	220

TABLE VIc. $(U|\chi(L)|40)$.

L	$(00\|\chi\|40)$	$(10\|\chi\|40)$	$(20\|\chi\|40)$	$(30\|\chi\|40)$	$(40\|\chi\|40)$
S	1	0	0	0	-1408
D	0	0	$-88(13)^{\frac{1}{2}}$	0	-44
F	0	1	0	$90(11)^{\frac{1}{2}}$	1078
G	0	0	$53(715/27)^{\frac{1}{2}}$	$-16(1001)^{\frac{1}{2}}$	‖ $-16720/9$ $-34(2618)^{\frac{1}{2}}/9$ ‖
G'	0	0	$7(15470/27)^{\frac{1}{2}}$	$64(442)^{\frac{1}{2}}$	‖ $-34(2618)^{\frac{1}{2}}/9$ $10942/9$ ‖
H	0	0	0	$-72(462)^{\frac{1}{2}}$	-704
I	0	0	$34(1045/31)^{\frac{1}{2}}$	$-9(21945/31)^{\frac{1}{2}}$	‖ $-2453/31$ $60(74613)^{\frac{1}{2}}/31$ ‖
I'	0	0	$-12(1785/31)^{\frac{1}{2}}$	$756(85/31)^{\frac{1}{2}}$	‖ $60(74613)^{\frac{1}{2}}/31$ $36088/31$ ‖
K	0	0	0	$-84(33)^{\frac{1}{2}}$	-132
L	0	0	0	0	‖ $-4268/31$ $924(1995)^{\frac{1}{2}}/31$ ‖
L'	0	0	0	0	‖ $924(1995)^{\frac{1}{2}}/31$ $11770/31$ ‖
M	0	0	0	$-99(15)^{\frac{1}{2}}$	-1067
N	0	0	0	0	528
Q	0	0	0	0	22

$$S(S+1)-3n/4 = [n/(n-2)]\sum_{v'S'}[S'(S'+1)-3(n-1)/4] \times (l^{n-1}v'S'+l|\}l^n vS)^2; \quad (50)$$

since S' may have only the two values $S-\frac{1}{2}$ and $S+\frac{1}{2}$, we obtain from (32) and (50) that

$$\begin{aligned}\textstyle\sum_{v'}(l^{n-1}v'S-\tfrac{1}{2}+l|\}l^n vS)^2 &= (n+2S+2)S/n(2S+1),\\ \textstyle\sum_{v'}(l^{n-1}v'S+\tfrac{1}{2}+l|\}l^n vS)^2 &= (n-2S)(S+1)/n(2S+1).\end{aligned} \quad (51)$$

Since, for $n=v$, v' may have only the value $v-1$, we get

$$(l^{v-1}v-1\ S-\tfrac{1}{2}+l|\}l^v vS)^2 = (v+2S+2)S/v(2S+1),$$
$$(l^{v-1}v-1\ S+\tfrac{1}{2}+l|\}l^v vS)^2 = (v-2S)(S+1)/v(2S+1),$$

and owing to (58) of III,

$$\begin{aligned}(l^{n-1}v-1\ S-\tfrac{1}{2}+l|\}l^n vS)^2 &= (4l+4-n-v)(v+2S+2)S/2n(2l+2-v)(2S+1),\\ (l^{n-1}v-1\ S+\tfrac{1}{2}+l|\}l^n vS)^2 &= (4l+4-n-v)(v-2S)(S+1)/2n(2l+2-v)(2S+1);\end{aligned} \quad (52a)$$

TABLE VII. $x((210), UU')$.

	(11)	(20)	(21)
(11)	0	0	$12(455)^{\frac{1}{2}}$
(20)	0	$-6/7$	$6(66)^{\frac{1}{2}}/7$
(21)	$12(455)^{\frac{1}{2}}$	$6(66)^{\frac{1}{2}}/7$	3/7, 0

TABLE VIII. $x((211), UU')$.

	(10)	(11)	(20)	(21)	(30)
(10)	0	0	0	0	$-20(143)^{\frac{1}{2}}$
(11)	0	0	0	$10(182)^{\frac{1}{2}}$	10
(20)	0	0	$-8/7$	$4(33)^{\frac{1}{2}}/7$	$4\sqrt{3}$
(21)	0	$10(182)^{\frac{1}{2}}$	$4(33)^{\frac{1}{2}}/7$	4/7, 3	2
(30)	$-20(143)^{\frac{1}{2}}$	10	$4\sqrt{3}$	2	2

subtracting (52a) from (51) we also get

$$\begin{aligned}(l^{n-1}v+1\,S-\tfrac{1}{2}+l|\}l^nvS)^2 &= (n-v)(4l+6-v+2S)S/2n(2l+2-v)(2S+1),\\ (l^{n-1}v+1\,S+\tfrac{1}{2}+l|\}l^nvS)^2 &= (n-v)(4l+4-v-2S)(S+1)/\\ &\qquad 2n(2l+2-v)(2S+1).\end{aligned} \tag{52b}$$

The phases of $(l^{n-1}v'S'+l|\}l^nvS)$ are independent of n and will be denoted by $\epsilon(v'S'|\}vS)$; they are arbitrary as long as the phases of $(W'\alpha'L'+l|W\alpha L)$ are not fixed. The latter are partially fixed by (47), and, comparing it with (61) of III, we have

$$\epsilon(v+1S'|\}vS)=(-1)^{l+S-S'+\frac{1}{2}}\epsilon(vS|\}v+1S'),$$

or, in a more general form,

$$\epsilon(v'S'|\}vS)=(-1)^{l+S-S'+(v'-v)/2}\epsilon(vS|\}v'S'). \tag{53}$$

Another partial limitation in the choice of $\epsilon(v'S'|\}vS)$ is given by the fact that, according to (20), every value of W corresponds to two couples of v and S, which are related by the equations,

$$v_1+2S_2=v_2+2S_1=2l+1; \tag{54}$$

it may be shown that from this fact follows the relation

$$\frac{\epsilon(v_1-2\,S_1|\}v_1-1\,S_1-\frac{1}{2})\epsilon(v_1-1\,S_1-\frac{1}{2}|\}v_1S_1)}{\epsilon(v_1-2\,S_1|\}v_1-1\,S_1+\frac{1}{2})\epsilon(v_1-1\,S_1+\frac{1}{2}|\}v_1S_1)} = \frac{\epsilon(v_2S_2+1|\}v_2+1\,S_2+\frac{1}{2})\epsilon(v_2+1\,S_2+\frac{1}{2}|\}v_2S_2)}{\epsilon(v_2S_2+1|\}v_2-1\,S_2+\frac{1}{2})\epsilon(v_2-1\,S_2+\frac{1}{2}|\}v_2S_2)}, \tag{55}$$

when v_1, S_1, v_2, and S_2 satify (54).

In order to satisfy (53) and (55), the following choice of phases was made for $l=3$:

$$\begin{aligned}\epsilon(v'S'|\}vS)&=(-1)^{S'} && \text{for } v \text{ odd},\\ \epsilon(v'S'|\}vS)&=(-1)^{S'+(v'-v)/2} && \text{for } v \text{ even}.\end{aligned} \tag{56}$$

3. The Calculation of $(W'U'+f|WU)$ and $(U'L'+f|UL)$

The coefficients of fractional parentage of f and f^2 are equal to unity; those of f^3 were calculated from (9) of III, and, in the cases where two doublets with the same L were allowed, $(f^2(^3F)f\,L|\}f^3\alpha\,L)$ was put equal to 0 in one of them, since for doublets of f^3 with $U\equiv(21)$ the eigenvalue of R vanishes according to (26), (27), and (28).

From the coefficients of fractional parentage of f, f^2, and f^3, several elements of $(W'U'+f|WU)$ and $(U'L'+f|UL)$ were obtained and are given in Tables III and IV. These tables were then extended by using (48), (49), (35), (36), and also (1) as it was pointed out in Section 2. In the very few cases where these equations were not sufficient for the determination of some elements, additional equations were obtained from (23) of III by requiring that $(f^nUvSL\|U^{(5)}\|f^nU'v'SL')$ should vanish unless $v=v'$ and $U=U'$, since the tensor $\mathbf{U}^{(5)}$ commutes with Q and R.

Owing to the present status of the experimental classification of the spectra of the rare earths, the terms of lower multiplicity are not yet interesting; we limited therefore Tables III and IV to those elements which are of use in the calculation of the coefficients of fractional parentage for f^4 and for the two highest multiplicities of f^5, f^6, and f^7.

6. THE SPECTRA OF f^n

1. The Choice of the Parameters

In Section 4 of II we considered the coefficients of Slater's integrals F^k as scalar products of tensors in the three-dimensional space; we shall now show that they may also be considered as particular components of tensors in the $(2l+1)$-dimensional space.

In full analogy to Section 3 of II it is possible to define as an irreducible tensor of the "type" W in the $(2l+1)$-dimensional space each operator whose components transform by a $(2l+1)$-dimensional rotation as the elements of the basis of the representation $\mathfrak{B}_W$ of $\mathfrak{d}_{2l+1}$.

In the three-dimensional space $\mathbf{u}^{(k)}$ was a tensor, and its components $u_q^{(k)}$ transformed as the spherical harmonics $Y(kq)$; in the $(2l+1)$-dimensional space $\mathbf{u}^{(k)}$ alone is no longer a tensor, but it may be shown that the quantities $(2k+1)^{\frac{1}{2}}u_q^{(k)}$ transform as the functions $\psi((20\cdots0)kq)$ if k is even, and as the functions $\psi((110\cdots0)kq)$ if k is odd, i.e., all the quantities $(4t+1)^{\frac{1}{2}}u_q^{(2t)}$ or $(4t-1)^{\frac{1}{2}}u_q^{(2t-1)}$ for $1\leqslant t\leqslant l$ are *together* the components of a sole tensor.

TABLE IX. $x((220), UU')$.

	(20)	(21)	(22)
(20)	3/14	$3(55)^{\frac{1}{2}}/7$	$-3(5/28)^{\frac{1}{2}}$
(21)	$3(55)^{\frac{1}{2}}/7$	−6/7, −3	$3/(7)^{\frac{1}{2}}$
(22)	$-3(5/28)^{\frac{1}{2}}$	$3/(7)^{\frac{1}{2}}$	3/2

TABLE X. $x((221), UU')$.

	(10)	(11)	(20)	(21)	(30)	(31)
(10)	0	0	0	0	$5(143)^{\frac{1}{2}}$	$-15(429)^{\frac{1}{2}}$
(11)	0	0	0	$14(910/11)^{\frac{1}{2}}$	$2(10)^{\frac{1}{2}}$	$2(39)^{\frac{1}{2}}/11$
(20)	0	0	2/7	$-10(6)^{\frac{1}{2}}/7$	$\sqrt{3}$	$9(3/7)^{\frac{1}{2}}$
(21)	0	$14(910/11)^{\frac{1}{2}}$	$-10(6)^{\frac{1}{2}}/7$	$-1/7$, $12/11$	$5(2/11)^{\frac{1}{2}}$	$3\sqrt{2}/11$
(30)	$5(143)^{\frac{1}{2}}$	$2(10)^{\frac{1}{2}}$	$\sqrt{3}$	$5(2/11)^{\frac{1}{2}}$	$-1/2$	$3/2(11)^{\frac{1}{2}}$
(31)	$-15(429)^{\frac{1}{2}}$	$2(39)^{\frac{1}{2}}/11$	$9(3/7)^{\frac{1}{2}}$	$3\sqrt{2}/11$	$3/2(11)^{\frac{1}{2}}$	1/22

TABLE XI. $x((222), UU')$.

	(00)	(10)	(20)	(30)	(40)
(00)	0	0	0	0	$-30(143)^{\frac{1}{2}}$
(10)	0	0	0	$-3(1430)^{\frac{1}{2}}$	$9(1430)^{\frac{1}{2}}$
(20)	0	0	6/11	$-3(42/11)^{\frac{1}{2}}$	$9\sqrt{2}/11$
(30)	0	$-3(1430)^{\frac{1}{2}}$	$-3(42/11)^{\frac{1}{2}}$	-3	$1/(11)^{\frac{1}{2}}$
(40)	$-30(143)^{\frac{1}{2}}$	$9(1430)^{\frac{1}{2}}$	$9\sqrt{2}/11$	$1/(11)^{\frac{1}{2}}$	3/11

In the seven-dimensional space the quantities

$$\sum_{kqk'q'} [(2k+1)(2k'+1)]^{\frac{1}{2}} u_1{}^{(k)}{}_q u_2{}^{(k')}{}_{q'} \times((200)(20)kq,\ (200)(20)k'q'|WUKQ) \quad (57)$$

will transform as $\psi(WUKQ)$, and, in particular, the quantities

$$\sum_{kq}(2k+1)u_1{}^{(k)}{}_q u_2{}^{(k)}{}_{-q}(kkq-q|kk00) \times((200)(20)k+(200)(20)k|WU0)$$
$$=\sum_k(2k+1)^{\frac{1}{2}}(\mathbf{u}_1{}^{(k)}\cdot\mathbf{u}_2{}^{(k)}) \times((200)(20)k+(200)(20)k|WU0) \quad (58)$$

will have the tensorial properties of $\psi(WUS)$.

Since, according to (45) of II,

$$f_k(f^2L)=(f^2LM|(\mathbf{C}_1{}^{(k)}\cdot\mathbf{C}_2{}^{(k)})|f^2LM),$$

or also

$$f_k(f^2L)=(3\|C^{(k)}\|3)^2(f^2LM|(\mathbf{u}_1{}^{(k)}\cdot\mathbf{u}_2{}^{(k)})|f^2LM), \quad (59)$$

we shall substitute to the f_k the linear combinations

$$N_i\sum_k(2k+1)^{\frac{1}{2}}(3\|C^{(k)}\|3)^{-2}f_k \times((200)(20)k+(200)(20)k|W_iU_iS), \quad (60)$$

where the N_i are convenient normalization factors and

$$W_1U_1\equiv(000)(00),\quad W_2U_2\equiv(400)(40),\quad W_3U_3\equiv(220)(22). \quad (61)$$

Since the only parents of the functions $\psi((000)(00)S)$, $\psi((400)(40)S)$, and $\psi((220)(22)S)$ are, respectively, the functions

$\psi((100)(10)F)$, $\psi((300)(30)F)$, and $\psi((210)(21)F)$,

the coefficients

$$((200)(20)k+(200)(20)k|W_iU_iS)$$

will be proportional to $((20)k+f|(10)F)$, $((20)k+f|(30)F)$ and $((20)k+f|(21)F)$, which are given in Table IV. Taking the values of $(3\|C^{(k)}\|3)$ from (51) of II and remembering that

$$f^k=D_kf_k, \quad (62)$$

where the D_k are the denominators of Table II[6] of TAS,* we define for the configurations f^n,

$$\begin{aligned} e_0&=f^0=n(n-1)/2,\\ e_1&=9f^0/7+f^2/42+f^4/77+f^6/462,\\ e_2&=143f^2/42-130f^4/77+35f^6/462,\\ e_3&=11f^2/42+4f^4/77-7f^2/462; \end{aligned} \quad (63)$$

the term $9f^0/7$ was added for convenience in e_1 without changing its tensorial properties, since both f^0 and e_1 are scalars in the seven-dimensional space.

The general expression of the energy matrices of f^n will be

$$e_0E^0+e_1E^1+e_2E^2+e_3E^3 \quad (64)$$

instead of

$$f^0F_0+f^2F_2+f^4F_4+f^6F_6; \quad (65)$$

the E^i are linear combinations of Slater's parameters, which are, however, different from those adopted empirically in (96) of II:

$$\begin{aligned} E^0&=F_0-10F_2-33F_4-286F_6,\\ E^1&=(70F_2+231F_4+2002F_6)/9,\\ E^2&=(F_2-3F_4+7F_6)/9,\\ E^3&=(5F_2+6F_4-91F_6)/3; \end{aligned} \quad (66)$$

TABLE XII. $c(WW'(220))$.

	(000)	(100)	(110)	(200)	(111)	(210)	(211)	(220)	(221)	(222)
(000)	0	0	0	0	0	0	0	1	0	0
(100)	0	0	0	0	0	1	0	0	1	0
(110)	0	0	1	0	0	0	1	1	1	0
(200)	0	0	0	1	0	0	1	1	0	1
(111)	0	0	0	0	1	1	1	1	1	0
(210)	0	1	0	0	1	2	1	0	2	1
(211)	0	0	1	1	1	1	3	1	2	1
(220)	1	0	1	1	1	0	1	2	1	1
(221)	0	1	1	0	1	2	2	1	3	1
(222)	0	0	0	1	0	1	1	1	1	1

* E. U. Condon and G. H. Shortley, *Theory of Atomic Spectra* (Cambridge University Press, London, 1935).

TABLE XIII. $c(UU'(22))$.

	(00)	(10)	(11)	(20)	(21)	(30)	(22)	(31)	(40)
(00)	0	0	0	0	0	0	1	0	0
(10)	0	0	0	0	1	0	0	1	0
(11)	0	0	1	0	0	1	1	1	0
(20)	0	0	0	1	1	1	1	1	1
(21)	0	1	0	1	2	1	0	2	2
(30)	0	0	1	1	1	2	1	2	1
(22)	1	0	1	1	0	1	2	1	1
(31)	0	1	1	1	2	2	1	3	2
(40)	0	0	0	1	2	1	1	2	2

the formulas for f^2 assume now the form

$$
\begin{aligned}
{}_0{}^1S&=E^0+9E^1 \\
{}_2{}^3P&=E^0+33E^3 \\
{}_2{}^1D&=E^0+2E^1+286E^2-11E^3 \\
{}_2{}^3F&=E^0 \qquad (67)\\
{}_2{}^1G&=E^0+2E^1-260E^2-4E^3 \\
{}_2{}^3H&=E^0-9E^3 \\
{}_2{}^1I&=E^0+2E^1+70E^2+7E^3
\end{aligned}
$$

For $n>2$, the e_i are matrices whose order equals the number of allowed states for a given SL; the elements of these matrices may be calculated by means of (1), but most of the calculations may be avoided by considering the tensorial properties of the e_i.

e_1 is a scalar also in the seven-dimensional space; it is therefore diagonal in the $vUSL$ scheme, and its eigenvalues are independent of L and U. We have from (67) that

$$e_1(f^2SL)=q_{12}+\tfrac{1}{2}-2(\mathbf{s}_1\cdot\mathbf{s}_2), \tag{68}$$

and owing to (50) of III and (3) we obtain that, in general, the eigenvalues of e_1 are

$$e_1(f^nvUSL)=9(n-v)/2+v(v+2)/4-S(S+1). \tag{69}$$

The matrices e_2 and e_3 are particular components of tensors in the seven-dimensional space. The dependence of their elements on U and L will be analogous to (28) of II, but the result is somewhat more complicated, since, in the decomposition of the external product of two irreducible representations of $\mathfrak{d}_7$, some representation may appear more than once. We now have

$$(f^nvWUSL|e_i|f^nv'W'U'SL)=\textstyle\sum_\alpha A_\alpha(W'U'L+W_iU_iS|\alpha WUL); \tag{70}$$

or, owing to an obvious extension of (46),

$$(f^nvWUSL|e_i|f^nv'W'U'SL)=\textstyle\sum_\beta B_\beta(WUL+W'U'L|\beta W_iU_iS)/(2L+1)^{\frac{1}{2}}. \tag{71}$$

The number of values which may be assumed by α and β equals the number of times that the representation $\mathfrak{B}_{W_i}$ appears in the decomposition of $\mathfrak{B}_W\times\mathfrak{B}_{W'}$ and will be denoted by $c(WW'W_i)$. A method for calculating these numbers is given by Weyl.[6]

TABLE XIVa. $(U|\varphi(L)|U')$ for $U, U'\neq(31), (40)$.

$(U\ U')$	S	P	D	F	G	H	I	K	L	M	N
$(11\|\varphi\|11)$	0	−11	0	0	0	3	0	0	0	0	0
$(20\|\varphi\|20)$	0	0	−11	0	−4	0	7	0	0	0	0
$(10\|\varphi\|21)$	0	0	0	1	0	0	0	0	0	0	0
$(20\|\varphi\|21)$	0	0	$6\sqrt{2}$	0	$(65)^{\frac{1}{2}}$	0	0	0	0	0	0
$(21\|\varphi\|21)$	0	0	−57	63	55	−105	0	−14	42	0	0
$(11\|\varphi\|30)$	0	$(11)^{\frac{1}{2}}$	0	0	0	$(39)^{\frac{1}{2}}$	0	0	0	0	0
$(20\|\varphi\|30)$	0	0	0	0	$2(5)^{\frac{1}{2}}$	0	3	0	0	0	0
$(21\|\varphi\|30)$	0	0	0	$(195)^{\frac{1}{2}}$	$-(143)^{\frac{1}{2}}$	$-2(42)^{\frac{1}{2}}$	0	$-4(17)^{\frac{1}{2}}$	0	0	0
$(30\|\varphi\|30)$	0	83	0	−72	20	−15	42	−28	0	6	0
$(00\|\varphi\|22)$	1	0	0	0	0	0	0	0	0	0	0
$(20\|\varphi\|22)$	0	0	$3(429)^{\frac{1}{2}}$	0	$4(65)^{\frac{1}{2}}$	0	$3(85)^{\frac{1}{2}}$	0	0	0	0
$(22\|\varphi\|22)$	144	0	69	0	−148	72	39	0	−96	0	56

TABLE XIVb. $(U|\varphi(L)|31)$.

L	$(10\|\varphi\|31)$	$(11\|\varphi\|31)$	$(21\|\varphi\|31)$	$(30\|\varphi\|31)$	$(31\|\varphi\|31)$
P	0	$(330)^{\frac{1}{2}}$	0	$17(143)^{\frac{1}{2}}$	209
D	0	0	$12(273)^{\frac{1}{2}}$	0	−200
F	1	0	$-36(5)^{\frac{1}{2}}$	$-16(39)^{\frac{1}{2}}$	$\Vert\ 624 \quad -80(143)^{\frac{1}{2}}\ \Vert$
F'	0	0	$-3(715)^{\frac{1}{2}}$	$24(33)^{\frac{1}{2}}$	$\Vert -80(143)^{\frac{1}{2}} \quad -616\ \Vert$
G	0	0	$11(7)^{\frac{1}{2}}$	$4(1001)^{\frac{1}{2}}$	836
H	0	$(85)^{\frac{1}{2}}$	$-2(1309/3)^{\frac{1}{2}}$	$(187/2)^{\frac{1}{2}}$	$\Vert -1353/2 \quad -5(6545)^{\frac{1}{2}}/2 \Vert$
H'	0	$(77)^{\frac{1}{2}}$	$-74(5/3)^{\frac{1}{2}}$	$31(35/2)^{\frac{1}{2}}$	$\Vert -5(6545)^{\frac{1}{2}}/2 \quad 703/2 \Vert$
I	0	0	0	$30(33)^{\frac{1}{2}}$	$\Vert -2662/5 \quad 528(34)^{\frac{1}{2}}/5 \Vert$
I'	0	0	0	0	$\Vert 528(34)^{\frac{1}{2}}/5 \quad -88/5 \Vert$
K	0	0	$-28(323/23)^{\frac{1}{2}}$	$4(437)^{\frac{1}{2}}$	$\Vert 6652/23 \quad 96(21318)^{\frac{1}{2}}/23 \Vert$
K'	0	0	$42(66/23)^{\frac{1}{2}}$	0	$\Vert 96(21318)^{\frac{1}{2}}/23 \quad -5456/23 \Vert$
L	0	0	$-6(190)^{\frac{1}{2}}$	0	−464
M	0	0	0	$-6(385)^{\frac{1}{2}}$	814
N	0	0	0	0	−616
O	0	0	0	0	352

[6] H. Weyl, *The Classical Groups* (Princeton University Press, Princeton, New Jersey, 1939), p. 229.

TABLE XIVc. $(U|\varphi(L)|40)$.

L	$(20\|\varphi\|40)$	$(21\|\varphi\|40)$	$(22\|\varphi\|40)$
S	0	0	$2(2145)^{\frac{1}{2}}$
D	$11(13)^{\frac{1}{2}}$	$-6(26)^{\frac{1}{2}}$	$9(33)^{\frac{1}{2}}$
F	0	$3(455)^{\frac{1}{2}}$	0
G	$-4(715/27)^{\frac{1}{2}}$	$-131(11/27)^{\frac{1}{2}}$	$-4(11/27)^{\frac{1}{2}}$
G'	$(15470/27)^{\frac{1}{2}}$	$17(238/27)^{\frac{1}{2}}$	$-17(238/27)^{\frac{1}{2}}$
H	0	$-12(21)^{\frac{1}{2}}$	$3(286)^{\frac{1}{2}}$
I	$7(1045/31)^{\frac{1}{2}}$	0	$3(3553/31)^{\frac{1}{2}}$
I'	$3(1785/31)^{\frac{1}{2}}$	0	$75(21/31)^{\frac{1}{2}}$
K	0	$-2(119)^{\frac{1}{2}}$	0
L	0	$22(105/31)^{\frac{1}{2}}$	$4(627/31)^{\frac{1}{2}}$
L'	0	$-84(19/31)^{\frac{1}{2}}$	$12(385/31)^{\frac{1}{2}}$
N	0	0	$-(2530)^{\frac{1}{2}}$

TABLE XV. $y(f^3, {}_v{}^2U, {}_3{}^2U')$.

	${}_3{}^2(11)$	${}_3{}^2(20)$	${}_3{}^2(21)$
${}_1{}^2(10)$	0	0	$-6(22)^{\frac{1}{2}}$
${}_3{}^2(11)$	2	0	0
${}_3{}^2(20)$	0	$10/7$	$2(66)^{\frac{1}{2}}/7$
${}_3{}^2(21)$	0	$2(66)^{\frac{1}{2}}/7$	$2/7$

2. The Calculation of e_2

For the values of W and W' which satisfy (16), $c(WW'(400))$ equals unity if $W=W'$ and $w_1=2$, and vanishes in any other case; it follows that e_2 is diagonal with respect to v and vanishes for $v=2S$, and also that for $v>2S$

$$(f^n vWUSL|e_2|f^n vWU'SL) = b(nvS)(WUL+WU'L|(400)(40)S)/(2L+1)^{\frac{1}{2}}. \quad (72)$$

By considerations which are very similar to the method used in Section 7 of II for calculating the energy matrices of d^n (and in particular for the proof that the relative positions of the quartets and sextets of d^5 are exactly opposed to those of the terms of d^2 with the same L), it may be shown that $b(nvS)$ is independent of n and that for two values of v and S which correspond to the same value of W the $b(nvS)$ differ only in the sign. We can therefore write

$$(f^n vUSL|e_2|f^n vU'SL) = \pm(WUL|e_2|WU'L), \quad (73)$$

where the upper sign holds for the values of v and S which appear in the first column of Table I, and the lower sign for the second column.

The actual calculation of $(WUL|e_2|WU'L)$ is simplified by the lemma (11): introducing it in (72) we have that

$$(WUL|e_2|WU'L) = \sum_\gamma x_\gamma(W, UU')(U|\chi_\gamma(L)|U'), \quad (74)$$

where x_γ is independent of L and $\chi_\gamma(L)$ is independent of W; the maximal number of independent $(U|\chi_\gamma(L)|U')$ is $c(UU'(40))$ and is given in Table V.

Not only $(WUL|e_2|WU'L)$, but also

$$\sum_{L_2}(UL|U_2L_2+f)(U_2|\chi_{\gamma 2}(L_2)|U_2') \times(U_2'L_2+f|U'L) \quad (75)$$

TABLE XVI. $y(f^4, {}_v{}^3U, {}_4{}^3U')$.

	${}_4{}^3(10)$	${}_4{}^3(11)$	${}_4{}^3(20)$	${}_4{}^3(21)$	${}_4{}^3(30)$
${}_2{}^3(10)$	0	0	0	$-12(33/5)^{\frac{1}{2}}$	0
${}_2{}^3(11)$	0	$6/5$	0	0	6
${}_4{}^3(10)$	0	0	0	$8(11/15)^{\frac{1}{2}}$	0
${}_4{}^3(11)$	0	$29/15$	0	0	$-1/3$
${}_4{}^3(20)$	0	0	$6/7$	$-8(11/147)^{\frac{1}{2}}$	$4/\sqrt{3}$
${}_4{}^3(21)$	$8(11/15)^{\frac{1}{2}}$	0	$-8(11/147)^{\frac{1}{2}}$	$-2/21$	$-4/3$
${}_4{}^3(30)$	0	$-1/3$	$4/\sqrt{3}$	$-4/3$	$1/3$

TABLE XVII. $y(f^4, {}_v{}^1U, {}_4{}^1U')$.

	${}_4{}^1(20)$	${}_4{}^1(21)$	${}_4{}^1(22)$
${}_0{}^1(00)$	0	0	$-12(22)^{\frac{1}{2}}$
${}_2{}^1(20)$	$3(3/175)^{\frac{1}{2}}$	$-4(33/35)^{\frac{1}{2}}$	$-(3/5)^{\frac{1}{2}}$
${}_4{}^1(20)$	$221/140$	$8(11/245)^{\frac{1}{2}}$	$-(7/80)^{\frac{1}{2}}$
${}_4{}^1(21)$	$8(11/245)^{\frac{1}{2}}$	$2/7$	0
${}_4{}^1(22)$	$-(7/80)^{\frac{1}{2}}$	0	$1/4$

TABLE XVIII. $y(f^5, {}_v{}^4U, {}_5{}^4U')$.

	${}_5{}^4(10)$	${}_5{}^4(11)$	${}_5{}^4(20)$	${}_5{}^4(21)$	${}_5{}^4(30)$
${}_3{}^4(00)$	0	0	0	0	0
${}_3{}^4(10)$	0	0	0	$9(11)^{\frac{1}{2}}$	0
${}_3{}^4(20)$	0	0	$3/(7)^{\frac{1}{2}}$	$(33/7)^{\frac{1}{2}}$	$-2(21)^{\frac{1}{2}}$
${}_5{}^4(10)$	0	0	0	$-(55/3)^{\frac{1}{2}}$	0
${}_5{}^4(11)$	0	$-1/3$	0	0	$-5/3$
${}_5{}^4(20)$	0	0	$5/7$	$5(11/147)^{\frac{1}{2}}$	$2/\sqrt{3}$
${}_5{}^4(21)$	$-(55/3)^{\frac{1}{2}}$	0	$5(11/147)^{\frac{1}{2}}$	$-4/21$	$-2/3$
${}_5{}^4(30)$	0	$-5/3$	$2/\sqrt{3}$	$-2/3$	$-1/3$

is expressible as linear combination of the $(U|\chi_\gamma(L)|U')$; it is therefore convenient to calculate at first the expressions (75), and then to assemble the results in the summation (1), where the coefficients of fractional parentage have the form (34). It is also possible to avoid at all the summations (75) for most of the values of L, after the different $(U|\chi_\gamma(L)|U')$ allowed by Table V are obtained from few simple $\chi_{\gamma 2}(L_2)$.

Although almost all the allowed χ_γ appear in the expressions (75), the linear combinations (74) are generally proportional to each other, and it is therefore possible to express the results by means of one $\chi(L)$ for every couple UU', with the sole exception of $U=U'\equiv(21)$, where both $\chi_\gamma(L)$ allowed by Table V are necessary for expressing the different $e_2(W)$. The functions $(U|\chi(L)|U')$ are tabulated in Tables VI, the values of $x(W, UU')$ in Tables VII–XI.

3. The Calculation of e_3

Together with e_3 it is useful to consider the operator

$$\Omega = -462^{\frac{1}{2}}\sum_k(2k+1)^{\frac{1}{2}}\mathbf{U}^{(k)2} \times((110)(11)k+(110)(11)k|(220)(22)S) = 33(\mathbf{U}^{(1)2}-\mathbf{U}^{(5)2}), \quad (76)$$

TABLE XIX. $y(f^5, {}_v{}^2U, {}_5{}^2U')$.

	${}_5{}^2(10)$	${}_5{}^2(11)$	${}_5{}^2(20)$	${}_5{}^2(21)$	${}_5{}^2(30)$	${}_5{}^2(31)$
${}_1{}^2(10)$	0	0	0	$36/(5)^{1/2}$	0	$-36\sqrt{2}$
${}_3{}^2(11)$	0	$3/\sqrt{2}$	0	0	$3(5)^{1/2}/2$	$-(39/8)^{1/2}$
${}_3{}^2(20)$	0	0	$3/7$	$-11(6)^{1/2}/7$	$-4\sqrt{3}$	0
${}_3{}^2(21)$	$3(33/10)^{1/2}$	0	$-3(33/98)^{1/2}$	$3/7(11)^{1/2}$	$-3/2\sqrt{2}$	$3/2(22)^{1/2}$
${}_5{}^2(10)$	0	0	0	$43/(30)^{1/2}$	0	$4\sqrt{3}$
${}_5{}^2(11)$	0	$-5/6$	0	0	$-5(5/72)^{1/2}$	$-(13/48)^{1/2}$
${}_5{}^2(20)$	0	0	$11/7$	$-11/7(6)^{1/2}$	$4/\sqrt{3}$	0
${}_5{}^2(21)$	$43/(30)^{1/2}$	0	$-11/7(6)^{1/2}$	$25/231$	$29/6(22)^{1/2}$	$1/22\sqrt{2}$
${}_5{}^2(30)$	0	$-5(5/72)^{1/2}$	$4/\sqrt{3}$	$29/6(22)^{1/2}$	$-1/12$	$1/4(11)^{1/2}$
${}_5{}^2(31)$	$4\sqrt{3}$	$-(13/48)^{1/2}$	0	$1/22\sqrt{2}$	$1/4(11)^{1/2}$	$1/44$

TABLE XX. $y(f^6, {}_4{}^5U, {}_6{}^5U')$.

	${}_6{}^5(11)$	${}_6{}^5(20)$	${}_6{}^5(21)$
${}_4{}^5(00)$	0	0	0
${}_4{}^5(10)$	0	0	$-6(11)^{1/2}$
${}_4{}^5(20)$	0	$-2(2/7)^{1/2}$	$2(33/7)^{1/2}$

which has the same tensorial properties as e_3; from (24) of III and (27) we have

$$\Omega=\tfrac{1}{2}L(L+1)-12G(G_2), \tag{77}$$

and therefore its matrix is diagonal in the UL scheme and has the eigenvalues

$$\omega(U,L)=\tfrac{1}{2}L(L+1)-12g(U). \tag{78}$$

We have from (67) that

$$e_3(f^2\,{}^3L)=-3\omega, \tag{79}$$

and since for every n

$$\Omega=66\sum_{i<j}((\mathbf{u}_i{}^{(1)}\cdot\mathbf{u}_j{}^{(1)})-(\mathbf{u}_i{}^{(5)}\cdot\mathbf{u}_j{}^{(5)})), \tag{80}$$

we obtain that for every term of f^n with maximal spin

$$e_3(f^n\,{}^{n+1}L)=-3\omega(U,L). \tag{81}$$

The values of $c(WW'(220))$ are given in Table XII' but the results are much simpler than could be expected from that table. The calculations show that

$$(f^n vUSL|e_3+\Omega|f^n vU'SL)=a(n,v)(f^v vUSL|e_3+\Omega|f^v vU'SL), \tag{82}$$

$$(f^6\,{}_6L|e_3+\Omega|f^6\,{}_6L)=(f^7\,{}_7L|e_3+\Omega|f^7\,{}_7L)=0; \tag{83}$$

for $n\leqslant 7$ it is

$$a(v+2,v)=(1-v)/(7-v),\quad a(v+4,v)=-4/(7-v), \tag{84}$$

and it may be noted that these equations satisfy the relation

$$\sum_{n=v}^{14-v}(f^n vUSL|e_3+\Omega|f^n vU'SL)=0. \tag{85}$$

The fact that $a(n,v)$ depends on v but not on S suggests that Eqs. (82) to (85) are connected with the properties of the symplectic group which leaves invariant the form (21), but the investigation of these properties is beyond the scope of this paper.

For $v\neq v'$ we found also

$$\begin{aligned}(f^5\,{}_1{}^2L|e_3|f^5\,{}_3{}^2L)&=(2/5)^{1/2}(f^3\,{}_1{}^2L|e_3|f^3\,{}_3{}^2L),\\(f^6\,{}_0{}^1L|e_3|f^6\,{}_4{}^1L)&=(9/5)^{1/2}(f^4\,{}_0{}^1L|e_3|f^4\,{}_4{}^1L),\\(f^6\,{}_2L|e_3|f^6\,{}_4L)&=(1/6)^{1/2}(f^4\,{}_2L|e_3|f^4\,{}_4L),\\(f^7\,{}_1{}^2L|e_3|f^7\,{}_5{}^2L)&=(3/2)^{1/2}(f^5\,{}_1{}^2L|e_3|f^5\,{}_5{}^2L).\end{aligned} \tag{86}$$

The values of $c(UU'(22))$ are given in Table XIII, but the calculations show that also when $c(UU'(22))>1$ we can write without any exception

$$(f^n vUSL|e_3+\Omega|f^n v'U'SL)=y(f^n,vSU,v'SU')(U|\varphi(L)|U'). \tag{87}$$

The functions $(U|\varphi(L)|U')$ are tabulated in Tables XIV, the values of $y(f^n,vSU,v'SU')$ which do not follow from (81), (82), (83), or (86) are given in Tables XV–XXIV.

TABLE XXI. $y(f^6, {}_v{}^3U, {}_6{}^3U')$.

	${}_6{}^3(10)$	${}_6{}^3(11)$	${}_6{}^3(20)$	${}_6{}^3(21)$	${}_6{}^3(30)$	${}_6{}^3(31)$
${}_2{}^3(10)$	0	0	0	$-48(2/5)^{1/2}$	0	-36
${}_2{}^3(11)$	0	$(6/5)^{1/2}$	0	0	$\sqrt{3}$	$3(13/10)^{1/2}$
${}_4{}^3(10)$	0	0	0	$46/(15)^{1/2}$	0	$-8(6)^{1/2}$
${}_4{}^3(11)$	0	$11/3(5)^{1/2}$	0	0	$-19/3\sqrt{2}$	$(13/60)^{1/2}$
${}_4{}^3(20)$	0	0	$-6\sqrt{2}/7$	$-22/7\sqrt{3}$	$8(2/3)^{1/2}$	0
${}_4{}^3(21)$	$-(110/3)^{1/2}$	0	$(22/147)^{1/2}$	$-16/21(11)^{1/2}$	$5/3\sqrt{2}$	$1/(22)^{1/2}$
${}_4{}^3(30)$	0	$-(5)^{1/2}/3$	$4(2/3)^{1/2}$	$4/3(11)^{1/2}$	$1/3\sqrt{2}$	$-1/(22)^{1/2}$

TABLE XXII. $y(f^8, {}_v{}^1U, {}_6{}^1U')$.

	${}_6{}^1(00)$	${}_6{}^1(10)$	${}_6{}^1(20)$	${}_6{}^1(30)$	${}_6{}^1(40)$
${}_2{}^1(20)$	0	0	$6/(55)^{\frac{1}{2}}$	$2(42/5)^{\frac{1}{2}}$	$6(2/55)^{\frac{1}{2}}$
${}_4{}^1(20)$	0	0	$-61/(770)^{\frac{1}{2}}$	$8(3/5)^{\frac{1}{2}}$	$-6/(385)^{\frac{1}{2}}$
${}_4{}^1(21)$	0	$3(22)^{\frac{1}{2}}$	$(2/7)^{\frac{1}{2}}$	$-\sqrt{3}$	$1/(7)^{\frac{1}{2}}$
${}_4{}^1(22)$	$-4(33/5)^{\frac{1}{2}}$	0	$-1/(22)^{\frac{1}{2}}$	0	$2/(11)^{\frac{1}{2}}$

TABLE XXIII. $y(f^7, {}_3{}^4U, {}_7{}^4U')$.

	${}_7{}^4(20)$	${}_7{}^4(21)$	${}_7{}^4(22)$
${}_3{}^4(00)$	0	0	$-12(11)^{\frac{1}{2}}$
${}_3{}^4(10)$	0	$6(33)^{\frac{1}{2}}$	0
${}_3{}^4(20)$	$-(5/7)^{\frac{1}{2}}$	$2(11/7)^{\frac{1}{2}}$	-1

TABLE XXIV. $y(f^7, {}_3{}^2U, {}_7{}^2U')$.

	${}_7{}^2(00)$	${}_7{}^2(10)$	${}_7{}^2(20)$	${}_7{}^2(30)$	${}_7{}^2(40)$
${}_3{}^2(11)$	0	0	0	$2(10)^{\frac{1}{2}}$	0
${}_3{}^2(20)$	0	0	$-16/(77)^{\frac{1}{2}}$	$-2(6)^{\frac{1}{2}}$	$6(2/77)^{\frac{1}{2}}$
${}_3{}^2(21)$	0	$-(66)^{\frac{1}{2}}$	$(6/7)^{\frac{1}{2}}$	1	$(3/7)^{\frac{1}{2}}$

From (71) and from the orthogonality between the functions $\psi((00)S)$, $\psi((40)S)$, and $\psi((22)S)$ follow the relations

$$\sum_L(2L+1)(U|\chi(L)|U)$$
$$=\sum_L(2L+1)(U|\varphi(L)|U)=0 \quad (88)$$

and

$$\sum_L(2L+1)(U|\chi(L)|U')(U'|\varphi(L)|U)=0, \quad (89)$$

which were useful for checking Tables VI and XIV.

Reprinted from the *Farkas Memorial Volume*,
Research Council of Israel, 1952

NUCLEAR LEVELS AND CASIMIR OPERATOR

GIULIO RACAH
The Hebrew University, Jerusalem

Received : May, 5, 1952

SECTION 1

The calculation of the levels of nuclear shells generally requires very long calculations, due to the great number of levels which belong to every nuclear configuration. It is our purpose to discuss some cases where the calculation may be greatly simplified by the use of convenient devices, and to stress in particular the advantage which may be obtained by the use of Casimir's operator[1].

In order to show the fundamental idea on which these calculations are based, we shall first consider the old example of the nuclear configuration p^n.[2]

If two nucleons belonging to a p-shell interact by spin-independent forces, their interaction energy can assume only three values, according to the three possible values of their resultant momentum. Therefore, by a convenient choice of the three constants a, b and c, it is always possible to express their interaction energy by the formula

$$a + bM_{12} + c(\mathbf{l}_1 \cdot \mathbf{l}_2), \tag{1}$$

where M_{12} is the "Majorana-operator" which has the eigenvalue 1 for space-symmetrical states and the eigenvalue -1 for space-antisymmetrical states, and $(\mathbf{l}_1 \cdot \mathbf{l}_2)$ is the scalar product of the angular momenta of the two nucleons.

If we have n nucleons in the p-shell, we need only to add the expressions (1) for every pair of nucleons, and obtain

$$\begin{aligned} V &= \frac{n(n-1)}{2}\, a + bM + \tfrac{1}{2} c\,[L(L+1) - nl\,(l+1)] = \\ &= \frac{n(n-1)}{2}\, a + bM + c\,[\tfrac{1}{2}L(L+1) - n]; \end{aligned} \tag{2}$$

here M is the eigenvalue of the Majorana operator $\sum_{i<k} M_{ik}$, which depends only on the partition $[n]$ characterizing the symmetry of the supermultiplet to which the level belongs, and has the expression[3]

$$M = \tfrac{1}{8}n(16-n) - \tfrac{1}{2}\,[P(P+4) + P'(P'+2) + P''^2]. \tag{3}$$

Calculating the three constants from the well-known energy values of the configuration p^2, we obtain for Wigner interaction the energy

$$V_W = \frac{n(n-1)}{2}(F^0 - F^2/5) + \left[\frac{n(n+1)}{2} + 2M - \tfrac{1}{2}L(L+1)\right](3F^2/25) \tag{4a}$$

and for Majorana interaction

$$V_M = M(F^0 - F^2/5) + \left[\frac{n(n+1)}{2} + 2M - \tfrac{1}{2}L(L+1)\right](3F^2/25); \tag{4b}$$

here F^0 and F^2 are the "generalized" parameters of Slater, which are defined by

$$F^k = \iint J_k(r_1, r_2)\ R^2(r_1)\ R^2(r_2)\ dr_1 dr_2, \tag{5}$$

where $R(r)/r$ is the radial wave function appropriate to the shell, and the $J_k(r_1, r_2)$ are defined by the expansion

$$J_k(r_1, r_2) = \Sigma_k J_k(r_1, r_2) P_k(\cos\omega_{12}). \tag{6}$$

It may be remarked that these generalized parameters are not necessarily decreasing functions of k, as they were in the classical case of Coulomb interaction; for short-range forces they may increase with k, and for the limiting case of very short range (δ-interaction) we have

$$F^k = (2k+1)F^0. \tag{7}$$

SECTION 2

If we want to extend this method to other shells where the two-particle configurations have more than three states, we must find some other operator which may be added to those appearing in (1) and (2) in order to have a larger number of free constants.

If we consider the two operators M and $L(L+1)$ which already appear in (2), we see that M is connected with the space-symmetry of the state, i.e. with the representation of the $(2l+1)$-dimensional unitary group to which the state belongs, and $L(L+1)$ is connected with the representation of the three-dimensional rotation group $R(3)$. It is therefore natural to search for a new group, which should be a subgroup of the $(2l+1)$-dimensional unitary group, and should still contain as a subgroup the elements of the $(2l+1)$-dimensional representation $\mathfrak{D}_l$ of the group $R(3)$. We shall then classify the states according to the representation of the new group to which they belong, and shall take as a new operator the operator whose eigenvalues characterize these representations.

Such a group is the $(2l+1)$-dimensional rotation group $R(2l+1)$, which leaves invariant the quadratic form

$$\Sigma_{m\ l}^{-l}(-1)^{l-m}\varphi(m)\,\varphi(-m). \tag{8}$$

Many properties of $R(2l+1)$ are very similar to the well known properties of $R(3)$. It is an $l(2l+1)$-dimensional continuous group, and is generated by $l(2l+1)$ infinitesmal operators $E_1 \ldots E_\varrho \ldots E_{l(2l+1)}$, which are the analogues of L_x, L_y and L_z. The analogues of the permutation relations

$$[L_x, L_y] = iL_z \tag{9a}$$

are the relations*

$$[E_\varrho, E_\sigma] = c_{\sigma\varrho}{}^{\tau} E_\tau \ , \tag{9b}$$

and the analogue of the operator

$$\Lambda = L_x^2 + L_y^2 + L_z^2 \tag{10a}$$

is Casimir's operator

$$G = (4l-2)\, g^{\varrho\sigma} E_\varrho E_\sigma, \tag{10b}$$

where the $g^{\varrho\sigma}$ are the elements of the inverse of Cartan's matrix

$$g_{\varrho\sigma} = c_{\varrho\mu}{}^{\nu} c_{\sigma\nu}{}^{\mu}. \tag{11}$$

The normalization factor $(4l-2)$ is not included in Casimir's original definition[1], but it is introduced here in order to avoid fractions in the following equations.

Every irreducible (single-valued) representation of $R(3)$ is characterized by an integer L, and to the representation $\mathfrak{D}_L$ corresponds the eigenvalue $L(L+1)$ of Λ; every irreducible (single-valued) representation of $R(2l+1)$ is characterized by a set $W = (w_1 w_2 \ldots w_l)$ of l integer numbers, and to the representation $\mathfrak{B}_W$ corresponds the eigenvalue

$$g(W) = w_1(w_1 + 2l - 1) + w_2(w_2 + 2l - 3) + \ldots + w_l(w_l + 1) \tag{12}$$

of Casimir's operator[4].

The state of one l-particle belongs to $(10\ldots0)$, or, more briefly, to (1), as it is customary to drop some or all zeros. The states of l^2 belong to three different representations: the S-state belongs to (00), the other states with an even L belong to (20), and the states with an odd L belong to (11).

If it is possible to express the energies of the levels of l^2 by

$$a + bM_{12} + c(\mathbf{l}_1 \cdot \mathbf{l}_2) + d(\mathbf{e}_1 \cdot \mathbf{e}_2), \tag{13}$$

where

$$(\mathbf{e}_1 \cdot \mathbf{e}_2) = (4l-2)\, g^{\varrho\sigma} e_\varrho^{(1)} e_\sigma^{(2)} \tag{14}$$

* The summation convention is used for Greek indices.

and $e_\sigma^{(i)}$ are the infinitesimal operators of $R(2l+1)$ which operate on the particle i, then the energy of a state of l^n will be

$$V = \frac{n(n-1)}{2} a + bM + \tfrac{1}{2}c\,[L(L+1) - nl(l+1)] + \tfrac{1}{2}d\,[g(W) - 2nl]; \tag{15}$$

it is also convenient to use instead of (13) the particular case of (15) for $n = 2$:

$$V = a + bM + c\,[\tfrac{1}{2}L(L+1) - l(l+1)] + d\,[\tfrac{1}{2}g(W) - 2l]. \tag{13'}$$

The levels of d^2 have for Wigner interaction the energies

$$\begin{aligned} S &= E^0 + 7E^1 \\ P &= E^0 \qquad\quad + 21E^2 \\ D &= E^0 + 2E^1 - \ \ 9E^2 \\ F &= E^0 \qquad\quad - \ \ 9E^2 \\ G &= E^0 + 2E^1 + \ \ 5E^2 \end{aligned} \tag{16}$$

where

$$\begin{aligned} E^0 &= F_0 - 7\,(F_2 + 9F_4)/2 = F^0 - (F^2 + F^4)/14, \\ E^1 &= 5\,(F_2 + 9F_4)/2 = 5\,(F^2 + F^4)/98 \\ E^2 &= (F_2 - 5F_4)/2 = (F^2 - 5F^4/9)/98; \end{aligned} \tag{17}$$

for Majorana interaction the signs have to be changed for P and F.

As (13') has four free parameters, it cannot be expected that the five values of V_W or V_M will fit it, but a convenient linear combination of them will certainly do it; it is easy to verify that $V_W + 2V_M$ fits (13') with

$$a = E^0 + 3E^1, \quad b = 2E^0 + 6E^1, \quad c = 6E^2, \quad d = -(3E^1 + 9E^2), \tag{18}$$

and therefore for every configuration d^n

$$V_W + 2V_M = \left[\frac{n(n-1)}{2} + 2M\right]E^0 + 3\left[\frac{n(n+3)}{2} + 2M - 1/2\,g(W)\right]E^1 + 3\left[\,L(L+1) - 3/2\,g(W)\,\right]E^2. \tag{19}$$

A more important application is the limiting case of very short range, as in this case E^0 and E^2 vanish according to (7); then (16) fit (13') with

$$a = E^1, \quad b = 2E^1, \quad c = 0, \quad d = -E^1, \tag{20}$$

and therefore for every configuration d^n with δ-interaction

$$V_W = V_M = \left[\frac{n(n+3)}{2} + 2M - \tfrac{1}{2}g(W)\right]E^1, \tag{21}$$

a result which was also given by Jahn.[5]

SECTION 3

The method described in the preceding section cannot be applied directly to configurations l^n with $l>2$, as the levels of l^2 do not fit (13′) even for a vanishing range of the interaction. If we wish to use this method also for such shells, we must now restrict the scope of our calculations.

Instead of considering separately the energy values of the levels of l^2, we shall consider the mean values* of the levels belonging to the same representation $\mathfrak{W}_W$ of $R(2l+1)$; as the levels of l^2 belong to only three representations $\mathfrak{W}_W$, these mean values may always be represented by

$$V = a + bM_{12} + d[\tfrac{1}{2}g(W) - 2l], \tag{22}$$

which is a particular case of (13′). Then we assert that the mean energy of the levels of l^n belonging to a given representation $\mathfrak{W}_W$ is

$$\overline{V} = \frac{n(n-1)}{2}a + bM + d[\tfrac{1}{2}g(W) - nl]. \tag{23}$$

In order to prove this fundamental theorem, we separate the two-particle interaction V into two parts,

$$V = V_1 + V_2, \tag{24}$$

where $V_1 = V$ is the mean value (22), and V_2 expresses the deviations of the different levels from the mean value. Then by considerations which are very similar to those used elsewhere for atomic spectra[6], it may be proved that the mean value of V_2 over the levels of l^n belonging to a given representation $\mathfrak{W}_W$ always vanishes; it follows that $V = V_1$ also for l^n, and according to our previous considerations the value of V_1 is given by (23).

If we write the energy levels of l^2 for Wigner interaction in the general form

$$V(L) = \Sigma_k f_k(l^2, L)F^k, \tag{25}$$

it has been shown[7] that

$$f_k(l^2, L) = (llLM|(\mathbf{C}_1^{(k)} \cdot \mathbf{C}_2^{(k)})|llLM) = \tag{26}$$

$$= (l||C^{(k)}||l)^2(-1)^{2l-L}W(llll; Lk);$$

from the general properties II(42)** and II(43) of the coefficients W and from the particular expression

$$W(aaaa; e0) = (-1)^{2a-e}/(2a+1) \tag{27}$$

* Every mean value must always be taken using the weights $2L+1$.

** Reference 7 will be referred to as II in the following.

we get

$$\Sigma_e (-1)^{2a-e}(2e+1)W(aaaa\,;ef) = (2a+1)\,\delta_{f0} \qquad (28a)$$

$$\Sigma_e (2e+1)W(aaaa\,;ef) = (-1)^{2a+f}(2a+1)\,W(aaaa;0f) = 1; \qquad (28b)$$

and from (25), (26), (28) and (27) we obtain

$$\begin{aligned}\Sigma_L (2L+1)\,V(L) &= (2l+1)^2 F^0,\\ \Sigma_L (-1)^L\,(2L+1)\,V(L) &= (2l+1)\,V(S).\end{aligned} \qquad (29)$$

If we define

$$\begin{aligned}E^0 &= [(2l+1)F_0 - V(S)]/2l,\\ E^1 &= [2l+1)[V(S) - F_0]/2l(2l+3),\end{aligned} \qquad (30)$$

we can write

$$\begin{aligned}V(00) &= V(S) = E_0 + (2l+3)\,E^1\\ V(11) &= E_0\\ V(20) &= E_0 + 2E^1.\end{aligned} \qquad (31)$$

Taking the values of a, b and d in (22) in order to fit (31), we get from (23).

$$V_W = \frac{n(n-1)}{2}E^0 + \left[\frac{n(n+2l-1)}{2} + 2M - \tfrac{1}{2}g(W)\right]E^1; \qquad (32a)$$

the analogous expression for Majorana interaction is obtained by changing the sign of $V(11)$ in (31) and proceeding in the same way:

$$\overline{V}_M = ME_0 + \left[\frac{n(n+2l-1)}{2} + 2M - \tfrac{1}{2}g(W)\right]E^1. \qquad (32b)$$

These formulae give in general only the baricenters of groups of levels, and not the exact position of the different levels; it should be remarked, however, that at least for short-range forces, the separation between levels belonging to the same $\mathfrak{B}_W$ are in general smaller than the distances between the baricenters, so that Eqs. (32) are sufficient for determining the order of the different groups of levels.

Furthermore, for those $\mathfrak{B}_W$ which contain only one level, Eqs. (32) give directly the binding energies of this level; and it is a most lucky chance that this happens exactly for the ground levels of most nuclei.

For even-even nuclei, the ground level is an S-level with W (0) belonging to the supermultiplet $(\frac{1}{2}\,\varepsilon, 0, 0)$, where $\varepsilon = |N_n - N_p|$; we have therefore for the binding energy of this ground level

$$\begin{aligned}E_W &= \frac{n(n-1)}{2}E^0 + \frac{n(n+4l+14) - \varepsilon(\varepsilon+8)}{4}E^1,\\ E_M &= \frac{n(16-n) - \varepsilon(\varepsilon+8)}{8}E^0 + \frac{n(n+4l+14) - \varepsilon(\varepsilon+8)}{4}E^1\end{aligned} \qquad (33)$$

For even-odd nuclei the ground level has $L = l$ and $W \equiv (1)$, belongs to the supermultiplet $(\frac{1}{2}\varepsilon, \frac{1}{2}, \pm\frac{1}{2})$, and its binding energies are

$$E_W = \frac{n(n-1)}{2} E^0 + \frac{n(n+4l+14) - \varepsilon(\varepsilon+8) - (4l+6)}{4} E^1,$$

$$E_M = \frac{n(16-n) - \varepsilon(\varepsilon+8) - 6}{8} E^0 + \\ + \frac{n(n+4l+14) - \varepsilon(\varepsilon+8) - (4l+6)}{4} E^1 \tag{34}$$

It may be remarked that for δ-interaction ($E^0 = 0$) between equivalent nucleons ($n = \varepsilon$) these expressions become

$$E = n(2l+3)E^1/2 \qquad \text{for even nuclei} \tag{35a}$$

and

$$E = (n-1)(2l+3)E^1/2 \qquad \text{for odd nuclei,} \tag{35b}$$

and are the LS-coupling analogues of the result of M. G. Mayer[8]; it follows that Mayer's result is independent of the particular assumption of jj-coupling.

For odd-odd nuclei the situation is somewhat more complicated. If $\varepsilon = 0$, the ground level is an S-level with $W \equiv (0)$ belonging to the supermultiplet (100), and its binding energies are

$$E_W = \frac{n(n-1)}{2} E^0 + \frac{n(n+4l+14) - 20}{4} E^1$$

$$E = \frac{n(16-n) - 20}{8} E^0 + \frac{n(n+4l+14) - 20}{4} E^1. \tag{36}$$

If $\varepsilon \neq 0$, the levels with the highest M belong to the supermultiplet $(\frac{1}{2}\varepsilon, 1, 0)$ and to $W \equiv (11)$, and their mean binding energies are

$$E_W = \frac{n(n-1)}{2} E^0 + \frac{n(n+4l+14) - \varepsilon(\varepsilon+8) - 8(l+1)}{4} E^1,$$

$$E_M = \frac{n(16-n) - (\varepsilon+2)(\varepsilon+6)}{8} E^0 + \\ + \frac{n(n+4l+14) - \varepsilon(\varepsilon+8) - 8(l+1)}{4} E^1. \tag{37}$$

For a long-range Majorana interaction ($E^0 >> E^1$), these levels are the lowest levels of the nucleus, but for a short-range interaction ($E^0 << E^1$), the S-level with $W \equiv (0)$ belonging to the supermultiplet $(\frac{1}{2}\varepsilon + 1, 0, 0)$ must also be considered, as its binding energies are

$$E_W = \frac{n(n-1)}{2} E^0 + \frac{n(n+4l+14) - (\varepsilon+2)(\varepsilon+10)}{4} E^1$$
$$E_M = \frac{n(16-n) - (\varepsilon+2)(\varepsilon+10)}{8} E^0 + \qquad (38)$$
$$+ \frac{n(n+4l+14) - (\varepsilon+2)(\varepsilon+10)}{4} E^1 .$$

Comparing (37 and (38), we see that for a δ-interaction ($E^0 = 0$), the S-level belonging to the supermultiplet ($\frac{1}{2}\varepsilon + 1$, 0, 0) is the ground level not only for $\varepsilon = 0$, but for every $\varepsilon < 2l - 3$; only for $\varepsilon > 2l - 3$ the ground level belongs to the supermultiplet ($\frac{1}{2}\varepsilon$, 1, 0). It may also be shown that for δ-interaction, the levels belonging to W (11) are degenerate and coincide with their baricenter (37).

Before concluding this section, it is necessary to point out that if the levels of l^2 cannot be fitted to (13′), but only to (22), for $n > 2$ the quantum numbers W are no longer "good" quantum numbers, and in the W-scheme the energy matrix has also non diagonal elements; but for short-range forces these non diagonal elements are generally small. For equivalent nucleons with δ-interaction they vanish identically, as it will be shown elsewhere.

Section 4

The methods developed in the preceding sections for LS-coupling may be applied without difficulty also to jj-coupling. In this case the situation is even simpler, as we do not have any more supermultiplets, but only charge-multiplets, and instead of the Majorana operator we shall use the isotopic spin T.

The most important difference is connected with the fact that the bilinear form

$$\sum_{m=-j}^{j} (-1)^{j-m} \varphi_1(m)\, \varphi_2(-m) \qquad (39)$$

is now skew-symmetric, and the linear group which leaves this form invariant is no more an orthogonal group, but the $(2j+1)$-dimensional "symplectic" group $Sp(2j+1)$[9]. The properties of $Sp(2j+1)$ are very similar to those of $R(2l+1)$: the irreducible representations are characterized by a set $W = (w_1, w_2 \ldots w_{j+\frac{1}{2}})$ of $j + \frac{1}{2}$ integer numbers, and to the representation $\mathfrak{B}_W$ corresponds now the eigenvalue

$$g(W) = w_1(w_1 + 2j + 1) + w_2(w_2 + 2j - 1) + \ldots$$
$$+ w_{j+\frac{1}{2}}(w_{j+\frac{1}{2}} + 2) \qquad (40)$$

of Casimir's operator.

In the configuration j^2 the state with $J = 0$ still belongs to W (00), but the other states with even J now belong to (11), and those with odd J belong to (20).

Instead of (13) we shall now write

$$a + b\,(\mathbf{t}_1 \cdot \mathbf{t}_2) + c(\mathbf{j}_1 \cdot \mathbf{j}_2) + d\,(\mathbf{e}_1 \cdot \mathbf{e}_2); \qquad \textbf{(41)}$$

instead of (15)

$$V = \frac{n(n-1)}{2} a + \tfrac{1}{2} b\,[T(T+1) - \tfrac{3}{4} n] + \tfrac{1}{2} c\,[J(J+1) - nj\,(j+1)] + \tfrac{1}{2} d\,[g(W) - 2n\,(j+1)], \qquad (42)$$

and instead of (13′)

$$V = a + b\,[\tfrac{1}{2} T\,(T+1) - \tfrac{3}{4}] + c\,[\tfrac{1}{2} J\,(J+1) - j\,(j+1)] + d\,[\tfrac{1}{2} g(W) - 2\,(j+1)]. \qquad (41')$$

These equations may be applied directly to the configurations $(3/2)^n$, as $(3/2)^2$ has just four levels. It may be remarked that for Wigner interaction, the levels with $J = 1$ and $J = 3$ coincide[10]; as they are the only couple of levels with the same quantum numbers T and W, c must vanish in (41′) and therefore also for $n > 2$ all levels belonging to the same representation $\mathfrak{B}_W$ must coincide; for $p^3_{3/2}$ this fact was already remarked by Inglis[11].

Another direct application of Eqs. (38) and (39) may be made to the configurations $(5/2)^n$ with equivalent nucleons, as in this last case $(5/2)^2$ has only three levels, which may be fitted to (41′) with $b = 0$.

For more complicated configurations, we are able to calculate only baricenters and ground levels as in Section 3, using Eqs. (41) and (42) with $c = 0$.

As the analogues of (26) hold also for jj-coupling, it follows from (28) that

$$\begin{aligned} \overline{V}_W(11) &= [(2j+1)^2 F^0 - (2j+3)\, V_W(0)]\,/2\,(j+1)(2j-1), \\ V_W(20) &= [(2j+1) F^0 + V_W(0)]\,/2(j+1). \end{aligned} \qquad (43a)$$

Fitting these values to (41′) we obtain from (42) that

$$V_W = \frac{n(n-1)}{2} \cdot \frac{(2j+1)F^0 - 2V_W(0)}{2j-1} + \left[\frac{n(n+2j+4)}{2} - 2T\,(T+1) - \tfrac{1}{2} g(W)\right] \frac{(2j+1)\,(V_W(0) - F^0)}{2(j+1)\,(2j-1)} \qquad (44a)$$

which is the analogue of (32a)

For Majorana interaction the calculation is more complicated, as the expression of its eigenvalues for j^2 is more involved. It will be shown in the appendix that

$$\begin{aligned} V(11) &= [2\,(2j+1)\, V_W(0) + 2\,(2j-1) V_M(0) - (2j+1)^2 F^0]\,/4(j+1)\,(2j-1) \\ V(20) &= [\,2V_W(0) + 2V_M(0) + (2j+1)\, F^0]\,/4\,(j+1), \end{aligned} \qquad (43b)$$

and proceeding as before we obtain

$$V_M = \frac{n(n-1)}{2} \cdot \frac{2V_W(0) - (2j+1)F^0}{2(2j-1)} +$$
$$+ \left[\frac{n(n+2j+4)}{2} - 2T(T+1) - \tfrac{1}{2}g(W)\right] \times$$
$$\times \frac{(4j-2)V_M(0) - 4V_W(0) + (2j+1)^2F^0}{8(j+1)(2j-1)} -$$
$$- [n(j+1) - \tfrac{1}{2}g(W)] \frac{2V_M(0) - (2j+1)F^0}{8(j+1)}. \tag{44b}$$

Applications to the calculation of ground levels may be made exactly as in the preceding section; we shall limit ourselves to the remark that for δ-interaction

$$V_W(0) = V_M(0) = (2j+1)F^0/2, \tag{45}$$

and for equivalent nucleons it is $T = n/2$; we obtain, therefore, as a particular case, the result of M. G. Mayer in its original form[8].

Appendix

We have already seen [Eqs. (25) and (26)] that the Wigner interaction between two particles is a sum of scalar products of tensors, and therefore the transformation from LS- to jj-coupling may be made by the general methods of the algebra of tensor matrices. But since the Majorana interaction is an exchange interaction, it must be treated as was treated in Section 5 of II that part of the electrostatic interaction of atomic spectra which depends on Slater's integrals G^k.

For Majorana interaction in LS-coupling we have

$$V_M(l^2, L) = \Sigma_k g_k(l^2, L)F^k, \tag{A1}$$

where, according to II(57) and (26),

$$g_k(l^2, L) = (-1)^L f_k(l^2, L). \tag{A2}$$

Introducing for g_k its expression II(59), we get

$$V_M(l^2, L) = \Sigma_{kr}(l||C^{(k)}||l)^2 (-1)^r (2r+1) W(llll; rk) F^k \times$$
$$\times (llLM|\,(\mathbf{u}_1^{(r)} . \mathbf{u}_2^{(r)})\,|llLM) = \tag{A3}$$
$$= \Sigma_r (2r+1) V_W(l^2, r)\ (llLM|(\mathbf{u}_1^{(r)} . \mathbf{u}_2^{(r)})|llLM)$$

The transformation to jj-coupling may now be performed by transforming the scalar products $(\mathbf{u}_1^{(r)} . \mathbf{u}_2^{(r)})$ to the jj-scheme. From the definition II(58) of $\mathbf{u}^{(r)}$ and from II(44b), we have

$$(\tfrac{1}{2}lj||u^{(r)}||\tfrac{1}{2}lj) = (-1)^{\frac{1}{2}+r-l-j}\,(2j+1) W(ljlj; \tfrac{1}{2}\ r) =$$
$$= [(2j+1)(2l+1) - r(r+1)]^{\frac{1}{2}} / (2l+1), \tag{A4}$$

and therefore, by II(38),

$$V_M(j^2, J) = \Sigma_r (2r+1) \frac{(2j+1)(2l+1) - r(r+1)}{(2l+1)^2} \times$$
$$\times \quad V_W(l^2, r)\ (-1)^{2j-J}\ W(jjjj; Jr). \tag{A5}$$

Applying (28a) we have

$$\Sigma_J (2J+1) V_M(j^2, J) = (2j+1)^2 V_W(S)/(2l+1); \tag{A6}$$

applying (28b) and the identity

$$(2j+1)(2l+1) = (2j+1)(2l+1) + 2(j-l)^2 - \tfrac{1}{2} = \tfrac{1}{2}(2j+1)^2 + 2l(l+1), \tag{A7}$$

we have

$$\Sigma_J (-1)^J (2J+1) V_M(j^2, J) =$$
$$= -\Sigma_r (2r+1) \frac{\frac{1}{2}(2j+1)^2 + 2l(l+1) - r(r+1)}{(2l+1)^2} V_W(l^2, r) =$$
$$= -\frac{(2j+1)^2}{2(2l+1)^2} \Sigma_r (2r+1) V_W(l^2, r) +$$
$$+ \Sigma_r (2r+1) \frac{r(r+1) - 2l(l+1)}{(2l+1)^2} V_W(l^2, r). \tag{A8}$$

Since

$$r(r+1) - 2l(l+1) = 2l(l+1)(2l+1)(-1)^{2l-r} W(llll; r1), \tag{A9}$$

and since $k = 1$ does not appear in (25), it follows from (26) and II(42) that the second sum in (A8) vanishes; the first sum is given by (29), and therefore

$$\Sigma_J (-1)^J (2J+1) V_M(j^2, J) = -(2j+1)^2 F^0/2 \tag{A10}$$

From (A6) and (A10), and from the fact that

$$V_W(j^2, 0) + V_M(j^2, 0) = (2j+1) V(l^2, S)/(2l+1), \tag{A11}$$

we easily obtain (43b).

REFERENCES

1. Casimir, H., 1931, *Proc. Kon. Acad. Amst.*, **34**, 844.
2. Racah, G., 1942, *Phys. Rev.*, **61**, 186.
3. Rosenfeld, L., 1948, *Nuclear Physics*, Amsterdam, Eq. (10.31—6).
4. For the general expression of the eigenvalues of Casimir's operator for every semisimple group, see Racah, G., 1950, *Rend. Lincei*, **8**, 108.
5. Jahn, H. A., 1950, *Proc. Roy. Soc.*, **A201**, 516.
6. Racah, G., 1949, *Phys. Rev.*, **76**, 1352, Eq. (88).
7. Racah, G., 1942, *Phys. Rev.*, **62**, 438.
8. Mayer, M. G., 1950, *Phys. Rev.*, **78**, 22.
9. Weyl, H., 1939, *Classical Groups*, Princeton, Chap. VI.
10. Inglis, D. R., 1931, *Phys. Rev.*, **38**, 862.
11. Inglis, D. R., 1950, *Phys. Rev.*, **77**, 724.

REVIEWS OF MODERN PHYSICS VOLUME 24, NUMBER 4 OCTOBER, 1952

Some Properties of the Racah and Associated Coefficients*

L. C. BIEDENHARN,§ J. M. BLATT,‡ AND M. E. ROSE†
Oak Ridge National Laboratory, Oak Ridge, Tennessee

I. INTRODUCTION

THE composition of angular momenta, one of the basic elements of quantum mechanics, is accomplished by means of the vector addition coefficients —known also as the Clebsch-Gordan or Wigner coefficients. This, in principle, constitutes a complete solution to problems with coupled angular momenta; but, in practice, say in the evaluation of matrix elements of composite systems, one is quite often led to involved summations of products of several vector addition coefficients that are carried out, if at all, with difficulty. A quite typical situation might involve the matrix elements, for a composite system, of operators acting on only one or more of the subsystems. An elementary example of this is the evaluation of the matrix elements $(lsjm|a\mathbf{L}+b\mathbf{S}|l's'jm')$ for the magnetic moment of a particle with spin. A direct evaluation—which, of course, in this example could be avoided—involves a cumbersome sum of the product of three vector addition coefficients. This direct procedure, in effect, computes the matrix elements in a "decoupled" scheme, and then seeks to relate them to the desired matrix elements in the "coupled" scheme. The complicated sum of vector addition coefficients automatically effects this transformation between coupling schemes, keeping proper account of the conservation of angular momentum.

Racah, in his work on complex atomic spectra, discussed in detail the properties of these transformations, and defined the coefficient $W(abcd;ef)$—since known as the Racah coefficient—as the transformation between the coupling schemes $(\mathbf{a}+\mathbf{b}=\mathbf{e};\ \mathbf{e}+\mathbf{d}=\mathbf{c})$ and $(\mathbf{b}+\mathbf{d}=\mathbf{f};\ \mathbf{a}+\mathbf{f}=\mathbf{c})$. We discuss this in more detail in Sec. III.

Since Racah's pioneer work, the W coefficients have been applied to a wide variety of problems, the angular correlation of successive radiations and the angular distribution of scattering and reaction cross sections being conspicuous examples. From the utilitarian point of view (effecting difficult summations of vector addition coefficients), the usefulness of the Racah coefficients for these angular correlation problems is apparent. The relationship of the "recoupling" approach to these problems is less obvious, but more fundamental, as has been clearly pointed out by Fano (reference 12).||

The coefficients that enter most naturally into the angular distribution of scattering and reaction cross sections are not the Racah coefficients themselves but a combination designated in reference 6 as the "Z coefficients" (defined by Eq. (23) below). For the specific application of the W and Z coefficients we shall, however, refer to the original literature. A bibliography for this purpose is included below. Besides giving algebraic tables of the W coefficients, we shall summarize the relevant algebraic properties of the coefficients.

The algebraic tables of the W coefficients are sufficiently complicated, especially for large values of the variables, that a numerical tabulation would be of great value. We have prepared such tables (reference 19) not only for the W but also the Z coefficients. However, space does not permit their inclusion here. Copies of these tables can be obtained from the Oak Ridge National Laboratory. The tabulation consists of 54 numerical tables of $W(l_1J_1l_2J_2;sL)$ for $s=\frac{1}{2}$ through 3 in steps of $\frac{1}{2}$ and $L=0\cdots 8$ in integer steps. The remaining parameters have the range $l_i=0$ through 4 or 5 (integer steps) and $J_i=0$ through 9/2 ($\frac{1}{2}$ integer steps) with the necessary restriction $J_i-s=$ integer [see discussion following Eq. (12)]. The Z coefficients are tabulated in 54 tables for the same range of parameters.

II. VECTOR ADDITION COEFFICIENTS

We shall not repeat here the definition of the vector addition coefficients, since this is extensively treated in references 1 and 2. Tables of these coefficients are given in reference 2. However, some of the symmetry relations for the vector addition coefficients are less well known, and we repeat the rules given independently by Racah (reference 3) and Eisenbud.¶

$$\begin{aligned}(ab\alpha\beta|abc\gamma)&=(ba-\beta-\alpha|bac-\gamma)\\&=(-1)^{a+b-c}(ba\beta\alpha|bac\gamma)\\&=(-1)^{a+b-c}(ab-\alpha-\beta|abc-\gamma)\\&=(-1)^{a-\alpha}\left[\frac{2c+1}{2b+1}\right]^{\frac{1}{2}}\cdot(ac\alpha-\gamma|acb-\beta)\\&=(-1)^{b+\beta}\left[\frac{2c+1}{2a+1}\right]^{\frac{1}{2}}\cdot(cb-\gamma\beta|cba-\alpha).\end{aligned}\qquad(1)$$

Other symmetry rules result from a combination of these basic symmetries.

* This paper is based in part on work performed for the AEC at the Oak Ridge National Laboratory.
† Oak Ridge National Laboratory.
‡ University of Illinois.
§ Now at Yale University.
|| Numbered references are given at the end of this article.

¶ L. Eisenbud, Ph.D. thesis, Princeton University (1948).

Racah also gives an explicit formula for the often-occurring coefficient $(ab00|abc0)$. This formula is most easily written in terms of the "triangle" coefficient $\Delta(abc)$ defined by

$$\Delta(abc)=\left[\frac{(a+b-c)!(a-b+c)!(-a+b+c)!}{(a+b+c+1)!}\right]^{\frac{1}{2}}. \quad (2)$$

$\Delta(abc)$ is clearly unchanged by any permutation of a, b, c. In general, we shall need the triangle coefficients only for values of a, b, c which satisfy the triangular inequalities

$$a+b\geqslant c,\quad b+c\geqslant a,\quad c+a\geqslant b. \quad (3)$$

Racah then defined the quantity g by

$$2g=a+b+c \quad (4)$$

and obtained

$$(ab00|abc0)=(-1)^{g+c}(2c+1)^{\frac{1}{2}}\Delta(abc)\frac{g!}{(g-a)!(g-b)!(g-c)!}\quad (a+b+c=\text{even})$$

$$(ab00|abc0)=0\quad (a+b+c=\text{odd}). \quad (5)$$

III. RACAH COEFFICIENTS

The physical significance of the W coefficients, as well as their properties, can be most easily obtained by a procedure due to Racah. Consider orthonormal wave functions, $\Psi(jm)$, of sharp angular momenta, that is to say, $\Psi(jm)$ is an eigenfunction of the rotation operator with eigenvalues $j(j+1)$ for J^2 and m for J_z. If we have two such wave functions, in different spaces, then a product wave function with sharp total angular momentum, is obtained in the usual way by application of the vector addition coefficients:**

$$\Phi[j_1j_2;J_{12}M_{12}]=\sum_{m_1}(j_1j_2m_1M_{12}-m_1|j_1j_2J_{12}M_{12})\Psi(j_1m_1)\Psi(j_2M_{12}-m_1). \quad (6)$$

The unitary property of the vector addition coefficients guarantees that $\Phi[j_1j_2;J_{12}M_{12}]$ is itself orthonormal in J_{12} and M_{12}. For the addition of three angular momentum vectors we consider a third wave function $\Psi(j_3m_3)$. A composite function with sharp total angular momentum can be obtained by combining $\Psi(j_3m_3)$ with $\Phi[j_1j_2;J_{12}M_{12}]$ using again the vector addition coefficients:

$$\Phi[j_1j_2(J_{12})j_3;JM]=\sum_{m_3}(J_{12}j_3m_3M-m_3|J_{12}j_3JM)\Psi(j_3m_3)\times\Phi[j_1j_2;J_{12}M-m_3]$$

$$=\sum_{m_1,m_3}(j_1j_2m_1M-m_3-m_1|j_1j_2J_{12}M-m_3)\cdot(J_{12}j_3M-m_3m_3|J_{12}j_3JM)\cdot\Psi(j_1m_1)\Psi(j_2M-m_3-m_1)\Psi(j_3m_3). \quad (7)$$

The wave functions thus formed are again orthonormal in J, M, and also in J_{12}. It is clear, though, that such wave functions are not unique, since we could just as well have combined first j_2 and j_3 to give J_{23} and then j_1 and J_{23} to give J. That is,

$$\Phi[j_1,j_2j_3(J_{23});JM]=\sum_{m_2,m_1}(j_2j_3m_2M-m_1-m_2|j_2j_3J_{23}M-m_1)\cdot(j_1J_{23}m_1M-m_1|j_1J_{23}JM)\cdot\Psi(j_1m_1)\Psi(j_2m_2)\Psi(j_3M-m_1-m_2). \quad (8)$$

To identify the composite wave functions we must indicate the coupling scheme, as in Eqs. (7) and (8). Composite wave functions for one coupling scheme are, however, linearly related to the wave functions for another scheme. Thus, we may write

$$\Phi[j_1,j_2j_3(J_{23});JM]=\sum_{J_{12}}\langle j_1,j_2j_3(J_{23})J|j_1j_2(J_{12})j_3J\rangle\cdot\Phi[j_1j_2(J_{12})j_3;JM]. \quad (9)$$

By using Eqs. (7) and (8), and the orthonormality of the $\Psi(j_im_i)$, one can obtain a relation for the transformation coefficient in terms of the vector addition coefficients:

$$\langle j_1,j_2j_3(J_{23})J|j_1j_2(J_{12})j_3J\rangle=\sum_{m,m_2}(j_1j_2M-mm_2|j_1j_2J_{12}M-m+m_2)\cdot(j_2j_3m_2m-m_2|j_2j_3J_{23}m)\cdot(j_1J_{23}M-mm|j_1J_{23}JM)\cdot(J_{12}j_3M-m+m_2\,m-m_2|J_{12}j_3JM). \quad (10)$$

It is this relation that Racah used to define the W coefficients originally. In his notation we have

$$(2e+1)^{\frac{1}{2}}(2f+1)^{\frac{1}{2}}W(abcd;ef)\equiv\langle ab(e)dc|a,bd(f)c\rangle. \quad (11)$$

Racah was able to perform the sum indicated on the right hand side of (10). His result is

$$W(abcd;ef)=\Delta(abe)\Delta(cde)\Delta(acf)\Delta(bdf)w(abcd;ef)$$

$$w(abcd;ef)=\sum_z\frac{(-1)^{z+a+b+c+d}\cdot(z+1)!}{(z-a-b-e)!(z-c-d-e)!\cdot(z-a-c-f)!(z-b-d-f)!}\cdot\frac{1}{(a+b+c+d-z)!(a+d+e+f-z)!(b+c+e+f-z)!}. \quad (12)$$

As the interpretation in terms of recoupling angular momenta indicates, the Racah function is defined for

** The vector addition $(ab\alpha\beta|abc\gamma)$ vanishes unless $\alpha+\beta=\gamma$. In the following, we shall explicitly satisfy this requirement, eliminating a formal sum over γ.

integral or half-integral values of the quantities a, b, c, d, e, f, with the limitation that each of the four triads

$$(a, b, e) \quad (c, d, e) \quad (a, c, f) \quad (b, d, f) \tag{13}$$

has an integral sum. The sum in (12) goes over integral values of z such that none of the factorials in the denominator has a negative argument.

According to its definition, W satisfies various selection rules, all of which can be summarized by saying that each of the four triads (13) must form a possible triangle, i.e., must satisfy the condition that any side of a triangle is smaller than or equal to the sum of the other two sides. If one or more of those four triangles degenerates into a straight line, the summation in (12) reduces to one term (see Eq. 29).

The Racah coefficients are highly symmetrical functions of the parameters a, b, c, d, e, f. The basic symmetry relations are

$$\begin{aligned} W(abcd; ef) &= W(badc; ef) = W(cdab; ef) = W(acbd; fe) \\ &= (-1)^{e+f-a-d} W(ebcf; ad) \\ &= (-1)^{e+f-b-c} W(aefd; bc). \end{aligned} \tag{14}$$

Additional symmetry relations follow from the ones stated here, so that there are altogether 24 different permutations of a, b, c, d, e, f [which correspond to all possible permutations between the four triads (13)], for which the corresponding W's differ at most by a sign.

One can readily obtain the properties of the W coefficients from the above "recoupling" technique. Since the composite wave functions $\Phi[\cdots; JM]$ are orthonormal, we know immediate from Eq. (9) that the transformation coefficients are unitary. In terms of the W functions the unitary property is expressed by

$$\sum_e (2e+1)(2f+1) W(abcd; ef) W(abcd; eg) = \delta_{fg}. \tag{15}$$

Another sum rule given by Racah, which can be obtained from this procedure, is

$$W(afgb; cd) = \sum_e (2e+1)(-1)^{a+b-e} \cdot W(abcd; ef) \cdot W(bacd; eg). \tag{16}$$

An extension of the recoupling procedure to four wave functions yields yet another sum rule for the Racah functions:††

$$W(a\alpha b\beta; c\gamma) W(a'\alpha b'\beta; c'\gamma) = \sum_\lambda (2\lambda+1) \cdot W(a'\lambda \alpha c; ac') W(b\lambda\beta c'; b'c) W(a'\lambda\gamma b; ab'). \tag{17}$$

Racah‡‡ has shown that, except for a phase, Eqs. (14), (15), (16), and (17) define the W functions completely. Hence, no further independent relations can exist.

The Racah coefficients are useful for the study of angular distributions since their application effects the summations over the magnetic quantum numbers. If we substitute Eqs. (7), (8), into (9), and use the orthonormality of the $\Psi(j_i m_i)$ we find the relation

$$\begin{aligned} &(ab\alpha\beta | abe\alpha+\beta)(ed\alpha+\beta\delta | edc\alpha+\beta+\delta) \\ &= \sum_f (2e+1)^{\frac{1}{2}}(2f+1)^{\frac{1}{2}} (bd\beta\delta | bdf\beta+\delta) \\ &\quad \cdot (af\alpha\beta+\delta | afc\alpha+\beta+\delta) \cdot W(abcd; ef). \end{aligned} \tag{18}$$

The usefulness of (18) may not be immediately apparent. In a fairly typical problem where one is faced with summing a product of several vector addition coefficients over a single magnetic quantum number, however, successive application of (18) will allow "recoupling" of the angular momenta involved in the vector addition coefficients until the magnetic quantum number sum can be carried out.

Using the unitary property of the vector addition coefficients Eq. (18) can be given a form that proves very useful for summations involving products of three vector addition coefficients:

$$\begin{aligned} &\sum_\beta (ab\alpha\beta | abe\alpha+\beta) \cdot (ed\alpha+\beta\gamma-\alpha-\beta | edc\gamma) \\ &\quad \cdot (bd\beta\gamma-\alpha-\beta | bdf\gamma-\alpha) = (2e+1)^{\frac{1}{2}}(2f+1)^{\frac{1}{2}} \\ &\quad \cdot (af\alpha\gamma-\alpha | afc\gamma) \cdot W(abcd; ef). \end{aligned} \tag{19}$$

The result given in Eq. (19) is more particularized than that in (18) above, since in the former we are restricted to a single Racah transformation.

For calculational purposes, a recursion formula for the W coefficients is desirable. Equation (17), it will be readily observed is, in fact, a generalized recursion relation. By specializing this formula various recursion relations can be derived. For example, take $c' = \frac{1}{2}$. The W's involving c' then have a simple algebraic form (see Sec. V) and we get as one case

$$\begin{aligned} W(a\alpha+\tfrac{1}{2}b\beta+\tfrac{1}{2}; c+\tfrac{1}{2}\gamma) &= (2c+1) \cdot \left[\frac{(\alpha+\beta+\gamma+2)(\alpha+\beta-\gamma+1)}{(\alpha+c+a+2)(\alpha+c+1-a)(\beta+b+c+2)(\beta+c+1-b)}\right]^{\frac{1}{2}} \cdot W(a\alpha b\beta; c\gamma) \\ &+ \left[\frac{(a+\alpha+1-c)(a+c-\alpha)(b+\beta+1-c)(b+c-\beta)}{(\alpha+a+c+2)(\alpha+c+1-a)(\beta+b+c+2)(\beta+c+1-b)}\right]^{\frac{1}{2}} \cdot W(a\alpha+\tfrac{1}{2}b\beta+\tfrac{1}{2}; c-\tfrac{1}{2}\gamma). \end{aligned} \tag{20}$$

†† L. C. Biedenharn, to appear in J. Math. Phys. (M.I.T.).

‡‡ Unpublished manuscript.

It will be observed that Eq. (20) is indeterminate if either $a=\alpha+c+1$ or $b=\beta+c+1$ or both. Although a recursion formula can be written for these cases, it is unnecessary to do so, since the desired W coefficient has a simple explicit form [see Eq. (29)].

Useful algebraic formulas for the W coefficients result from giving numerical values to one variable (say the e) in Eq. (12). Two triangle conditions then restrict the remaining variables:

$$\begin{aligned}&(1)\quad |a-e|\leqslant b\leqslant a+e,\\&(2)\quad |c-e|\leqslant d\leqslant c+e.\end{aligned}\tag{21}$$

The formulas can thus be conveniently tabulated in a square array of $(2e+1)^2$ entries, similar to the tabulation of the vector addition coefficients. We give such tables for $e=\frac{1}{2}$, 1, $\frac{3}{2}$, 2 in Sec. V. If algebraic formulas for higher values of e are desired, the recursion formulas can be used to generate them.

In the correlation of successive radiations involving pure multipoles the Racah coefficients that enter have two repeated indices, in the form $W(a\nu cd;ad)$. Only integer values of ν can occur. A recursion relation for these W's takes a particularly simple and useful form:

$$\begin{aligned}W(a\nu+1\,cd;ad)&=\frac{2\nu+1}{\nu+1}\cdot\frac{2a(a+1)+2d(d+1)-2c(c+1)-\nu(\nu+1)}{[(2a+2+\nu)(2a-\nu)(2d+2+\nu)(2d-\nu)]^{\frac{1}{2}}}\\&\cdot W(a\nu cd;ad)-\frac{\nu}{\nu+1}\\&\cdot\left[\frac{(2a+\nu+1)(2a+1-\nu)(2d+\nu+1)(2d+1-\nu)}{(2a+\nu+2)(2a-\nu)(2d+2+\nu)(2d-\nu)}\right]^{\frac{1}{2}}\\&\cdot W(a\nu-1\,cd;ad).\end{aligned}\tag{22}$$

Equation (22) is by far the simplest way to obtain algebraic forms for the $W(a\nu cd;ad)$. A short tabulation of these W's is given in Sec. V.

IV. Z COEFFICIENTS

Coefficients which are more appropriate for the angular distribution problem, as mentioned earlier, are not the Racah coefficients themselves, but the combination

$$Z(abcd;ef)=i^{f-a+c}[(2a+1)(2b+1)(2c+1)(2d+1)]^{\frac{1}{2}}\cdot W(abcd;ef)(ac00|acf0).\tag{23}$$

Since there exists the simple explicit formula (7) for the relevant Clebsch-Gordan coefficient, the computation of Z is a simple matter once the tabulation of the W coefficients has been carried out.

The Z coefficients obey all the selection rules for the Racah coefficients [see (12) and the discussion there] as well as the selection rule

$$Z=0 \text{ unless } a+c+f=\text{even},\tag{24}$$

which follows from (5). This restriction has the consequence that the phase factor i^{f-a+c} in (23) is always real and equal to ±1. Of the various symmetry relations for the Racah coefficients, only one is needed for the angular distribution problem, namely,

$$Z(l_1J_1l_2J_2;sL)=(-1)^LZ(l_2J_2l_1J_1;sL).\tag{25}$$

A sum rule for Z coefficients follows from Racah's sum rule (15). This becomes for the Z coefficients

$$\sum_b Z(abcd;ef)Z(abc'd;ef)=\delta_{cc'}(2a+1)(2d+1)[(ac00|acf0)]^2.\tag{26}$$

Finally, we give here the values of Z when either e or f vanish. $e=0$ corresponds to vanishing channel spin; $f=0$ is related to the total cross section.

$$Z(abcd;0f)=\delta_{ab}\delta_{cd}(-1)^{2f}i^{f-a+c}\cdot[(2a+1)(2c+1)]^{\frac{1}{2}}(ac00|acf0),\tag{27}$$

$$Z(abcd;e0)=\delta_{ac}\delta_{bd}(-1)^{b-e}(2b+1)^{\frac{1}{2}}.\tag{28}$$

In the application to angular distributions in nuclear reactions $f=L$ is integral, so that the factor $(-1)^{2f}$ in (27) is always equal to $+1$.

Tables I, II, III, and IV for the W coefficients combined with Eqs. (5) and (23) suffice to determine the Z coefficients explicitly.

V. ALGEBRAIC FORMULAS FOR THE W COEFFICIENTS

The summation in Eq. (12) reduces to a single term if any one of the triangles formed from the triads (abe) (cde) (acf) (bdf) reduces to a line. In all such cases the symmetry conditions (16) allow the coefficient in question to be permuted to the form

$$W(abcd;a+bf)=\left[\frac{2a!2b!(a+b+c+d+1)!(a+b+c-d)!(a+b+d-c)!(c+f-a)!(d+f-b)!}{(2a+2b+1)!(c+d-a-b)!(a+c-f)!(a+f-c)!(a+c+f+1)!(b+d-f)!(b+f-d)!(b+f+d+1)!}\right]^{\frac{1}{2}}.\tag{29}$$

For any one variable equal to zero we have a special case of Eq. (29), and the result is [after using symmetry conditions (14)]

$$W(abcd;0f)=(-1)^{b+c-f}(2b+1)^{-\frac{1}{2}}(2c+1)^{-\frac{1}{2}}\delta_{ab}\delta_{cd}.\tag{30}$$

TABLE I. $W(l_1J_1l_2J_2;\frac{1}{2},L)$.

	$l_1=J_1+\frac{1}{2}$	$l_1=J_1-\frac{1}{2}$
$l_2=J_2+\frac{1}{2}$	$(-1)^{J_1+J_2-L}\left[\dfrac{(J_1+J_2+L+2)(J_1+J_2-L+1)}{(2J_1+1)(2J_1+2)(2J_2+1)(2J_2+2)}\right]^{\frac{1}{2}}$	$(-1)^{J_1+J_2-L}\left[\dfrac{(L-J_1+J_2+1)(L+J_1-J_2)}{(2J_1)(2J_1+1)(2J_2+1)(2J_2+2)}\right]^{\frac{1}{2}}$
$l_2=J_2-\frac{1}{2}$	$(-1)^{J_1+J_2-L}\left[\dfrac{(L+J_1-J_2+1)(L-J_1+J_2)}{(2J_1+1)(2J_1+2)(2J_2)(2J_2+1)}\right]^{\frac{1}{2}}$	$(-1)^{J_1+J_2-L-1}\left[\dfrac{(J_1+J_2+L+1)(J_1+J_2-L)}{2J_1(2J_1+1)(2J_2)(2J_2+1)}\right]^{\frac{1}{2}}$

TABLE II. $W(l_1J_1l_2J_2;1,L)$.

	$l_2=J_2+1$
$l_1=J_1+1$	$(-1)^{J_1+J_2-L}\left[\dfrac{(L+J_1+J_2+3)(L+J_1+J_2+2)(-L+J_1+J_2+2)(-L+J_1+J_2+1)}{4(2J_1+3)(J_1+1)(2J_1+1)(2J_2+3)(J_2+1)(2J_2+1)}\right]^{\frac{1}{2}}$
$l_1=J_1$	$(-1)^{J_1+J_2-L}\left[\dfrac{(L+J_1+J_2+2)(-L+J_1+J_2+1)(L-J_1+J_2+1)(L+J_1-J_2)}{4J_1(2J_1+1)(J_1+1)(2J_2+1)(J_2+1)(2J_2+3)}\right]^{\frac{1}{2}}$
$l_1=J_1-1$	$(-1)^{J_1+J_2-L}\left[\dfrac{(L+J_1-J_2)(L+J_1-J_2-1)(L-J_1+J_2+2)(L-J_1+J_2+1)}{4(2J_1+1)(2J_1-1)(J_1)(J_2+1)(2J_2+1)(2J_2+3)}\right]^{\frac{1}{2}}$
	$l_2=J_2$
$l_1=J_1+1$	$(-1)^{J_1+J_2-L}\left[\dfrac{(L+J_1+J_2+2)(L+J_1-J_2+1)(J_1+J_2-L+1)(L-J_1+J_2)}{4(2J_1+1)(J_1+1)(2J_1+3)(J_2)(J_2+1)(2J_2+1)}\right]^{\frac{1}{2}}$
$l_1=J_1$	$(-1)^{J_1+J_2-L-1}\left[\dfrac{J_1(J_1+1)+J_2(J_2+1)-L(L+1)}{[4J_1(J_1+1)(2J_1+1)(J_2)(J_2+1)(2J_2+1)]^{\frac{1}{2}}}\right]$
$l_1=J_1-1$	$(-1)^{J_1+J_2-L-1}\left[\dfrac{(L+J_1+J_2+1)(-L+J_1+J_2)(L+J_1-J_2)(L-J_1+J_2+1)}{4(2J_1+1)(J_1)(2J_1-1)(J_2)(2J_2+1)(J_2+1)}\right]^{\frac{1}{2}}$
	$l_2=J_2-1$
$l_1=J_1+1$	$(-1)^{J_1+J_2-L}\left[\dfrac{(L-J_1+J_2)(L-J_1+J_2-1)(L+J_1-J_2+2)(L+J_1-J_2+1)}{4(2J_1+1)(J_1+1)(2J_1+3)(2J_2-1)(J_2)(2J_2+1)}\right]^{\frac{1}{2}}$
$l_1=J_1$	$(-1)^{J_1+J_2-L-1}\left[\dfrac{(L+J_1+J_2+1)(L+J_1-J_2+1)(L+J_2-J_1)(J_1+J_2-L)}{4J_1(2J_1+1)(J_1+1)(J_2)(2J_2+1)(2J_2-1)}\right]^{\frac{1}{2}}$
$l_1=J_1-1$	$(-1)^{J_1+J_2-L}\left[\dfrac{(L+J_1+J_2+1)(L+J_1+J_2)(-L+J_1+J_2)(-L+J_1+J_2-1)}{4(2J_1+1)(J_1)(2J_1-1)(2J_2+1)(J_2)(2J_2-1)}\right]^{\frac{1}{2}}$

The Racah Coefficient $W(a\nu cd;ad)$

In order to remove irrational normalizing factors, define

$$W(a\nu cd;ad)\equiv\left[\frac{(2a-\nu)!(2d-\nu)!}{(2a+\nu+1)!(2d+\nu+1)!}\right]^{\frac{1}{2}}Y_\nu(acd). \quad (31)$$

In addition, define the variable x by

$$x\equiv c(c+1)-a(a+1)-d(d+1), \quad (32)$$

and introduce the convention that $\bar{a}=a(a+1)$, etc.

The Y_ν are then rational polynomials of the νth order in x with coefficients involving $\bar{a}$, $\bar{d}$ rationally. The lowest polynomials are

$$Y_0=1,$$

$$Y_1=-2x,$$

$$Y_2=6x^2+6x-8\bar{a}\bar{d},$$

$$Y_3=-20x^3-80x^2+16x[3\bar{a}\bar{d}-\bar{a}-\bar{d}-3]+80\bar{a}\bar{d},$$

$$\begin{aligned}Y_4=70x^4+700x^3&+x^2[1560-240\bar{a}\bar{d}+200\bar{a}+200\bar{d}]\\&+x[720+480\bar{a}+480\bar{d}-1360\bar{a}\cdot\bar{d}]\\&+48\bar{a}\bar{d}[2\bar{a}\bar{d}-4\bar{a}-4\bar{d}-27].\end{aligned}$$

The Y_ν for higher values of ν may be generated from the recursion relation

$$Y_{\nu+1}=\left(\frac{2\nu+1}{\nu+1}\right)Y_1Y_\nu-(2\nu+1)Y_\nu-\left(\frac{\nu}{\nu+1}\right)(4\bar{a}+1-\nu^2)(4\bar{d}+1-\nu^2)Y_{\nu-1}. \quad (33)$$

TABLE III. $W(l_1J_1l_2J_2;\frac{3}{2},L)$.

	$l_1=J_1+3/2$
$l_2=J_2+\frac{3}{2}$	$(-1)^{J_1+J_2-L}\left[\dfrac{(L+J_1+J_2+4)(L+J_1+J_2+3)(L+J_1+J_2+2)(-L+J_1+J_2+3)(-L+J_1+J_2+2)(-L+J_1+J_2+1)}{(2J_1+4)(2J_1+3)(2J_1+2)(2J_1+1)(2J_2+4)(2J_2+3)(2J_2+2)(2J_2+1)}\right]^{\frac{1}{2}}$
$l_2=J_2+\frac{1}{2}$	$(-1)^{J_1+J_2-L}\left[\dfrac{3(L+J_1+J_2+3)(L+J_1+J_2+2)(L+J_1-J_2+1)(L-J_1+J_2)(-L+J_1+J_2+2)(-L+J_1+J_2+1)}{(2J_1+4)(2J_1+3)(2J_1+2)(2J_1+1)(2J_2+3)(2J_2+2)(2J_2+1)(2J_2)}\right]^{\frac{1}{2}}$
$l_2=J_2-\frac{1}{2}$	$(-1)^{J_1+J_2-L}\left[\dfrac{3(L+J_1+J_2+2)(L+J_1-J_2+2)(L+J_1-J_2+1)(L-J_1+J_2)(L-J_1+J_2-1)(J_1+J_2-L+1)}{(2J_1+4)(2J_1+3)(2J_1+2)(2J_1+1)(2J_2+2)(2J_2+1)(2J_2)(2J_2-1)}\right]^{\frac{1}{2}}$
$l_2=J_2-\frac{3}{2}$	$(-1)^{J_1+J_2-L}\left[\dfrac{(L-J_1+J_2)(L-J_1+J_2-1)(L-J_1+J_2-2)(L+J_1-J_2+3)(L+J_1-J_2+2)(L+J_1-J_2+1)}{(2J_1+4)(2J_1+3)(2J_1+2)(2J_1+1)(2J_2+1)(2J_2)(2J_2-1)(2J_2-2)}\right]^{\frac{1}{2}}$
	$l_1=J_1+\frac{1}{2}$
$l_2=J_2+\frac{3}{2}$	$(-1)^{J_1+J_2-L}\left[\dfrac{3(L+J_1+J_2+3)(L+J_1+J_2+2)(-L+J_1+J_2+2)(-L+J_1+J_2+1)(L-J_1+J_2+1)(L+J_1-J_2)}{(2J_1+3)(2J_1+2)(2J_1+1)(2J_1)(2J_2+4)(2J_2+3)(2J_2+2)(2J_2+1)}\right]^{\frac{1}{2}}$
$l_2=J_2+\frac{1}{2}$	$(-1)^{J_1+J_2-L-1}\dfrac{[(L+J_1+J_2+2)(-L+J_1+J_2+1)]^{\frac{1}{2}}[(L+J_1+J_2+3)(-L+J_1+J_2)-2(L-J_1+J_2)(L+J_1-J_2)]}{[(2J_1+3)(2J_1+2)(2J_1+1)(2J_1)(2J_2+3)(2J_2+2)(2J_2+1)(2J_2)]^{\frac{1}{2}}}$
$l_2=J_2-\frac{1}{2}$	$(-1)^{J_1+J_2-L-1}\dfrac{[(L+J_1-J_2+1)(L-J_1+J_2)]^{\frac{1}{2}}[2(L+J_1+J_2+2)(-L+J_1+J_2)-(L-J_1+J_2-1)(L+J_1-J_2)]}{[(2J_1+3)(2J_1+2)(2J_1+1)(2J_1)(2J_2+2)(2J_2+1)(2J_2)(2J_2-1)]^{\frac{1}{2}}}$
$l_2=J_2-\frac{3}{2}$	$(-1)^{J_1+J_2-L-1}\left[\dfrac{3(L+J_1+J_2+1)(L-J_1+J_2)(L-J_1+J_2-1)(L+J_1-J_2+2)(L+J_1-J_2+1)(-L+J_1+J_2)}{(2J_1+3)(2J_1+2)(2J_1+1)(2J_1)(2J_2+1)(2J_2)(2J_2-1)(2J_2-2)}\right]^{\frac{1}{2}}$
	$l_1=J_1-\frac{1}{2}$
$l_2=J_2+\frac{3}{2}$	$(-1)^{J_1+J_2-L}\left[\dfrac{3(L+J_1+J_2+2)(-L+J_1+J_2+1)(L-J_1+J_2+2)(L-J_1+J_2+1)(L+J_1-J_2)(L+J_1-J_2-1)}{(2J_1+2)(2J_1+1)(2J_1)(2J_1-1)(2J_2+4)(2J_2+3)(2J_2+2)(2J_2+1)}\right]^{\frac{1}{2}}$
$l_2=J_2+\frac{1}{2}$	$(-1)^{J_1+J_2-L}\dfrac{[(L-J_1+J_2+1)(L+J_1-J_2)]^{\frac{1}{2}}[(L-J_1+J_2)(L+J_1-J_2-1)-2(L+J_1+J_2+2)(-L+J_1+J_2)]}{[(2J_1+2)(2J_1+1)(2J_1)(2J_1-1)(2J_2+3)(2J_2+2)(2J_2+1)(2J_2)]^{\frac{1}{2}}}$
$l_2=J_2-\frac{1}{2}$	$(-1)^{J_1+J_2-L}\dfrac{[(L+J_1+J_2+1)(-L+J_1+J_2)]^{\frac{1}{2}}[(L+J_1+J_2+2)(-L+J_1+J_2-1)-2(L+J_1-J_2)(L-J_1+J_2)]}{[(2J_1+2)(2J_1+1)(2J_1)(2J_1-1)(2J_2+2)(2J_2+1)(2J_2)(2J_2-1)]^{\frac{1}{2}}}$
$l_2=J_2-\frac{3}{2}$	$(-1)^{J_1+J_2-L}\left[\dfrac{3(L+J_1+J_2+1)(L+J_1+J_2)(L-J_1+J_2)(L+J_1-J_2+1)(-L+J_1+J_2)(-L+J_1+J_2-1)}{(2J_1+2)(2J_1+1)(2J_1)(2J_1-1)(2J_2+1)(2J_2)(2J_2-1)(2J_2-2)}\right]^{\frac{1}{2}}$
	$l_1=J_1-3/2$
$l_2=J_2+\frac{3}{2}$	$(-1)^{J_1+J_2-L}\left[\dfrac{(L+J_1-J_2)(L+J_1-J_2-1)(L+J_1-J_2-2)(L-J_1+J_2+3)(L-J_1+J_2+2)(L-J_1+J_2+1)}{(2J_1+1)(2J_1)(2J_1-1)(2J_1-2)(2J_2+4)(2J_2+3)(2J_2+2)(2J_2+1)}\right]^{\frac{1}{2}}$
$l_2=J_2+\frac{1}{2}$	$(-1)^{J_1+J_2-L-1}\left[\dfrac{3(L+J_1+J_2+1)(-L+J_1+J_2)(L+J_1-J_2)(L+J_1-J_2-1)(L-J_1+J_2+2)(L-J_1+J_2+1)}{(2J_1+1)(2J_1)(2J_1-1)(2J_1-2)(2J_2+3)(2J_2+2)(2J_2+1)(2J_2)}\right]^{\frac{1}{2}}$
$l_2=J_2-\frac{1}{2}$	$(-1)^{J_1+J_2-L}\left[\dfrac{3(L+J_1+J_2+1)(L+J_1+J_2)(-L+J_1+J_2)(-L+J_1+J_2-1)(L+J_1-J_2)(L-J_1+J_2+1)}{(2J_1+1)(2J_1)(2J_1-1)(2J_1-2)(2J_2+2)(2J_2+1)(2J_2)(2J_2-1)}\right]^{\frac{1}{2}}$
$l_2=J_2-\frac{3}{2}$	$(-1)^{J_1+J_2-L-1}\left[\dfrac{(L+J_1+J_2+1)(L+J_1+J_2)(L+J_1+J_2-1)(-L+J_1+J_2)(-L+J_1+J_2-1)(-L+J_1+J_2-2)}{(2J_1+1)(2J_1)(2J_1-1)(2J_1-2)(2J_2+1)(2J_2)(2J_2-1)(2J_2-2)}\right]^{\frac{1}{2}}$

Algebraic formulas that result from taking the variable e in Eq. (12) to have the values $\frac{1}{2}$, 1, $\frac{3}{2}$, and 2 are given in Tables I through IV below. Table I gives the Racah coefficient $W(l_1J_1l_2J_2;\frac{1}{2}L)$; Table II, $W(l_1J_1l_2J_2;1L)$; Table III, $W(l_1J_1l_2J_2;\frac{3}{2}L)$; and Table IV, $W(l_1J_1l_2J_2;2L)$.

TABLE IV. $W(l_1J_1l_2J_2; 2, L)$.

	$l_1 = J_1+2$
$l_2=J_2+2$	$(-1)^{L-J_1-J_2}\left[\dfrac{(L+J_1+J_2+5)(L+J_1+J_2+4)(L+J_1+J_2+3)(L+J_1+J_2+2)(-L+J_1+J_2+4)(-L+J_1+J_2+3)(-L+J_1+J_2+2)(-L+J_1+J_2+1)}{(2J_1+5)(2J_1+4)(2J_1+3)(2J_1+2)(2J_1+1)\cdot(2J_2+5)(2J_2+4)(2J_2+3)(2J_2+2)(2J_2+1)}\right]^{1/2}$
$l_2=J_2+1$	$(-1)^{L-J_1-J_2}\left[\dfrac{4(L+J_1+J_2+4)(L+J_1+J_2+3)(L+J_1+J_2+2)(L+J_1-J_2+1)(-L+J_1+J_2+3)(-L+J_1+J_2+2)(-L+J_1+J_2+1)(L-J_1+J_2)}{(2J_1+5)(2J_1+4)(2J_1+3)(2J_1+2)(2J_1+1)\cdot(2J_2+4)(2J_2+3)(2J_2+2)(2J_2+1)(2J_2)}\right]^{1/2}$
$l_2=J_2$	$(-1)^{L-J_1-J_2}\left[\dfrac{6(L+J_1+J_2+3)(L+J_1+J_2+2)(L+J_1-J_2+2)(L+J_1-J_2+1)(-L+J_1+J_2+2)(-L+J_1+J_2+1)(L-J_1+J_2)(L-J_1+J_2-1)}{(2J_1+5)(2J_1+4)(2J_1+3)(2J_1+2)(2J_1+1)\cdot(2J_2+3)(2J_2+2)(2J_2+1)(2J_2)(2J_2-1)}\right]^{1/2}$
$l_2=J_2-1$	$(-1)^{L-J_1-J_2}\left[\dfrac{4(L+J_1+J_2+2)(L-J_1+J_2)(L-J_1+J_2-1)(L-J_1+J_2-2)(L+J_1-J_2+3)(L+J_1-J_2+2)(L+J_1-J_2+1)(-L+J_1+J_2+1)}{(2J_1+5)(2J_1+4)(2J_1+3)(2J_1+2)(2J_1+1)\cdot(2J_2+2)(2J_2+1)(2J_2)(2J_2-1)(2J_2-2)}\right]^{1/2}$
$l_2=J_2-2$	$(-1)^{L-J_1-J_2}\left[\dfrac{(L-J_1+J_2)(L-J_1+J_2-1)(L-J_1+J_2-2)(L-J_1+J_2-3)(L+J_1-J_2+4)(L+J_1-J_2+3)(L+J_1-J_2+2)(L+J_1-J_2+1)}{(2J_1+5)(2J_1+4)(2J_1+3)(2J_1+2)(2J_1+1)\cdot(2J_2+1)(2J_2)(2J_2-1)(2J_2-2)(2J_2-3)}\right]^{1/2}$
	$l_1 = J_1+1$
$l_2=J_2+2$	$(-1)^{L-J_1-J_2}\left[\dfrac{4(L+J_1+J_2+4)(L+J_1+J_2+3)(L+J_1+J_2+2)(L-J_1+J_2+1)(L+J_1-J_2)(-L+J_1+J_2+3)(-L+J_1+J_2+2)(-L+J_1+J_2+1)}{(2J_1+4)(2J_1+3)(2J_1+2)(2J_1+1)(2J_1)\cdot(2J_2+5)(2J_2+4)(2J_2+3)(2J_2+2)(2J_2+1)}\right]^{1/2}$
$l_2=J_2+1$	$(-1)^{L-J_1-J_2-1}\left[\dfrac{(L+J_1+J_2+3)(L+J_1+J_2+2)(-L+J_1+J_2+2)(-L+J_1+J_2+1)}{(2J_1+4)(2J_1+3)(2J_1+2)(2J_1+1)(2J_1)\cdot(2J_2+4)(2J_2+3)(2J_2+2)(2J_2+1)(2J_2)}\right]^{1/2}\cdot 4\cdot[(J_1+1)(J_1-J_2)-L(L+1)+J_2(J_2+2)]$
$l_2=J_2$	$(-1)^{L-J_1-J_2-1}\left[\dfrac{6(L+J_1+J_2+2)(L-J_1+J_2)(L+J_1-J_2+1)(-L+J_1+J_2+1)}{(2J_1+4)(2J_1+3)(2J_1+2)(2J_1+1)(2J_1)\cdot(2J_2+3)(2J_2+2)(2J_2+1)(2J_2)(2J_2-1)}\right]^{1/2}\cdot 2\cdot[J_1(J_1+2)+J_2(J_2+1)-L(L+1)]$
$l_2=J_2-1$	$(-1)^{L-J_1-J_2-1}\left[\dfrac{(L-J_1+J_2)(L-J_1+J_2-1)(L+J_1-J_2+2)(L+J_1-J_2+1)}{(2J_1+4)(2J_1+3)(2J_1+2)(2J_1+1)(2J_1)\cdot(2J_2+2)(2J_2+1)(2J_2)(2J_2-1)(2J_2-2)}\right]^{1/2}\cdot 4\cdot[J_1(J_1+2)+J_2(J_2+1)+J_1J_2-L(L+1)]$
$l_2=J_2-2$	$(-1)^{L-J_1-J_2-1}\left[\dfrac{4(L+J_1+J_2+1)(-L+J_1+J_2)(L-J_1+J_2)(L-J_1+J_2-1)(L-J_1+J_2-2)(L+J_1-J_2+3)(L+J_1-J_2+2)(L+J_1-J_2+1)}{(2J_1+4)(2J_1+3)(2J_1+2)(2J_1+1)(2J_1)\cdot(2J_2+1)(2J_2)(2J_2-1)(2J_2-2)(2J_2-3)}\right]^{1/2}$
	$l_1 = J_1$
$l_2=J_2+2$	$(-1)^{L-J_1-J_2}\left[\dfrac{6(L+J_1+J_2+3)(L+J_1+J_2+2)(L+2-J_1+J_2)(L+1-J_1+J_2)(L+J_1-J_2)(L+J_1-J_2-1)(-L+J_1+J_2+2)(-L+J_1+J_2+1)}{(2J_1+3)(2J_1+2)(2J_1+1)(2J_1)(2J_1-1)\cdot(2J_2+5)(2J_2+4)(2J_2+3)(2J_2+2)(2J_2+1)}\right]^{1/2}$
$l_2=J_2+1$	$(-1)^{L-J_1-J_2-1}\left[\dfrac{6(L+J_1+J_2+2)(L+J_1-J_2)(L-J_1+J_2+1)(-L+J_1+J_2+1)}{(2J_1+3)(2J_1+2)(2J_1+1)(2J_1)(2J_1-1)\cdot(2J_2+4)(2J_2+3)(2J_2+2)(2J_2+1)(2J_2)}\right]^{1/2}\cdot 2\cdot[J_1(J_1+1)+J_2(J_2+2)-L(L+1)]$

TABLE IV.—*Continued.*

$l_2=J_2$	$(-1)^{L-J_1-J_2}\left[\dfrac{1}{(2J_1+3)(2J_1+2)(2J_1+1)(2J_1)(2J_1-1)\cdot(2J_2+3)(2J_2+2)(2J_2+1)(2J_2)(2J_2-1)}\right]^{\frac{1}{2}}\cdot 6\cdot[A(A+1)-\frac{4}{3}J_1(J_1+1)J_2(J_2+1)]$
	where $A\equiv L(L+1)-J_1(J_1+1)-J_2(J_2+1)$
$l_2=J_2-1$	$(-1)^{L-J_1-J_2}\left[\dfrac{6(L+J_1+J_2+1)(L-J_1+J_2)(L+J_1-J_2+1)(-L+J_1+J_2)}{(2J_1+3)(2J_1+2)(2J_1+1)(2J_1)(2J_1-1)\cdot(2J_2+2)(2J_2+1)(2J_2)(2J_2-1)(2J_2-2)}\right]^{\frac{1}{2}}\cdot 2\cdot[J_1(J_1+1)-L(L+1)+J_2^2-1]$
$l_2=J_2-2$	$(-1)^{L-J_1-J_2}\left[\dfrac{6(L+J_1+J_2+1)(L+J_1+J_2)(L-J_1+J_2)(L-J_1+J_2-1)(-L+J_1+J_2)(-L+J_1+J_2-1)(L+J_1-J_2+2)(L+J_1-J_2+1)}{(2J_1+3)(2J_1+2)(2J_1+1)(2J_1)(2J_1-1)\cdot(2J_2+1)(2J_2)(2J_2-1)(2J_2-2)(2J_2-3)}\right]^{\frac{1}{2}}$
	$l_1=J_1-1$
$l_2=J_2+2$	$(-1)^{L-J_1-J_2}\left[\dfrac{4(L+J_1+J_2+2)(-L+J_1+J_2+1)(L-J_1+J_2+3)(L-J_1+J_2+2)(L-J_1+J_2+1)(L+J_1-J_2)(L+J_1-J_2-1)(L+J_1-J_2-2)}{(2J_1-2)(2J_1-1)(2J_1)(2J_1+1)(2J_1+2)\cdot(2J_2+1)(2J_2+2)(2J_2+3)(2J_2+4)(2J_2+5)}\right]^{\frac{1}{2}}$
$l_2=J_2+1$	$(-1)^{L-J_1-J_2-1}\left[\dfrac{(L-J_1+J_2+2)(L-J_1+J_2+1)(L+J_1-J_2)(L+J_1-J_2-1)}{(2J_1-2)(2J_1-1)(2J_1)(2J_1+1)(2J_1+2)\cdot(2J_2)(2J_2+1)(2J_2+2)(2J_2+3)(2J_2+4)}\right]^{\frac{1}{2}}\cdot 4\cdot[(J_1-1)(J_1+J_2+2)-(L+J_2+2)(L-J_2-1)]$
$l_2=J_2$	$(-1)^{L-J_1-J_2}\left[\dfrac{6(L+J_1+J_2+1)(L+J_1-J_2)(-L+J_1+J_2)(L-J_1+J_2+1)}{(2J_1-2)(2J_1-1)(2J_1)(2J_1+1)(2J_1+2)\cdot(2J_2-1)(2J_2)(2J_2+1)(2J_2+2)(2J_2+3)}\right]^{\frac{1}{2}}\cdot 2\cdot[J_1^2-1-(L+J_2+1)(L-J_2)]$
$l_2=J_2-1$	$(-1)^{L-J_1-J_2-1}\left[\dfrac{(L+J_1+J_2+1)(L+J_1+J_2)(-L+J_1+J_2)(-L+J_1+J_2-1)}{(2J_1-2)(2J_1-1)(2J_1)(2J_1+1)(2J_1+2)\cdot(2J_2-2)(2J_2-1)(2J_2)(2J_2+1)(2J_2+2)}\right]^{\frac{1}{2}}\cdot 4\cdot[(J_1-1)(J_1-J_2+1)-(L+J_2)(L-J_2+1)]$
$l_2=J_2-2$	$(-1)^{L-J_1-J_2-1}\left[\dfrac{4(L+J_1+J_2+1)(L+J_1+J_2)(L+J_1+J_2-1)(L+J_1-J_2+1)(-L+J_1+J_2)(-L+J_1+J_2-1)(-L+J_1+J_2-2)(L-J_1+J_2)}{(2J_1-2)(2J_1-1)(2J_1)(2J_1+1)(2J_1+2)\cdot(2J_2-3)(2J_2-2)(2J_2-1)(2J_2)(2J_2+1)}\right]^{\frac{1}{2}}$
	$l_1=J_1-2$
$l_2=J_2+2$	$(-1)^{L-J_1-J_2}\left[\dfrac{(L+4-J_1+J_2)(L+3-J_1+J_2)(L+2-J_1+J_2)(L+1-J_1+J_2)(L+J_1-J_2)(L+J_1-J_2-1)(L+J_1-J_2-2)(L+J_1-J_2-3)}{(2J_1-3)(2J_1-2)(2J_1-1)(2J_1)(2J_1+1)\cdot(2J_2+5)(2J_2+4)(2J_2+3)(2J_2+2)(2J_2+1)}\right]^{\frac{1}{2}}$
$l_2=J_2+1$	$(-1)^{L-J_1-J_2-1}\left[\dfrac{4(L+1+J_1+J_2)(L+3-J_1+J_2)(L+2-J_1+J_2)(L+1-J_1+J_2)(L+J_1-J_2)(L+J_1-J_2-1)(L+J_1-J_2-2)(-L+J_1+J_2)}{(2J_1-3)(2J_1-2)(2J_1-1)(2J_1)(2J_1+1)\cdot(2J_2+4)(2J_2+3)(2J_2+2)(2J_2+1)(2J_2)}\right]^{\frac{1}{2}}$
$l_2=J_2$	$(-1)^{L-J_1-J_2}\left[\dfrac{6(L+1+J_1+J_2)(L+J_1+J_2)(L+2-J_1+J_2)(L+1-J_1+J_2)(L+J_1-J_2)(L+J_1-J_2-1)(-L+J_1+J_2)(-L+J_1+J_2-1)}{(2J_1-3)(2J_1-2)(2J_1-1)(2J_1)(2J_1+1)\cdot(2J_2+3)(2J_2+2)(2J_2+1)(2J_2)(2J_2-1)}\right]^{\frac{1}{2}}$
$l_2=J_2-1$	$(-1)^{L-J_1-J_2-1}\left[\dfrac{4(L+1+J_1+J_2)(L+J_1+J_2)(L+J_1+J_2-1)(L+1-J_1+J_2)(L+J_1-J_2)(-L+J_1+J_2)(-L+J_1+J_2-1)(-L+J_1+J_2-2)}{(2J_1-3)(2J_1-2)(2J_1-1)(2J_1)(2J_1+1)\cdot(2J_2+2)(2J_2+1)(2J_2)(2J_2-1)(2J_2-2)}\right]^{\frac{1}{2}}$
$l_2=J_2-2$	$(-1)^{L-J_1-J_2}\left[\dfrac{(L+1+J_1+J_2)(L+J_1+J_2)(L+J_1+J_2-1)(L+J_1+J_2-2)(-L+J_1+J_2)(-L+J_1+J_2-1)(-L+J_1+J_2-2)(-L+J_1+J_2-3)}{(2J_1-3)(2J_1-2)(2J_1-1)(2J_1)(2J_1+1)\cdot(2J_2+1)(2J_2)(2J_2-1)(2J_2-2)(2J_2-3)}\right]^{\frac{1}{2}}$

REFERENCES

General:

1. E. Wigner, *Gruppentheorie* (Vieweg, 1931).
2. E. U. Condon and G. H. Shortley, *Theory of Atomic Spectra* (Cambridge University Press, Cambridge, England, 1935).
3. G. Racah, Phys. Rev. **62**, 438 (1942).
4. G. Racah, Phys. Rev. **63**, 367 (1943).
5. G. Racah, "Lectures on Group Theory" (Institute for Advanced Study, Princeton, New Jersey, 1951).

Angular Dependence of Scattering and Reaction Cross Sections:

6. J. M. Blatt and L. C. Biedenharn, Revs. Modern Phys. **23**, 256 (1952).

Atomic and Molecular Spectra:

7. R. Bersohn, J. Chem. Phys. **18**, 1124 (1950).
8. R. E. Trees, Phys. Rev. **82**, 683 (1951).

Directional Correlation of Successive Radiations:

9. J. W. Gardner, Proc. Phys. Soc. (London) **62A**, 763 (1949).
10. S. P. Lloyd, Ph.D. thesis, University of Illinois (1951).
11. G. Racah, Phys. Rev. **82**, 309 (1951).
12. U. Fano, Natl. Bur. Standards No. 1214, Washington, D. C. (1951).
13. L. C. Biedenharn, G. B. Arfken, and M. E. Rose, Phys. Rev. **83**, 586 (1951).
14. K. Alder, Phys. Rev. **83**, 1266 (1951); Phys. Rev. **84**, 369 (1951); and forthcoming article in Helv. Phys. Acta.
15. Rose, Biedenharn, and Arfken, Phys. Rev. **84**, 5 (1952).
16. L. C. Biedenharn and M. E. Rose (to be submitted to the Revs. Modern Phys.).

Nuclear Structure:

17. H. A. Jahn, Proc. Roy. Soc. (London) **A201**, 516 (1950); Proc. Roy. Soc. (London) **A205**, 192 (1951).
18. G. Racah, Chapters 5 and 6 of reference 5.

Numerical Tables:

19. L. C. Biedenharn, Oak Ridge National Laboratory Report No. 1098 (1952).

Reprinted from JOURNAL OF MATHEMATICS AND PHYSICS
Vol. XXXI, No. 4, January, 1953

AN IDENTITY SATISFIED BY THE RACAH COEFFICIENTS*

BY L. C. BIEDENHARN

1. In discussing the theory of complex spectra[1], Racah found it expedient to introduce the function, subsequently called the Racah coefficient,

$$W(abcd; ef) = \Delta(abe)\,\Delta(cde)\,\Delta(acf)\,\Delta(bdf)$$
$$\cdot \sum_z (-1)^z \frac{(a+b+c+d+1-z)!}{(a+b-e-z)!\,(c+d-e-z)!\,(a+c-f-z)!}$$
$$\cdot \frac{1}{(b+d-f-z)!\,z!\,(e+f-a-d+z)!\,(e+f-b-c+z)!}$$
$$\Delta(abe) \equiv \left[\frac{(a+b-e)!\,(a-b+e)!\,(-a+b+e)!}{(a+b+e+1)!}\right]^{\frac{1}{2}} \tag{1}$$

In this sum z takes on all integer values, including zero. W is defined for all integer and half integer values of its arguments subject to the restriction that each of the triads (abe), (cde), (acf), (bdf) has an integer sum. The arguments have the significance of angular momentum vectors and the Racah coefficient can be intepreted as the transformation coefficient[2] from the coupling scheme $(\mathbf{a} + \mathbf{b} = \mathbf{e}, \mathbf{e} + \mathbf{d} = \mathbf{c})$ to the scheme $(\mathbf{b} + \mathbf{d} = \mathbf{f}, \mathbf{a} + \mathbf{f} = \mathbf{c})$. This point will be discussed more fully below, along with other properties of the coefficients, using the procedure of ref. 2.

Recently it has been found that the Racah coefficients are useful in discussing many other topics of current physical interest: the theory of the angular correlation of successive radiations[3–7], the angular dependence of scattering and reaction cross sections[8], nuclear structure[9], to mention a few. A need for more explicit algebraic formulae for the W coefficients (specifying values for one or more of the variables) as well as numerical tabulation became apparent. Jahn[9] recently published tables of algebraic formulae for the W's similar in form to the well-known tabulation of the vector addition coefficients[10]. The numerical tabulation was undertaken by Blatt and the author[11], in connection with a review paper on

* This document is based on work performed for the AEC at the Oak Ridge National Laboratory.

[1] G. Racah, Phys. Rev. **62,** 438 (1942).
[2] G. Racah, Phys. Rev. **63,** 367 (1943).
[3] J. W. Gardner, Proc. Phys. Soc., **62A,** 763 (1949).
[4] S. P. Lloyd, Phys. Rev. **83,** 716 (1951); Ph.D. Thesis, University of Ill. (1951).
[5] L. C. Biedenharn, G. B. Arfken, and M. E. Rose, Phys. Rev. **83,** 586 (1951).
[6] U. Fano, National Bureau of Standards Report #1214 (1951).
[7] G. Racah, Phys. Rev. **84,** 910 (1951).
[8] J. M. Blatt and L. C. Biedenharn, Phys. Rev. **82,** 123(L) (1951). An expanded version of this work has been submitted to the Reviews of Modern Physics.
[9] H. A. Jahn, Proc. Roy. Soc., **A201,** 516 (1950); **A205,** 192 (1951).
[10] E. P. Wigner, "Gruppentheorie" Vieweg, (1931). E. U. Condon and G. H. Shortley, "Theory of Atomic Spectra," Cambridge, (1935).
[11] L. C. Biedenharn, "Tables of the Racah Coefficients," ORNL Report No. 1098, (1951).

the angular dependence of nuclear and reaction cross sections. In setting up the task of numerical calculation it became clear that neither (1) nor the more explicit algebraic forms deduced from it[9] was the best way to proceed. A more desirable procedure would involve computation from recursion formulae, that should exist, of course. In developing such recursion formulae, a quite general identity for the Racah coefficients was found, which is the subject of this note. The significance of this identity for the theory of the W coefficients has been discussed by Racah[12], and this note will therefore be confined to a simple demonstration of the identity only.

Besides their application to problems of physical interest, the Racah coefficients are of considerable mathematical interest for group theoretical problems. In particular, Racah has discussed the application to an algebra of tensor operators and the construction of matrices that decompose the Kronecker product.[13]

2. Let us consider normalized wave functions of the form $\psi(jm)$ which are eigenfunctions of the operators $\mathbf{J}^2 = J_x^2 + J_y^2 + J_z^2$ and J_z with eigenvalues $j(j+1)$ and m respectively. $\mathbf{J}$ is here the total angular momentum operator, and satisfies the commutation relations $[J_x, J_y] = iJ_z$ etc. If we have two such wave functions, for different particles, we can form a composite wave function that is the simultaneous eigenfunction of the operators $\mathbf{J}_1^2$; $\mathbf{J}_2^2$; $(\mathbf{J}_1 + \mathbf{J}_2)^2 = \mathbf{J}^2$; $J_z = J_{1z} + J_{2z}$ in the usual way using the vector addition coefficients[10]:

$$\Phi(JM) = \sum_{m_1} (j_1 j_2 m_1 M - m_1 \mid j_1 j_2 JM) \psi(j_1 m_1) \psi(j_2 M - m_1) \tag{2}$$

The vector addition coefficient $(ab\alpha\beta \mid abc\gamma)$ vanishes unless $\gamma = \alpha + \beta$. In (2), and in subsequent equations, we have explicitly satisfied this requirement, eliminating a formal sum over γ. The vector addition coefficients are unitary (and real), satisfying the relations[10]

$$\sum_\alpha (ab\alpha\gamma - \alpha \mid abc\gamma)(ab\alpha\gamma - \alpha \mid abc'\gamma) = \delta_{cc'}$$
$$\sum_c (ab\alpha\gamma - \alpha \mid abc\gamma)(ab\alpha'\gamma - \alpha' \mid abc\gamma) = \delta_{\alpha\alpha'} .$$

It follows from this, and the orthonormality of the $\psi(j_i m_i)$, that the $\Phi(JM)$ are also orthonormal.

If we now consider three particles, we can define composite wave functions of sharp total angular momentum, $\mathbf{J} = \mathbf{J}_1 + \mathbf{J}_2 + \mathbf{J}_3$, in a similar way. The composite wave functions so defined are, however, not unique but depend upon the coupling scheme. If, for example, we couple first j_1 and j_2 to form J_a and then couple J_a and j_3 to get J we get the composite wave functions:

$$\begin{aligned}\Phi(j_1 j_2 (J_a) j_3 \,;\, JM) \\ = \sum_m \Big[\sum_{m_1} (j_1 j_2 m_1 m - m_1 \mid j_1 j_2 J_a m) \psi(j_1 m_1) \psi(j_2 m - m_1)\Big] \\ \cdot (J_a j_3 m M - m \mid J_a j_3 JM) \psi(j_3 M - m)\end{aligned} \tag{3}$$

[12] G. Racah, (unpublished). Dr. U. Fano kindly let us see his copy of a preliminary form of the manuscript.

[13] G. Racah, "Lectures on Group Theory and Spectroscopy", Institute for Advanced Study, Princeton, N. J. (1951).

Here the good quantum numbers for the composite wave function are J, M, J_a, and, of course, j_1, j_2, j_3. The notation for Φ is chosen to indicate the coupling scheme. On the other hand we might have coupled first j_2 and j_3 to form J_b and then coupled j_1 and J_b to form J. This yields the wave function:

$$\Phi(j_1, j_2 j_3(J_b); JM) = \sum_m \left[\sum_{m_2} (j_2 j_3 m_2 m - m_2 \mid j_2 j_3 J_b m)\psi(j_2 m_2)\psi(j_3 m - m_2)\right] \cdot (j_1 J_b M - m m \mid j_1 J_b JM)\psi(j_1 M - m) \quad (4)$$

The two sets of wave functions defined by (3) and (4) are easily shown to be orthonormal, that is to say:

$$\langle \Phi(j_1 j_2(J_a') j_3 ; J'M') \mid \Phi(j_1 j_2(J_a) j_3 ; JM) \rangle = \delta_{JJ'} \delta_{MM'} \delta_{J_a J_a'} \quad (5)$$

Similarly for the wave functions (4).

Since both these sets of wave functions span the same space, it is clear they must be related by equations of the form*:

$$\Phi(j_1, j_2 j_3(J_b); JM) = \sum_{J_a} \langle j_1, j_2 j_3(J_b) J \mid j_1 j_2(J_a) j_3 J \rangle \Phi(j_1 j_2(J_a) j_3 ; JM) \quad (6)$$

Using the orthonormality of the two sets of wave functions we deduce that the transformation is unitary:

$$\sum_{J_a} \langle j_1, j_2 j_3(J_b) J \mid j_1 j_2(J_a) j_3 J \rangle^* \langle j_1, j_2 j_3(J_b') J \mid j_1 j_2(J_a) j_3 J \rangle = \delta_{J_b J_b'} \quad (7)$$

Furthermore, by substituting (3) and (4) into (6) using the orthonormality of the $\psi(j_i m_i)$ we find:

$$(j_2 j_3 m_2 m - m_2 \mid j_2 j_3 J_b m)(j_1 J_b M - m m \mid j_1 J_b JM) = \sum_{J_a} \langle j_1, j_2 j_3(J_b) J \mid j_1 j_2(J_a) j_3 J \rangle \cdot (j_1 j_2 M - m m_2 \mid j_1 j_2 J_a M - m + m_2)(J_a j_3 M + m_2 - m m - m_2 \mid J_a j_3 JM) \quad (8)$$

The unitarity of the vector addition coefficients then allows one to write:

$$\langle j_1, j_2 j_3(J_b) J \mid j_1 j_2(J_a) j_3 J \rangle = \sum_{m_1, m_2} (j_1 j_2 m_1 m_2 \mid j_1 j_2 J_a m_1 + m_2) \cdot (j_2 j_3 m_2 M - m_1 - m_2 \mid j_2 j_3 J_b M - m_1)(j_1 J_b m_1 M - m_1 \mid j_1 J_b JM) \cdot (J_a j_3 m_1 + m_2 M - m_1 - m_2 \mid J_a j_3 JM) \quad (9)$$

These last relations show incidentally that the transformation coefficients are real. Comparing (9) with Racah's work[1, 2] one finds:

$$\langle j_1, j_2 j_3(J_b) J \mid j_1 j_2(J_a) j_3 J \rangle = [(2J_a + 1)(2J_b + 1)]^{\frac{1}{2}} \cdot W(j_1 j_2 J j_3 ; J_a J_b) \quad (10)$$

* In (6) we used the fact that the transformation coefficient is independent of M. This can be seen by using the Wigner-Eckart theorem, Wigner, l. c. p. 264.

By using the properties of the vector addition coefficients* and (9), we find that the W coefficients have the symmetries:

$$\begin{aligned} W(abcd;ef) &= W(badc;ef) = W(cdab;ef) = W(acbd;fe) \\ &= (-1)^{a+d-e-f}W(ebcf;ad) = (-1)^{b+c-e-f}W(aefd;bc) \end{aligned} \tag{11}$$

From index balance we can deduce in (9) that the sum is independent of M (M is, of course, restricted to lie between $-J$ and J). Using this latter fact, Racah was able to perform the impressive algebraic feat of summing the right hand side of (9) to get (1).

Using (10) to rewrite (7) we find:

$$\sum_{J_a} (2J_a + 1)(2J_b + 1)W(j_1j_2Jj_3;\, J_aJ_b)W(j_1j_2Jj_3;\, J_aJ_b') = \delta_{J_bJ_b'} \tag{12}$$

In this notation (6) reads:

$$\begin{aligned} &\Phi(j_1, j_2j_3(J_b); JM) \\ &\quad = \sum_{J_a} [(2J_a+1)(2J_b+1)]^{\frac{1}{2}}W(j_1j_2Jj_3;\, J_aJ_b)\Phi(j_1j_2(J_a)j_3;\, JM) \end{aligned} \tag{6'}$$

Interchange now j_1 and j_2.

$$\begin{aligned} &\Phi(j_2, j_1j_3(J_b); JM) \\ &\quad = \sum_{J_a} [(2J_a+1)\,(2J_b+1)]^{\frac{1}{2}}W(j_2j_1Jj_3;\, J_aJ_b)\Phi(j_2j_1(J_a)j_3;\, JM) \\ &= \sum_{J_a} [(2J_a+1)(2J_b+1)\,]^{\frac{1}{2}}W(j_2j_1Jj_3;\, J_aJ_b)(-1)^{j_1+j_2-J_a}\Phi(j_1j_2(J_a)j_3;\, JM) \end{aligned} \tag{13}$$

In getting (13) we used (3) and the fact that:

$$(j_1j_2m_1m - m_1 \mid j_1j_2J_am) = (-1)^{j_1+j_2-J_a}(j_2j_1m - m_1m_1 \mid j_2j_1J_am \tag{14}$$

Taking the inner product of (6′) and (13) yields:

$$\begin{aligned} &\langle\Phi(j_1, j_2j_3(J_b); JM) \mid \Phi(j_2, j_1j_3(J_b'); JM)\rangle \\ &\quad = \sum_{J_a} (2J_a+1)[(2J_b+1)(2J_b'+1)]^{\frac{1}{2}}\cdot(-1)^{j_1+j_2-J_a}W(j_1j_2Jj_3;\, J_aJ_b) \\ &\qquad \cdot W(j_2j_1Jj_3;\, J_aJ_b') \end{aligned} \tag{15}$$

This inner product, however, is itself a Racah coefficient and using (9) and (10), we find another relation[1] for the W's:

$$\begin{aligned} &W(j_1J_bJ_b'j_2;\, Jj_3) \\ &\quad = \sum_{J_a} (2J_a+1)(-1)^{j_1+j_1-J_a}W(j_1j_2Jj_3;\, J_aJ_b)\cdot W(j_2j_1Jj_3;\, J_aJ_b') \end{aligned} \tag{16}$$

To proceed further it is a natural generalization to consider the coupling of more than three systems. For four wave functions, $\psi(j_im_i)$, the composite func-

* These are the symmetry properties of the vector addition coefficients:

$$\begin{aligned} (ab\alpha\beta \mid abc\gamma) &= (-1)^{a+b-c}(ba\beta\alpha) \mid bac\gamma) \\ &= (-1)^{a+b-c}(ab - \alpha - \beta \mid abc - \gamma) \\ &= (-1)^{a-\alpha}[(2c+1)/(2a+1)]^{\frac{1}{2}}(ac\alpha - \gamma \mid acb - \beta) \end{aligned}$$

and all other symmetries resulting from combinations of these basic symmetries.

tions of sharp J, M will involve two intermediate coupling vectors, as, for example, in the coupling scheme ($\mathbf{j}_1 + \mathbf{j}_2 = \mathbf{J}_a$, $\mathbf{j}_3 + \mathbf{j}_4 = \mathbf{J}_b$, $\mathbf{J}_a + \mathbf{J}_b = \mathbf{J}$) symbolized by $(j_1j_2(J_a), j_3j_4(J_b), J)$. There are 15 different types of coupling that are possible and, just as before, there exist unitary (real) transformations between the composite wave functions defined by each coupling scheme. One can write out the explicit form of these transformations in terms of six vector addition coefficients, in complete analogy to (9). Such transformations are naturally more general than the Racah coefficients, and are not without their own interest—a familiar example is the passage from (LS) to (jj) coupling for two electron wave functions. Fano[6] has shown that they are the transformations that occur in the general problem of the coupling of tensor parameters, or, as it applies to the particular problem of angular correlations, the correlation of successive radiations from a non-random initial state. The triple correlation problem[5] is another example of the occurrence of such transformations.

As one might expect, however, these more general transformations are expressible in terms of the Racah coefficients. If the transformation is between the coupling schemes $(j_1j_2(J_a), j_3j_4(J_b), J)$ to $(j_1j_2(J_a), j_3(J_d), j_4, J)$, or more generally between coupling schemes where only one of the intermediate couplings is unchanged, it is clear that only a single W coefficient is involved.

Consider next the transformation between the wave functions defined by the coupling scheme $(j_1j_2(J_a), j_3j_4(J_b), J)$ and the coupling scheme ($\mathbf{j}_2 + \mathbf{j}_3 = \mathbf{J}_c$, $\mathbf{j}_1 + \mathbf{J}_c = \mathbf{J}_d$, $\mathbf{J}_d + \mathbf{j}_4 = \mathbf{J}$) symbolized by $(j_1, j_2j_3(J_c), (J_d), j_4, J)$, that is:

$$\begin{aligned}\Phi(j_1, j_2j_3(J_c), (J_d)j_4; JM) &= \textstyle\sum_{J_aJ_b} \langle j_1, j_2j_3(J_c), (J_d), j_4, J \mid j_1j_2(J_a), j_3j_4(J_b), J\rangle \\ &\quad\cdot\Phi(j_1j_2(J_a), j_3j_4(J_b); JM)\end{aligned} \tag{17}$$

We can accomplish this transformation in two steps, each one of which involves a re-coupling of the type given by (5). First we change the coupling of j_1, j_2 and j_3 in the wave function on the left hand side of (17):

$$\begin{aligned}\Phi(j_1, j_2j_3(J_c), (J_d)j_4; JM) &\\ = \textstyle\sum_{J_a} \langle j_1, j_2j_3(J_c)J_d \mid j_1j_2(J_a)j_3\, J_d\rangle &\Phi(j_1j_2(J_a), j_3(J_d), j_4; JM)\end{aligned} \tag{18}$$

and secondly re-couple J_aj_3 and j_4 in the wave functions appearing on the right hand side of (18) using (5) once again:

$$\begin{aligned}\Phi(j_1j_2(J_a), j_3(J_d), j_4; JM) &\\ = \textstyle\sum_{J_b} \langle J_aj_3(J_d)j_4J \mid J_a, j_3j_4(J_b)J\rangle &\cdot\Phi(j_1j_2(J_a), j_3j_4(J_b); JM)\end{aligned} \tag{19}$$

Substituting (18) and (19) in (17) yields the result:

$$\begin{aligned}\langle j_1, j_2j_3(J_c), (J_d), j_4J \mid j_1j_2(J_a), j_3j_4(J_b), J\rangle &\\ = \langle j_1, j_2j_3(J_c)J_d \mid j_1\, j_2(J_a)\, j_3\, J_d\rangle\cdot &\langle J_aj_3(J_d), j_4J \mid J_a, j_3j_4(J_b)J\rangle\end{aligned} \tag{20}$$

This procedure is, however, not unique. We might equally well have applied (5) three times in the sequence:

(a) Recouple j_1, J_c and j_4 in the wave function on the left hand side of (17):

$$\Phi(j_1, j_2j_3(J_c), (J_d), j_4; JM) = \sum_\alpha \langle j_1 J_c(J_d) j_4 J \mid j_1, J_c j_4(\alpha) J\rangle \Phi(j_1; j_2j_3(J_c), j_4(\alpha); JM) \tag{21}$$

(b) Recouple j_2j_3 and j_4 in the wave functions on the right hand side of (21):

$$\Phi(j_1; j_2j_3(J_c)j_4(\alpha); JM) = \sum_{J_b} \langle j_2j_3(J_c)j_4\alpha \mid j_2, j_3j_4(J_b)\alpha\rangle \Phi(j_1; j_2, j_3j_4(J_b), \alpha; JM) \tag{22}$$

(c) Finally, recouple j_1j_2 and J_b in the wave functions on the right hand side of (22):

$$\Phi(j_1; j_2, j_3j_4(J_b), \alpha; JM) = \sum_{J_a} \langle j_1, j_2 J_b(\alpha) J \mid j_1j_2(J_a) J_b J\rangle \Phi(j_1j_2(J_a), j_3j_4(J_b); JM) \tag{23}$$

Putting these steps together we now obtain:

$$\begin{aligned}\langle j_1, j_2j_3(J_c), (J_d), j_4 J \mid j_1j_2(J_a), j_3j_4(J_b), J\rangle &= \sum_\alpha \langle j_1 J_c(J_d) j_4 J \mid j_1, J_c j_4(\alpha) J\rangle \cdot \langle j_2j_3(J_c) j_4\alpha \mid j_2, j_3j_4(J_b)\alpha\rangle \\ &\quad \cdot \langle j_1, j_2 J_b(\alpha) J \mid j_1j_2(J_a) J_b J\rangle\end{aligned} \tag{24}$$

By comparing (20) and (24), we find the desired identity for the Racah coefficients. In the notation of (10), and written more symmetrically by means of (11), the identity reads:

$$W(a\alpha b\beta; c\gamma) W(\bar{a}\alpha\bar{b}\beta; \bar{c}\gamma) = \sum_\lambda (2\lambda + 1) W(\bar{a}\lambda\alpha c; a\bar{c}) W(b\lambda\beta\bar{c}; \bar{b}c) W(\bar{a}\lambda\gamma b; a\bar{b}) \tag{25}$$

Since our interest is primarily in exhibiting this result, suffice it to say that all the remaining transformations between coupling schemes for four wave functions can be expressed as sums of products of Racah functions.

It has been pointed out by Racah[12] that, aside from an indeterminate phase, (11) (12) (16) and (25) determine the W coefficients completely. It follows from this important result that further relations independent of these do not exist.

3. The application of the identity (25) to give a recursion relation is immediate. Take, for example, $\bar{c} = \frac{1}{2}$. Then in the sum λ takes on but two values $\lambda = c \pm \frac{1}{2}$. Solving for the term with $\lambda = c + \frac{1}{2}$ yields:

$$\begin{aligned}W(\bar{a}c + \tfrac{1}{2}\gamma b; a\bar{b}) \cdot [2(c+1) W(\bar{a}c + \tfrac{1}{2}\alpha c; a\tfrac{1}{2}) W(bc + \tfrac{1}{2}\beta\tfrac{1}{2}; \bar{b}c)] &= W(a\alpha b\beta; c\gamma) \cdot [W(\bar{a}\alpha\bar{b}\beta; \tfrac{1}{2}\gamma)] - W(\bar{a}c - \tfrac{1}{2}\gamma b; a\bar{b}) \\ &\quad \cdot [2c W(\bar{a}c - \tfrac{1}{2}\alpha c; a\tfrac{1}{2}) W(bc - \tfrac{1}{2}\beta\tfrac{1}{2}; \bar{b}c)]\end{aligned} \tag{26}$$

Since the W coefficients with one variable equal to $\frac{1}{2}$ are simple algebraic functions[9, 11], the terms in brackets can be regarded as known. (26) is then a recursion formula, in that it relates a Racah coefficient with one variable equal to $c + \frac{1}{2}$ to coefficients with variables c and $c - \frac{1}{2}$. Hence, from a table of W coefficients with one variable equal to $\frac{1}{2}$, we can generate all desired tables. In practice, this procedure proved of great value.

Further specialization of (26) is, of course, advisable in numerical calculations. Reference (11) contains such formulae.

Other relations result from different choices of the variable $\bar{c}$. In particular, the choice $\bar{c} = 1$ yields useful relations.

In the angular correlation problem involving pure multipole radiations only the Racah coefficient with two repeated parameters, $W(a\nu cd; ad)$, occurs. This coefficient bears an interesting relation to the Legendre polynomials[13]. In the limit of large a, c, d we have:

$$[(2a + 1)(2d + 1)]^{\frac{1}{2}} W(a\nu cd; ad) \to (-1)^{\nu} P_{\nu} (\cos \vartheta)$$

$$\cos \vartheta \equiv \frac{c(c+1) - d(d+1) - a(a+1)}{2ad} \tag{27}$$

The recursion relation when particularized for these Racah functions becomes:

$$W(a\nu cd; ad)W(a\bar{\nu}cd; ad) = \sum_{\lambda} [(2\lambda + 1)W(a\lambda a\nu; a\bar{\nu})W(d\lambda d\nu; d\bar{\nu})]W(a\lambda cd; ad) \tag{28}$$

The identity in (28) is thus seen to be the analogon to the familiar expansion for a product of Legendre functions. In the limit of large a, d we find, incidentally, that the bracketed terms are related to the vector addition coefficients, that is,

$$[(2\lambda + 1)(2a + 1)]^{\frac{1}{2}} W(a\lambda a\nu; a\bar{\nu}) \to (-1)^{\bar{\nu}} (\nu\bar{\nu}00 \mid \nu\bar{\nu}\lambda 0) + 0(1/a) \tag{29}$$

Particularizing (28) further, by letting $\bar{\nu} = 1$, yields the recursion relation:

$$\begin{aligned} W(a\nu + 1cd; ad) = {} & \frac{2\nu + 1}{(\nu + 1)} \frac{2a(a+1) + 2d(d+1) - 2c(c+1) - \nu(\nu+1)}{[(2a + 2 + \nu)(2a - \nu)(2d + 2 + \nu)(2d - \nu)]^{\frac{1}{2}}} \\ & \cdot W(a\nu\, cd; ad \\ & - \frac{\nu}{\nu + 1}\left[\frac{(2a + \nu + 1)(2a + 1 - \nu)(2d + \nu + 1)(2d + 1 - \nu)}{(2a + 2 + \nu)(2a - \nu)(2d + 2 + \nu)(2d - \nu)}\right]^{\frac{1}{2}} \\ & \cdot W(a\nu - 1cd; ad) \end{aligned} \tag{30}$$

This relation proves quite convenient for obtaining algebraic forms and numerical values for the $W(a\nu cd; ad)$.

Acknowledgment. The author wishes to express his appreciation for the considerable aid and encouragement extended to him by Professor John M. Blatt.

It has come to the author's attention that Professor E. P. Wigner in an unpublished manuscript (1941) defined and discussed functions equivalent to the Racah coefficients.

OAK RIDGE NATIONAL LABORATORY
OAK RIDGE, TENNESSEE

(Received March 10, 1952)

Reprinted from the
Proceedings of the Royal Society, A, *volume* 218, p. 370, 1953

Theoretical studies in nuclear structure

V. The matrix elements of non-central forces with an application to the $2p$-shell

By J. P. Elliott*

Department of Mathematics, University of Southampton

(*Communicated by R. E. Peierls, F.R.S.—Received* 4 *January* 1952
—*Rewritten* 16 *February* 1953)

Appendix

In finding the matrix elements of Q in a system of more than one shell in § 1·3 we met U function summations of the type (16) in which one U function came from (11) and the other two from the extraction of the wave functions of the 'chosen' particles from each of the two shells and their recoupling to form the two-particle states. The summation variable is the T, S or L value of the recoupled wave function of the remaining particles.

Using the symmetry properties of the U functions, (16) may be rewritten

$$\sum_{J_{124}} U(J_{124}j_4j_2j_1;\ J_{12}J_{14})\,U(J_{12}j_4j_3J;\ J_{124}J_{123})\,U(j_3Jj_2J_{14};\ J_{124}J_{23})$$
$$= U(j_1j_2J_{123}j_3;\ J_{12}J_{23})\,U(J_{23}j_1Jj_4;\ J_{123}J_{14}), \qquad (25)$$

and illustrated by the figure of three tetrahedra with a common edge J_{124}, or two tetrahedra with a common base j_1, J_{23}, J_{123}.

We may prove (25) by relating each side of the equation to the transformation coefficient

$$T = \{(j_1j_2)J_{12}, j_3, J_{123}, j_4, J \mid (j_2j_3)J_{23}, (j_1j_4)J_{14}, J\}. \qquad (26)$$

By using the property

$$\{(j_1j_2)J_{12}, j_3, J_{123} \mid j_1, (j_2j_3)J_{23}, J_{123}\} = U(j_1j_2J_{123}j_3;\ J_{12}J_{23})$$

several times, we may break down the left-hand side of (26) by extracting j_2 from J_{12} and recoupling it with j_3, and then coupling the remaining j_1 and j_4. This gives

$$T = (-1)^{j_1+J_{23}-J_{123}}U(j_1j_2J_{123}j_3;\ J_{12}J_{23})\,U(J_{23}j_1Jj_4;\ J_{123}J_{14}).$$

Alternatively, we may extract j_2 from J_{23} and recouple it with J_{14} in the right-hand side of (26). Now extract j_1 from J_{14} recoupling it with j_2, and finally extract J_{12} from J_{124}, recoupling it with j_3. We are then left with the dummy suffix J_{124} in the relation

$$T = (-1)^{j_1+J_{23}-J_{123}} \sum_{J_{124}} U(J_{124}j_4j_2j_1;\ J_{12}J_{14})\,U(J_{12}j_4j_3J;\ J_{124}J_{123})\,U(j_3Jj_2J_{14};\ J_{124}J_{23}),$$

which proves (25).

* Present address: The Atomic Energy Research Establishment, Harwell.

ON ANGULAR MOMENTUM

Julian Schwinger

The commutation relations of an arbitrary angular momentum vector can be reduced to those of the harmonic oscillator. This provides a powerful method for constructing and developing the properties of angular momentum eigenvectors. In this paper many known theorems are derived in this way, and some new results obtained. Among the topics treated are the properties of the rotation matrices; the addition of two, three, and four angular momenta; and the theory of tensor operators.

1. Introduction

One of the methods of treating a general angular momentum in quantum mechanics is to regard it as the superposition of a number of elementary "spins," or angular momenta with $j = \frac{1}{2}$. Such a spin assembly, considered as a Bose-Einstein system, can be usefully discussed by the method of second quantization. We shall see that this procedure unites the compact symbolism of the group theoretical approach with the explicit operator techniques of quantum mechanics.

We introduce spin creation and annihilation operators associated with a given spatial reference system, $a_\zeta^+ = (a_+^+, a_-^+)$ and $a_\zeta = (a_+, a_-)$, which satisfy

$$[a_\zeta, a_{\zeta'}] = 0, \qquad [a_\zeta^+, a_{\zeta'}^+] = 0,$$
$$[a_\zeta, a_{\zeta'}^+] = \delta_{\zeta\zeta'}. \tag{1.1}$$

The number of spins and the resultant angular momentum are then given by

$$n = \sum_\zeta a_\zeta^+ a_\zeta = n_+ + n_-,$$
$$\mathbf{J} = \sum_{\zeta,\zeta'} a_\zeta^+ (\zeta|\tfrac{1}{2}\boldsymbol{\sigma}|\zeta') a_{\zeta'}. \tag{1.2}$$

With the conventional matrix representation for $\boldsymbol{\sigma}$, the components of $\mathbf{J}$ appear as

$$J_+ = J_1 + iJ_2 = a_+^+ a_-, \; J_- = J_1 - iJ_2 = a_-^+ a_+,$$
$$J_3 = \tfrac{1}{2}(a_+^+ a_+ - a_-^+ a_-) = \tfrac{1}{2}(n_+ - n_-). \tag{1.3}$$

Of course, this realization of the angular momentum commutation properties in terms of those of harmonic oscillators can be introduced without explicit reference to the composition of spins.

To evaluate the square of the total angular momentum

$$\mathbf{J}^2 = \sum_{\zeta\zeta'\zeta''\zeta'''} a_\zeta^+ a_{\zeta'} a_{\zeta''}^+ a_{\zeta'''} (\zeta|\tfrac{1}{2}\boldsymbol{\sigma}|\zeta')\cdot(\zeta''|\tfrac{1}{2}\boldsymbol{\sigma}|\zeta'''), \tag{1.4}$$

we employ the matrix elements of the spin permutation operator

$$P^{(12)} = \tfrac{1}{2}(1 + \boldsymbol{\sigma}^{(1)}\cdot\boldsymbol{\sigma}^{(2)}). \tag{1.5}$$

Thus

$$(\zeta|\boldsymbol{\sigma}|\zeta')\cdot(\zeta''|\boldsymbol{\sigma}|\zeta''') = 2\delta_{\zeta\zeta'''}\delta_{\zeta'\zeta''} - \delta_{\zeta\zeta'}\delta_{\zeta''\zeta'''}, \tag{1.6}$$

and

$$\mathbf{J}^2 = \tfrac{1}{2}\sum_{\zeta\zeta''} a_\zeta^+ a_{\zeta'} a_{\zeta'}^+ a_\zeta - \tfrac{1}{4}n^2. \tag{1.7}$$

According to the commutation relations (1.1),

$$\sum_{\zeta,\zeta'} a_\zeta^+ a_{\zeta'} a_{\zeta'}^+ a_\zeta = \sum_\zeta a_\zeta^+ (n+2) a_\zeta = n(n+1), \tag{1.8}$$

whence

$$\mathbf{J}^2 = \tfrac{1}{2}n(\tfrac{1}{2}n + 1); \tag{1.9}$$

a given number of spins, $n = 0, 1, 2, \ldots$, possesses a definite angular momentum quantum number,

$$j = \tfrac{1}{2}n = 0, \tfrac{1}{2}, 1, \ldots . \tag{1.10}$$

We further note that, according to (1.3), a state with a fixed number of positive and negative spins also has a definite magnetic quantum number,

$$m = \tfrac{1}{2}(n_+ - n_-), \qquad j = \tfrac{1}{2}(n_+ + n_-). \tag{1.11}$$

Therefore, from the eigenvector of a state with prescribed occupation numbers,

$$\Psi(n_+ n_-) = \frac{(a_+^+)^{n_+}}{(n_+!)^{1/2}} \frac{(a_-^+)^{n_-}}{(n_-!)^{1/2}} \Psi_0, \tag{1.12}$$

$$a_\pm \Psi_0 = 0,$$

we obtain the angular momentum eigenvector[1]

$$\Psi(jm) = \frac{(a_+^+)^{j+m}(a_-^+)^{j-m}}{[(j+m)!(j-m)!]^{1/2}} \Psi_0. \tag{1.13}$$

Familiar as a symbolic expression of the transformation properties of angular momentum eigenvectors[2], this form is here a precise operator construction of the eigenvector.

[1] A direct proof is given in Appendix A.

[2] See, for example, H. Weyl, "The Theory of Groups and Quantum Mechanics" (E. P. Dutton and Company, Inc., New York, 1931), p. 189.

On multiplying (1.13) with an analogous monomial constructed from the components of the arbitrary spinor $x_\zeta = (x_+, x_-)$

$$\phi_{jm}(x) = \frac{x_+^{j+m} x_-^{j-m}}{[(j+m)!(j-m)!]^{1/2}} \tag{1.14}$$

we obtain, after summation with respect to m, and then with respect to j,

$$\sum_{m=-j}^{j} \phi_{jm}(x)\Psi(jm) = \frac{(xa^+)^{2j}}{(2j)!}\Psi_0, \tag{1.15}$$

and

$$\sum_{jm} \phi_{jm}(x)\Psi(jm) = e^{(xa^+)}\Psi_0, \tag{1.16}$$

in which we have written

$$(xa^+) = \sum_\zeta x_\zeta a_\zeta^+. \tag{1.17}$$

To illustrate the utility of (1.16), conceived of as an eigenvector generating function, we shall verify the orthogonality and normalization of the eigenvectors (1.13). Consider, then,

$$\begin{aligned}(e^{(xa^+)}\Psi_0, e^{(ya^+)}\Psi_0) &= \sum \phi_{jm}(x^*)(\Psi(jm), \Psi(j'm'))\phi_{j'm'}(y) \\ &= (\Psi_0, e^{(x^*a)}e^{(ya^+)}\Psi_0).\end{aligned} \tag{1.18}$$

According to the commutation relations (1.1), and $a_\zeta\Psi_0 = 0$, we have

$$a_\zeta f(a^+)\Psi_0 = \left(\frac{\partial f(a^+)}{\partial a_\zeta^+}\right)\Psi_0, \tag{1.19}$$

whence

$$\begin{aligned}(\Psi_0, e^{(x^*a)}\, e^{(ya^+)}\Psi_0) &= e^{(x^*y)}(e^{(y^*a)}\Psi_0, \Psi_0) = e^{(x^*y)} \\ &= \sum_{jm} \phi_{jm}(x^*)\phi_{jm}(y).\end{aligned} \tag{1.20}$$

We have thus proved that

$$(\Psi(jm), \Psi(j'm')) = \delta_{jj'}\,\delta_{mm'}. \tag{1.21}$$

As a second elementary example, we shall obtain the matrix elements of powers of $J_\pm$ by considering the effect of the operators $e^{\lambda J_\pm}$ on (1.16). We have

$$\begin{aligned}\sum_{jm} \phi_{jm}(x)e^{\lambda J_+}\Psi(jm) &= e^{\lambda a_+^+ a_-}\, e^{(xa^+)}\Psi_0 = e^{\lambda x_- a_+^+}\, e^{(xa^+)}\Psi_0 \\ &= e^{(x_+ + \lambda x_-)a_+^+ + x_- a_-^+}\Psi_0 \\ &= \sum_{jm} \phi_{jm}(x_+ + \lambda x_-, x_-)\Psi(jm),\end{aligned} \tag{1.22}$$

and therefore

$$\sum_{j'm'} (jm|e^{\lambda J_+}|j'm')\phi_{j'm'}(x) = \phi_{jm}(x_+ + \lambda x_-, x_-), \tag{1.23}$$

which, on expansion, yields the nonvanishing matrix element

$$(jm|J_+^{m-m'}|jm') = \left[\frac{(j+m)!}{(j+m')!}\frac{(j-m')!}{(j-m)!}\right]^{1/2}, \qquad m - m' > 0. \tag{1.24}$$

Similarly

$$\sum_{j'm'} (jm|e^{\lambda J_-}|j'm')\phi_{j'm'}(x) = \phi_{jm}(x_+, x_- + \lambda x_+), \tag{1.25}$$

and

$$(jm|J_-^{m'-m}|jm') = \left[\frac{(j+m')!}{(j+m)!}\frac{(j-m)!}{(j-m')!}\right]^{1/2}, \qquad m' - m > 0. \tag{1.26}$$

A particular consequence of (1.24) and (1.26) is

$$\begin{aligned}\Psi(jm) &= \left[\frac{1}{(2j)!}\frac{(j-m)!}{(j+m)!}\right]^{1/2} \cdot J_+^{j+m}\Psi(j, -j) \\ &= \left[\frac{1}{(2j)!}\frac{(j+m)!}{(j-m)!}\right]^{1/2} \cdot J_-^{j-m}\Psi(jj),\end{aligned} \tag{1.27}$$

which details the construction of an arbitrary eigenvector from those possessing the maximum values of $|m|$ compatible with a given j.

It is also possible to exhibit an operator which permits the construction of an arbitrary eigenvector from that possessing the minimum value of j compatible with a given m. Indeed, (1.13), written in the form

$$\Psi(jm) = \frac{(a_+^+ a_-^+)^{j-|m|}}{[(j+|m|)!(j-|m|)!]^{1/2}}(a_+^+)^{|m|+m}(a_-^+)^{|m|-m}\Psi_0, \tag{1.28}$$

states that

$$\Psi(jm) = \left[\frac{(2|m|)!}{(j+|m|)!(j-|m|)!}\right]^{1/2} K_+^{j-|m|}\Psi(|m|, m), \tag{1.29}$$

where K_+ and two associated operators are defined by

$$\begin{aligned}K_+ = a_+^+ a_-^+, \qquad K_- = a_+ a_-, \\ K_3 = \tfrac{1}{2}(n_+ + n_- + 1).\end{aligned} \tag{1.30}$$

It is easily seen that

$$[J_3, K_\pm] = [J_3, K_3] = 0, \tag{1.31}$$

and that

$$\begin{aligned}[K_3, K_+] = K_+, \qquad [K_3, K_-] = -K_-, \\ [K_+, K_-] = -2K_3.\end{aligned} \tag{1.32}$$

The latter are analogous to the commutation properties of J, save for the algebraic sign of the commutator $[K_+, K_-]$. In keeping with this qualified analogy we also have

$$J_3^2 - \tfrac{1}{4} = K_3(K_3 - 1) - K_+K_- = K_3(K_3 + 1) - K_-K_+ \qquad (1.33)$$

as compared with

$$J^2 = J_3(J_3 - 1) + J_+J_- = J_3(J_3 + 1) + J_-J_+. \qquad (1.34)$$

Noting that the eigenvalue of K_3 is $j + \frac{1}{2}$, we see that the roles of j and m are essentially interchanged in K. The hyperbolic nature of the space in which the latter operates is thus related to the restriction $|m| \leq j$.

If (1.29) is multiplied by a similar numerical quantity, and then summed with respect to j, one obtains

$$\sum_{j=|m|}^{\infty} \left[\frac{(2|m|)!}{(j+|m|)!(j-|m|)!}\right]^{1/2} \lambda^{j-|m|}\Psi(jm) = F_{2|m|}(\lambda K_+)\Psi(|m|, m), \quad (1.35)$$

where

$$F_r(z) = r! z^{-r/2} I_r(2z^{1/2}) = \sum_{n=0}^{\infty} \frac{r!}{n!(r+n)!} z^n, \qquad (1.36)$$

and I_r is the cylinder function of imaginary argument. A simpler generating function is given by

$$\sum_j \left[\frac{1}{(2|m|)!} \frac{(j+|m|)!}{(j-|m|)!}\right]^{1/2} \lambda^{j-|m|}\Psi(jm) = e^{\lambda K_+}\Psi(|m|, m). \qquad (1.37)$$

2. Rotations

A significant interpretation is obtained for (1.15) by introducing the operators

$$a'^+_+ = (xa^+)\ , \qquad a'_+ = (x^*a), \qquad (2.1)$$

$$a'^+_- = [x^*a^+], \qquad a'_- = [xa],$$

where

$$[xy] = x_+y_- - x_-y_+. \qquad (2.2)$$

With the restriction

$$(x^*x) = 1, \qquad (2.3)$$

these operators also obey the commutation relations (1.1), and must therefore constitute spin creation and annihilation operators associated with an altered spatial reference system. Accordingly, (1.15) can be viewed as the expression of the state $m = j$, in a rotated coordinate system, as a linear combination of the eigenvectors in a fixed coordinate system,

$$\Psi'(jj) = \frac{(a_+'^+)^{2j}}{((2j)!)^{1/2}} \Psi_0 = ((2j)!)^{1/2} \sum_{m=-j}^{j} \phi_{jm}(x) \Psi(jm). \tag{2.4}$$

The unitary nature of this transformation is here easily verified,

$$(2j)! \sum_m \phi_{jm}(x^*) \phi_{jm}(x) = (x^* x)^{2j} = 1. \tag{2.5}$$

In general

$$\Psi'(jm') = \phi_{jm'}(a'^+) \Psi_0 = \sum_m \Psi(jm) U_{mm'}^{(j)}, \tag{2.6}$$

where the coefficients are to be inferred from

$$\sum_m \phi_{jm}(a^+) U_{mm'}^{(j)} = \phi_{jm'}(x_+ a_+^+ + x_- a_-^+, \; - x_-^* a_+^+ + x_+^* a_-^+). \tag{2.7}$$

It is useful to introduce the unitary operator that generates $\Psi'(jm')$ from $\Psi(jm)$,

$$\Psi'(jm') = U \Psi(jm), \tag{2.8}$$

which permits an alternative construction of the coefficients in (2.6),

$$U_{mm'}^{(j)} = (jm|U|jm'). \tag{2.9}$$

In terms of the successive rotations characterized by Eulerian angles, ϕ, θ, ψ, U is given explicitly by

$$U = e^{-i\psi J_3''} e^{-i\theta J_2'} e^{-i\phi J_3}, \tag{2.10}$$

where

$$J' = e^{-i\phi J_3} J e^{i\phi J_3},$$

$$J'' = e^{-i\theta J_2'} J' e^{i\theta J_2'} \tag{2.11}$$

are the operators appropriate to the coordinate systems produced by the previous rotations. The resulting expression for $U(\phi\theta\psi)$ is

$$U = e^{-i\phi J_3} e^{-i\theta J_2} e^{-i\psi J_3}, \qquad U^{-1} = e^{i\psi J_3} e^{i\theta J_2} e^{i\phi J_3}. \tag{2.12}$$

The angular momentum operators associated with the new coordinate system,

$$J' = UJU^{-1}, \tag{2.13}$$

can be constructed from the transformed creation and annihilation operators,

$$a_+'^+ = U a_+^+ U^{-1} = e^{-(i/2)(\psi+\phi)} \cos \tfrac{1}{2}\theta \, a_+^+ + e^{-(i/2)(\psi-\phi)} \sin \tfrac{1}{2}\theta \, a_-^+ \tag{2.14}$$

$$a_-'^+ = U a_-^+ U^{-1} = -e^{-(i/2)(\psi-\phi)} \sin \tfrac{1}{2}\theta \, a_+^+ + e^{(i/2)(\psi+\phi)} \cos \tfrac{1}{2}\theta \, a_-^+.$$

In evaluating (2.14), we have made use of the relations

$$e^{-i\psi J_3}\, a_\pm^+\, e^{i\psi J_3} = e^{\mp(i/2)\psi}\, a_\pm^+,$$

$$e^{-i\theta J_2}\, a_\pm^+\, e^{i\theta J_2} = \cos\tfrac{1}{2}\theta\; a_\pm^+ \pm \sin\tfrac{1}{2}\theta\; a_\mp^+, \tag{2.15}$$

of which the former follows immediately from the significance of $a_\pm^+$ as a positive (negative) spin creation operator, while the latter may be verified by differentiation with respect to θ, in conjunction with the commutation relations

$$[a_\pm^+, J_2] = \mp\, (i/2)\, a_\mp^+. \tag{2.16}$$

The form of (2.14) is in agreement with (2.1) and (2.3), where

$$x_+ = e^{-(i/2)(\psi+\phi)} \cos\tfrac{1}{2}\theta,\; x_- = e^{-(i/2)(\psi-\phi)} \sin\tfrac{1}{2}\theta. \tag{2.17}$$

To construct the matrix of U, we consider

$$(e^{(xa^+)}\Psi_0,\, Ue^{(ya^+)}\Psi_0) = \sum_{jm} \phi_{jm}(x^*)\, U_{mm'}^{(j)} \phi_{jm'}(y)$$

$$= (\Psi_0,\, e^{(x^*a)} e^{(ya'^+)}\Psi_0), \tag{2.18}$$

in which the a'^+ are the operators (2.14). On writing

$$(ya'^+) = (a^+uy) \tag{2.19}$$

where u is the matrix

$$u = \begin{pmatrix} e^{-(i/2)(\phi+\psi)} \cos\tfrac{1}{2}\theta, & -e^{-(i/2)(\phi-\psi)} \sin\tfrac{1}{2}\theta \\ e^{-(i/2)(\phi-\psi)} \sin\tfrac{1}{2}\theta, & e^{(i/2)(\phi+\psi)} \cos\tfrac{1}{2}\theta \end{pmatrix} \tag{2.20}$$

we immediately obtain

$$\sum_{jm} \phi_{jm}(x^*)\, U_{mm'}^{(j)}\, \phi_{jm'}(y) = e^{(x^*uy)}. \tag{2.21}$$

Since (2.12) implies that

$$U_{mm'}^{(j)}(\phi\theta\psi) = e^{-im\phi}\, U_{mm'}^{(j)}(\theta) e^{-im'\psi}, \tag{2.22}$$

where

$$U_{mm'}^{(j)}(\theta) = (jm|e^{-i\theta J_2}|jm'), \tag{2.23}$$

we may simplify (2.21) by placing $\phi = \psi = 0$, thereby obtaining

$$\sum_{jm} \phi_{jm}(x^*)\, U_{mm'}^{(j)}(\theta) \phi_{jm'}(y) = \exp\{\cos\tfrac{1}{2}\theta (x^*y) - \sin\tfrac{1}{2}\theta [x^*y]\}. \tag{2.24}$$

The matrix u is unitary and unimodular, that is, possesses a unit determinant. Its representation in terms of spin matrices has, as it must, the form of (2.12),

$$u = e^{-(i/2)\phi\sigma_3}\, e^{-(i/2)\theta\sigma_2}\, e^{-(i/2)\psi\sigma_3}. \tag{2.25}$$

Any such unitary matrix can be presented as

$$u = e^{-i\mathscr{H}} \tag{2.26}$$

where $\mathscr{H}$ is a Hermitian matrix. Since

$$\det u = e^{-i\,\mathrm{tr}\mathscr{H}}, \tag{2.27}$$

$\mathscr{H}$ must be a traceless Hermitian matrix and, accordingly, is a linear combination of the spin matrices, with real coefficients. Hence u can be written as

$$u = e^{-(i/2)\gamma\mathbf{n}\cdot\boldsymbol{\sigma}} \tag{2.28}$$

where $\mathbf{n}$ is a unit vector, specified by two angles, α and β. The fact that (2.28) is the matrix describing a rotation through the angle γ about the axis $\mathbf{n}$ affirms the well-known equivalence between an arbitrary rotation and a simple rotation about a suitably chosen axis. The rotation angle γ is easily obtained by comparing the trace of u, in its two versions,

$$\tfrac{1}{2}\mathrm{tr}\, u = \cos\tfrac{1}{2}\gamma = \cos\tfrac{1}{2}\theta\cos\tfrac{1}{2}(\phi+\psi). \tag{2.29}$$

More generally, the trace of U for a given j depends only upon the rotation angle γ. We define[3]

$$\chi^{(j)} = \sum_{m=-j}^{j} U^{(j)}_{mm} = \mathrm{tr}\, P_j U, \tag{2.30}$$

in which P_j is the projection operator for the states with quantum number j. If we remark that U must also have the form of (2.28),

$$U = e^{-i\gamma\mathbf{n}\cdot\mathbf{J}} \tag{2.31}$$

we immediately obtain

$$\chi^{(j)} = \sum_{m=-j}^{j} e^{-im\gamma} = \frac{\sin(j+\frac{1}{2})\gamma}{\sin\frac{1}{2}\gamma}. \tag{2.32}$$

However, we can also derive this directly from the generating function (2.21).

For simplicity we shall assume the reference system to be so chosen that u is a diagonal matrix, with eigenvalues $e^{\pm(i/2)\gamma}$. We replace x_ζ^* with $t(\partial/\partial y_\zeta)$ and evaluate the derivatives at $y_\zeta = 0$. According to

$$\phi_{jm}(\partial/\partial y)\phi_{jm'}(y)]_{y_\zeta=0} = \delta_{m,m'}, \tag{2.33}$$

we then have

$$\sum_j t^{2j}\chi^{(j)} = \exp\left(te^{-(i/2)\gamma}\frac{\partial}{\partial y_+}; y_+\right)\cdot\exp\left(te^{(i/2)\gamma}\frac{\partial}{\partial y_-}; y_-\right)\Big]_{y_\zeta=0} \tag{2.34}$$

in which the notation reflects the necessity of placing the derivatives to the left of the powers of y_ζ. Now

[3] This trace is the character of group theory.

$$\exp\left(\lambda\frac{\partial}{\partial y}; y\right) = \sum_{n=0}^{\infty}\frac{\lambda^n}{n!}\left(\frac{\partial}{\partial y}\right)^n y^n = \sum_{n=0}^{\infty}\lambda^n = \frac{1}{1-\lambda}, \tag{2.35}$$

and therefore

$$\sum_j t^{2j}\chi^{(j)} = \frac{1}{1-t\cdot\exp\left(-\frac{i}{2}\gamma\right)}\,\frac{1}{1-t\cdot\exp\left(\frac{i}{2}\gamma\right)}$$

$$= \frac{1}{1-2t\cdot\cos\frac{1}{2}\gamma + t^2}, \tag{2.36}$$

which is a generating function for the $\chi^{(j)}$. On writing

$$\frac{1}{1-t\cdot\exp\left(-\frac{i}{2}\gamma\right)}\,\frac{1}{1-t\cdot\exp\left(\frac{i}{2}\gamma\right)}$$

$$= \frac{1}{2it\cdot\sin\frac{1}{2}\gamma}\left[\frac{1}{1-t\cdot\exp\left(\frac{i}{2}\gamma\right)} - \frac{1}{1-t\cdot\exp\left(-\frac{i}{2}\gamma\right)}\right], \tag{2.37}$$

and expanding in powers of t, one obtains

$$\chi^{(j)}(\gamma) = \frac{\sin(j+\frac{1}{2})\gamma}{\sin\frac{1}{2}\gamma}. \tag{2.38}$$

Symmetry properties of $U^{(j)}_{mm'}(\phi\theta\psi)$ are easily inferred from (2.21). According to the invariance of (x^*uy) under the substitutions $\phi \leftrightarrow \psi + \pi$, $x^* \leftrightarrow y$, and $\phi \rightarrow \phi - \pi$, $\theta \rightarrow \pi - \theta$, $\psi \rightarrow -\psi$, $y_\pm \rightarrow iy_\mp$, we have

$$U^{(j)}_{mm'}(\phi\theta\psi) = U^{(j)}_{m'm}(\psi+\pi, \theta, \phi-\pi) = i^{2j}U^{(j)}_{m,-m'}(\phi-\pi, \pi-\theta, -\psi). \tag{2.39}$$

Among the additional equivalent forms produced by successive application of these transformations are

$$i^{2j}U^{(j)}_{-mm'}(-\phi, \pi-\theta, \psi+\pi) = U^{(j)}_{-m-m'}(\pi-\phi, \theta, -\pi-\psi)$$

$$= U^{(j)}_{-m'-m}(-\psi, \theta, -\phi). \tag{2.40}$$

We also note that

$$U^{(j)*}_{mm'}(\phi\theta\psi) = U^{(j)}_{mm'}(-\phi, \theta, -\psi) = U^{(j)}_{-m-m'}(\phi+\pi, \theta, \psi-\pi). \tag{2.41}$$

On removing the angles ϕ and ψ with the aid of (2.22), we find that the content of (2.39) and (2.40) is

$$U^{(j)}_{mm'}(\theta) = (-1)^{j-m}U^{(j)}_{m-m'}(\pi-\theta) = (-1)^{j-m'}U^{(j)}_{-mm'}(\pi-\theta)$$

$$= (-1)^{m-m'}U^{(j)}_{-m-m'}(\theta) = (-1)^{m-m'}U^{(j)}_{m'm}(\theta) = U^{(j)}_{-m'-m}(\theta). \tag{2.42}$$

In view of these relations, it is sufficient to exhibit $U^{(j)}_{mm'}(\theta)$ for non-negative values of m and m'.

On expanding the generating function (2.24) in terms of $\phi_{jm}(x^*)$, or of $\phi_{jm'}(y)$, we obtain the equivalent expressions

$$\sum_{m'} U^{(j)}_{mm'}(\theta)\phi_{jm'}(y) = \phi_{jm}(\cos\tfrac{1}{2}\theta\, y_+ - \sin\tfrac{1}{2}\theta\, y_-, \sin\tfrac{1}{2}\theta\, y_+ + \cos\tfrac{1}{2}\theta\, y_-), \quad (2.43a)$$

$$\sum_{m} \phi_{jm}(x^*)U^{(j)}_{mm'}(\theta) = \phi_{jm'}(\cos\tfrac{1}{2}\theta\, x^*_+ + \sin\tfrac{1}{2}\theta\, x^*_-, -\sin\tfrac{1}{2}\theta\, x^*_+ + \cos\tfrac{1}{2}\theta x^*_-), \quad (2.43b)$$

of which the latter is the counterpart of (2.7). As a convenient means of constructing $U^{(j)}_{mm'}(\theta)$, we place

$$x^*_+ = \sin\tfrac{1}{2}\theta \cos\tfrac{1}{2}\theta, \qquad x^*_- = t - \cos^2\tfrac{1}{2}\theta,$$

so that (2.43b) reads

$$\sum_m \frac{(\sin\frac{1}{2}\theta\cos\frac{1}{2}\theta)^{j+m}}{[(j+m)!(j-m)!]^{1/2}}(t-\cos^2\tfrac{1}{2}\theta)^{j-m}\, U^{(j)}_{mm'}(\theta)$$
$$= (-1)^{j-m'}\left[\frac{(\sin\frac{1}{2}\theta)^{j+m}(\cos\frac{1}{2}\theta)^{j-m'}}{[(j+m')!(j-m')!]^{1/2}}\right] t^{j+m'}(1-t)^{j-m'}. \quad (2.44)$$

Thus

$$U^{(j)}_{mm'}(\theta) = (-1)^{j-m'}\left[\frac{(j+m)!}{(j-m)!}\frac{1}{(j+m')!(j-m')!}\right]^{1/2}$$
$$\cdot[(\sin\tfrac{1}{2}\theta)^{-m+m'}(\cos\tfrac{1}{2}\theta)^{-m-m'}]$$
$$\cdot\left[\left(\frac{d}{dt}\right)^{j-m} t^{j+m'}(1-t)^{j-m'}\right]_{t=\cos^2\frac{1}{2}\theta}. \quad (2.45)$$

The structure of the right side will be recognized as that of the Jacobi polynomial,

$$\mathscr{F}_n(a,b;t) = F(-n, a+n, b; t) = \frac{(b-1)!}{(b+n-1)!}t^{1-b}(1-t)^{b-a}$$
$$\cdot\left(\frac{d}{dt}\right)^n t^{b+n-1}(1-t)^{a-b+n}, \quad (2.46)$$

whence[4]

$$U^{(j)}_{mm'}(\theta) = \frac{(-1)^{j-m'}}{(m+m')!}\left[\frac{(j+m)!}{(j-m)!}\frac{(j+m')!}{(j-m')!}\right]^{1/2}(\sin\tfrac{1}{2}\theta)^{m-m'}(\cos\tfrac{1}{2}\theta)^{m+m'}$$
$$\cdot\mathscr{F}_{j-m}(2m+1, m+m'+1; \cos^2\tfrac{1}{2}\theta). \quad (2.47)$$

[4] This is equivalent to the result obtained by P. Güttinger, *Z. Phys.* **73**, 169 (1931).

Other forms can be obtained from (2.43), corresponding to the variety of transformations permissible to hypergeometric functions. Thus the known relation

$$F(a, b, c; x) = (1-x)^{-a} F\left(a, c-b, c; -\frac{x}{1-x}\right), \tag{2.48}$$

applied to (2.47), gives

$$U^{(j)}_{mm'}(\theta) = \frac{(-1)^{j-m'}}{(m+m')!}\left[\frac{(j+m)!}{(j-m)!}\frac{(j+m')!}{(j-m')!}\right]^{1/2} (\sin \tfrac{1}{2}\theta)^{2j} (\cot \tfrac{1}{2}\theta)^{m+m'}$$
$$\cdot F(m-j, m'-j, m+m'+1; -\cot^2 \tfrac{1}{2}\theta). \tag{2.49}$$

Another aspect of reference system transformation is best discussed in terms of

$$U^{(j)*}_{mm'}(\phi\theta\psi) = e^{im\phi} U^{(j)}_{mm'}(\theta) e^{im'\psi} = (jm'|U^{-1}|jm). \tag{2.50}$$

This quantity is the transformation function

$$(\Psi'(jm'), \Psi(jm)) = (\omega, jm'|jm), \tag{2.51}$$

in which we have used ω to designate collectively the angles $\phi\theta\psi$, relating the new reference system to the fixed one. We shall be interested in the differential characterization of this transformation function, in its dependence upon the Eulerian angles. Now

$$\begin{aligned} \frac{1}{i}\frac{\partial}{\partial\phi} U^{-1} &= U^{-1} J_3 \\ \frac{1}{i}\frac{\partial}{\partial\psi} U^{-1} &= J_3 U^{-1} = U^{-1} J_3' \\ \frac{1}{i}\frac{\partial}{\partial\theta} U^{-1} &= U^{-1} e^{-i\phi J_3} J_2 e^{i\phi J_3} = U^{-1} J_\theta, \end{aligned} \tag{2.52}$$

where

$$\begin{aligned} J_3' &= J_3 \cos\theta + \tfrac{1}{2}\sin\theta (J_+ e^{-i\phi} + J_- e^{i\phi}), \\ J_\theta &= \frac{1}{2i}(J_+ e^{-i\phi} - J_- e^{i\phi}), \end{aligned} \tag{2.53}$$

and, therefore

$$\begin{aligned} \frac{1}{i}\frac{\partial}{\partial\phi}(\omega|\;) &= (\omega|J_3|\;) \\ e^{i\phi}\left[\frac{\partial}{\partial\theta} + \frac{1}{\sin\theta}\left(\frac{1}{i}\frac{\partial}{\partial\psi} - \cos\theta\frac{1}{i}\frac{\partial}{\partial\phi}\right)\right](\omega|\;) &= (\omega|J_+|\;) \\ e^{-i\phi}\left[-\frac{\partial}{\partial\theta} + \frac{1}{\sin\theta}\left(\frac{1}{i}\frac{\partial}{\partial\psi} - \cos\theta\frac{1}{i}\frac{\partial}{\partial\phi}\right)\right](\omega|\;) &= (\omega|J_-|\;). \end{aligned} \tag{2.54}$$

This is a differential operator representation of an arbitrary angular momentum vector. The familiar differential operators associated with an orbital angular momentum emerge if the transformation function is independent of ψ. Since this corresponds to $m' = 0$, the quantum number j must then be an integer.[5]

The differential operators (2.54) are well-known in connection with angular momentum of a rigid body, and, accordingly, the eigenvalue equation for J^2 in this representation will be identical with the symmetrical top wave equation. To construct this equation directly, we remark that

$$\begin{aligned} J^2 &= J_3^2 + (\tfrac{1}{2}J_+e^{-i\phi} + \tfrac{1}{2}J_-e^{i\phi})^2 - (\tfrac{1}{2}J_+e^{-i\phi} - \tfrac{1}{2}J_-e^{i\phi})^2 \\ &= J_3^2 + \left[\frac{J_3' - J_3\cos\theta}{\sin\theta}\right]^2 + J_\theta^2 \qquad (2.55) \\ &= \frac{J_3'^2 - 2J_3'J_3\cos\theta + J_3^2}{\sin^2\theta} + J_\theta^2 + \cot\theta\frac{1}{i}J_\theta, \end{aligned}$$

since

$$[J_3', J_3] = \sin\theta\frac{1}{i}J_\theta. \qquad (2.56)$$

On referring to (2.52), we immediately obtain

$$-\left[\frac{\partial^2}{\partial\theta^2} + \cot\theta\frac{\partial}{\partial\theta} + \frac{1}{\sin^2\theta}\left(\frac{\partial^2}{\partial\psi^2} - 2\cos\theta\frac{\partial}{\partial\psi}\frac{\partial}{\partial\phi} + \frac{\partial^2}{\partial\phi^2}\right)\right]U^{-1} = U^{-1}J^2, \qquad (2.57)$$

and the analogous differential equation for $(\omega|\quad)$, including the eigenvalue equation

$$\left[\frac{\partial^2}{\partial\theta^2} + \cot\theta\frac{\partial}{\partial\theta} + j(j+1) - \frac{m^2 - 2mm'\cos\theta + m'^2}{\sin^2\theta}\right](\omega, jm'|jm) = 0. \qquad (2.58)$$

An integral theorem concerning the angular dependence of U, or U^{-1}, is stated by

$$\int U\, d\omega = P_0, \qquad (2.59)$$

where P_0 is the projection operator for the state $j = 0$, and

$$d\omega = \tfrac{1}{2}\sin\theta\cdot d\theta\cdot\frac{1}{4\pi}d\phi\cdot\frac{1}{4\pi}d\psi, \qquad (2.60)$$

$$\int d\omega = 1\,.$$

[5] The fact that the general differential operators (2.54) admit half-integral values of j has been noticed by F. Bopp and R. Haag, *Z. Naturforsch.* **5a**, 644 (1950).

The integration domain is here understood to be

$$0 \leq \phi < 4\pi, \qquad 0 \leq \psi < 4\pi, \qquad 0 \leq \theta \leq \pi. \tag{2.61}$$

To prove this theorem we subject (2.57) to the angular integrations contained in $d\omega$. In virtue of the periodicity possessed by U^{-1} over 4π intervals of ϕ and ψ, we obtain

$$\int U^{-1}\, d\omega\, J^2 = -\tfrac{1}{2}\left[\sin\theta \frac{\partial}{\partial\theta}\int U^{-1}\frac{d\phi}{4\pi}\frac{d\psi}{4\pi}\right]_{\theta=0}^{\pi} = 0. \tag{2.62}$$

This result asserts the vanishing of $\int U^{-1}\, d\omega$, and the Hermitian conjugate $\int U\, d\omega$, except for the state with $j = 0$. The fact that the rotation operator U reduces to unity for this spherically symmetrical state completes the proof of (2.59). We shall defer application of this theorem to the next section.

3. Addition of Two Angular Momenta

Two kinematically independent angular momenta, $\mathbf{J}_1$ and $\mathbf{J}_2$, can be expressed by

$$\mathbf{J}_1 = \sum_{\zeta,\zeta'} a_\zeta^+ (\zeta|\tfrac{1}{2}\boldsymbol{\sigma}|\zeta')a_{\zeta'},$$

$$\mathbf{J}_2 = \sum_{\zeta,\zeta'} b_\zeta^+ (\zeta|\tfrac{1}{2}\boldsymbol{\sigma}|\zeta')b_{\zeta'}, \tag{3.1}$$

where the a and b operators individually obey (1.1), but are mutually commutative. In studying the eigenvectors of the total angular momentum,

$$\mathbf{J} = \mathbf{J}_1 + \mathbf{J}_2, \tag{3.2}$$

the following scalar operators play an important role:

$$\mathcal{J}_+ = (a^+b), \qquad \mathcal{J}_- = (b^+a),$$

$$\mathcal{J}_3 = \tfrac{1}{2}[(a^+\, a) - (b^+\, b)] = \tfrac{1}{2}(n_1 - n_2), \tag{3.3}$$

and

$$\mathcal{K}_+ = [a^+b^+], \qquad \mathcal{K}_- = [ab],$$

$$\mathcal{K}_3 = \tfrac{1}{2}[(a^+\, a) + (b^+\, b)] + 1 = \tfrac{1}{2}n + 1\,. \tag{3.4}$$

As one can easily verify by direct calculation, the operators $\mathcal{J}$ and $\mathcal{K}$ commute with each other (as well as with J), and obey

$$[\mathcal{J}_3, \mathcal{J}_\pm] = \pm\,\mathcal{J}_\pm, \qquad [\mathcal{J}_+, \mathcal{J}_-] = 2\mathcal{J}_3,$$

$$[\mathcal{K}_3, \mathcal{K}_\pm] = \pm\,\mathcal{K}_\pm, \quad [\mathcal{K}_+, \mathcal{K}_-] = -2\mathcal{K}_3. \tag{3.5}$$

It will be noted that the commutation properties of the $\mathcal{J}$ operators are those of a conventional angular momentum, while the $\mathcal{K}$ operators are analogous to the hyperbolic angular momentum K, which was discussed in the first section. We shall denote the eigenvalues of $\mathcal{J}_3$ and $\mathcal{K}_3$ by μ and ν, respectively. These quantities have the following significance,

$$\mu = j_1 - j_2, \qquad \nu = j_1 + j_2 + 1. \tag{3.6}$$

In evaluating the square of the resultant angular momentum, we encounter

$$\begin{aligned} 2\,\mathbf{J}_1 \cdot \mathbf{J}_2 &= \tfrac{1}{2} \sum_{\zeta\zeta'\zeta''\zeta'''} a_\zeta^+ a_{\zeta'} b_{\zeta''}^+ b_{\zeta'''} (\zeta|\boldsymbol{\sigma}|\zeta') \cdot (\zeta''|\boldsymbol{\sigma}|\zeta''') \\ &= \sum_{\zeta\zeta'} a_\zeta^+ a_{\zeta'} b_{\zeta'}^+ b_\zeta - \tfrac{1}{2} n_1 n_2. \end{aligned} \tag{3.7}$$

This can be expressed either in terms of the $\mathcal{J}$ operators, or of the $\mathcal{K}$ operators, since

$$\mathcal{J}_- \mathcal{J}_+ = \sum_{\zeta\zeta'} b_{\zeta'}^+ a_{\zeta'} a_\zeta^+ b_\zeta = n_2 + \sum_{\zeta\zeta'} a_\zeta^+ a_{\zeta'} b_{\zeta'}^+ b_\zeta, \tag{3.8}$$

and

$$\mathcal{K}_+ \mathcal{K}_- = \sum_{\zeta\zeta'} a_\zeta^+ b_{\zeta'}^+ (a_\zeta b_{\zeta'} - a_{\zeta'} b_\zeta) = n_1 n_2 - \sum_{\zeta\zeta'} a_\zeta^+ a_{\zeta'} b_{\zeta'}^+ b_\zeta. \tag{3.9}$$

Indeed,

$$\mathbf{J}^2 = \mathcal{J}_3(\mathcal{J}_3 + 1) + \mathcal{J}_- \mathcal{J}_+ = \mathcal{J}_3(\mathcal{J}_3 - 1) + \mathcal{J}_+ \mathcal{J}_-, \tag{3.10}$$

and

$$\mathbf{J}^2 = \mathcal{K}_3(\mathcal{K}_3 - 1) - \mathcal{K}_+ \mathcal{K}_- = \mathcal{K}_3(\mathcal{K}_3 + 1) - \mathcal{K}_- \mathcal{K}_+. \tag{3.11}$$

From the first, conventional, representation of $\mathbf{J}^2$ in terms of the angular momentum $\mathcal{J}$, we infer that

$$j \geq |\mu|, \tag{3.12}$$

or

$$j \geq |j_1 - j_2|, \tag{3.13}$$

while the hyperbolic representation implies that

$$\nu - 1 \geq j, \tag{3.14}$$

or

$$j_1 + j_2 \geq j. \tag{3.15}$$

We have thus arrived at

$$j_1 + j_2 \geq j \geq |j_1 - j_2|, \tag{3.16}$$

the familiar restriction on the composition of two angular momenta.

An eigenvector of $\mathbf{J}^2$ is conveniently labelled by the eigenvalues of J_3, $\mathcal{J}_3$, and $\mathcal{K}_3$. In virtue of (3.6), the resulting eigenvector $\Psi(jm\mu\nu)$ is equivalently designated as $\Psi(j_1 j_2 jm)$. In particular, the state with $\nu = j + 1$ corresponds to $j_1 + j_2 = j$, and $2j_1 = j + \mu$, $2j_2 = j - \mu$. The special state of this type with $m = j$ can be realized in only one way, since $m = j_1 + j_2$ requires that $m_1 = j_1$, $m_2 = j_2$. Thus

$$\Psi(jj\mu j+1) = \frac{(a_+^+)^{j+\mu}}{((j+\mu)!)^{1/2}} \frac{(b_+^+)^{j-\mu}}{((j-\mu)!)^{1/2}} \Psi_0. \tag{3.17}$$

With an arbitrary reference system, this result becomes

$$((2j)!)^{1/2} \sum_{m=-j}^{j} \phi_{jm}(x)\Psi(jm\mu j+1) = \frac{(xa^+)^{j+\mu}(xb^+)^{j-\mu}}{[(j+\mu)!\,(j-\mu)!]^{1/2}} \Psi_0, \tag{3.18}$$

according to (2.4). We multiply this $\mathscr{J}$ analogue of (1.13) with $\phi_{j\mu}(\xi)$, and sum with respect to μ,

$$((2j)!)^{1/2} \sum_{m\mu} \phi_{jm}(x)\phi_{j\mu}(\xi)\Psi(jm\mu j+1) = \frac{(\xi_+(xa^+) + \xi_-(xb^+))^{2j}}{(2j)!} \Psi_0. \tag{3.19}$$

Further summation with respect to j then yields

$$\sum_{jm\mu} ((2j)!)^{1/2}\phi_{jm}(x)\phi_{j\mu}(\xi)\Psi(jm\mu j+1) = e^{\xi_+(xa^+)+\xi_-(xb^+)}\,\Psi_0. \tag{3.20}$$

To complete the determination of the eigenvector $\Psi(jm\mu\nu)$, we need the analogue of (1.29), specifying the eigenvector with arbitrary ν in terms of that with the minimum value, $j+1$. For this purpose, we examine the operator[6]

$$V = t^{2\mathscr{K}_3 - 1} \tag{3.21}$$

which has the following significant properties,

$$t\frac{\partial}{\partial t}V = (2\mathscr{K}_3 - 1)V, \qquad \left(t\frac{\partial}{\partial t}\right)^2 V = (2\mathscr{K}_3 - 1)^2 V, \tag{3.22}$$

and

$$V^{-1}\mathscr{K}_- V = t^2\mathscr{K}_-, \qquad \mathscr{K}_- V = t^2 V\mathscr{K}_-. \tag{3.23}$$

In conjunction with

$$4\mathbf{J}^2 + 1 = (2\mathscr{K}_3 - 1)^2 - 4\mathscr{K}_+\mathscr{K}_-, \tag{3.24}$$

we obtain

$$\left(\frac{\partial^2}{\partial t^2} + \frac{1}{t}\frac{\partial}{\partial t} - \frac{4\mathbf{J}^2+1}{t^2}\right)V - 4\mathscr{K}_+ V\mathscr{K}_- = 0, \tag{3.25}$$

an ordered operator form of Bessel's equation. The solution is

$$V = t^{(4\mathbf{J}^2+1)^{\frac{1}{2}}} F_{(4\mathbf{J}^2+1)^{\frac{1}{2}}}(t^2\mathscr{K}_+; P; \mathscr{K}_-), \tag{3.26}$$

where P is an integration constant, and the notation is intended to indicate that P is inserted between the powers of $\mathscr{K}_+$ and $\mathscr{K}_-$ in the ordered operator

[6] Our procedure here is based upon the general method of Appendix A.

expansion of the function F defined in (1.36). The second solution of the Bessel equation has been rejected in order to conform with the fact that $t^{2\mathscr{K}_3-1}$ must vanish as $t\to 0$, in view of the non-negative character of $\mathscr{K}_3-1$. The operator (3.26) can also be written as

$$V=\sum_j t^{2j+1}F_{2j+1}(t^2\mathscr{K}_+;P_{j,j+1};\mathscr{K}_-)$$

$$=\sum_j\sum_{\nu=j+1}^{\infty}t^{2\nu-1}P_{j\nu}, \tag{3.27}$$

where $P_{j\nu}$ is the projection operator for the state with the indicated eigenvalues. According to the well-known Bessel function power series we then have

$$P_{j\nu}=\omega_{j\nu}(\mathscr{K}_+)P_{j,j+1}\omega_{j\nu}(\mathscr{K}_-), \tag{3.28}$$

where

$$\omega_{j\nu}(\lambda)=\left[\frac{(2j+1)!}{(\nu+j)!\,(\nu-j-1)!}\right]^{1/2}\lambda^{\nu-j-1}. \tag{3.29}$$

This yields the desired eigenvector relation,

$$\Psi(jm\mu\nu)=\omega_{j\nu}(\mathscr{K}_+)\Psi(jm\mu j+1). \tag{3.30}$$

It will be noted that, with respect to j and ν, Eq. (3.30) is converted into (1.29) by the substitutions

$$j\to|m|-\tfrac{1}{2},\qquad \nu\to j+\tfrac{1}{2}, \tag{3.31}$$

which are in accord with the significance of K. Corresponding, then, to the generating functions (1.35) and (1.37), we have

$$\sum_{\nu=j+1}^{\infty}\omega_{j\nu}(\lambda)\Psi(jm\mu\nu)=F_{2j+1}(\lambda\mathscr{K}_+)\Psi(jm\mu j+1), \tag{3.32}$$

and

$$((2j+1)!)^{-1/2}\sum_{\nu=j+1}^{\infty}\chi_{j\nu}(\lambda)\Psi(jm\mu\nu)=e^{\lambda\mathscr{K}_+}\Psi(jm\mu j+1), \tag{3.33}$$

in which

$$\chi_{j\nu}(\lambda)=\left[\frac{(\nu+j)!}{(\nu-j-1)!}\right]^{1/2}\lambda^{\nu-j-1}. \tag{3.34}$$

The application of the operator $e^{\lambda\mathscr{K}_+}$ to (3.20) thus produces

$$\sum_{jm\mu\nu}(2j+1)^{-1/2}\phi_{jm}(x)\phi_{j\mu}(\xi)\chi_{j\nu}(\lambda)\Psi(jm\mu\nu)=e^{\lambda[a^+b^+]+\xi_+(xa^+)+\xi_-(xb^+)}\Psi_0. \tag{3.35}$$

The eigenvectors are exhibited somewhat more explicitly[7] in the result

[7] The normalization constant does not automatically appear in the corresponding group theory formula. B. L. van der Waerden, "Die gruppentheoretische Methode in der Quantenmechanik" (Berlin, 1932).

obtained by applying $\omega_{j_\nu}(\mathscr{K}_+)$ to (3.18),

$$\sum_{m=-j}^{j} \phi_{jm}(x)\Psi(j_1j_2jm) = \left[\frac{2j+1}{(j_1+j_2+j+1)!}\right]^{1/2} \cdot\left[\frac{[a^+b^+]^{j_1+j_2-j}\cdot(xa^+)^{j+j_1-j_2}\cdot(xb^+)^{j_2+j-j_1}}{[(j_1+j_2-j)!(j+j_1-j_2)!(j_2+j-j_1)!]^{1/2}}\right]\Psi_0, \quad (3.36)$$

in which we have employed j_1 and j_2, rather than μ and ν. For the purpose of converting (3.36) into a convenient expression for the transformation function

$$(j_1j_2jm|j_1m_1j_2m_2) = (\Psi(j_1j_2jm), \Psi(j_1m_1j_2m_2)), \quad (3.37)$$

we make the replacement $x_+ \to z_-^*$, $x_- \to -z_+^*$, and take the scalar product with the generating function of the $\Psi(j_1m_1j_2m_2)$,

$$\sum_{j_1m_1j_2m_2} \phi_{j_1m_1}(x)\phi_{j_2m_2}(y)\Psi(j_1m_1j_2m_2) = e^{(xa^+)+(yb^+)}\Psi_0. \quad (3.38)$$

The ensuing formula can be written

$$\sum_{m_1m_2m_3} \phi_{j_1m_1}(x)\phi_{j_2m_2}(y)\phi_{j_3m_3}(z)X(j_1j_2j_3;\, m_1m_2m_3)$$
$$= [(j_1+j_2+j_3+1)!]^{-1/2}\cdot\frac{[yz]^{j_2+j_3-j_1}\cdot[zx]^{j_3+j_1-j_2}\cdot[xy]^{j_1+j_2-j_3}}{[(j_2+j_3-j_1)!(j_3+j_1-j_2)!(j_1+j_2-j_3)!]^{1/2}}, \quad (3.39)$$

in virtue of the definition[8]

$$(j_1j_2jm|j_1m_1j_2m_2) = (2j+1)^{1/2}(-1)^{j_1-j_2+m}X(j_1j_2j;\, m_1m_2-m). \quad (3.40)$$

Multiplication with

$$\Phi_{j_1j_2j_3}(\alpha\beta\gamma) = [(J+1)!]^{1/2}\frac{\alpha^{J-2j_1}\beta^{J-2j_2}\gamma^{J-2j_3}}{[(J-2j_1)!(J-2j_2)!(J-2j_3)!]^{1/2}}$$
$$J = j_1+j_2+j_3, \quad (3.41)$$

and summation with respect to j_1, j_2, and j_3, then yields the generating function

$$\sum_{jm} \phi_{j_1m_1}(x)\phi_{j_2m_2}(y)\phi_{j_3m_3}(z)\Phi_{j_1j_2j_3}(\alpha\beta\gamma)X(j_1j_2j_3;\, m_1m_2m_3) = e^{\alpha[yz]+\beta[zx]+\gamma[xy]}. \quad (3.42)$$

[8] This X coefficient is related to the V coefficient of G. Racah, *Phys. Rev.* **62,** 438 (1942), by $X = (-1)^{j_2+j-j_1}V$. We have introduced the X coefficient by virtue of its greater symmetry: compare Eqs. (3.44), (3.45) with Eq. (19a) of Racah's paper (henceforth referred to as R). Editors' note: This X coefficient is identical to the Wigner 3-j symbol; i.e., $X(j_1j_2j;\, m_1m_2m) = \begin{pmatrix} j_1 & j_2 & j \\ m_1 & m_2 & m \end{pmatrix}$.

Symmetry properties of the X coefficients can be easily inferred from the invariance of the generating function to particular substitutions. Thus, the null effect of multiplying x_+, y_+, z_+ by $e^{(i/2)\varphi}$, and x_-, y_-, z_- by $e^{-(i/2)\varphi}$, indicates that X vanishes unless

$$m_1 + m_2 + m_3 = 0. \tag{3.43}$$

The invariance of the generating function for simultaneous cyclic permutations of x, y, z and α, β, γ implies the corresponding property for X:

$$X(j_1j_2j_3;\, m_1m_2m_3) = X(j_2j_3j_1;\, m_2m_3m_1) = X(j_3j_1j_2;\, m_3m_1m_2). \tag{3.44}$$

The interchange of x and y, combined with the substitutions $\alpha \leftrightarrow -\beta$, $\gamma \to -\gamma$, discloses the behavior of the X coefficients with respect to non-cyclic permutations,

$$\begin{aligned} X(j_2j_1j_3;\, m_2m_1m_3) &= X(j_1j_3j_2;\, m_1m_3m_2) = X(j_3j_2j_1;\, m_3m_2m_1) \\ &= (-1)^J X(j_1j_2j_3;\, m_1m_2m_3), \end{aligned} \tag{3.45}$$

while the exchange of x_+, y_+, z_+ with x_-, y_-, z_-, in conjunction with sign reversals for α, β, γ, leads to

$$X(j_1j_2j_3;\, -m_1-m_2-m_3) = (-1)^J X(j_1j_2j_3;\, m_1m_2m_3). \tag{3.46}$$

Among the implied properties of the transformation function (3.37) are

$$\begin{aligned} (j_2j_1jm|j_2m_2j_1m_1) &= (j_1j_2j-m|j_1-m_1\, j_2-m_2) \\ &= (-1)^{j_1+j_2-j}(j_1j_2jm|j_1m_1j_2m_2). \end{aligned} \tag{3.47}$$

The expression for $X(j_1j_2j_3;\, m_1m_2m_3)$, obtained by expanding (3.39), is

$$X(j;\, m) = [(J+1)!]^{-1/2} \sum_n (-1)^n \prod_{i=1}^{3} \frac{[(j_i+m_i)!(j_i-m_i)!(J-2j_i)!]^{1/2}}{(J-2j_i-n_i)!n_i!} \tag{3.48}$$

in which

$$n = n_1 + n_2 + n_3, \tag{3.49}$$

and the summation is to be extended over all n_i subject to

$$J - 2j_i \geq n_i \geq 0, \tag{3.50}$$

and

$$n_2 - n_3 = m_1 - j_2 + j_3,\ n_3 - n_1 = m_2 - j_3 + j_1,\ n_1 - n_2 = m_3 - j_1 + j_2. \tag{3.51}$$

The latter conditions can also be written as

$$\begin{aligned} J - 2j_1 - n_1 &= j_2 + m_2 - n_3 = j_3 - m_3 - n_2 \\ J - 2j_2 - n_2 &= j_3 + m_3 - n_1 = j_1 - m_1 - n_3 \\ J - 2j_3 - n_3 &= j_1 + m_1 - n_2 = j_2 - m_2 - n_1. \end{aligned} \tag{3.52}$$

It follows from the non-negative character of these quantities that the n_i are uniquely determined if one of the nine integers $J-2j_i, j_i+m_i, j_i-m_i$ is equal to zero. In general, the number of terms in the sum (3.48) exceeds by unity the smallest of these nine integers. It is a matter of convenience which of the n_i is chosen as the summation parameter.

The X coefficient can also be exhibited in closed form whenever the $|m_i|$ have the minimum values compatible with the given j_i. The simplest illustration of this is provided by $X(j_1j_2j_3;000)$ corresponding to integral values of j_1, j_2, and j_3. Note that this quantity vanishes, according to (3.46), if $\frac{1}{2}J$ is not an integer. Our procedure here is to place $x_- = \partial/\partial x_+$, with analogous substitutions for y_- and z_-, and to evaluate the derivatives at $x_+ = y_+ = z_+ = 0$. Since

$$[(j_1+m_1)!(j_1-m_1)!]^{-1/2}\cdot(\partial/\partial x_+)^{j_1-m_1}\, x_+^{j_1+m_1}]_{x_+=0} = \delta_{m_1,0}, \tag{3.53}$$

this effectively isolates the $m=0$ terms in (3.42). The reduction of the generating function can be performed with the aid of the following theorem concerning ordered operators, which will be proved in Appendix B. If a and a^+ are two operators satisfying $[a, a^+]=1$, and $f(a^+)$ is an arbitrary function, we have

$$e^{za;\,a^+} f(a^+) = \frac{1}{1-z} f\left(\frac{a^+}{1-z}\right) e^{\frac{z}{1-z}a^+;\,a}. \tag{3.54}$$

The differential operator realization of this, with $a=\partial/\partial a^+$, is the form actually employed.

The result of the calculation is

$$\sum_j \Phi_{j_1j_2j_3}(\alpha\beta\gamma) X(j_1j_2j_3;\,000) = (1+\alpha^2+\beta^2+\gamma^2)^{-1}, \tag{3.55}$$

which is a generating function for $X(j;\,0)$. On writing

$$\begin{aligned}(1+\alpha^2+\beta^2+\gamma^2)^{-1} &= \sum_{J=0,2,\cdots} (-1)^{\frac{1}{2}J}(\alpha^2+\beta^2+\gamma^2)^{\frac{1}{2}J} \\ &= \sum_j (-1)^{\frac{1}{2}J}(\tfrac{1}{2}J)!\,\frac{\alpha^{J-2j_1}\,\beta^{J-2j_2}\,\gamma^{J-2j_3}}{(\frac{1}{2}J-j_1)!(\frac{1}{2}J-j_2)!(\frac{1}{2}J-j_3)!},\end{aligned} \tag{3.56}$$

we obtain the explicit formula[9]

$$X(j;\,0) = (-1)^{\frac{1}{2}J}\frac{(\frac{1}{2}J)!}{[(J+1)!]^{1/2}}\prod_{i=1}^{3}\frac{[(J-2j_i)!]^{1/2}}{(\frac{1}{2}J-j_i)!}. \tag{3.57}$$

We extend this argument by making the substitutions $x_-\to\partial/\partial x_+$, $y_-\to\partial/\partial y_+$, $z_+\to\partial/\partial z_-$, and evaluating the derivatives for arbitrary x_+, y_+,

[9] This result is contained in *R*, Eq. (22′).

and z_-. In view of

$$\phi_{j_1m_1}(x) \to \left[\frac{(j_1+m_1)!}{(j_1-m_1)!}\right]^{1/2} \frac{x_+^{2m_1}}{(2m_1)!}, \qquad m_1 \geq 0$$
$$0, \qquad m_1 < 0, \tag{3.58}$$

and

$$\phi_{j_3m_3}(z) \to \left[\frac{(j_3+|m_3|)!}{(j_3-|m_3|)!}\right]^{1/2} \frac{z_-^{2|m_3|}}{(2|m_3|)!}, \qquad m_3 \leq 0$$
$$0, \qquad m_3 > 0, \tag{3.59}$$

we shall thereby obtain the X coefficient for $m_1 \geq 0, m_2 \geq 0, -m_3 = m_1 + m_2$. The values of X when two of the m_i are negative can then be inferred from (3.46). The generating function now becomes

$$e^{\alpha[yz]+\beta[zx]+\gamma[xy]} \to (1+\alpha^2+\beta^2+\gamma^2)^{-1}$$
$$\cdot \exp\left\{\frac{z_-}{1+\alpha^2+\beta^2+\gamma^2}[(\alpha\gamma-\beta)x_+ + (\beta\gamma-\alpha)y_+]\right\} \tag{3.60}$$

and, on expanding in powers of x_+, y_+, and z_-, we find that

$$\sum_j \Phi_{j_1j_2j_3}(\alpha\beta\gamma) \prod_i \left[\frac{(j_i+|m_i|)!}{(j_i-|m_i|)!}\right]^{1/2} X(j;m)$$
$$= (2|m_3|)! \frac{(\alpha\gamma-\beta)^{2m_1}(\beta\gamma-\alpha)^{2m_2}}{(1+\alpha^2+\beta^2+\gamma^2)^{2|m_3|+1}}. \tag{3.61}$$

The result attained by further expansion of (3.61) is

$$\left[(J+1)! \prod_i \frac{(j_i+|m_i|)!}{(j_i-|m_i|)!} \frac{1}{(J-2j_i)!}\right]^{1/2} X(j;m)$$
$$= \sum_{n_1n_2} (-1)^{\frac{1}{2}J_2-2|m_3|} \frac{(\frac{1}{2}J_3)!}{\prod_i (\frac{1}{2}J_i-j_i-|m_i|)!} \frac{(2m_1)!}{(2m_1-n_1)!n_1!} \frac{(2m_2)!}{(2m_2-n_2)!n_2!} \tag{3.62}$$

where

$$J_1 = J+n_1-n_2, \qquad J_2 = J-n_1+n_2, \qquad J_3 = J+n_1+n_2. \tag{3.63}$$

The double summation is to be extended over such non-negative integers that satisfy

$$J-2j_1-n_2 \geq 2m_1-n_1 \geq 0$$
$$J-2j_2-n_1 \geq 2m_2-n_2 \geq 0$$
$$J-2j_3 \geq 2|m_3|-n_1-n_2 \geq 0, \tag{3.64}$$

and for which $J + n_1 + n_2$ is an even integer. The sum consists of a single term if one of the $J - 2j_i$ vanishes, or if $m_1 = m_2 = 0$. This simplification may also result from the evenness requirement on J_3. Thus

$$\left[\frac{(j_1+\frac{1}{2})(j_3+\frac{1}{2})(J+1)!}{\prod_i (J-2j_i)!}\right]^{\frac{1}{2}} X(j_1j_2j_3;\tfrac{1}{2}0-\tfrac{1}{2})$$

$$= (-1)^{\frac{1}{2}J-1}\cdot\frac{(\frac{1}{2}J)!}{(\frac{1}{2}J-j_1-\frac{1}{2})(\frac{1}{2}J-j_2)!(\frac{1}{2}J-j_3-\frac{1}{2})!}, \qquad J \text{ even}$$

$$= (-1)^{\frac{1}{2}J+\frac{1}{2}}\cdot\frac{(\frac{1}{2}J+\frac{1}{2})!}{(\frac{1}{2}J-j_1)!(\frac{1}{2}J-j_2-\frac{1}{2})!(\frac{1}{2}J-j_3)!}, \qquad J \text{ odd}, \quad (3.65)$$

which are the X coefficients with the minimum $|m_i|$ corresponding to half-integral values for two of the j_i.

The orthogonality and normalization of the eigenvectors $\Psi(jm\mu\nu)$ can be verified, with the aid of (3.35), by an extension of the procedure leading to (1.21). According to Eq. (C7) of Appendix C, we have

$$(\exp\{\lambda[a^+b^+] + \xi_+(xa^+) + \xi_-(xb^+)\}\,\Psi_0,$$

$$\exp\{\kappa[a^+b^+] + \eta_+(ya^+) + \eta_-(yb^+)\}\,\Psi_0) = \frac{1}{(1-\lambda^*\kappa)^2}\exp\left[\frac{(\xi^*\eta)(x^*y)}{1-\lambda^*\kappa}\right], \quad (3.66)$$

and the expansion

$$\frac{1}{(1-\lambda^*\kappa)^2}\exp\frac{[(\xi^*\eta)(x^*y)}{1-\lambda^*\kappa}$$

$$= \sum_{jm\mu\nu}\frac{1}{2j+1}\phi_{jm}(x^*)\phi_{jm}(y)\phi_{j\mu}(\xi^*)\phi_{j\mu}(\eta)\chi_{j\nu}(\lambda^*)\chi_{j\nu}(\kappa) \quad (3.67)$$

establishes that

$$(\Psi(jm\mu\nu), \Psi(j'm'\mu'\nu')) = \delta_{jj'}\,\delta_{mm'}\,\delta_{\mu\mu'}\,\delta_{\nu\nu'}. \quad (3.68)$$

The unitary nature of the transformation $\Psi(j_1m_1j_2m_2) \to \Psi(j_1j_2jm)$, and of its inverse, imposes the following conditions upon the X coefficients,

$$\sum_{m_1m_2} X(j_1j_2j_3, m_1m_2m_3)X(j_1j_2j_3'; m_1m_2m_3') = \frac{1}{2j_3+1}\delta_{j_3j'_3}\,\delta_{m_3m'_3} \quad (3.69)$$

and

$$\sum_{j_3m_3}(2j_3+1)X(j_1j_2j_3; m_1m_2m_3)X(j_1j_2j_3; m_1'm_2'm_3) = \delta_{m_1m'_1}\,\delta_{m_2m'_2}. \quad (3.70)$$

As a particular consequence of (3.69), we have

$$\sum_m [X(j; m)]^2 = 1\,. \quad (3.71)$$

The Rotation Matrices

The results of this section can be applied in developing further the properties of the matrices $U_{mm'}^{(j)}(\phi\theta\psi)$, which were introduced in Section 2. If U is the operator generating a reference system rotation for the composite system with angular momentum $\mathbf{J} = \mathbf{J}_1 + \mathbf{J}_2$, while U_1 and U_2 are the corresponding operators for the individual angular momenta, we have

$$U = U_1 U_2, \tag{3.72}$$

according to the exponential form (2.31). In particular, theorem (2.59) states that

$$\int U_1 U_2 d\omega = P_0, \tag{3.73}$$

where P_0 is the projection operator for the $j = 0$ state of the resultant angular momentum. On taking matrix elements of the latter equation, we find

$$\begin{aligned} \int U_{m_1 m_1'}^{(j_1)}(\omega) U_{m_2 m'_2}^{(j_2)}(\omega) d\omega &= (j_1 m_1 j_2 m_2 | P_0 | j_1 m_1' j_2 m_2') \\ &= (j_1 m_1 j_2 m_2 | j_1 j_2 00)(j_1 j_2 00 | j_1 m_1' j_2 m_2') \\ &= \frac{1}{2j_1 + 1} \delta_{j_1 j_2} \delta_{-m_1 m_2} \delta_{-m'_1 m'_2} (-1)^{m_1 - m'_2} \end{aligned} \tag{3.74}$$

since

$$(j_1 j_2 00 | j_1 m_1 j_2 m_2) = (2j_1 + 1)^{-1/2} (-1)^{j_1 - m_1} \delta_{j_1 j_2} \delta_{-m_1 m_2}. \tag{3.75}$$

In view of (2.41), it is also possible to write (3.74) as

$$\int U_{m_1 m'_1}^{(j_1)*}(\omega) U_{m_2 m'_2}^{(j_2)}(\omega) d\omega = \frac{1}{2j_1 + 1} \delta_{j_1 j_2} \delta_{m_1 m_2} \delta_{m'_1 m'_2}, \tag{3.76}$$

which expresses the orthogonality properties of the rotation matrices, in their dependence upon the rotation parameters.

The orthogonality relation of the trace $\chi^{(j)}$, derived from (3.76), is

$$\int \chi^{(j_1)*} \chi^{(j_2)} \, d\omega = \delta_{j_1 j_2}. \tag{3.77}$$

This integral can be simplified, since the $\chi^{(j)}$ depend only upon the rotation angle γ. We write

$$d\omega = \int_0^{2\pi} \tfrac{1}{2} d\gamma \sin \tfrac{1}{2}\gamma \; \delta\!\left(\cos \tfrac{1}{2}\gamma - \cos \tfrac{1}{2}\theta \cdot \cos \frac{\phi + \psi}{2}\right) d\omega, \tag{3.78}$$

and, after first performing the $d\omega$ integration, obtain

$$\int_0^{2\pi} \chi^{(j_1)}(\gamma)^* \chi^{(j_2)}(\gamma) \frac{1}{\pi} \sin^2 \frac{\gamma}{2} \, d\gamma = \delta_{j_1 j_2}, \tag{3.79}$$

which can be verified directly.

We return to (3.72) and observe that its matrix element is

$$U^{(j_1)}_{m_1m'_1}(\omega)U^{(j_2)}_{m_2m'_2}(\omega) = \sum_{jmm'} (j_1m_1j_2m_2|j_1j_2jm)U^{(j)}_{mm'}(\omega)(j_1j_2jm|j_1m'_1j_2m'_2)$$

$$= \sum_{jmm'} (2j+1)X(j_1j_2j;\, m_1m_2-m)(-1)^{m-m'}U^{(j)}_{mm'}(\omega)X(j_1j_2j;\, m'_1m'_2-m'), \tag{3.80}$$

or

$$U^{(j_1)}_{m_1m'_1}(\omega)U^{(j_2)}_{m_2m'_2}(\omega)$$
$$= \sum_{jmm'} (2j+1)X(j_1j_2j;\, m_1m_2m)U^{(j)}_{mm'}(\omega)^*X(j_1j_2j;\, m'_1m'_2m'). \tag{3.81}$$

With the use of the orthogonality relation (3.76), this can be presented in the symmetrical form

$$\int U^{(j_1)}_{m_1m_1'}U^{(j_2)}_{m_2m'_2}U^{(j_3)}_{m_3m'_3}d\omega = X(j_1j_2j_3;\, m_1m_2m_3)X(j_1j_2j_3;\, m'_1m'_2m'_3). \tag{3.82}$$

Specializations of this integral are provided by

$$U^{(l)}_{m0} = \left(\frac{4\pi}{2l+1}\right)^{1/2} Y^*_{lm}(\theta\phi), \tag{3.83}$$

and

$$U^{(l)}_{00} = P_l(\cos\theta), \tag{3.84}$$

where Y_{lm} is the spherical harmonic associated with integral l, and $P_l(\cos\theta)$ is the Legendre polynomial.
Thus

$$\int Y_{l_1m_1}Y_{l_2m_2}Y_{l_3m_3}\, \tfrac{1}{2}\sin\theta\, d\theta\, \frac{1}{2\pi}\, d\phi = \left[\prod_i \left(\frac{2l_i+1}{4\pi}\right)\right]^{1/2} X(l;0)\, X(l;m), \tag{3.85}$$

and

$$\int_0^\pi P_{l_1}(\cos\theta)\, P_{l_2}(\cos\theta)\, P_{l_3}(\cos\theta)\, \tfrac{1}{2}\sin\theta\, d\theta = [X(l;0)]^2. \tag{3.86}$$

The multiplication property of the trace, as derived from (3.80) is

$$\chi^{(j_1)}(\gamma)\chi^{(j_2)}(\gamma) = \sum_{j=|j_1-j_2|}^{j=j_1+j_2} \chi^{(j)}(\gamma), \tag{3.87}$$

which can also be expressed in the form

$$\int_0^{2\pi} \chi^{(j_1)}\chi^{(j_2)}\chi^{(j_3)}\, \frac{1}{\pi}\sin^2 \tfrac{1}{2}\gamma\, d\gamma = \begin{cases} 1, & J-2j_i \geq 0 \\ 0, & \text{otherwise.} \end{cases} \tag{3.88}$$

One can regard this as a realization of the projection operator statement of the angular momentum composition law,

$$P_{j_1}P_{j_2} = \sum_{j=|j_1-j_2|}^{j=j_1+j_2} P_j, \tag{3.89}$$

since (3.87) is the trace of the equation obtained by multiplying (3.89) with $U_1U_2 = U$.

We shall conclude this discussion by deriving the completeness relations for the functions $\chi^{(j)}(\omega)$ and $U^{(j)}_{mm'}(\omega)$. Referring to (2.36), the generating function of the $\chi^{(j)}$, we replace t therein with $te^{(i/2)\gamma'}$ and obtain

$$\sum_j t^{2j}\chi^{(j)}(\omega)e^{i(j+\frac{1}{2})\gamma'} = \frac{1}{(1+t^2)\cos\frac{\gamma'}{2} - 2t\cos\frac{\gamma}{2} - i\sin\frac{\gamma'}{2}(1-t^2)}, \tag{3.90}$$

the imaginary part of which can be written

$$\sum_j t^{2j}\chi^{(j)}(\omega)\chi^{(j)}(\omega') = \frac{(1-t^2)}{(1-t^2)^2\left(1-\frac{4t}{(1+t)^2}\cos\frac{\gamma}{2}\cos\frac{\gamma'}{2}\right)+4t^2\left(\cos\frac{\gamma}{2}-\cos\frac{\gamma'}{2}\right)^2}. \tag{3.91}$$

We now consider the limit $t\to 1$, and infer from the known result

$$\lim_{\varepsilon\to 0}\frac{1}{\pi}\frac{\varepsilon}{x^2+\varepsilon^2} = \delta(x), \tag{3.92}$$

that

$$\sum_j \chi^{(j)}(\omega)\chi^{(j)}(\omega') = \frac{\pi}{2}\frac{1}{\sin(\gamma/2)}\,\delta\left(\cos\frac{\gamma}{2}-\cos\frac{\gamma'}{2}\right). \tag{3.93}$$

However

$$\int\frac{\pi}{2}\frac{1}{\sin\gamma/2}\,\delta\left(\cos\frac{\gamma}{2}-\cos\frac{\gamma'}{2}\right)d\omega = \int_0^{2\pi}\delta\left(\cos\frac{\gamma}{2}-\cos\frac{\gamma'}{2}\right)\sin\frac{\gamma}{2}\,\tfrac{1}{2}\,d\gamma = 1, \tag{3.94}$$

so that (3.93) can be written

$$\sum_j \chi^{(j)}(\omega)\chi^{(j)}(\omega') = \delta(\omega-\omega'), \tag{3.95}$$

which is the completeness relation of the $\chi^{(j)}$. As a specialization of (3.95), we place $\gamma' = 0$ and find

$$\sum_j (2j+1)\chi^{(j)}(\omega) = \delta(\omega). \tag{3.96}$$

An operator expression for the composition of successive rotations is given by

$$U(\omega)U^{-1}(\omega') = U(\omega - \omega'). \tag{3.97}$$

We take the trace of this equation for the states with quantum number j, and, in virtue of the unitary property of U, obtain

$$\sum_{mm'} U^{(j)}_{mm'}(\omega)U^{(j)}_{mm'}(\omega')^* = \chi^{(j)}(\omega - \omega'), \tag{3.98}$$

which is in the nature of an addition theorem. The completeness relation for the $U^{(j)}_{mm'}(\omega)$ is reached on multiplying (3.98) with $2j + 1$ and summing with respect to j. In view of (3.96), we have

$$\sum_{jmm'} (2j + 1)U^{(j)}_{mm'}(\omega)U^{(j)}_{mm'}(\omega')^* = \delta(\omega - \omega'). \tag{3.99}$$

On integration of (3.98) and (3.99) with respect to the Eulerian angle ψ, there emerges the addition theorem and the completeness relation of the spherical harmonics.

4. Three and Four Angular Momenta

Eigenvectors for the resultant of three angular momenta can be built up in several ways, as symbolized by

$$\mathbf{J} = \mathbf{J}_1 + (\mathbf{J}_2 + \mathbf{J}_3) = \mathbf{J}_2 + (\mathbf{J}_3 + \mathbf{J}_1) = \mathbf{J}_3 + (\mathbf{J}_1 + \mathbf{J}_2). \tag{4.1}$$

Thus, according to the first procedure, we construct $\Psi(j_1m_1j_2j_3j_{23}m_{23})$ and then $\Psi(j_1[j_2j_3]j_{23}jm)$, while the last method yields $\Psi(j_3[j_1j_2]j_{12}jm)$. The notation $[j_2j_3]$, for example, is intended to indicate that these angular momenta are not involved explicitly in the composition of j_1 and j_{23} to form j. Similarly, four angular momenta can be combined in various pairs,

$$\mathbf{J} = (\mathbf{J}_1 + \mathbf{J}_2) + (\mathbf{J}_3 + \mathbf{J}_4) = (\mathbf{J}_2 + \mathbf{J}_3) + (\mathbf{J}_4 + \mathbf{J}_1) = (\mathbf{J}_1 + \mathbf{J}_3) + (\mathbf{J}_2 + \mathbf{J}_4), \tag{4.2}$$

in which the first method, say, yields $\Psi([j_1j_2]j_{12}[j_3j_4]j_{34}jm)$ through the intermediary of $\Psi(j_1j_2j_{12}m_{12}j_3j_4j_{34}m_{34})$. Our problem in this section is the evaluation of the transformation function connecting two such schemes of adding four angular momenta. The analogous question for three angular momenta can be regarded as a specialization of this more symmetrical problem.

To facilitate the addition of angular momenta in pairs, we observe that the generating function (3.35), written as

$$\sum_{j_1j_2jm} (2j + 1)^{-1/2}\phi_{jm}(x)\Phi_{j_1j_2j}(\alpha_1\alpha_2\alpha_3)\Psi(j_1j_2jm)$$
$$= \exp(\alpha_3[a^+b^+] + \alpha_2(xa^+) + \alpha_1(xb^+))\Psi_0, \tag{4.3}$$

can be obtained from

$$\sum_{j_1m_1j_2m_2} \phi_{j_1m_1}(t_1)\phi_{j_2m_2}(t_2)\Psi(j_1m_1j_2m_2) = \exp((t_1a^+) + (t_2b^+))\Psi_0 \qquad (4.4)$$

by the application of the differential operator

$$\exp\left(\alpha_3\left[\frac{\partial}{\partial t_1}\frac{\partial}{\partial t_2}\right] + \alpha_2\left(x\frac{\partial}{\partial t_1}\right) + \alpha_1\left(x\frac{\partial}{\partial t_2}\right)\right) \qquad (4.5)$$

with the understanding that the derivatives are to be evaluated at $t_1 = t_2 = 0$. Accordingly, if we apply (4.5) and

$$\exp\left(\beta_3\left[\frac{\partial}{\partial t_3}\frac{\partial}{\partial t_4}\right] + \beta_2\left(y\frac{\partial}{\partial t_3}\right) + \beta_1\left(y\frac{\partial}{\partial t_4}\right)\right), \qquad (4.6)$$

to the generating function of the $\Psi(j_1m_1j_2m_2j_3m_3j_4m_4)$, namely,

$$\exp((t_1a^+) + (t_2b^+) + (t_3c^+) + (t_4d^+))\Psi_0, \qquad (4.7)$$

we shall obtain a function generating $\Psi(j_1j_2j_{12}m_{12}j_3j_4j_{34}m_{34})$. The further application of the operator

$$\exp\left(\gamma_3\left[\frac{\partial}{\partial x}\frac{\partial}{\partial y}\right] + \gamma_2\left(z\frac{\partial}{\partial x}\right) + \gamma_1\left(z\frac{\partial}{\partial y}\right)\right) \qquad (4.8)$$

then produces

$$\sum_{j_1j_2j_3j_4j_{12}j_{34}jm} [(2j_{12}+1)(2j_{34}+1)(2j+1)]^{-1/2}\phi_{jm}(z)\Phi_{j_1j_2j_{12}}(\alpha) \cdot\Phi_{j_3j_4j_{34}}(\beta)\Phi_{j_{12}j_{34}j}(\gamma)\Psi([j_1j_2]j_{12}[j_3j_4]j_{34}jm) = e^Q\Psi_0, \qquad (4.9)$$

in which

$$Q = \alpha_3[a^+b^+] + \beta_3[c^+d^+] + \gamma_3\alpha_1\beta_1[b^+d^+] + \gamma_3\alpha_1\beta_2[b^+c^+] + \gamma_3\alpha_2\beta_1[a^+d^+] + \gamma_3\alpha_2\beta_2[a^+c^+] + \gamma_2\alpha_2(za^+) + \gamma_2\alpha_1(zb^+) + \gamma_1\beta_2(zc^+) + \gamma_1\beta_1(zd^+). \qquad (4.10)$$

As an important specialization of (4.9), yielding the eigenvectors with $j = 0$, we place $\gamma_1 = \gamma_2 = 0$, and $\gamma_3 = 1$, with the result

$$\sum_{j_1j_2j_3j_4j'} (2j'+1)^{-1/2}\Phi_{j_1j_2j'}(\alpha)\Phi_{j_3j_4j'}(\beta)\Psi([j_1j_2]j'[j_3j_4]j'00) = e^R\,\Psi_0,$$

$$R = \alpha_3[a^+b^+] + \beta_3[c^+d^+] + \alpha_1\beta_1[b^+d^+] + \alpha_1\beta_2[b^+c^+] + \alpha_2\beta_1[a^+d^+] + \alpha_2\beta_2[a^+c^+], \qquad (4.11)$$

where $j' = j_{12} = j_{34}$. An analogous equation for a different mode of addition is

$$\sum_{j_1j_2j_3j_4j'} (2j''+1)^{-1/2}\Phi_{j_1j_3j''}(\alpha')\Phi_{j_2j_4j''}(\beta')\Psi([j_1j_3]j''[j_2j_4]j''00) = e^{R'}\Psi_0,$$

$$R' = \alpha_3'[a^+c^+] + \beta_3'[b^+d^+] + \alpha_1'\beta_1'[c^+d^+] + \alpha_1'\beta_2'[c^+b^+] + \alpha_2'\beta_1'[a^+d^+] + \alpha_2'\beta_2'[a^+b^+]. \qquad (4.12)$$

The transformation function connecting the two schemes is determined by[10]

$$\sum_{j_1j_2j_3j_4j'j''}(-1)^{j'+j''-j_1-j_4}\Phi_{j_1j_2j'}(\alpha)\Phi_{j_3j_4j'}(\beta)\Phi_{j_1j_3j''}(\alpha')\Phi_{j_2j_4j''}(\beta')W(j_1j_2j_3j_4;j'j'') = (e^{R}\Psi_0, e^{R'}\Psi_0), \qquad (4.13)$$

in which we have written[11]

$$([j_1j_2]j'[j_3j_4]j'0|[j_1j_3]j''[j_2j_4]j''0) = (-1)^{j'+j''-j_1-j_4} \cdot[(2j'+1)(2j''+1)]^{1/2}W(j_1j_2j_3j_4;j'j''). \qquad (4.14)$$

We now employ the theorem [Eq. (C28)]

$$(\exp\{\tfrac{1}{2}\sum_{\mu,\nu=1}^{4}\lambda^*_{\mu\nu}[A^+_\mu A^+_\nu]\}\Psi_0,\ \exp\{\tfrac{1}{2}\sum_{\mu,\nu=1}^{4}\kappa_{\mu\nu}[A^+_\mu A^+_\nu]\}\Psi_0) = [1-\tfrac{1}{2}\sum\lambda^*_{\mu\nu}\kappa_{\mu\nu}+|\lambda^*|^{1/2}|\kappa|^{1/2}]^{-2}, \qquad (4.15)$$

in which the $A_{\zeta\mu}$ are four sets of two component operators, obeying

$$[A_{\zeta\mu}, A^+_{\zeta'\nu}] = \delta_{\mu\nu}\,\delta_{\zeta\zeta'}, \qquad (4.16)$$

and $|\lambda|$, $|\kappa|$ are the determinants of the antisymmetrical matrices $\lambda_{\mu\nu}$ and $\kappa_{\mu\nu}$. For the application in question,

$$|\lambda|^{1/2}|\kappa|^{1/2} = -\,\alpha_3\beta_3\alpha_3'\beta_3',$$

$$\tfrac{1}{2}\sum_{\mu\nu}\lambda_{\mu\nu}\kappa_{\mu\nu} = \alpha_3\alpha_2'\beta_2'+\beta_3\alpha_1'\beta_1'+\alpha_3'\alpha_2\beta_2+\beta_3'\alpha_1\beta_1-\alpha_1\beta_2\alpha_1'\beta_2'+\alpha_2\beta_1\alpha_2'\beta_1'. \qquad (4.17)$$

On changing the signs of α_1 and β_3', we obtain for the generating function of the W coefficients,

$$\sum_{j_1j_2j_3j_4j'j''}\Phi_{j_1j_2j'}(\alpha)\Phi_{j_3j_4j'}(\beta)\Phi_{j_1j_3j''}(\alpha')\Phi_{j_2j_4j''}(\beta')W(j_1j_2j_3j_4;j'j'')$$
$$= [1-\alpha_3\alpha_2'\beta_2'-\beta_3\alpha_1'\beta_1'-\alpha_3'\alpha_2\beta_2-\beta_3'\alpha_1\beta_1-\alpha_1\beta_2\alpha_1'\beta_2'-\alpha_2\beta_1\alpha_2'\beta_1'+\alpha_3\beta_3\alpha_3'\beta_3']^{-2}. \qquad (4.18)$$

The symmetry properties expressed by

$$W(j_1j_2j_3j_4;j'j'') = W(j_2j_1j_4j_3;j'j'') = W(j_3j_4j_1j_2;j'j'') = W(j_1j_3j_2j_4;j''j') \qquad (4.19)$$

follow from the invariance of (4.18) under the respective substitutions:

$\alpha_1\leftrightarrow\alpha_2$, $\beta_1\leftrightarrow\beta_2$, $\alpha'\leftrightarrow\beta'$; $\alpha_1'\leftrightarrow\alpha_2'$, $\beta_1'\leftrightarrow\beta_2'$, $\alpha\leftrightarrow\beta$; $\alpha\leftrightarrow\alpha'$, $\beta\leftrightarrow\beta'$, while the more complicated transformation $(\alpha_1\alpha_2\alpha_3)\rightarrow(-\alpha_3\alpha_2\alpha_1)$, $(\alpha_1'\alpha_2'\alpha_3')\leftrightarrow(\beta_3\beta_1\beta_2)$, $(\beta_1'\beta_2'\beta_3')\rightarrow(\beta_1'\beta_3'-\beta_2')$ yields

$$W(j_1j_2j_3j_4;j'j'') = (-1)^{j'+j''-j_1-j_4}W(j'j_2j_3j'';j_1j_4). \qquad (4.20)$$

[10] For simplicity we have assumed that the parameters α, β are real. The generating function (4.18) is valid without this restriction.

[11] The W coefficient thereby defined is the same as that discussed in R.

Twenty-four equivalent forms for W are obtained by repeated use of (4.19) and (4.20).

Further characteristics of W follow from the composition properties of the transformation function (4.14), which we shall temporarily indicate by $(12, 34j'|13, 24j'')$. Thus

$$\sum_{j''} (12, 34j'|13, 24j'')(13, 24j''|12, 34j''') = \delta_{j'j'''} \tag{4.21}$$

and

$$\sum_{j''} (12, 34j'|13, 24j'')(13, 24j''|14, 23j''') = (12, 34j'|14, 23j'''). \tag{4.22}$$

All of these quantities can be expressed in terms of W. The interchange of 2 and 4, and of 3 and 4 in (4.14) yields, with the aid of (3.47),

$$(13, 24j''|14, 23j''') = (-1)^{j_2+j_3+j_4-j_1}[(2j''+1)(2j'''+1)]^{1/2} W(j_1j_4j_3j_2; j'''j''), \tag{4.23}$$

and

$$(12, 34j'|14, 23j''') = (-1)^{j'''+j_4-j_1}[(2j'+1)(2j'''+1)]^{1/2} W(j_1j_2j_4j_3; j'j'''). \tag{4.24}$$

Therefore

$$\sum_{j''} (2j''+1) W(j_1j_2j_3j_4; j'j'') W(j_1j_2j_3j_4; j'''j'') = \frac{1}{2j'+1}\delta_{j'j'''} \tag{4.25}$$

and

$$\sum_{j''} (-1)^{j'+j''+j'''+j_1+j_2+j_3+j_4}(2j''+1) W(j_1j_2j_3j_4; j'j'') W(j_1j_4j_3j_2; j'''j'') = W(j_1j_2j_4j_3; j'j'''). \tag{4.26}$$

These formulae can be combined by placing $j_2 = j_4$, $j' = j'''$ in (4.26) and, after multiplication with $2j'+1$, performing the summation with respect to j' by means of (4.25). We obtain

$$\sum_{j'} (2j'+1) W(j_1j_2j_2j_3; j'j') = \sum_{j''} (-1)^{j''-j_2+j_3}, \tag{4.27}$$

in which the values assumed by j'' are those compatible with the existence of $W(j_1j_2j_3j_2; j'j'')$, namely, $j'' \geq |j_1 - j_3|$, $j'' \leq j_1 + j_3$, $2j_2$. Accordingly,

$$\sum_{j'} (2j'+1) W(j_1j_2j_2j_3; j'j') = \begin{cases} 1, & k \text{ even} \\ 0, & k \text{ odd} \end{cases} \tag{4.28}$$

where k is the smaller of the two integers $j_1 + j_3 - |j_1 - j_3|$, $2j_2 - |j_1 - j_3|$. One of the consequences of (4.28),

$$W(j_1j_2j_20; j_2j_2) = \frac{1}{2j_2+1}, \qquad j_1 \leq 2j_2, \tag{4.29}$$

is a particular example of

$$W(j_1 j_2 j_3 0; j_3 j_2) = [(2j_2+1)(2j_3+1)]^{-1/2}, \quad |j_2 - j_3| \leq j_1 \leq j_2 + j_3, \tag{4.30}$$

which follows from (4.14) on remarking that, with $j_4 = 0$, the interchange of j_2 and j_3 simply multiplies the eigenvector with $(-1)^{j_2+j_3-j_1}$.

The relation between the W and X coefficients can be inferred from (4.14) by writing

$$\begin{aligned}
\Psi([j_1 j_2] j' [j_3 j_4] j' 0) \\
&= \sum_{m_{12}} (2j'+1)^{-1/2} (-1)^{j'-m_{12}} \Psi(j_1 j_2 j' m_{12} j_3 j_4 j' - m_{12}) \\
&= (2j'+1)^{1/2} (-1)^{j_1+j_3-j_2-j_4} \sum_{m_{12}} X(j_1 j_2 j'; m_1 m_2 - m_{12}) (-1)^{j'-m_{12}} \\
&\qquad \cdot X(j_3 j_4 j'; m_3 m_4 m_{12}) \Psi(j_1 m_1 j_2 m_2 j_3 m_3 j_4 m_4),
\end{aligned} \tag{4.31}$$

which, with the similar representation of $\Psi([j_1 j_3] j'' [j_2 j_4] j'' 0)$, yields

$$\begin{aligned}
\sum_m X(j_1 j_2 j'; m_1 m_2 - m_{12}) (-1)^{j'-m_{12}} X(j_3 j_4 j'; m_3 m_4 m_{12}) \\
\cdot X(j_1 j_3 j''; m_1 m_3 - m_{13}) (-1)^{j''-m_{13}} X(j_2 j_4 j''; m_2 m_4 m_{13}) \\
= (-1)^{j'+j''+j_1+j_4} W(j_1 j_2 j_3 j_4; j' j'').
\end{aligned} \tag{4.32}$$

The general expression obtained for W by expanding the generating function (4.18) can be cast into the form

$$\begin{aligned}
W(j_1 j_2 j_3 j_4; j' j'') = \prod_r [(n+p+1-n_r)!]^{-1/2} \prod_{r,s} [(n_r+p_s)!]^{1/2} \\
\cdot \sum (-1)^{p_3} \frac{(n+p+1)!}{\prod_{r,s} n_r! p_s!},
\end{aligned} \tag{4.33}$$

where

$$n = \sum_{r=1}^{4} n_r, \qquad p = \sum_{s=1}^{3} p_s, \tag{4.34}$$

and the summation is to be extended over the non-negative integers, n_r, p_s, for which

$$\begin{aligned}
j_2 + j' - j_1 - p_1 &= j_4 + j' - j_3 - p_2 = j_2 + j_4 - j'' - p_3 = n_1, \\
j_3 + j' - j_4 - p_1 &= j_1 + j' - j_2 - p_2 = j_1 + j_3 - j'' - p_3 = n_2, \\
j_3 + j'' - j_1 - p_1 &= j_4 + j'' - j_2 - p_2 = j_3 + j_4 - j' - p_3 = n_3, \\
j_2 + j'' - j_4 - p_1 &= j_1 + j'' - j_3 - p_2 = j_1 + j_2 - j' - p_3 = n_4.
\end{aligned} \tag{4.35}$$

The number of terms in the sum exceeds by unity the smallest of the twelve quantum number combinations, $j_2 + j' - j_1$, etc.; the sum reduces to a single term if one such combination vanishes. The choice of summation parameter is a matter of convenience.

We now return to the general problem, that of evaluating the transformation function

$$([j_1j_2]j_{12}[j_3j_4]j_{34}jm|[j_1j_3]j_{13}[j_2j_4]j_{24}jm) \equiv (-1)^{j_{12}+j_{24}-j_1-j_4}$$
$$\cdot[(2j_{12}+1)(2j_{34}+1)(2j_{13}+1)(2j_{24}+1)]^{1/2}S(j_1j_2j_3j_4;j_{12}j_{34}j_{13}j_{24};j). \quad (4.36)$$

A generating function for the S coefficient is given by[12]

$$\sum\Phi_{j_1j_2j_{12}}(\alpha)\Phi_{j_3j_4j_{34}}(\beta)\Phi_{j_{12}j_{34}j}(\gamma)\Phi_{j_1j_3j_{13}}(\alpha')$$
$$\cdot\Phi_{j_2j_4j_{24}}(\beta')\Phi_{j_{13}j_{24}j}(\gamma')S(j_1j_2j_3j_4;j_{12}j_{34}j_{13}j_{24};j)$$
$$=[1+\alpha_3\beta_3\alpha_3'\beta_3'-\gamma_3(\alpha_3'\alpha_2\beta_2+\beta_3'\alpha_1\beta_1)-\gamma_3'(\alpha_3\alpha_2'\beta_2'+\beta_3\alpha_1'\beta_1')$$
$$-\gamma_3\gamma_3'(\alpha_2\beta_1\alpha_2'\beta_1'+\alpha_1\beta_2\alpha_1'\beta_2')-\gamma_2\gamma_2'(\alpha_2\alpha_2'+\beta_3\beta_3'\alpha_1\alpha_1')-\gamma_1\gamma_1'(\beta_1\beta_1'+\alpha_3\alpha_3'\beta_2\beta_2')$$
$$-\gamma_1\gamma_2'(\beta_2\alpha_1'+\alpha_3\beta_3'\beta_1\alpha_2')+\gamma_2\gamma_1'(\alpha_1\beta_2'+\alpha_3'\beta_3\alpha_2\beta_1')]^{-2}, \quad (4.37)$$

where the sum is over all j's. The connection with the X coefficients is contained in

$$\sum X(j_1j_2j_{12};m_1m_2-m_{12})X(j_3j_4j_{34};m_3m_4-m_{34})X(j_{12}j_{34}j;m_{12}m_{34}-m)$$
$$\cdot X(j_1j_3j_{13};m_1m_3-m_{13})X(j_2j_4j_{24};m_2m_4-m_{24})X(j_{13}j_{24}j;m_{13}m_{24}-m)$$
$$=(-1)^{j_{34}+j_{13}+j_1+j_4+2j}S(j_1j_2j_3j_4;j_{12}j_{34}j_{13}j_{24};j) \quad (4.38)$$

(the sum is over all m's) and the W coefficient appears as a special example,

$$S(j_1j_2j_3j_4;j'j''j'j'';0)=[(2j'+1)(2j''+1)]^{-1/2}W(j_1j_2j_3j_4;j'j''). \quad (4.39)$$

In view of the complexity of the S coefficient we shall be content to record here only those cases that can be expressed in terms of W. This occurs whenever one of the nine quantum numbers involved in the S coefficient equals zero, which is a consequence of (4.39) and the fact that the symmetry of S is such that any of the other quantum numbers can appear in the position of j. Thus, it follows from either (4.37) or (4.38) that

$$S(j_1j_2j_3j_4;j_{12}j_{34}j_{13}j_{24};j)$$
$$=(-1)^{j_{13}+j_{24}-j_1+j_2-j_3-j_4}S(j_{12}j_1j_{34}j_3;j_2j_4jj_{13};j_{24})$$
$$=(-1)^{j_{24}+j_{34}-j_{12}-j_2-j_4-j}S(jj_{12}j_{13}j_1;j_{34}j_3j_{24}j_2;j_4), \quad (4.40)$$

which are representative of the eight permutations of this type. We obtain from (4.39) that

$$S(j_1j_2j_3j_2;j_{12}j_{34}j_{13}0;j_{13})$$
$$=(-1)^{j_{13}-j_1-j_3}[(2j_2+1)(2j_{13}+1)]^{-1/2}W(j_1j_{12}j_3j_{34};j_2j_{13}), \quad (4.41)$$

[12] This is obtained with the aid of Eq. (C30).

and

$$S(j_1j_2j_30;j_{12}j_3j_{13}j_2;j)$$
$$=(-1)^{j_3-j_{12}-j}[(2j_2+1)(2j_3+1)]^{-1/2}W(jj_{12}j_{13}j_1;j_3j_2). \quad (4.42)$$

The latter result contains the solution to the problem of three angular momenta. Expressed in terms of a transformation function, without explicit reference to the angular momentum with zero quantum number, (4.42) states that

$$([j_1j_2]j_{12}j_3jm|[j_1j_3]j_{13}j_2jm)$$
$$=(-1)^{j_{12}+j_{13}-j_1-j}[(2j_{12}+1)(2j_{13}+1)]^{1/2}W(j_1j_2j_3j;j_{12}j_{13}). \quad (4.43)$$

A slightly simpler form[13] is obtained on permuting the indices 1 and 2, together with a change in sense of addition for j_1 and j_{23},

$$([j_1j_2]j_{12}j_3jm|j_1[j_2j_3]j_{23}jm)$$
$$=[(2j_{12}+1)(2j_{23}+1)]^{1/2}W(j_1j_2jj_3;j_{12}j_{23}). \quad (4.44)$$

As a particular consequence of this result, note that, according to (4.30),

$$([j_1j_2]j_3j_30|j_1[j_2j_3]j_10)=1, \quad (4.45)$$

that is, the eigenvector for the null resultant of three angular momenta is independent of the mode of addition, provided that the order of the angular momenta is preserved. As one representation of this eigenvector we have

$$\Psi(j_1j_2j_30)=\sum_{m_3}[(2j_3+1)]^{-1/2}(-1)^{j_3+m_3}\Psi([j_1j_2]j_3-m_3j_3m_3), \quad (4.46)$$

and therefore

$$(j_1m_1j_2m_2j_3m_3|j_1j_2j_30)=[(2j_3+1)]^{-1/2}(-1)^{j_3+m_3}(j_1m_1j_2m_2|j_1j_2j_3-m_3)$$
$$=(-1)^{j_1+j_3-j_2}X(j_1j_2j_3;m_1m_2m_3), \quad (4.47)$$

in virtue of (3.40). Thus, the X coefficient, originally defined in terms of the addition of two angular momenta, now appears as characterizing three angular momenta with a null resultant.

This possibility, of replacing $\mathbf{J}_1+\mathbf{J}_2=\mathbf{J}$ with $\mathbf{J}_1+\mathbf{J}_2+\mathbf{J}_3=0$, depends upon the circumstance that the negative of an angular momentum operator is, in a certain sense, also an angular momentum operator. The commutation relations

$$\mathbf{J}\times\mathbf{J}=i\mathbf{J} \quad (4.48)$$

imply that

$$(-\mathbf{J})\times(-\mathbf{J})=-i(-\mathbf{J}), \quad (4.49)$$

[13] G. Racah, *Phys. Rev.* **63**, 367(1943).

which reassume the form (4.48) on changing the sign of i (complex, not Hermitian conjugation). Therefore

$$\mathbf{J}' = -\mathbf{J}^* \tag{4.50}$$

is an angular momentum operator. To find the eigenvectors of $\mathbf{J}'$, we notice that a rotation operator U is a function of $i\mathbf{J}$ and real angles. Therefore

$$U' = U^* \tag{4.51}$$

is the same function of $\mathbf{J}'$ that U is of $\mathbf{J}$. On taking the complex conjugate of the equation

$$U\Psi(jm') = \sum_m \Psi(jm)U^{(j)}_{mm'} \tag{4.52}$$

we obtain

$$U'\Psi^*(jm') = \sum_m \Psi^*(jm)(-1)^{m-m'}U^{(j)}_{-m-m'}, \tag{4.53}$$

with the aid of (2.41). Hence

$$\Psi'(jm) = (-1)^{j+m}\Psi^*(j-m) \tag{4.54}$$

are the eigenvectors associated with $\mathbf{J}'$.

Now observe that the following dyadic, formed from the eigenvectors of a single angular momentum,

$$(2j+1)^{-1/2}\sum_m \Psi(jm)\Psi^*(jm), \tag{4.55}$$

is unchanged by a rotation of the reference system, since

$$\sum_{mm'm''} \Psi(jm')(jm'|U|jm)(jm|U^{-1}|jm'')\Psi^*(jm'') = \sum_m \Psi(jm)\Psi^*(jm). \tag{4.56}$$

Therefore, on employing (4.54) we infer that the *vector*

$$(2j+1)^{-1/2} \sum_m \Psi(jm)(-1)^{j-m}\Psi'(j-m) \tag{4.57}$$

describes the spherically symmetrical state of two angular momenta, which is in agreement with (3.75). This is the basic example of the relationship involved in (4.47).

5. Tensor Operators

An irreducible tensor operator of rank $j(=0, \frac{1}{2}, 1, \ldots)$ is a set of $2j+1$ operators, $T(jm)$, which transforms in the following manner under a change in coordinate system,

$$UT(jm')U^{-1} = \sum_{m=-j}^{j} T(jm)U^{(j)}_{mm'}. \tag{5.1}$$

On taking the Hermitian conjugate of this equation and employing (2.41), we find that $i^{2m}T(j-m)^\dagger$ transforms in the same manner as $T(jm)$. We therefore define the Hermitian conjugate tensor $T^\dagger$ according to

$$T^\dagger(jm) = i^{2m}(Tj-m)^\dagger. \tag{5.2}$$

The tensor that is conjugate to $T^\dagger$ is then described by

$$T^{\dagger\dagger}(jm) = i^{2m}(T^\dagger(j-m))^\dagger = i^{2m}(i^{-2m}T(jm)^\dagger)^\dagger = (-1)^{2m}T(jm), \tag{5.3}$$

or

$$T^{\dagger\dagger} = (-1)^{2j}T. \tag{5.4}$$

This shows that Hermitian tensors, $T\dagger = T$, exist only for integral j,[14] and satisfy

$$T(jm) = (-1)^m T(j-m)^\dagger. \tag{5.5}$$

The product of two tensor operators transforms under coordinate system rotations according to

$$\begin{aligned} UT_1(j_1m_1')T_2(j_2m_2')U^{-1} &= (UT_1(j_1m_1')U^{-1})(UT_2(j_2m_2')U^{-1}) \\ &= \sum_{m_1m_2} T_1(j_1m_1)T_2(j_2m_2)U^{(j_1)}_{m_1m'_1}U^{(j_2)}_{m_2m'_2}. \end{aligned} \tag{5.6}$$

It follows from (3.80) that

$$\sum_{m_1m_2} T_1(j_1m_1)T_2(j_2m_2)(j_1m_1j_2m_2|j_1j_2jm) \equiv T(j_1j_2jm) \tag{5.7}$$

obeys

$$UT(j_1j_2jm')U^{-1} = \sum_m T(j_1j_2jm)U^{(j)}_{mm'}, \tag{5.8}$$

and is therefore an irreducible tensor of rank j.

For a tensor operator applied to an angular momentum eigenvector we have, analogously,

$$\begin{aligned} U(T(j_1m_1')\Psi(j_2m_2')) &= (UT(j_1m_1')U^{-1})(U\Psi(j_2m_2')) \\ &= \sum_{m_1m_2} T(j_1m_1)\Psi(j_2m_2)U^{(j_1)}_{m_1m'_1}U^{(j_2)}_{m_2m'_2} \end{aligned} \tag{5.9}$$

so that

$$\sum_{m_1m_2} T(j_1m_1)\Psi(j_2m_2)(j_1m_1j_2m_2|j_1j_2jm) \equiv \Phi(j_1j_2jm) \tag{5.10}$$

obeys

$$U\Phi(j_1j_2jm') = \sum_m \Phi(j_1j_2jm)U^{(j)}_{mm'} \tag{5.11}$$

[14] It is similarly impossible to identify the $\Psi'(jm)$ of (4.54) with $\Psi(jm)$, for all m, if j is half-integral.

and is therefore an angular momentum eigenvector with quantum numbers j and m.

The magnetic quantum number dependence of tensor operator matrix elements is contained in the last statement. On introducing explicitly the additional quantum numbers necessary to form a complete set, we are led to write

$$\sum_{qm'} T(kq)\Psi(\gamma'j'm')(kqj'm'|kj'jm) = \sum_{\gamma} \Psi(\gamma jm)(2j+1)^{-1/2}[\gamma j|T^{(k)}|\gamma'j'], \quad (5.12)$$

where we have employed different letters for the tensor operator indices in order to simplify the notation. It follows from (5.12) that[15]

$$\begin{aligned}(\gamma jm|T(kq)|\gamma'j'm') &= (2j+1)^{-1/2}[\gamma j|T^{(k)}|\gamma'j'](kj'jm|kqj'm') \\ &= (-1)^{k-j'+m}[\gamma j|T^{(k)}|\gamma'j']X(jkj';\,-mqm'). \quad (5.13)\end{aligned}$$

As an alternative derivation of the latter result,[16] we remark that

$$\begin{aligned}(\gamma jm|T(kq)|\gamma'j'm') &= (U\Psi(\gamma jm),\, UT(kq)U^{-1}U\Psi(\gamma'j'm')) \\ &= \sum_{m''q'm'''} (\gamma jm''|T(kq')|\gamma'j'm''')(-1)^{m-m''}U^{(j)}_{-m''-m}U^{(k)}_{q'q}U^{(j')}_{m'''m'}. \quad (5.14)\end{aligned}$$

An integration with respect to ω then yields, according to (3.82),

$$\begin{aligned}(\gamma jm|T(kq)|\gamma'j'm') = \sum_{m''q'm'''} &(-1)^{m-m''}X(jkj';\,-mqm')X(jkj';\,-m''q'm''') \\ &\cdot(\gamma jm''|T(kq')|\gamma'j'm'''), \quad (5.15)\end{aligned}$$

which is (5.13), with

$$[\gamma j|T^{(k)}|\gamma'j'] = \sum_{mqm'} (-1)^{j'-k-m}X(jkj';\,-mqm')(\gamma jm|T(kq)|\gamma'j'm'). \quad (5.16)$$

According to the definition of the Hermitian conjugate tensor, we have

$$\begin{aligned}(\gamma jm|T^{\dagger}(kq)|\gamma'j'm') &= i^{2q}(\gamma'j'm'|T(k-q)|\gamma jm)^* \\ &= i^{2q}(-1)^{k-j+m'}[\gamma'j'|T^{(k)}|\gamma j]^*X(jkj';\,-mqm'), \quad (5.17)\end{aligned}$$

or

$$[\gamma j|T^{(k)\dagger}|\gamma'j'] = i^{2j'-2j}[\gamma'j'|T^{(k)}|\gamma j]^*, \quad (5.18)$$

in which use has been made of the X coefficient properties contained in (3.45) and (3.46). For a Hermitian tensor, this result reads

$$[\gamma j|T^{(k)}|\gamma'j'] = (-1)^{j-j'}[\gamma'j'|T^{(k)}|\gamma j]^*. \quad (5.19)$$

[15] The relation between the rectangular bracket symbol and the analogous quantity defined in R is

$$[\gamma j|T^{(k)}|\gamma'j'] = (-1)^{k+j-j'}(\gamma j\|T^{(k)}\|\gamma'j').$$

[16] This is the method employed by E. Wigner, "Gruppentheorie und ihre Anwendung auf die Quantenmechanik der Atomspektrem" (Braunschweig, 1931), p. 263.

If the tensor operators T_1 and T_2 of (5.7) refer to the same dynamical variables, we may write

$$(\gamma jm|T(k_1k_2kq)|\gamma' j'm') = (-1)^{k-j'+m}[\gamma j|T^{(k)}(k_1k_2)|\gamma' j']X(jkj';\ -mqm'), \tag{5.20}$$

where in view of (5.16),

$$[\gamma j|T^{(k)}(k_1k_2)|\gamma' j'] = \sum_{mqm'} (-1)^{j'-k-m}X(jkj';\ -mqm')(k_1q_1k_2q_2|k_1k_2kq) \cdot \sum_{\gamma'' j'' m''} (\gamma jm|T_1(k_1q_1)|\gamma'' j'' m'')(\gamma'' j'' m''|T_2(k_2q_2)|\gamma' j'm'). \tag{5.21}$$

The resulting magnetic quantum number summation, involving four X coefficients, can be identified with a W coefficient,

$$[\gamma j|T^{(k)}(k_1k_2)|\gamma' j'] = (2k+1)^{\frac{1}{2}} \sum_{\gamma'' j''} W(k_1k_2jj';\ kj'') \cdot [\gamma j|T_1^{(k_1)}|\gamma'' j''][\gamma'' j''|T_2^{(k_2)}|\gamma' j']. \tag{5.22}$$

When T_1 and T_2 are tensor operators associated with different dynamical variables, so that

$$[T_1, J_2] = [T_2, J_1] = 0, \tag{5.23}$$

we have

$$(\gamma j_1j_2jm|T(k_1k_2kq)|\gamma' j_1'j_2'j'm') = (-1)^{k-j'+m}[\gamma j_1j_2j|T^{(k)}(k_1k_2)|\gamma' j_1'j_2'j'] \cdot X(jkj';\ -mqm'). \tag{5.24}$$

Here

$$[\gamma j_1j_2j|T^{(k)}(k_1k_2)|\gamma' j_1'j_2'j'] = \sum (-1)^{j'-k-m}X(jkj';\ -mqm') \cdot (k_1q_1k_2q_2|k_1k_2kq)(j_1j_2jm|j_1m_1j_2m_2)(j_1'j_2'j'm'|j_1'm_1'j_2'm_2') \cdot (\gamma j_1m_1|T_1(k_1q_1)|\gamma'' j_1'm_1')(\gamma'' j_2m_2|T_2(k_2q_2)|\gamma' j_2'm_2'), \tag{5.25}$$

where the sum is over all m variables. This magnetic quantum number summation, involving six X coefficients, can be identified with an S coefficient,

$$[\gamma j_1j_2j|T^{(k)}(k_1k_2)|\gamma' j_1'j_2'j'] = [(2j+1)(2j'+1)(2k+1)]^{1/2}(-1)^{j_2+j'_1-j'-k_1+k}S(j_1j_2j_1'j_2';\ jj'k_1k_2;\ k) \cdot \sum_{\gamma''} [\gamma j_1|T_1^{(k_1)}|\gamma'' j_1'][\gamma'' j_2|T_2^{(k_2)}|\gamma' j_2']. \tag{5.26}$$

Special examples which require only the W coefficient are

$$[\gamma j_1j_2j|T^{(0)}(k_1k_1)|\gamma' j_1'j_2'j] = \left[\frac{2j+1}{2k_1+1}\right]^{1/2}(-1)^{j_2+j'_1-j-k_1}W(j_1j_2j_1'j_2';\ jk_1) \cdot \sum_{\gamma''} [\gamma j_1|T_1^{(k_1)}|\gamma'' j_1'][\gamma'' j_2|T_2^{(k_1)}|\gamma' j_2'], \tag{5.27}$$

$$[\gamma j_1 j_2 j | T_1^{(k)} | \gamma' j_1' j_2 j']$$
$$= [(2j+1)(2j'+1)]^{1/2}(-1)^{j_2+k-j_1-j'} W(j_1 j j_1' j'; j_2 k)[\gamma j_1 | T_1^{(k)} | \gamma' j_1'], \quad (5.28)$$

and

$$[\gamma j_1 j_2 j | T_2^{(k)} | \gamma' j_1 j_2' j']$$
$$= [(2j+1)(2j'+1)]^{1/2}(-1)^{j_1+k-j'_2-j} W(j_2 j j_2' j'; j_1 k)[\gamma j_2 | T_2^{(k)} | \gamma' j_2']. \quad (5.29)$$

Further relations connecting the S and W coefficients can be deduced from these results. We shall illustrate this for the simpler situation in which only W is involved. We multiply the two scalar operators[17]

$$T^{(0)}(k_1 k_1) = \sum_{q_1} (2k_1+1)^{-1/2} T_1(k_1 q_1)(-1)^{k_1-q_1} T_2(k_1 - q_1), \quad (5.30)$$

and

$$T^{(0)}(k_2 k_2) = \sum_{q_2} (2k_2+1)^{-1/2} T_1(k_2 q_2)(-1)^{k_2-q_2} T_2(k_2 - q_2), \quad (5.31)$$

to obtain

$$T^{(0)}(k_1 k_1) T^{(0)}(k_2 k_2) = \sum_{q_1 q_2} [(2k_1+1)(2k_2+1)]^{-1/2} T_1(k_1 q_1) T_1(k_2 q_2)$$
$$\cdot (-1)^{k_1+k_2-q_1-q_2} T_2(k_1 - q_1) T_2(k_2 - q_2). \quad (5.32)$$

On writing

$$T_1(k_1 q_1) T_1(k_2 q_2) = \sum_{kq} T_1(k_1 k_2 k q)(k_1 k_2 k q | k_1 q_1 k_2 q_2) \quad (5.33)$$

and

$$T_2(k_1 - q_1) T_2(k_2 - q_2) = \sum_{kq} T_2(k_1 k_2 k - q)(-1)^{k_1+k_2-k}(k_1 k_2 k q | k_1 q_1 k_2 q_2), \quad (5.34)$$

this becomes

$$T^{(0)}(k_1 k_1) T^{(0)}(k_2 k_2)$$
$$= \sum_{kq} [(2k_1+1)(2k_2+1)]^{-1/2} T_1(k_1 k_2 k q)(-1)^{k-q} T_2(k_2 k_2 k - q)$$
$$\equiv \sum_k \left[\frac{2k+1}{(2k_1+1)(2k_2+1)}\right]^{1/2} T^{(0)}([k_1 k_2]k[k_1 k_2]k). \quad (5.35)$$

A matrix element of this equation, when evaluated with the aid of (5.22) and (5.27), yields the information that

$$W(j_1 j_2 j_1'' j_2''; j k_1) W(j_1'' j_2'' j_1' j_2'; j k_2)$$
$$= (-1)^{j''_1+j''_2-j} \sum_k (2k+1)(-1)^{k_1+k_2-k} W(j_1 j_2 j_1' j_2'; jk)$$
$$\cdot W(k_1 k_2 j_1 j_1'; k j_1'') W(k_1 k_2 j_2 j_2'; k j_2''). \quad (5.36)$$

[17] Here T_1 and T_2 are functions of different dynamical variables.

Tensor operators can be constructed from the spin creation and annihilation operators. Thus, consider the operator

$$e^{\zeta_+(za^+)+\zeta_-[za]} = \sum_{kq\alpha} \phi_{kq}(z)\phi_{k\alpha}(\zeta)t(kq\alpha), \tag{5.37}$$

formed from the commuting quantities (za^+) and $[za]$. On subjecting this to a unitary transformation, we find

$$e^{\zeta_+(za'^+)+\zeta_-[za']} = \sum_{kq\alpha} \phi_{kq}(z)\phi_{k\alpha}(\zeta)Ut(kq\alpha)U^{-1}, \tag{5.38}$$

where the transformed creation and annihilation operators are described by (2.14). Now, according to (2.19), we have

$$(za'^+) = (z'a^+), \qquad [za'] = [z'a], \qquad z' = uz, \tag{5.39}$$

in which the second statement stems from the fact that a_- and a_+ transform in the same way as a_+^+ and $-a_-^+$. Therefore,

$$\begin{aligned} e^{\zeta_+(za'^+)+\zeta_-[za']} &= \sum_{kq\alpha} \phi_{kq}(uz)\phi_{k\alpha}(\zeta)t(kq\alpha) \\ &= \sum_{kqq'\alpha} U_{qq'}^{(k)}\phi_{kq'}(z)\phi_{k\alpha}(\zeta)t(kq\alpha), \end{aligned} \tag{5.40}$$

on employing (2.21). We have thereby shown that

$$Ut(kq'\alpha)U^{-1} = \sum_q t(kq\alpha)U_{qq'}^{(k)}. \tag{5.41}$$

On taking the Hermitian conjugate of (5.37) and making the substitution $z_+^* \to z_-$, $z_-^* \to -z_+$, $\zeta_-^* \to \zeta_+$, $\zeta_+^* \to -\zeta_-$, which restores this generating operator to its original form, we find that

$$t(kq\alpha) = (-1)^{q+\alpha}t(k-q-\alpha)^\dagger. \tag{5.42}$$

Accordingly, the adjoint tensor is given by

$$t^\dagger(kq\alpha) = i^{2\alpha}t(kq-\alpha). \tag{5.43}$$

The significance of α can be appreciated from

$$\frac{(za^+)^{k+\alpha}[za]^{k-\alpha}}{[(k+\alpha)!(k-\alpha)!]^{1/2}} = \sum_q \phi_{kq}(z)t(kq\alpha), \tag{5.44}$$

namely, 2α is the excess of creation with respect to annihilation operators. Therefore, if $t(kq\alpha)$ is applied to an angular momentum eigenvector with quantum number j', it will produce an eigenvector with quantum number j, such that

$$\alpha = j - j'. \tag{5.45}$$

To evaluate the matrix elements of $t(kq\alpha)$, we examine

$$(e^{(xa^+)}\Psi_0, e^{\zeta_+(za^+)+\zeta_-[za]}\cdot e^{(ya^+)}\Psi_0)$$
$$= \sum_{jmj'm'kq\alpha} \phi_{jm}(x^*)(jm|t(kq\alpha)|j'm')\phi_{j'm'}(y)\phi_{kq}(z)\phi_{k\alpha}(\zeta) = e^{(x^*y)+\zeta_+(x^*z)+\zeta_-[zy]}. \tag{5.46}$$

The substitution $x_+^* \to x_-$, $x_-^* \to -x_+$ places this in the form

$$\sum_{jmj'm'kq\alpha} (-1)^{j-m}\phi_{j-m}(x)(jm|t(kq\alpha)|j'm')\phi_{j'm'}(y)\phi_{kq}(z)\phi_{k\alpha}(\zeta) = e^{\zeta_-[zy]+[yx]-\zeta_+[xz]}, \tag{5.47}$$

and comparison with (3.42) shows that

$$e^{\zeta_-[zy]+[yx]-\zeta_+[xz]} = \sum_{jmj'm'kq} \phi_{j-m}(x)\phi_{kq}(z)\phi_{j'm'}(y)\phi_{k,\,j-j'}(\zeta)$$
$$\cdot(-1)^{k+j-j'}\left[\frac{(j+j'+k+1)!}{(j+j'-k)!}\right]^{1/2} X(jkj';\ -mqm'). \tag{5.48}$$

Therefore

$$(jm|t(kq\alpha)|j'm) = \delta_{\alpha,\,j-j'}(-1)^{k-j'+m}\left[\frac{(j+j'+k+1)!}{(j+j'-k)!}\right]^{1/2} X(jkj';\ -mqm), \tag{5.49}$$

or

$$[j|t^{(k)}(\alpha)|j'] = \delta_{\alpha,\,j-j'}\left[\frac{(j+j'+k+1)!}{(j+j'-k)!}\right]^{1/2}. \tag{5.50}$$

Of particular interest are the operators with $\alpha = 0$ (k integral),

$$\frac{(za)^k[za]^k}{k!} = \sum_q \phi_{kq}(z)t(kq0). \tag{5.51}$$

Indeed

$$-(za^+)[za] = \mathfrak{a}\cdot\mathbf{J}, \tag{5.52}$$

where $\mathfrak{a}$ is a null vector,

$$\mathfrak{a}\cdot\mathfrak{a} = 0, \tag{5.53}$$

with the components

$$\mathfrak{a}_1 = -z_+^2 + z_-^2, \qquad \mathfrak{a}_2 = -i(z_+^2 + z_-^2), \qquad \mathfrak{a}_3 = 2z_+z_-. \tag{5.54}$$

It is well known that if $\mathbf{r}$ is a position vector, $(\mathfrak{a}\cdot\mathbf{r})^k$ is a spherical harmonic of order k,

$$\frac{(\mathfrak{a}\cdot\mathbf{r})^k}{2^k k!} = \left[\frac{4\pi}{2k+1}\right]^{1/2} \sum_q \phi_{kq}(z)Y_{kq}(\mathbf{r}), \tag{5.55}$$

where $Y_{kq}(\mathbf{r})$, which usually designates a surface spherical harmonic, here includes the factor r^k. Accordingly, we write

$$\frac{(\mathfrak{a}\cdot\mathbf{J})^k}{2^k k!} = \left[\frac{4\pi}{2k+1}\right]^{1/2} \sum_q \phi_{kq}(z)Y_{kq}(\mathbf{J}), \tag{5.56}$$

in which $Y_{kq}(\mathbf{J})$ differs from the analogous $Y_{kq}(\mathbf{r})$ only in that the order of factors is significant. With this notation, we have

$$t(kq0) = \left[\frac{4\pi}{2k+1}\right]^{1/2} (-2)^k Y_{kq}(\mathbf{J}), \tag{5.57}$$

and

$$[j|Y^{(k)}(\mathbf{J})|j] = \left[\frac{2k+1}{4\pi}\right]^{1/2} \cdot (-\tfrac{1}{2})^k \cdot \left[\frac{(2j+k+1)!}{(2j-k)!}\right]^{1/2}. \tag{5.58}$$

Notice also that the tensor $t(kq0)$ is Hermitian, according to (5.43), so that the operator harmonics satisfy

$$Y_{kq}(\mathbf{J})^{\dagger} = (-1)^q Y_{k-q}(\mathbf{J}). \tag{5.59}$$

The matrix elements of the tensor operator

$$Y(k_1k_2kq) = \sum_{q_1q_2} Y_{k_1q_1}(\mathbf{J}) Y_{k_2q_2}(\mathbf{J})(k_1q_1k_2q_2|k_1k_2kq) \tag{5.60}$$

are described by

$$[j|Y^{(k)}(k_1k_2)|j] = (2k+1)^{1/2} W(k_1k_2jj;kj)[j|Y^{(k_1)}|j][j|Y^{(k_2)}|j], \tag{5.61}$$

in view of (5.22). With respect to their effect on an eigenvector with quantum number j, one can assert that

$$Y(k_1k_2kq) = Y_{kq}(\mathbf{J}) \frac{[j|Y^{(k)}(k_1k_2|j]}{[j|Y^{(k)}|j]}, \tag{5.62}$$

which becomes a generally valid operator equation on replacing $j(j+1)$ with $\mathbf{J}^2$. Hence

$$\sum_{q_1q_2} Y_{k_1q_1}(\mathbf{J}) Y_{k_2q_2}(\mathbf{J})(k_1q_1k_2q_2|k_1k_2kq) = Y_{kq}(\mathbf{J})(2k+1)^{1/2} W(k_1k_2jj;kj) \frac{[j|Y^{(k_1)}|j][j|Y^{(k_2)}|j]}{[j|Y^{(k)}|j]}. \tag{5.63}$$

The example of this result for $k=0$ can be written

$$\sum_q Y_{kq}(\mathbf{J}) Y_{kq}(\mathbf{J})^{\dagger} = \frac{1}{2j+1}[j|Y^{(k)}|j]^2 = \frac{2k+1}{4\pi}\frac{1}{4^k}\frac{(2j+k+1)!}{(2j+1)(2j-k)!} \tag{5.64}$$

in which we have employed

$$W(kkjj;0j) = (-1)^k[(2j+1)(2k+1)]^{-1/2}. \tag{5.65}$$

One can easily exhibit the right side of (5.64) as a function of $j(j+1)$, and thus obtain the operator equation

$$\sum_q Y_{kq}(\mathbf{J}) Y_{kq}(\mathbf{J})^{\dagger} = \frac{2k+1}{4\pi}\{\mathbf{J}^2\}^k,$$

$$\{\mathbf{J}^2\}^k \equiv \prod_{n=0}^{k-1}\left[\mathbf{J}^2 - \frac{n}{2}\left(\frac{n}{2}+1\right)\right]. \tag{5.66}$$

The structure of the operator $\{\mathbf{J}^2\}^k$ can also be inferred from the two requirements that it annihilate any eigenvector with $j < \frac{1}{2}k$, and that it simplify to the kth power of $\mathbf{J}^2$ as j becomes very large.

We return to (5.63), displayed in the form

$$Y_{k_1q_1}(\mathbf{J})Y_{k_2q_2}(\mathbf{J}) = \left[\frac{(2k_1+1)(2k_2+1)}{4\pi}\right]^{1/2} \sum_{kq} Y_{kq}(\mathbf{J}) f_{k_1k_2k}(\mathbf{J}^2)(k_1k_2kq|k_1q_1k_2q_2), \tag{5.67}$$

where

$$\left[\frac{\{\mathbf{J}^2\}^k}{\{\mathbf{J}^2\}^{k_1}\{\mathbf{J}^2\}^{k_2}}\right]^{1/2} f_{k_1k_2k}(\mathbf{J}^2) = (2j+1)^{1/2} W(k_1k_2jj;\,kj). \tag{5.68}$$

The analogous equation for $Y_{k_2q_2}(\mathbf{J})Y_{k_1q_1}(\mathbf{J})$ differs from (5.67) only by the inclusion of the factor $(-1)^{k_1+k_2-k}$, as follows from (3.47). The addition and subtraction of these two equations then yields

$$\{Y_{k_1q_1}(\mathbf{J}),\, Y_{k_2q_2}(\mathbf{J})\} = \left[\frac{(2k_1+1)(2k_2+1)}{4\pi}\right]^{1/2} \cdot \sum_{\text{even},\,kq} Y_{kq}(\mathbf{J}) f_{k_1k_2k}(\mathbf{J}^2)(k_1k_2kq|k_1q_1k_2q_2) \tag{5.69}$$

and

$$[Y_{k_1q_1}(\mathbf{J}),\, Y_{k_2q_2}(\mathbf{J})] = \left[\frac{(2k_1+1)(2k_2+1)}{4\pi}\right]^{1/2} \cdot \sum_{\text{odd},\,kq} Y_{kq}(\mathbf{J}) f_{k_1k_2k}(\mathbf{J}^2)(k_1k_2kq|k_1q_1k_2q_2), \tag{5.70}$$

where the parity referred to is that of $k_1 + k_2 - k$. In the latter equation we have the commutation properties of these operator functions of $\mathbf{J}$.

As an elementary application of (5.70), we take its trace for the states with quantum number j. In view of the null trace possessed by a commutator, we infer that the trace of $Y_{kq}(\mathbf{J})$ vanishes for every k that can occur in (5.70). Since these k values are $|k_1 - k_2| + 1,\ |k_1 - k_2| + 3, \ldots, k_1 + k_2 - 1$, we obtain[18]

$$\operatorname{tr}^{(j)} Y_{kq}(\mathbf{J}) = 0, \qquad k > 0, \tag{5.71}$$

[18] This theorem is easily proved for an arbitrary tensor operator by taking the trace of (5.1) for states with a given j, and integrating with respect to ω

$$\sum_{m=-j}^{j} (\gamma jm|T(kq)|\gamma' jm) = 0, \qquad k > 0.$$

Of course, k must be integral if the individual matrix elements are not to vanish.

or

$$\frac{1}{2j+1}\,\mathrm{tr}^{(j)} Y_{kq}(\mathbf{J}) = \delta_{k0}. \tag{5.72}$$

With the aid of this result, the trace of (5.67) is evaluated as

$$\frac{1}{2j+1}\,\mathrm{tr}^{(j)} Y_{k_1q_1}(\mathbf{J})^\dagger Y_{k_2q_2}(\mathbf{J}) = \frac{1}{4\pi}\{j(j+1)\}^{k_1}\,\delta_{k_1k_2}\delta_{q_1q_2}, \tag{5.73}$$

which expresses the orthogonality of the operator harmonics. The multiplication of (5.67) with $Y_{k_3q_3}(\mathbf{J})$ then yields

$$\frac{1}{2j+1}\,\mathrm{tr}^{(j)} Y_{k_1q_1}(\mathbf{J}) Y_{k_2q_2}(\mathbf{J}) Y_{k_3q_3}(\mathbf{J}) = \left[\prod_i \frac{2k_i+1}{4\pi}\{j(j+1)\}^{k_i}\right]^{1/2}$$
$$\cdot X(k;q)(-1)^{k_1-k_2}(2j+1)^{1/2}W(k_1k_2jj;k_3j). \tag{5.74}$$

A comparison with (3.85) shows that, in the limit of large j,

$$(-1)^{k_1-k_2}(2j+1)^{1/2}W(k_1k_2jj;k_3j) \to X(k_1k_2k_3;000). \tag{5.75}$$

Turning to tensor operators formed from two angular momenta, we remark that, for matrix elements diagonal in j,

$$\sum_{q_1q_2} Y_{k_1q_1}(\mathbf{J}_1) Y_{k_2q_2}(\mathbf{J}_2)(k_1q_1k_2q_2|k_1k_2kq)$$
$$= Y_{kq}(\mathbf{J})(2j+1)(2k+1)^{1/2}(-1)^{j_1+j_2-j-k_1+k}S(j_1j_2j_1j_2;jjk_1k_2;k)$$
$$\cdot\frac{[j_1|Y^{(k_1)}|j_1][j_2|Y^{(k_2)}|j_2]}{[j|Y^{(k)}|j]}. \tag{5.76}$$

No such restriction is required for the special example

$$\sum_q Y_{kq}(\mathbf{J}_1)^\dagger Y_{kq}(\mathbf{J}_2) = (-1)^{j_1+j_2-j}W(j_1j_2j_1j_2;jk)[j_1|Y^{(k)}|j_1][j_2|Y^{(k)}|j_2]. \tag{5.77}$$

In terms of the Legendre polynomial operator defined by

$$\sum_q Y_{kq}(\mathbf{J}_1)^\dagger Y_{kq}(\mathbf{J}_2) = \frac{2k+1}{4\pi}P_k(\mathbf{J}_1,\mathbf{J}_2), \tag{5.78}$$

the latter equation can be written

$$[\{\mathbf{J}_1^2\}^k\{\mathbf{J}_2^2\}^k]^{-1/2}P_k(\mathbf{J}_1,\mathbf{J}_2) = (-1)^{j_1+j_2-j}(2j_1+1)^{1/2}(2j_2+1)^{1/2}W(j_1j_2j_1j_2;jk), \tag{5.79}$$

which indicates the limiting form of the right side for large j_1, j_2, and j. The simple result obtained for $j=0$ can be expressed as

$$P_k(\mathbf{J},-\mathbf{J}) = (-1)^k\{\mathbf{J}^2\}^k. \tag{5.80}$$

A multiplication theorem for the Legendre operator is obtained from the observation that

$$\frac{2k_1+1}{4\pi}\frac{2k_2+1}{4\pi}P_{k_1}(\mathbf{J}_1, \mathbf{J}_2)P_{k_2}(\mathbf{J}_1, \mathbf{J}_2) = \sum_{k,q}(Y_{k_2q_2}(\mathbf{J}_1)Y_{k_1q_1}(\mathbf{J}_1))^\dagger Y_{k_1q_1}(\mathbf{J}_2)Y_{k_2q_2}(\mathbf{J}_2), \quad (5.81)$$

namely,[19]

$$P_{k_1}(\mathbf{J}_1, \mathbf{J}_2)P_{k_2}(\mathbf{J}_1, \mathbf{J}_2) = \sum_k (2k+1)P_k(\mathbf{J}_1, \mathbf{J}_2)(-1)^{k_1+k_2-k}f_{k_1k_2k}(\mathbf{J}_1^2)f_{k_1k_2k}(\mathbf{J}_2^2). \quad (5.82)$$

On placing $k_2 = 1$, we obtain a simple recurrence relation from which the Legendre operators can be constructed successively, starting with

$$P_0(\mathbf{J}_1, \mathbf{J}_2) = 1. \quad (5.83)$$

The coefficients in the recurrence relation can be computed from

$$W(k1jj; k+1j) = -W(k+1\,1jj; kj) = \left[\frac{4j(j+1)-k^2-2k}{4j(j+1)(2j+1)}\frac{k+1}{(2k+1)(2k+3)}\right]^{1/2}, \quad (5.84)$$

and

$$\sum_k (2k+1)(2j+1)[W(k_11jj; kj)]^2 = 1. \quad (5.85)$$

Thus

$$f_{k1k+1}(\mathbf{J}^2) = \left[\frac{k+1}{(2k+1)(2k+3)}\right]^{1/2},$$

$$f_{k1k-1}(\mathbf{J}^2) = -\left[\frac{k}{(2k-1)(2k+1)}\right]^{1/2}\left(\mathbf{J}^2 - \frac{k^2-1}{4}\right),$$

$$(f_{k1k}(\mathbf{J}^2))^2 = \tfrac{1}{4}\frac{k(k+1)}{2k+1}, \quad (5.86)$$

and therefore,

$$\left(\mathbf{J}_1\cdot\mathbf{J}_2 + \frac{k(k+1)}{4}\right)P_k(\mathbf{J}_1, \mathbf{J}_2) = \frac{k+1}{2k+1}P_{k+1}(\mathbf{J}_1, \mathbf{J}_2) + \frac{k}{2k+1}\left(\mathbf{J}_1^2 - \frac{k^2-1}{4}\right)\left(\mathbf{J}_2^2 - \frac{k^2-1}{4}\right)P_{k-1}(\mathbf{J}_1, \mathbf{J}_2). \quad (5.87)$$

[19] This is a particular example of the theorem on the product of two W coefficients [Eq. (5.36)].

As the first few Legendre operators, obtained in succession from (5.87) with $k=0, 1, 2$, we have

$$\begin{aligned}
&P_1(\mathbf{J}_1, \mathbf{J}_2) = \mathbf{J}_1 \cdot \mathbf{J}_2, \\
&P_2(\mathbf{J}_1, \mathbf{J}_2) = \tfrac{3}{2}\mathbf{J}_1 \cdot \mathbf{J}_2(\mathbf{J}_1 \cdot \mathbf{J}_2 + \tfrac{1}{2}) - \tfrac{1}{2}\mathbf{J}_1^2\mathbf{J}_2^2, \\
&P_3(J_1, J_2) = \tfrac{5}{2}\mathbf{J}_1 \cdot \mathbf{J}_2(\mathbf{J}_1 \cdot \mathbf{J}_2 + \tfrac{1}{2})(\mathbf{J}_1 \cdot \mathbf{J}_2 + \tfrac{3}{2}) - \tfrac{5}{6}(\mathbf{J}_1 \cdot \mathbf{J}_2 + \tfrac{3}{2})\mathbf{J}_1^2\mathbf{J}_2^2 \\
&\qquad\qquad - \tfrac{2}{3}\mathbf{J}_1 \cdot \mathbf{J}_2(\mathbf{J}_1^2 - \tfrac{3}{4})(\mathbf{J}_2^2 - \tfrac{3}{4}).
\end{aligned} \tag{5.88}$$

A useful check upon these results is afforded by (5.80).

A statement analogous to (5.62) can be made for an arbitrary tensor operator; as far as matrix elements diagonal in j are concerned,

$$T(kq) = Y_{kq}(\mathbf{J}) \frac{[j|T^{(k)}|j]}{[j|Y^{(k)}|j]}. \tag{5.89}$$

The coefficient in this relation can be expressed in other ways. Thus, we have

$$\sum_q Y_{kq}(\mathbf{J})^\dagger T(kq) = \sum_q Y_{kq}(\mathbf{J})^\dagger Y_{kq}(\mathbf{J}) \frac{[j|T^{(k)}|j]}{[j|Y^{(k)}|j]}, \tag{5.90}$$

which leads to the projection rule

$$T(kq) \rightarrow \frac{4\pi}{2k+1} Y_{kq}(\mathbf{J}) \frac{1}{\{\mathbf{J}^2\}^k} \sum_{q'} Y_{kq'}(\mathbf{J})^\dagger T(kq'), \tag{5.91}$$

for isolating the part of a tensor operator that contributes to matrix elements diagonal in j. Alternatively, we consider the particular matrix element

$$(jj|T(k0)|jj) = (jj|Y_{k0}(\mathbf{J})|jj) \frac{[j|T^{(k)}|j]}{[j|Y^{(k)}|j]}. \tag{5.92}$$

Now

$$(jj|Y_{k0}(\mathbf{J})|jj) = (-1)^k[j|Y^{(k)}|j]X(jkj; -j0j) = \left[\frac{2k+1}{4\pi}\right]^{1/2} \frac{1}{2^k} \frac{(2j)!}{(2j-k)!}, \tag{5.93}$$

so that, for matrix elements diagonal in j,

$$T(kq) = \left[\frac{4\pi}{(2k+1)}\right]^{1/2} Y_{kq}(\mathbf{J})\, 2^k \frac{(2j-k)!}{(2j)!} (jj|T(k0)|jj). \tag{5.94}$$

Appendix A

We shall describe a method which produces simultaneously the eigenvalues and eigenvectors of the angular momentum operators. Consider for this purpose the unitary operator

$$V = \exp(i\chi\tfrac{1}{2}n + i\phi J_3), \tag{A1}$$

which has the eigenvalues $\exp(ij\chi + im\phi)$. The operator V can be interpreted as

$$V = \sum_{jm} [\exp(ij\chi + im\phi)]P(jm), \tag{A2}$$

where $P(jm)$, the projection operator for the state with the indicated eigenvalues, is represented in terms of the corresponding eigenvector by the dyadic

$$P(jm) = \Psi(jm)\Psi(jm)^*. \tag{A3}$$

Accordingly, if V can be constructed and displayed in the form (A2), we shall have achieved our goal.

We write

$$V = \exp(\tfrac{1}{2}i(\gamma_+ n_+ + \gamma_- n_-)),$$

$$\gamma_+ = \chi + \phi, \qquad \gamma_- = \chi - \phi, \tag{A4}$$

and deduce the differential equations

$$\frac{\partial}{\partial \gamma_\zeta} V = \tfrac{1}{2} i a_\zeta^+ a_\zeta V$$

$$= \tfrac{1}{2} i [\exp(\tfrac{1}{2} i \gamma_\zeta)] a_\zeta^+ V a_\zeta, \tag{A5}$$

with the aid of

$$V^{-1} a_\zeta V = (\exp(\tfrac{1}{2} i \gamma_\zeta)) a_\zeta. \tag{A6}$$

The latter can be verified by differentiation,

$$\frac{\partial}{\partial \gamma_\zeta} V^{-1} a_\zeta V = \tfrac{1}{2} i V^{-1} [a_\zeta, n_\zeta] V = \tfrac{1}{2} i V^{-1} a_\zeta V, \tag{A7}$$

or from the general theorem

$$a_\zeta f(n_\zeta) = f(n_\zeta + 1) a_\zeta. \tag{A8}$$

In virtue of the operator ordering in (A5), the solution of these equations which reduces to unity for $\gamma_\zeta = 0$ is given by

$$V = \exp\{(e^{\frac{1}{2}i\gamma_+} - 1)a_+^+; a_+ + (e^{\frac{1}{2}i\gamma_-} - 1)a_-^+; a_-\} \tag{A9}$$

where

$$\exp(\lambda a^+; a) = \sum \frac{\lambda^n}{n!} (a^+)^n (a)^n \tag{A10}$$

is a correspondingly ordered form of the exponential. We write this solution as

$$V = \exp(\sum_\zeta e^{\frac{1}{2}i\gamma_\zeta} a_\zeta^+; P_0; a_\zeta), \tag{A11}$$

which is intended to indicate that

$$P_0 = \exp(-a_+^+; a_+ - a_-^+; a_-) = \exp(-(a^+; a)) \tag{A12}$$

is to be inserted between the powers of a_ζ^+ and a_ζ in the ordered operator expansion of V:

$$V = \sum_{n_+, n_- = 0}^{\infty} \{\exp[\tfrac{1}{2}i(n_+\gamma_+ + n_-\gamma_-)]\} \frac{(a_+^+)^{n_+}(a_-^+)^{n_-}}{(n_+!n_-!)^{1/2}} P_0 \frac{(a_+)^{n_+}(a_-)^{n_-}}{(n_+!n_-!)^{1/2}}. \tag{A13}$$

We have thus obtained the form (A2), with

$$j = \tfrac{1}{2}(n_+ + n_-), \qquad m = \tfrac{1}{2}(n_+ - n_-), \qquad n_+, n_- = 0, 1, 2, \ldots, \tag{A14}$$

and

$$P(jm) = \phi_{jm}(a^+) P_0 \phi_{jm}(a), \tag{A15}$$

in which we have employed the notation

$$\phi_{jm}(a^+) = \frac{(a_+^+)^{j+m}(a_-^+)^{j-m}}{[(j+m)!(j-m)!]^{1/2}}. \tag{A16}$$

In terms of the eigenvector Ψ_0, defined by

$$P_0 = \Psi_0 \Psi_0^*, \tag{A17}$$

the angular momentum eigenvectors are exhibited as

$$\Psi(jm) = \phi_{jm}(a^+)\Psi_0. \tag{A18}$$

The fundamental property of $\Psi_0 = \Psi(00)$ is deduced from

$$[a_\zeta, P_0] = (\partial/\partial a_\zeta^+)P_0 = -P_0 a_\zeta, \tag{A19}$$

or

$$a_\zeta P_0 = 0, \tag{A20}$$

namely

$$a_\zeta \Psi_0 = 0. \tag{A21}$$

The simple generating function for the eigenvectors, (1.16), can also be obtained by noting that

$$\begin{aligned}(\Psi(jm), e^{(xa^+)}\Psi_0) &= (\Psi_0, \phi_{jm}(a) e^{(xa^+)}\Psi_0) \\ &= \phi_{jm}(x).\end{aligned} \tag{A22}$$

Indeed,

$$\begin{aligned}e^{(xa^+)}\Psi_0 &= \sum_{jm} \Psi(jm)(\Psi(jm), e^{(xa^+)}\Psi_0) \\ &= \sum_{jm} \phi_{jm}(x)\Psi(jm).\end{aligned} \tag{A23}$$

Appendix B

The ordered operator

$$A = \exp(za; a^+), \qquad [a, a^+] = 1, \tag{B1}$$

satisfies

$$[a, A] = (\partial/\partial a^+)A = zaA,$$

$$[a^+, A] = -(\partial/\partial a)A = -Aza^+, \tag{B2}$$

or

$$(1-z)aA = Aa, \qquad a^+A = (1-z)Aa^+. \tag{B3}$$

Therefore

$$\frac{\partial}{\partial z}A = aAa^+ = \frac{1}{1-z}Aaa^+ = \frac{1}{1-z}A + \frac{1}{1-z}Aa^+a$$

$$= \frac{1}{1-z}A + \frac{1}{(1-z)^2}a^+Aa, \tag{B4}$$

the solution of which implies the ordered operator identity,

$$\exp(za; a^+) = \frac{1}{1-z}\exp\left(\frac{z}{1-z}a^+; a\right). \tag{B5}$$

A particular consequence of this relation

$$\exp(za; a^+)\cdot\Psi_0 = \frac{1}{1-z}\Psi_0, \qquad a\Psi_0 = 0, \tag{B6}$$

is derived directly in the text (Eq. (2.35)). The properties of A contained in (B3) are also displayed in the generalizations

$$\exp(za; a^+)\cdot f(a^+) = \frac{1}{1-z}f\left(\frac{a^+}{1-z}\right)\exp\left(\frac{z}{1-z}a^+; a\right),$$

$$f(a)\exp(za; a^+) = \exp\left(\frac{z}{1-z}a^+; a\right)\frac{1}{1-z}f\left(\frac{a}{1-z}\right). \tag{B7}$$

The particular examples of these identities provided by

$$\exp(za; a^+)\cdot(a^+)^r = \frac{(a^+)^r}{(1-z)^{r+1}}\exp\left(\frac{z}{1-z}a^+; a\right) \tag{B8}$$

and

$$a^r\exp(za; a^+) = \exp\left(\frac{z}{1-z}a^+; a\right)\frac{a^r}{(1-z)^{r+1}} \tag{B9}$$

are operator forms of the Laguerre polynomial generating functions. Thus, if we place $a^+ = x$, $a = \partial/\partial x$, and let both sides of (B8) operate upon e^{-x}, we obtain

$$\sum_{n=0}^{\infty} \frac{z^n}{n!}\left(\frac{\partial}{\partial x}\right)^n x^{n+r}e^{-x} = \frac{x^r}{(1-z)^{r+1}} \exp\left(\frac{z}{1-z}x; \frac{\partial}{\partial x}\right) e^{-x}$$

$$= \frac{x^r}{(1-z)^{r+1}} \exp\left(-\frac{z}{1-z}x\right) e^{-x}, \qquad \text{(B10)}$$

or

$$\frac{\exp\left(-\frac{zx}{1-z}\right)}{(1-z)^{r+1}} = \sum_{n=0}^{\infty} z^n L_n^{(r)}(x), \qquad \text{(B11)}$$

where

$$L_n^{(r)}(x) = \frac{1}{n!} x^{-r}e^x\left(\frac{d}{dx}\right)^n (x^{n+r}e^{-x}). \qquad \text{(B12)}$$

A similar procedure applied to (B9) yields

$$\sum_{n=0}^{\infty} \frac{z^n}{n!}\left(\frac{\partial}{\partial x}\right)^{n+r} x^n e^{-x} = \exp\left(\frac{z}{1-z}x; \frac{\partial}{\partial x}\right)\frac{\left(\frac{\partial}{\partial x}\right)^r}{(1-z)^{r+1}} e^{-x}$$

$$= \frac{(-1)^r}{(1-z)^{r+1}} \exp\left(-\frac{z}{1-z}x\right) e^{-x}, \qquad \text{(B13)}$$

which proves the equivalence between (B12) and

$$L_n^{(r)}(x) = \frac{(-1)^r}{n!} e^x \left(\frac{d}{dx}\right)^{n+r} (x^n e^{-x}). \qquad \text{(B14)}$$

Another example of an ordered operator identity involves the cylinder function [Eq. (1.36)]

$$F_r(z) = r! z^{-r/2} I_r(2z^{1/2}) = \frac{r!}{2\pi i}\oint dt \frac{\exp\left(t + \frac{z}{t}\right)}{t^{r+1}}. \qquad \text{(B15)}$$

We have

$$a^r F_r(za; a^+) = \frac{r!}{2\pi i}\oint dt \frac{e^t}{t^{r+1}} a^r \exp\left(\frac{z}{t}a; a^+\right)$$

$$= \frac{r!}{2\pi i} e^z \oint dt \frac{e^{t-z}}{(t-z)^{r+1}} \exp\left(\frac{z}{t-z}a^+; a\right) a^r$$

$$= e^z F_r(za^+; a) a^r, \qquad \text{(B16)}$$

and similarly

$$F_r(za; a^+)(a^+)^r = e^z (a^+)^r F_r(za^+; a). \qquad \text{(B17)}$$

From these identities we obtain the Laguerre polynomial generating function

$$\sum_{n=0}^{\infty} \frac{r!}{(n+r)!} z^n L_n^{(r)}(x) = e^z F_r(-zx). \tag{B18}$$

Appendix C

It is our purpose in this section to evaluate a class of scalar products, the simplest illustration of which is

$$T^{(2)} = (\exp(\lambda[a^+b^+] + \xi_+(xa^+) + \xi_-(xb^+))\Psi_0,$$
$$\exp(\kappa[a^+b^+] + \eta_+(ya^+) + \eta_-(yb^+))\Psi_0). \tag{C1}$$

Differentiation with respect to ξ_+^* yields

$$(\partial/\partial\xi_+^*)T^{(2)} = (e\cdots\Psi_0, (x^*a)e\cdots\Psi_0)$$
$$= \eta_+(x^*y)T^{(2)} + \kappa([xb]e\cdots\Psi_0, e\cdots\Psi_0), \tag{C2}$$

or

$$(1-\lambda^*\kappa)(\partial/\partial\xi_+^*)T^{(2)} = \eta_+(x^*y)T^{(2)}. \tag{C3}$$

The solution of this, and analogous equations, is

$$T^{(2)} = \exp\left(\frac{(\xi^*\eta)(x^*y}{1-\lambda^*\kappa}\right)\cdot T_0^{(2)}, \tag{C4}$$

where

$$T_0^{(2)} = (\exp(\lambda[a^+b^+])\Psi_0, \exp(\kappa[a^+b^+])\Psi_0)$$
$$= (\Psi_0, \exp(\lambda^*\kappa(a^+;a))\Psi_0) = \frac{1}{(1-\lambda^*\kappa)^2}, \tag{C5}$$

in view of the simple generalization of (B6)

$$\exp(z(a;a^+))\Psi_0 = \frac{1}{(1-z)^2}\Psi_0. \tag{C6}$$

Therefore

$$T^{(2)} = \frac{1}{(1-\lambda^*\kappa)^2}\exp\left(\frac{(\xi^*\eta)(x^*y)}{1-\lambda^*\kappa}\right). \tag{C7}$$

One can prove, in a similar manner, that

$$(\exp(\lambda[a^+b^+] + (x_1a^+) + (x_2b^+))\Psi_0, \exp(\kappa[a^+b^+] + (y_1a^+) + (y_2b^+))\Psi_0)$$
$$= \frac{1}{(1-\lambda^*\kappa)^2}\exp\left(\frac{1}{1-\lambda^*\kappa}\{(x_1^*y_1) + (x_2^*y_2) + \kappa[x_1^*x_2^*] + \lambda^*[y_1y_2]\}\right). \tag{C8}$$

The general member of the class exemplified by (C1) is

$$T^{(n)} = (\exp(\tfrac{1}{2}\sum_{\mu\nu}\lambda_{\mu\nu}[A_\mu^+ A_\nu^+] + \sum_\mu \xi_\mu(xA_\mu^+))\Psi_0,$$
$$\exp(\tfrac{1}{2}\sum_{\mu\nu}\kappa_{\mu\nu}[A_\mu^+ A_\nu^+] + \sum_\mu \eta_\mu(yA_\mu^+))\Psi_0), \quad \text{(C9)}$$

where the A_μ are n sets of two-component operators, obeying

$$[A_{\zeta\mu}, A_{\zeta'\nu}^+] = \delta_{\mu\nu}\delta_{\zeta\zeta'}, \quad \text{(C10)}$$

while $\lambda_{\mu\nu}$ and $\kappa_{\mu\nu}$ form antisymmetrical matrices. Following the same procedure, we evaluate

$$(\partial/\partial\xi_\mu^*)T^{(n)} = (e\cdots\Psi_0, (x^*A_\mu)e\cdots\Psi_0)$$
$$= \eta_\mu(x^*y)T^{(n)} + \sum_\nu \kappa_{\mu\nu}([xA_\nu]e\cdots\Psi_0, e\cdots\Psi_0), \quad \text{(C11)}$$

whence

$$(\partial/\partial\xi_\mu^*)T^{(n)} + \sum_\beta \kappa_{\mu\beta}\lambda_{\beta\nu}^*(\partial/\partial\xi_\nu^*)T^{(n)} = \eta_\mu(x^*y)T^{(n)}. \quad \text{(C12)}$$

The solution of this equation can be expressed in a matrix notation as

$$T^{(n)} = \exp\left[(x^*y)\sum_{\mu\nu}\xi_\mu^*\left(\frac{1}{1+\kappa\lambda^*}\right)_{\mu\nu}\eta_\nu\right]T_0^{(n)}, \quad \text{(C13)}$$

where

$$T_0^{(n)} = (\exp(\tfrac{1}{2}\sum_{\mu\nu}\lambda_{\mu\nu}[A_\mu^+ A_\nu^+])\Psi_0, \exp(\tfrac{1}{2}\sum_{\mu\nu}\kappa_{\mu\nu}[A_\mu^+ A_\nu^+])\Psi_0)$$
$$= (\Psi_0, Q\Psi_0), \quad \text{(C14)}$$

and

$$Q = \exp(\tfrac{1}{2}\sum_{\mu\nu}\lambda_{\mu\nu}^*[A_\mu A_\nu])\cdot\exp(\tfrac{1}{2}\sum_{\mu\nu}\kappa_{\mu\nu}[A_\mu^+ A_\nu^+]). \quad \text{(C15)}$$

To evaluate $T_0^{(n)}$, we employ the following properties of Q,

$$(\partial/\partial\lambda_{\mu\nu}^*)Q = [A_\mu A_\nu]Q, \quad \text{(C16)}$$

and

$$[xA_\mu]Q - Q[xA_\mu] = -Q\sum_\nu \kappa_{\mu\nu}(xA_\nu^+), \quad \text{(C17a)}$$

$$Q(xA_\mu^+) - (xA_\mu^+)Q = \sum_\nu \lambda_{\mu\nu}^*[xA_\nu]Q, \quad \text{(C17b)}$$

in which x is an arbitrary constant spinor. One can combine (C17a, b) into

$$\sum_\nu (1+\kappa\lambda^*)_{\mu\nu}[xA_\nu]Q = Q[xA_\mu] - \sum_\beta \kappa_{\mu\beta}(xA_\beta^+)Q, \quad \text{(C18)}$$

or

$$[xA_\nu]Q = \sum_{\nu\beta}\left(\frac{1}{1+\kappa\lambda^*}\right)_{\nu\beta}Q[xA_\beta] - \sum_{\nu\beta}\left(\frac{1}{1+\kappa\lambda^*}\kappa\right)_{\nu\beta}(xA_\beta^+)Q. \quad \text{(C19)}$$

Therefore

$$[A_\mu A_\nu]Q = \sum_\beta \left(\frac{1}{1+\kappa\lambda^*}\right)_{\nu\beta} [A_\mu Q A_\beta] - \sum_\beta \left(\frac{1}{1+\kappa\lambda^*}\right)_{\nu\beta} (A_\beta^+ A_\mu)Q$$

$$-2\left(\frac{1}{1+\kappa\lambda^*}\kappa\right)_{\nu\mu} Q, \tag{C20}$$

from which we obtain $(A_\mu \Psi_0 = 0)$

$$(\partial/\partial\lambda_{\mu\nu}^*)T_0^{(n)} = -2\left(\frac{1}{1+\kappa\lambda^*}\kappa\right)_{\nu\mu} T_0^{(n)}. \tag{C21}$$

Thus, with respect to changes in the matrix λ^*, we have

$$\delta \log T_0^{(n)} = \tfrac{1}{2}\sum_{\mu\nu} \delta\lambda_{\mu\nu}^*(\partial/\partial\lambda_{\mu\nu}^*) \log T_0^{(n)} = -\mathrm{tr}\left(\frac{1}{1+\kappa\lambda^*}\kappa\delta\lambda^*\right). \tag{C22}$$

On comparing this with the theorem on differentiation of a determinant,

$$\delta \log |A| = \mathrm{tr}(A^{-1}\delta A), \tag{C23}$$

we obtain the desired general evaluation,

$$T_0^{(n)} = \frac{1}{|1+\kappa\lambda^*|}. \tag{C24}$$

A recurrence relation for $T_0^{(n)}$ can also be established with the aid of (C13). Thus, we have

$$T_0^{(n)} = (\exp(\tfrac{1}{2}\sum_{\mu\nu=1}^{n-1} \lambda_{\mu\nu}[A_\mu^+ A_\nu^+] + \sum_{\mu\nu=1}^{n-1} \lambda_{n\mu}[A_n^+ A_\mu^+])\Psi_0,$$

$$\exp(\tfrac{1}{2}\sum_{\mu\nu=1}^{n-1} \kappa_{\mu\nu}[A_\mu^+ A_\nu^+] + \sum_{\mu\nu=1}^{n-1} \kappa_{n\mu}[A_n^+ A_\mu^+])\Psi_0)$$

$$= \left(\Psi_0, \exp\left((A_n; A_n^+)\sum_{\mu\nu}\lambda_{n\mu}^*\left(\frac{1}{1+\kappa'\lambda'^*}\right)_{\mu\nu}\kappa_{n\nu}\right)\Psi_0\right)T_0^{(n-1)}$$

$$= \left[1 - \sum_{\mu\nu}\lambda_{n\mu}^*\left(\frac{1}{1+\kappa'\lambda'^*}\right)_{\mu\nu}\kappa_{n\nu}\right]^{-2} T_0^{(n-1)}, \tag{C25}$$

in which κ' and λ' designate the matrices of dimensionality $n-1$.

The actual construction of the $T_0^{(n)}$ can be performed without detailed calculations. It follows from (C24) and (C25) that $T_0^{(n)}$ has the form of the inverse square of a power series in the components of λ^* and κ, where the last terms of the series, $(-1)^{\frac{1}{2}n}|\lambda^*|^{1/2}|\kappa|^{1/2}$, vanishes for n odd. Thus, beginning with

$$T_0^{(2)} = [1 - \lambda_{12}^*\kappa_{12}]^{-2} = [1 - |\lambda^*|^{1/2}|\kappa|^{1/2}]^{-2}, \tag{C26}$$

We infer that $T_0^{(3)}$ has the same structure, suitably extended for the additional dimension,

$$T_0^{(3)} = [1 - (\lambda_{12}^*\kappa_{12} + \lambda_{23}^*\kappa_{23} + \lambda_{31}^*\kappa_{31})]^{-2}$$
$$= [1 - \tfrac{1}{2} \sum_{\mu\nu=1}^{3} \lambda_{\mu\nu}^*\kappa_{\mu\nu}]^{-2}, \tag{C27}$$

and therefore

$$T_0^{(4)} = [1 - \tfrac{1}{2} \sum_{\mu\nu=1}^{4} \lambda_{\mu\nu}^*\kappa_{\mu\nu} + |\lambda^*|^{1/2}|\kappa|^{1/2}]^{-2}, \tag{C28}$$

where

$$|\lambda|^{1/2} = \lambda_{12}\lambda_{34} + \lambda_{23}\lambda_{14} + \lambda_{31}\lambda_{24}$$
$$= \tfrac{1}{8} \sum_{\mu\nu\sigma\tau=1}^{4} \varepsilon_{\mu\nu\sigma\tau}\lambda_{\mu\nu}\lambda_{\sigma\tau}, \tag{C29}$$

and ε is the completely antisymmetric symbol. For the last indication of this general procedure we remark that, as the extension of (C28), we have

$$T_0^{(5)} = [1 - \tfrac{1}{2} \sum_{\mu\nu=1}^{5} \lambda_{\mu\nu}^*\kappa_{\mu\nu} + \sum_{\mu\nu=1}^{5} (\lambda^*)_\alpha(\kappa)_\alpha]^{-2}, \tag{C30}$$

in which

$$(\lambda)_\alpha = \tfrac{1}{8} \sum_{\mu\nu\sigma\tau=1}^{5} \varepsilon_{\alpha\mu\nu\sigma\tau}\lambda_{\mu\nu}\lambda_{\sigma\tau}. \tag{C31}$$

Symmetry Properties of the Wigner $9j$ Symbol

H. A. JAHN AND J. HOPE
Mathematics Department, The University, Southampton, England
(Received July 9, 1953)

The 72 symmetry relations of the Wigner $9j$ symbol

$$\begin{Bmatrix} j_1 j_2 j_3 \\ j_4 j_5 j_6 \\ j_7 j_8 j_9 \end{Bmatrix} = \frac{\langle (j_1j_2)j_3, (j_4j_5)j_6, j_9m \,|\, (j_1j_4)j_7, (j_2j_5)j_8, j_9m\rangle}{\{(2j_3+1)(2j_6+1)(2j_7+1)(2j_8+1)\}^{\frac{1}{2}}}$$

are given. The group of symmetry may be generated by the following elements: (i) an odd permutation of the rows or columns of the symbol multiples it by $(-1)^R$, where $R=j_1+j_2+j_3+j_4+j_5+j_6+j_7+j_8+j_9$; (ii) a reflection of the symbol in either of the two diagonals leaves it invariant. These results are conveniently deduced from the Schwinger generating function for the $9j$ symbol. In an addendum a $12j$ symbol is defined and some of its properties discussed.

AN important concept in the theory of complex spectra is the Wigner $9j$ symbol.[1] This is defined (Wigner) by

$$\begin{Bmatrix} j_1 j_2 j_3 \\ j_4 j_5 j_6 \\ j_7 j_8 j_9 \end{Bmatrix} = \frac{\langle (j_1j_2)j_3, (j_4j_5)j_6, j_9m \,| \; (j_1j_4)j_7, (j_2j_5)j_8, j_9m\rangle}{\{(2j_3+1)(2j_6+1)(2j_7+1)(2j_8+1)\}^{\frac{1}{2}}}, \quad (1)$$

the relation with the χ function[2] of Hope or the S function[3] of Schwinger being

$$\chi(abcd, ef, gh, k) = \{(2e+1)(2f+1)(2g+1)(2h+1)\}^{\frac{1}{2}} \begin{Bmatrix} a & b & e \\ c & d & f \\ g & h & k \end{Bmatrix}, \quad (1a)$$

$$S(j_1j_2j_3j_4;\, j_{12}j_{34}j_{13}j_{24};\, j) = (-1)^{j_1+j_4-j_{12}-j_{24}} \begin{Bmatrix} j_1 j_2 j_{12} \\ j_3 j_4 j_{34} \\ j_{13} j_{24} j \end{Bmatrix}. \quad (1b)$$

Its numerical evaluation is best carried out from the following expression (Jahn,[4] Hope,[2] Wigner[1]) in terms of Wigner $6j$ symbols (or Racah W functions):

$$\begin{Bmatrix} j_1 j_2 j_3 \\ j_4 j_5 j_6 \\ j_7 j_8 j_9 \end{Bmatrix} = \sum_x (-1)^{2x}(2x+1) \times \begin{Bmatrix} j_1 j_4 j_7 \\ j_8 j_9 x \end{Bmatrix} \begin{Bmatrix} j_2 j_5 j_8 \\ j_4 x\; j_6 \end{Bmatrix} \begin{Bmatrix} j_3 j_6 j_9 \\ x\; j_1 j_2 \end{Bmatrix}, \quad (2)$$

the relation between the Wigner $6j$ symbol and the Racah W function being

$$\begin{Bmatrix} j_1 j_2 j_3 \\ j_4 j_5 j_6 \end{Bmatrix} = (-1)^{j_1+j_2+j_4+j_5} W(j_1j_2j_5j_4;\, j_3j_6). \quad (3)$$

Related useful formulas,[4] involving Wigner $3j$ symbols [see (7) below], are:

$$\sum_{\substack{m_1m_2m_3m_4 \\ M_{13}M_{24}}} \langle j_1m_1j_2m_2|J_{12}M_{12}\rangle\langle j_3m_3j_4m_4|J_{34}M_{34}\rangle \times\langle J_{13}M_{13}J_{24}M_{24}|JM\rangle\langle j_1m_1j_3m_3|J_{13}M_{13}\rangle \times\langle j_2m_2j_4m_4|J_{24}M_{24}\rangle = \langle J_{12}M_{12}J_{34}M_{34}|JM\rangle \times\chi(j_1j_2j_3j_4, J_{12}J_{34}, J_{13}J_{24}, J), \quad (3a)$$

$$\sum_{\substack{m_1m_2m_3m_4 \\ M_{12}M_{34}M_{13}M_{24}}} \langle j_1m_1j_2m_2|J_{12}M_{12}\rangle\langle j_3m_3j_4m_4|J_{34}M_{34}\rangle \times\langle J_{12}M_{12}J_{34}M_{34}|JM\rangle\langle j_1m_1j_3m_3|J_{13}M_{13}\rangle \times\langle j_2m_2j_4m_4|J_{24}M_{24}\rangle\langle J_{13}M_{13}J_{24}M_{24}|J'M'\rangle = \delta(J,J')\delta(M,M')\chi(j_1j_2j_3j_4, J_{12}J_{34}, J_{13}J_{24}, J). \quad (3b)$$

The $9j$ symbol, apart from its immediate application to the transformation from one coupling scheme to another, i.e., from ls to jj coupling (see Hope[2]) or to the theory of fractional parentage coefficients (see Elliott[5,6]), has its most important application to the evaluation of the matrix elements of tensor products of tensor operators taken between vector-coupled states. This application is contained in the formula (Hope, Wigner, Schwinger):

$$\langle\gamma j_1j_2j\|T^{(k)}(k_1k_2)\|\gamma'j_1'j_2'j'\rangle = \{(2j+1)(2j'+1)(2k+1)\}^{\frac{1}{2}}\sum_{\gamma''}\langle\gamma j_1\|T_1^{(k_1)}\|\gamma''j_1'\rangle \times\langle\gamma''j_2\|T_2^{(k_2)}\|\gamma'j_2'\rangle \begin{Bmatrix} j_1 & j_2 & j \\ j_1' & j_2' & j' \\ k_1 & k_2 & k \end{Bmatrix}, \quad (4)$$

for the amplitude matrix of the tensor product

$$T(k_1k_2kq) = \sum_{q_1q_2}\langle k_1q_1k_2q_2|kq\rangle T_1(k_1q_1)T_2(k_2q_2), \quad (5)$$

of tensor operators $T_1(k_1q_1)$, $T_2(k_2q_2)$, where the operators T_1 do not act on that part of the system characterized by j_2 and similarly the T_2 operators do not act on the part of the system characterized by j_1 (the variables γ'' enter if the operators act on other parts

---

[1] E. P. Wigner, "On the matrices which reduce the Kronecker products of representations of simply reducible groups," hectographed paper, Princeton, 1951 (unpublished).

[2] J. Hope, Ph.D. thesis, London University, 1952 (unpublished).

[3] J. Schwinger, "On Angular Momentum," Report NYO-3071, Nuclear Development Associates, Inc., White Plains, New York, 1952 (unpublished).

[4] The senior author (H.A.J.) derived formula (2) and the corresponding expressions (3a), (b) in terms of six Wigner $3j$ symbols in Birmingham in 1949 in connection with a β-decay problem of Dr. T. H. R. Skyrme. (For an account of the derivation see reference 2, pp. 15–17.)

[5] J. P. Elliott, Ph.D. thesis, London University, 1952 (unpublished).

[6] J. P. Elliott, Proc. Roy. Soc. (London) **A218**, 345 (1953).

of the system). We are using here the tensor amplitude or double-barred matrix elements, as defined by Racah,[7] rather than the slightly different definition of Schwinger (Wigner's definition is identical with that of Racah), *viz.*,

$$\begin{aligned}\langle \alpha jm|T(k_1k_2kq)|\alpha'j'm'\rangle &= (-1)^{2k}\langle j'm'kq|jm\rangle(2j+1)^{-\frac{1}{2}}\langle\alpha j\|T^{(k)}(k_1k_2)\|\alpha'j'\rangle \quad (6)\\ &= (-1)^{j-m}\begin{pmatrix} j & k & j' \\ -m & q & m'\end{pmatrix}\langle \alpha j\|T^{(k)}(k_1k_2)\|\alpha'j'\rangle, \quad (6a)\end{aligned}$$

where the Wigner $3j$ symbol $\begin{pmatrix} j_1 j_2 j_3 \\ m_1m_2m_3\end{pmatrix}$ is defined (Wigner, Schwinger) by

$$\begin{aligned}\langle j_1m_1j_2m_2|jm\rangle &\equiv \langle j_1m_1j_2m_2|j_1j_2jm\rangle\\ &= (-1)^{j_1-j_2+m}(2j+1)^{\frac{1}{2}}\begin{pmatrix} j_1 & j_2 & j\\ m_1 & m_2 & -m\end{pmatrix}, \quad (7)\end{aligned}$$

and has the simple symmetry relations described by Wigner and Schwinger [odd permutation of the columns multiplies the symbol by $(-1)^J$, where $J=j_1+j_2+j_3$; a simultaneous change in sign of m_1, m_2, and m_3 has the same effect]. In Hope's notation the above tensor product is denoted by $(T^{k_1}{}_{(1)}\odot_q{}^kT^{k_2}{}_{(2)})$; Talmi[8] uses the notation $[T_1^{(k_1)}\times T_2^{(k_2)}]_q^{(k)}$; the notation used here is that of Schwinger.

Some of the symmetry relations of the $9j$ symbol have been considered by Hope and also by Schwinger. It is the purpose of this note to point out that the complete group of symmetry (consisting of 72 symmetry relations) may be generated by the following elements: (i) An odd permutation of the rows or columns of the symbol multiples it by $(-1)^R$, where $R=\sum_{i=1}^{9} j_i$; i.e., using the abbreviation

$$\begin{Bmatrix}123\\456\\789\end{Bmatrix} \equiv \begin{Bmatrix}j_1j_2j_3\\j_4j_5j_6\\j_7j_8j_9\end{Bmatrix},$$

we have

$$\begin{aligned}(-1)^R\begin{Bmatrix}123\\456\\789\end{Bmatrix} &= \begin{Bmatrix}213\\546\\879\end{Bmatrix} = \begin{Bmatrix}321\\654\\987\end{Bmatrix} = \begin{Bmatrix}132\\465\\798\end{Bmatrix}\\ &= \begin{Bmatrix}456\\123\\789\end{Bmatrix} = \begin{Bmatrix}789\\456\\123\end{Bmatrix} = \begin{Bmatrix}123\\789\\456\end{Bmatrix}. \quad (8)\end{aligned}$$

(ii) A reflection of the symbol in either of the two diagonals leaves it invariant; i.e.,

$$\begin{Bmatrix}123\\456\\789\end{Bmatrix} = \begin{Bmatrix}147\\258\\369\end{Bmatrix} = \begin{Bmatrix}963\\852\\741\end{Bmatrix}. \quad (9)$$

The full group of permutations (subgroup of the symmetric group S_9), with the elements arranged in classes is as follows, the star on the symbol for the class denoting that the permutations in the starred class multiply the $9j$ symbol by $(-1)^R$. [*Note*: in the following symbols for permutations the decomposition into cycles is indicated by the separate rows of the symbol, i.e., $\begin{pmatrix}1452\\3768\end{pmatrix}$ means the permutation (1452) (3768) in the usual notation.]

C_1: I,

C_2: $\begin{Bmatrix}15\\38\\67\end{Bmatrix}, \begin{Bmatrix}18\\35\\49\end{Bmatrix}, \begin{Bmatrix}16\\29\\57\end{Bmatrix}, \begin{Bmatrix}27\\34\\59\end{Bmatrix}, \begin{Bmatrix}19\\26\\48\end{Bmatrix}, \begin{Bmatrix}24\\37\\68\end{Bmatrix},$

C_3: $\begin{Bmatrix}12\\48\\57\\69\end{Bmatrix}, \begin{Bmatrix}13\\49\\58\\67\end{Bmatrix}, \begin{Bmatrix}14\\26\\35\\89\end{Bmatrix}, \begin{Bmatrix}15\\24\\36\\78\end{Bmatrix}, \begin{Bmatrix}16\\25\\34\\79\end{Bmatrix}, \begin{Bmatrix}17\\29\\38\\56\end{Bmatrix}, \begin{Bmatrix}18\\27\\39\\45\end{Bmatrix}, \begin{Bmatrix}19\\28\\37\\46\end{Bmatrix}, \begin{Bmatrix}23\\47\\59\\68\end{Bmatrix},$

C_4: $\begin{Bmatrix}123\\456\\789\end{Bmatrix}, \begin{Bmatrix}132\\465\\798\end{Bmatrix}, \begin{Bmatrix}147\\258\\369\end{Bmatrix}, \begin{Bmatrix}174\\285\\396\end{Bmatrix},$

C_5: $\begin{Bmatrix}159\\267\\348\end{Bmatrix}, \begin{Bmatrix}195\\276\\384\end{Bmatrix}, \begin{Bmatrix}168\\249\\357\end{Bmatrix}, \begin{Bmatrix}186\\294\\375\end{Bmatrix},$

C_6: $\begin{pmatrix}168\\239745\end{pmatrix}, \begin{pmatrix}348\\179652\end{pmatrix}, \begin{pmatrix}267\\139854\end{pmatrix}, \begin{pmatrix}294\\136587\end{pmatrix}, \begin{pmatrix}159\\287463\end{pmatrix}, \begin{pmatrix}357\\128964\end{pmatrix}, \begin{pmatrix}186\\254793\end{pmatrix}, \begin{pmatrix}384\\125697\end{pmatrix}, \begin{pmatrix}276\\145893\end{pmatrix}, \begin{pmatrix}249\\178563\end{pmatrix}, \begin{pmatrix}195\\236478\end{pmatrix}, \begin{pmatrix}375\\146982\end{pmatrix},$

$C_7{}^*$: $\begin{Bmatrix}12\\45\\78\end{Bmatrix}, \begin{Bmatrix}13\\46\\79\end{Bmatrix}, \begin{Bmatrix}23\\56\\89\end{Bmatrix}, \begin{Bmatrix}14\\25\\36\end{Bmatrix}, \begin{Bmatrix}17\\28\\39\end{Bmatrix}, \begin{Bmatrix}47\\58\\69\end{Bmatrix},$

$C_8{}^*$: $\begin{pmatrix}123\\486759\end{pmatrix}, \begin{pmatrix}456\\183729\end{pmatrix}, \begin{pmatrix}789\\153426\end{pmatrix}, \begin{pmatrix}147\\268359\end{pmatrix}, \begin{pmatrix}258\\167349\end{pmatrix}, \begin{pmatrix}369\\157248\end{pmatrix}, \begin{pmatrix}132\\495768\end{pmatrix}, \begin{pmatrix}465\\192738\end{pmatrix}, \begin{pmatrix}798\\162435\end{pmatrix}, \begin{pmatrix}174\\295386\end{pmatrix}, \begin{pmatrix}285\\194376\end{pmatrix}, \begin{pmatrix}396\\184275\end{pmatrix},$

$C_9{}^*$: $\begin{pmatrix}1254\\3867\end{pmatrix}, \begin{pmatrix}1364\\2957\end{pmatrix}, \begin{pmatrix}1948\\2365\end{pmatrix}, \begin{pmatrix}1287\\3594\end{pmatrix}, \begin{pmatrix}1397\\2684\end{pmatrix}, \begin{pmatrix}1675\\2398\end{pmatrix}, \begin{pmatrix}1926\\4785\end{pmatrix}, \begin{pmatrix}1835\\4796\end{pmatrix}, \begin{pmatrix}2734\\5896\end{pmatrix},$

$C_{10}{}^*$: $\begin{pmatrix}1452\\3768\end{pmatrix}, \begin{pmatrix}1463\\2759\end{pmatrix}, \begin{pmatrix}1849\\2563\end{pmatrix}, \begin{pmatrix}1782\\3495\end{pmatrix}, \begin{pmatrix}1793\\2486\end{pmatrix}, \begin{pmatrix}1576\\2893\end{pmatrix}, \begin{pmatrix}1629\\4587\end{pmatrix}, \begin{pmatrix}1538\\4697\end{pmatrix}, \begin{pmatrix}2437\\5698\end{pmatrix}.$

The elements (i), forming the class $C_7{}^*$, generate a self-conjugate subgroup of 36 elements consisting of

[7] G. Racah, Phys. Rev. **62**, 438 (1942).
[8] I. Talmi, Phys. Rev. **89**, 1065 (1953).

the classes C_1, C_3, C_4, C_5, C_7^*, C_8^*. Addition of either of the elements (ii) in the class C_2 then gives rise to the whole group of 72 elements.

These symmetry relations may be used, as pointed out by Hope and Schwinger, to bring a j from any one place to any other place in the symbol; thus when one of the j's is zero (as in the important application to the matrix elements of the scalar product of two tensor operators of equal rank), we have the relations:

$$\begin{Bmatrix} a & b & e \\ c & d & e \\ f & f & 0 \end{Bmatrix} = \begin{Bmatrix} 0 & e & e \\ f & d & b \\ f & c & a \end{Bmatrix} = \begin{Bmatrix} e & 0 & e \\ c & f & a \\ d & f & b \end{Bmatrix} = \begin{Bmatrix} f & f & 0 \\ d & c & e \\ b & a & e \end{Bmatrix} = \begin{Bmatrix} f & b & d \\ 0 & e & e \\ f & a & c \end{Bmatrix}$$
$$= \begin{Bmatrix} a & f & c \\ e & 0 & e \\ b & f & d \end{Bmatrix} = \begin{Bmatrix} b & a & e \\ f & f & 0 \\ d & c & e \end{Bmatrix} = \begin{Bmatrix} e & d & c \\ e & b & a \\ 0 & f & f \end{Bmatrix} = \begin{Bmatrix} c & e & d \\ a & e & b \\ f & 0 & f \end{Bmatrix}$$
$$= (-1)^{b+c+e+f}\{(2e+1)(2f+1)\}^{-\frac{1}{2}} \begin{Bmatrix} a & b & e \\ d & c & f \end{Bmatrix}.$$

The simple symmetry relations of the $6j$ symbol given by Wigner [invariance with respect to permutations of the columns and invariance also with respect to simultaneous inversion of any two columns: $\begin{Bmatrix} a & b & e \\ d & c & f \end{Bmatrix} = \begin{Bmatrix} b & a & e \\ c & d & f \end{Bmatrix}$, etc., $\begin{Bmatrix} a & b & e \\ d & c & f \end{Bmatrix} = \begin{Bmatrix} d & c & e \\ a & b & f \end{Bmatrix}$, etc.] may also be deduced from the above relations for the $9j$ symbol.

The above symmetry relations are most simply deduced from the Schwinger generating function for the Wigner $9j$ symbol as described in Schwinger's report. For many applications of this symbol to the evaluation of matrix elements of noncentral forces in nuclei, see Hope's thesis[2] and forthcoming publication.

ADDENDUM: DEFINITION OF THE $12j$ SYMBOL AND SOME OF ITS PROPERTIES[9]

A natural generalization of the Wigner $9j$ symbol occurs in the direct evaluation of $\langle n|n-2, 2\rangle$ orbital or charge-spin fractional parentage coefficients required in the theory of atomic spectra or nuclear structure. One meets there with the problem of evaluating the effect of the permutation $P_{n-1,n-2}$ of the coordinates of particles $n-1$ and $n-2$ on the following vector-coupled orbital state of n particles:

$$\Phi[\{(j_1{}^\alpha j_2{}^\beta \cdots j_{n-4}{}^\gamma)J_r,\ (j_{n-3}{}^a j_{n-2}{}^b)J_{ab}\}J';\ \times(j_{n-1}{}^c j_n{}^d)J_{cd}]J. \quad \text{(A1)}$$

Here the angular momenta j_α, j_β, $\cdots$, j_γ of the first $n-4$ particles are vector-coupled to a resultant J_r in some definite way uniquely specifying the state of these $n-4$ particles. This resultant J_r is in turn coupled to the resultant J_{ab} of the respective angular momenta j_a, j_b of particles $n-3$ and $n-2$ to an intermediate resultant J'; this intermediate resultant J' is coupled finally to the resultant J_{cd} of the respective angular momenta j_c, j_d of particles $n-1$ and n to form the total resultant J for the n particle system. The new state obtained by permuting the coordinates of particles $n-1$ and $n-2$ in this state is linearly expressible in terms of states of the same kind (with J_{ac}, J_{bd} replacing, respectively, J_{ab} and J_{cd} and a new intermediate resultant J'' replacing J', but with the same fixed values of J_r and J). The coefficients of the linear expression, depending on the twelve angular momenta,

$$j_a j_b j_c j_d,\ J_{ab}J_{cd}J_{ac}J_{bd},\ J'J'',\ J_rJ,$$

may be used to define a $12j$ symbol as follows:

$$\begin{aligned} P_{n-2,n-1}\Phi[\{(j_1{}^\alpha j_2{}^\beta \cdots j_{n-4}{}^\gamma)J_r,\ (j_{n-3}{}^a j_{n-2}{}^b)J_{ab}\}J'; \\ \times(j_{n-1}{}^c j_n{}^d)J_{cd}]J \\ = \sum_{J_{ac}J_{bd}J''} \{[J_{ab}][J_{cd}][J_{ac}][J_{bd}][J'][J'']\}^{\frac{1}{2}} \\ \times\begin{Bmatrix} j_a & j_b & J_{ab} & J' \\ j_c & j_d & J_{cd} & J'' \\ J_{ac} & J_{bd} & J_r & J \end{Bmatrix} \Phi[\{(j_1{}^\alpha j_2{}^\beta \cdots j_{n-4}{}^\gamma)J_r, \\ \times(j_{n-3}{}^a j_{n-2}{}^c)J_{ac}\}J'';\ (j_{n-1}{}^b j_n{}^d)J_{bd}]J, \end{aligned} \quad \text{(A2)}$$

where, following Wigner,[1] we use the abbreviation

$$[J] \equiv 2J+1. \quad \text{(A3)}$$

Defining thus the $12j$ symbol by

$$\begin{aligned} \begin{Bmatrix} a & b & e & p \\ c & d & f & q \\ g & h & r & s \end{Bmatrix} = \{[e][f][g][h][p][q]\}^{-\frac{1}{2}} \\ \times\langle[\{r, (ab)e\}p, (cd)f]sm| \\ \times[\{r, (ac)g\}q, (bd)h]sm\rangle, \end{aligned} \quad \text{(A4)}$$

one obtains easily the explicit expression

$$\begin{Bmatrix} a & b & e & p \\ c & d & f & q \\ g & h & r & s \end{Bmatrix} = (-1)^{e+f+g+h} \sum_x (-1)^{2x}(2x+1) \begin{Bmatrix} a & b & e \\ c & d & f \\ g & h & x \end{Bmatrix} \begin{Bmatrix} r & e & p \\ f & s & x \end{Bmatrix} \begin{Bmatrix} r & g & q \\ h & s & x \end{Bmatrix}, \quad \text{(A5)}$$

in terms of Wigner $9j$ and $6j$ symbols. The following symmetry relations also may be simply deduced:

$$\begin{Bmatrix} a & b & e & p \\ c & d & f & q \\ g & h & r & s \end{Bmatrix} = \begin{Bmatrix} a & c & g & q \\ b & d & h & p \\ e & f & r & s \end{Bmatrix} = \begin{Bmatrix} d & c & f & p \\ b & a & e & q \\ h & g & s & r \end{Bmatrix} = \begin{Bmatrix} d & b & h & q \\ c & a & g & p \\ f & e & s & r \end{Bmatrix}. \quad \text{(A6)}$$

Important special cases with one argument zero are the following:

$$\begin{Bmatrix} a & b & e & f \\ c & d & f & h \\ g & h & r & 0 \end{Bmatrix} = \begin{Bmatrix} a & c & g & h \\ b & d & h & f \\ e & f & r & 0 \end{Bmatrix} = \begin{Bmatrix} d & c & f & f \\ b & a & e & h \\ h & g & 0 & r \end{Bmatrix} = \begin{Bmatrix} d & b & h & h \\ c & a & g & f \\ f & e & 0 & r \end{Bmatrix}$$
$$= \{[f][h]\}^{-\frac{1}{2}} \begin{Bmatrix} a & b & e \\ c & d & f \\ g & h & r \end{Bmatrix}. \quad \text{(A7)}$$

[9] The work reported in this Addendum was performed by H. A. Jahn.

The complete set of special values of the $12j$ symbol when any one of its twelve arguments is zero is as follows:

$$\left\{\begin{matrix} a & 0 & a & p \\ c & d & f & q \\ g & d & r & s \end{matrix}\right\}=\left\{\begin{matrix} d & d & 0 & a \\ f & s & p & g \\ c & q & a & r \end{matrix}\right\}=\left\{\begin{matrix} r & p & a & 0 \\ q & s & d & c \\ g & f & a & d \end{matrix}\right\}=\left\{\begin{matrix} a & c & g & q \\ 0 & d & d & p \\ a & f & r & s \end{matrix}\right\}$$

$$\left\{\begin{matrix} r & g & q & d \\ a & a & 0 & f \\ p & c & s & d \end{matrix}\right\}=\left\{\begin{matrix} r & q & g & c \\ p & s & f & 0 \\ a & d & a & d \end{matrix}\right\}=\left\{\begin{matrix} d & f & c & g \\ d & s & q & a \\ 0 & p & a & r \end{matrix}\right\}=\left\{\begin{matrix} r & a & p & f \\ g & a & c & d \\ q & 0 & s & d \end{matrix}\right\}$$

$$=\frac{(-1)^{a+2c+d+p+q+r+s}}{\{[a][d]\}^{\frac{1}{2}}}\left\{\begin{matrix} p & c & q \\ d & s & f \end{matrix}\right\}\left\{\begin{matrix} p & c & q \\ g & r & a \end{matrix}\right\}, \quad \text{(A8)}$$

together with

$$\left\{\begin{matrix} 0 & b & b & p \\ c & d & f & q \\ c & h & r & s \end{matrix}\right\}=\left\{\begin{matrix} d & c & f & p \\ b & 0 & b & q \\ h & c & s & r \end{matrix}\right\}=\left\{\begin{matrix} r & q & c & c \\ p & s & f & b \\ b & h & 0 & d \end{matrix}\right\}=\left\{\begin{matrix} s & q & h & b \\ p & r & b & c \\ f & c & d & 0 \end{matrix}\right\}$$

$$=\{[b][c]\}^{-\frac{1}{2}}\left\{\begin{matrix} r & b & p \\ c & d & f \\ q & h & s \end{matrix}\right\}. \quad \text{(A9)}$$

A group of 16 symmetry relations of the $12j$ symbol, together with a convenient new notation, has been found by R. J. Ord-Smith and will be reported upon shortly.

The $12j$ symbol is being used in Southampton in nuclear structure calculations on light nuclei with interconfigurational mixing.

Progress of Theoretical Physics, Vol. 11, No. 2, February 1954

Generalized Racah Coefficient and its Applications

Akito ARIMA, Hisashi HORIE and Yukito TANABE

Department of Physics, University of Tokyo

(Received November 15, 1953)

Generalized Racah coefficient, designated as U coefficient in this paper, has been defined as the transformation function between two different coupling schemes in pairs of any four angular momenta, corresponding to the Racah coefficient defined as the transformation function between two different coupling schemes of any three angular momenta. Several simple properties of the U coefficient have been derived, and the method of tensor operators made to be extended to more general problems. Transformation coefficients between LS- and jj-coupling schemes in a many particle system can be evaluated by making use of these coefficients.

§ 1. Introduction

The Racah coefficient has proved to play a very important role in detailed theories of the atomic and nuclear spectroscopy[1-3], and also to be useful for the studies of the nuclear radiations and reactions.[4),5)] It is defined as the transformation function between two different coupling schemes of any three angular momenta j_1, j_2 and j_3 by

$$[(2J_{12}+1)(2J_{23}+1)]^{1/2}W(j_1j_2\,Jj_3;\,J_{12}\,J_{23})=(j_1j_2(J_{12})j_3\,J\,|\,j_1,j_2j_3(J_{23})J),\quad(1)$$

where $\boldsymbol{j}_1+\boldsymbol{j}_2=\boldsymbol{J}_{12}$, $\boldsymbol{j}_2+\boldsymbol{j}_3=\boldsymbol{J}_{23}$ and $\boldsymbol{J}_{12}+\boldsymbol{j}_3=\boldsymbol{J}_1+\boldsymbol{J}_{23}=\boldsymbol{J}$. In a similar way, we can define the generalized Racah coefficient which is designated as the U coefficient in this paper, as the transformation function between two different coupling orders in pairs of any four angular momenta j_1, j_2, j_3 and j_4 by

$$[(2J_{12}+1)(2J_{34}+1)(2J_{13}+1)(2J_{24}+1)]^{1/2}U\begin{pmatrix} j_1 & j_2 & J_{12}\\ j_3 & j_4 & J_{34}\\ J_{13} & J_{24} & J\end{pmatrix}$$

$$=(j_1j_2(J_{12})j_3j_4(J_{34})J\,|\,j_1j_3(J_{13})j_2j_4(J_{24})J),\qquad(2)$$

where $\boldsymbol{j}_1+\boldsymbol{j}_2=\boldsymbol{J}_{12}$, $\boldsymbol{j}_3+\boldsymbol{j}_4=\boldsymbol{J}_{34}$, $\boldsymbol{j}_2+\boldsymbol{j}_4=\boldsymbol{J}_{24}$, $\boldsymbol{j}_1+\boldsymbol{j}_3=\boldsymbol{J}_{13}$ and $\boldsymbol{J}_{12}+\boldsymbol{J}_{34}=\boldsymbol{J}_{13}+\boldsymbol{J}_{24}=\boldsymbol{J}$, and the nine angular monenta as the arguments in the U coefficients are arranged in three rows and columns in natural order. It is, therefore, expressed by a sum of the products of six Clebsch-Gordan coefficients of the vector additions as

$$(j_1j_2(J_{12})j_3j_4(J_{34})J\,|\,j_1j_3(J_{13})j_2j_4(J_{24})J)=\sum(j_1j_2\,m_1m_2\,|\,j_1j_2J_{12}M)$$
$$\cdot(j_3j_4\,m_3m_4\,|\,j_3j_4J_{34}\,M_{34})(J_{12}J_{34}\,M_{12}\,M_{34}\,|\,J_{12}J_{34}\,JM)(j_1j_3\,m_1m_3\,|\,j_1j_3J_{13}M_{13})$$
$$\cdot(j_2j_4\,m_2\,m_4\,|\,j_2j_4\,J_{24}\,M_{24})(J_{13}J_{24}\,M_{13}\,M_{24}\,|\,J_{13}\,J_{24}JM),\qquad(3)$$

where the summation is extended over all possible values of m_1, m_2, m_3, m_4, M_{12}, M_{34}, M_{13} and M_{24}, restricted by obvious relations $m_1+m_2=M_{12}$, $m_3+m_4=M_{34}$, $m_1+m_3=M_{13}$, $m_2+m_4=M_{24}$, and $M_{12}+M_{34}=M_{13}+M_{24}=M$.

It is shown in the next section, that the U coefficient can be expressed in terms of a sum of the products of three Racah coefficients and that the Racah coefficient is a special case of the U coefficient. Some other properties of the coefficient, also, will be derived there. In sec. 3, the method of tensor operators are extended to more general operators which are constructed as tensor products of two tensor operators. It enables us to treat the spin-dependent interactions in a more general way. And in sec. 4, the method of the calculation of the transformation function between LS- and jj-coupling schemes is derived which seems very important in the treatment of nuclear shell model, especially for light nuclei. Finally some recurrence formulae for the coefficients are given in the Appendix.

§ 2. Properties of the U coefficients

The U coefficient in (2) is defined for integral and half-integral values of the nine parameters, with the limitation that each of the six triads

$$(j_1, j_2, J_{12}),\ (j_3, j_4, J_{34}),\ (J_{12}, J_{34}, J),\ (j_1, j_3, J_{13}),\ (j_2, j_4, J_{24}),\ (J_{13}, J_{24}, J) \tag{4}$$

has an integral sum, and vanishes unless the elements of each triad (4) satisfy the triangular inequalities according to the definition (2) and (3).

The summation in (3) can be carried out by making use of the Racah coefficients if we introduce the following intermediate state characterised by $\boldsymbol{J}_{123}=\boldsymbol{J}_{12}+\boldsymbol{j}_3$, that is

$$\begin{aligned}&(j_1j_2(J_{12})j_3j_4(J_{34})J\,|\,j_1j_3(J_{13})j_2j_4(J_{24})J)\\&\quad=\sum_{J_{123}}(J_{12},j_3j_4(J_{34})J\,|\,J_{12}j_3(J_{123})j_4J)\,(j_1j_2(J_{12})j_3J_{123}\,|\,j_1j_3(J_{13})j_2J_{123})\\&\qquad\cdot(J_{13}j_2(J_{123})\,j_4J\,|\,J_{13},j_3j_4(J_{34})J),\end{aligned} \tag{5}$$

where the summation over J_{123} is extended over all possible values compatible with the condition $\boldsymbol{J}_{123}=\boldsymbol{J}_{12}+\boldsymbol{j}_3=\boldsymbol{J}_{13}+\boldsymbol{j}_2$. Therefore the U coefficient can be expressed in terms of Racah coefficients with RIII (4) and (5). In abbreviated notations for arguments, it is given by

$$U\begin{pmatrix}a & b & e\\ c & d & e'\\ f & f' & g\end{pmatrix}=\sum_{\lambda}(2\lambda+1)\,W(fgbd\,;\,f'\lambda)\,W(egcd\,;\,e'\lambda)\,W(fcbe\,;\,a\lambda). \tag{6}$$

It is easily seen that the Racah coefficient can be obtained as a special case of the U coefficient in which any one of the six arguments b, c, d, e, f and g appearing in any two W's in the right hand side is equal to zero. For example, if $g=0$, $e=e'$ and $f=f'$ result for non-vanishing U, which is given by

$$U\begin{pmatrix} a & b & e \\ c & d & e \\ f & f & 0 \end{pmatrix} = (-1)^{e+f-a-d}\, W(abcd;\, ef)/[(2e+1)(2f+1)]^{1/2}. \tag{7}$$

Owing to the symmetry properties of the U coefficient which we can show in the following, the coefficient reduces always to the W coefficient if any one of its nine arguments is equal to zero.

We can immediately derive the following symmetry properties from those of the Racah coefficients (see RII (40a) and (40b)) and the relations given by RII (43) and Biedenharn, Blatt and Rose's[6] (17).

i) Transposition of " rows " and " columns " :

$$U\begin{pmatrix} a & b & e \\ c & d & e' \\ f & f' & g \end{pmatrix} = U\begin{pmatrix} a & c & f \\ b & d & f' \\ e & e' & g \end{pmatrix}. \tag{8}$$

ii) Interchanges of two " rows " or " columns " ;

$$U\begin{pmatrix} a & b & e \\ c & d & e' \\ f & f' & g \end{pmatrix} = (-1)^{\sigma} U\begin{pmatrix} c & d & e' \\ a & b & e \\ f & f' & g \end{pmatrix} = (-1)^{\sigma} U\begin{pmatrix} f & f' & g \\ c & d & e' \\ a & b & e \end{pmatrix}, \tag{9}$$

where $\sigma = a+b+c+d+e+e'+f+f'+g$ (=integer).

Combining (8) and (9), we obtain 72 different arrangements of the nine parameters. For example, we can rewrite formula (6) into the following more symmetrical form :

$$U\begin{pmatrix} a & b & e \\ c & d & e' \\ f & f' & g \end{pmatrix} = (-1)^{\sigma} \sum_{\lambda} (2\lambda+1)\, W(bcef;\, \lambda a)\, W(bcf'e';\, \lambda d)\, W(efe'f';\, \lambda g), \tag{10}$$

where the " diagonal " elements of U appear as the last arguments of the three W coefficients. We shall prefer this form as the standard formula connecting the U and the W coefficients. Furthermore, it is easy to see that, if $a=c$, $b=d$ and $e=e'$, according to (9),

$$U\begin{pmatrix} a & b & e \\ a & b & e \\ f & f' & g \end{pmatrix} = 0, \qquad (f+f'+g=\text{odd}). \tag{11}$$

Some identities can be derived from the definition of the U coefficient as a transformation function between two different couplings in pairs of four angular momenta. Four angular momenta a, b, c and d can be combined into various pairs, as follows :

$$\boldsymbol{g} = (\boldsymbol{a}+\boldsymbol{b}) + (\boldsymbol{c}+\boldsymbol{d}) = (\boldsymbol{a}+\boldsymbol{c}) + (\boldsymbol{b}+\boldsymbol{d}) = (\boldsymbol{a}+\boldsymbol{d}) + (\boldsymbol{b}+\boldsymbol{c}). \tag{12}$$

It follows at once that

$$\sum_{ff'}(ab(e)cd(e')g \mid ac(f)bd(f')g)(ac(f)bd(f')g \mid ab(e_1)cd(e_1')g)$$
$$=\delta(e, e_1)\delta(e', e_1'). \tag{13}$$

In terms of U's, this gives us from (2), the following orthogonality relation between them :

$$\sum_{ff'}(2f+1)(2f'+1)U\begin{pmatrix} a & b & e \\ c & d & e' \\ f & f' & g \end{pmatrix}U\begin{pmatrix} a & b & e_1 \\ c & d & e_1' \\ f & f' & g \end{pmatrix}$$
$$=\delta(e, e_1)\delta(e', e_1')/[(2e+1)(2e'+1)]. \tag{14}$$

Since the transformation function between the first and third coupling orders in pairs of (12) can be expressed, through the second one, as

$$\sum_{ff'}(ab(e)cd(e')g \mid ac(f)bd(f')g)(ac(f)bd(f')g \mid ad(h)bc(h')g)$$
$$=(ab(e)cd(e')g \mid ad(h)bc(h')g), \tag{15}$$

we obtain, another useful relation between U's :*

$$\sum_{ff'}(-1)^{e'-f'-2c+h'}(2f+1)(2f'+1)U\begin{pmatrix} a & b & e \\ c & d & e' \\ f & f' & g \end{pmatrix}U\begin{pmatrix} a & c & f \\ d & b & f' \\ h & h' & g \end{pmatrix}=U\begin{pmatrix} a & b & e \\ d & c & e' \\ h & h' & g \end{pmatrix}. \tag{16}$$

We can see without difficulties that (15) and (16) are the generalization of RII (42) and (43), if we put g equal to zero. Beside these relations we have obtained several other relations between U's and between U's and W's, some of which will be given in Appendix. We shall also show there how recurrence formulae for U coefficients are obtained from one of them.

§ 3. Application to the calculation of matrix elements of tensor operators

(a) Tensor product of two tensor operators

The tensor product of two tensor operators $\boldsymbol{T}^{(k_1)}$ and $\boldsymbol{U}^{(k_2)}$ is defined in the usual way by an irreducible form

---

*) The following relations are easily proved :

$$(ab(e)cd(e')g \mid ad(f)bc(f')g)=(-1)^{c+d-e'}[(2e+1)(2e'+1)(2f+1)(2f'+1)]^{1/2}U\begin{pmatrix} a & b & e \\ d & c & e' \\ f & f' & g \end{pmatrix},$$

and

$$(ab(e)cd(e')g \mid ad(f)cb(f')g)=(-1)^{e-e'-b+f'}[(2e+1)(2e'+1)(2f+1)(2f'+1)]^{1/2}U\begin{pmatrix} a & b & e \\ d & c & e' \\ f & f' & g \end{pmatrix}.$$

$$[\boldsymbol{T}^{(k_1)} \times \boldsymbol{U}^{(k_2)}]_Q^{(K)} = \sum_{q_1 q_2} T_{q_1}^{(k_1)} \cdot U_{q_2}^{(k_2)} (k_1 k_2 q_1 q_2 \mid k_1 k_2 KQ). \tag{17}$$

In practical applications the most important tensor products are those in which two tensor operators operate on different parts of a composite system. The operator of this type appears in many problems, for example, in the calculation of matrices of spin-dependent interactions[7-9], of multipole moments of radiations in the nuclear shell model[10], and of polarization of emerging particles in nuclear reactions[11].*

When $\boldsymbol{T}^{(k_1)}$ operates on system 1 and $\boldsymbol{U}^{(k_2)}$ operates on system 2, the matrix element of a tensor product of $\boldsymbol{T}^{(k_1)}$ and $\boldsymbol{U}^{(k_2)}$ in $(j_1 j_2 JM)$ scheme is given by

$$(j_1 j_2 JM \mid [\boldsymbol{T}^{(k_1)} \times \boldsymbol{U}^{(k_2)}]_Q^{(K)} \mid j_1' j_2' J' M') = \sum (j_1 j_2 JM \mid j_1 j_2 m_1 m_2)$$

$$\cdot (j_1 m_1 \mid T_{q_1}^{(k_1)} \mid j_1' m_1') (j_2 m_2 \mid U_{q_2}^{(k_2)} \mid j_2' m_2')$$

$$\cdot (j_1' j_2' m_1' m_2' \mid j_1' j_2' J' M') (k_1 k_2 q_1 q_2 \mid k_1 k_2 KQ). \tag{18}$$

And, if the double-barred element is defined in accordance with RII (29) by

$$(\alpha jm \mid T_q^{(k)} \mid \alpha' j' m') = (\alpha j \parallel T^{(k)} \parallel \alpha' j') (j' k m' q \mid j' k j m) / (2j+1)^{1/2}, \tag{19}$$

it is easy to be shown that the double-barred elements of the tensor product $[\boldsymbol{T}^{(k_1)} \times \boldsymbol{U}^{(k_2)}]^{(K)}$ are expressed in terms of those of $\boldsymbol{T}^{(k_1)}$ and $\boldsymbol{U}^{(k_2)}$ and a U coefficient as follows:

$$(j_1 j_2 J \parallel [T^{(k_1)} \times U^{(k_2)}]^{(K)} \parallel j_1' j_2' J') = (j_1 \parallel T^{(k_1)} \parallel j_1') (j_2 \parallel U^{(k_2)} \parallel j_2')$$

$$\cdot [(2J+1)(2J'+1)(2K+1)]^{1/2} U \begin{pmatrix} j_1 & j_1' & k_1 \\ j_2 & j_2' & k_2 \\ J & J' & K \end{pmatrix}. \tag{20}$$

This simple and symmetrical formula is a natural generalization of RII (38), (44a) and (44b). First of all, noting that

$$(\boldsymbol{T}^{(k)} \cdot \boldsymbol{U}^{(k)}) = (-1)^k (2k+1)^{1/2} [T^{(k)} \times U^{(k)}]^{(0)}{}_0 \tag{21}$$

where in the left-hand side $(\boldsymbol{T}^{(k)} \cdot \boldsymbol{U}^{(k)}) = \sum_q (-1)^q T_q^{(k)} U_{\bar{q}}^{(k)}$ represents the scalar product of the two tensor operators $\boldsymbol{T}^{(k)}$ and $\boldsymbol{U}^{(k)}$, and putting $k_1 = k_2 = k$ and $K = 0$ in (20), we obtain the relation RII (38)

$$(j_1 j_2 JM \mid (\boldsymbol{T}^{(k)} \cdot \boldsymbol{U}^{(k)}) \mid j_1' j_2' JM) = (-1)^{j_1 + j_2' - J}$$

$$\cdot (j_1 \parallel T^{(k)} \parallel j_1') (j_2 \parallel U^{(k)} \parallel j_2') W(j_1 j_2 j_1' j_2'; Jk). \tag{22}$$

Putting further $k_2 = 0$ and $k_1 = K = k$ in (20), and noting that $(j \parallel 1 \parallel j) = (2j+1)^{1/2}$, we get the relation given by RII (44a)

*) For the same coefficient as our U's, U. Fano and G. Racah seem to have given the notation $X(abc; cde'; ff'g)$ in their unpublished paper (cf. ref. 11).

$$(j_1j_2J \| T^{(k)} \| j_1'j_2J') = (-1)^{k+j_2-j_1'-J}$$
$$\cdot (j_1 \| T^{(k)} \| j_1')[(2J+1)(2J'+1)]^{1/2} W(j_1Jj_1'J' ; j_2k). \qquad (23)$$

The relation RII (44b) can be obtained in a similar way by putting $k_1=0$ and $k_2=K=k$.

It must be observed that for the double-barred elements of the tensor product,

$$(j_1j_2J \| [T^{(k_1)} \times U^{(k_2)}]^{(K)} \| j_1'j_2'J') = (-1)^{J-J'+k_1+k_2-K}$$
$$\cdot (j_1'j_2'J' \| [T^{(k_1)} \times U^{(k_2)}]^{(K)} \| j_1j_2J), \qquad (24)$$

corresponding to RII (31). As a special case of (24), the following formula is obtained, by putting $j_1=j_1'$, $j_2=j_2'$ and $J=J'$,

$$(j_1j_2J \| [T^{(k_1)} \times U^{(k_2)}]^{(K)} \| j_1j_2J) = 0 \qquad (k_1+k_2-K=\text{odd}), \qquad (25)$$

which can also be derived from (11) immediately.

(b) Matrix elements of the scalar product of two irreducible composite tensors

In order to calculate the matrix of spin-dependent interactions, it is necesary to treat the scalar product of two irreducible composite tensors. Hence we consider this quantity more in detail. Let $\boldsymbol{T}^{(k_1k_2;K)}$ be an irreducible composite tensor of degree K with respect to $\boldsymbol{J}=\boldsymbol{j}_1+\boldsymbol{j}_2$ which behaves as an irreducible tensor of degree k_1 and k_2 with respect to j_1 and j_2 respectively (This may be considered as an abbreviation of the tensor product given in (17)), and $\boldsymbol{U}^{(k_1k_2;K)}$ have a similar meaning. By making use of (20) and BBR (1), the matrix elements of this scalar product in (j_1j_2JM) scheme are given by

$$(\gamma j_1j_2JM| (\boldsymbol{T}^{(k_1k_2;K)} \cdot \boldsymbol{U}^{(k_1k_2;K)}) | \gamma j_1'j_2'JM)$$
$$= (2K+1)\sum_{\gamma''j_1''j_2''}(\gamma j_1j_2 \| T^{(k_1,k_2)} \| \gamma''j_1''j_2'')(\gamma''j_1''j_2'' \| U^{(k_1,k_2)} \| \gamma' j_1'j_2')$$
$$\cdot \sum_{J''}(-1)^{J-J''}(2J''+1)\, U\begin{pmatrix} j_1 & j_1'' & k_1 \\ j_2 & j_2'' & k_2 \\ J & J'' & K \end{pmatrix} U\begin{pmatrix} j_1'' & j_1' & k_1' \\ j_2'' & j_2' & k_2' \\ J'' & J & K \end{pmatrix}. \qquad (26)$$

The summation over J'' in (26) can be carried out, using the relation between U and W coefficients and with RII (43) and BBR (17), so that (26) is written in an expected form as

$$(\gamma j_1j_2JM| (T^{(k_1k_2;K)} \cdot U^{(k_1'k_2';K)}) | \gamma' j_1'j_2'JM)$$
$$= (2K+1)\sum_{\lambda}(-1)^{j_1+j_2'-J} W(j_1j_2j_1'j_2' ; J\lambda)$$
$$\cdot \sum_{\gamma''j_1''j_2''}(-1)^{k_1'+k_2}(\gamma j_1j_2 \| T^{(k_1,k_2)} \| \gamma''j_1''j_2'')(\gamma''j_1''j_2'' \| U^{(k_1',k_2')} \| \gamma' j_1'j_2')$$
$$\cdot W(j_1j_2'k_1k_2' ; \lambda j_1'') W(j_2j_2'k_2k_3' ; \lambda j_2'') W(k_1k_2k_1'k_2' ; K\lambda). \qquad (27)$$

This formula is useful, especially, in the treatment of the spin-dependent interactions[7-9], in which this reduction has been done in a more straightforward way. For example, in the case of the spin-spin interaction between electrons, the angular momenta j_1 and j_2 are the total spin and the total orbital angular momenta respectively, and $k_1=k_1'=1$ and $k_2'=k_2+2$ ($k=0, 2, \cdots$), so that only $\lambda=2$ appears in this equation.

(c) Coefficients of the exchange integrals of a many particle system

Racah has given a general method for obtaining the coefficient of exchange integrals in the case of electrostatic interactions RII, sec. 5. We shall extend this method to spin-dependent interactions, making use of the result in this section. The spin-dependent interaction can be represented as a scalar product of two irreducible tensors which have degree K ($K\neq 0$) in the spin and the ordinary space respectively; for example, in the tensor or the spin-spin interaction $K=2$ and in the spin-orbit interaction $K=1$.

First, we consider only the orbital part and assume that the irreducible tensor is a tensor product of two tensor operators of degree k_1 and k_2, the former operating on particle 1 and the latter on particle 2. Then we obtain the orbital part of coefficients of exchange integrals in terms of double-barred elements as

$$(-1)^{l_1+l_2-L'}(l_1 l_2 L \| [T_1^{(k_1)}\times U_2^{(k_2)}]^{(K)} \| l_2 l_1 L')$$
$$=(-1)^{l_1+l_2-L'}(l_1 \| T^{(k_1)} \| l_2)(l_2 \| U^{(k_2)} \| l_1)[(2L+1)(2L'+1)(2K+1)]^{1/2} \cdot U\begin{pmatrix} l_1 & l_2 & L \\ l_1 & l_2 & L' \\ k_1 & k_2 & K \end{pmatrix},$$

and owing to (15) and RII (31), it follows that

$$(-1)^{l_1+l_2-L}(l_1 l_2 L \| [T^{(k_1)}\times U^{(k_2)}]^{(K)} \| l_2 l_1 L')$$
$$=(l_1 \| T^{(k_1)} \| l_2)(l_1 \| U^{(k_2)} \| l_2)[(2L+1)(2L'+1)(2K+1)]^{1/2}$$
$$\cdot\sum_{r,s}(-1)^{k_2-s}\cdot(2r+1)(2S+1)U\begin{pmatrix} l_1 & l_1 & r \\ l_2 & l_2 & s \\ k_1 & k_2 & K \end{pmatrix}U\begin{pmatrix} l_1 & l_2 & L \\ l_1 & l_2 & L' \\ r & s & K \end{pmatrix}. \tag{28}$$

Futhermore, if we define the unit tensor $u^{(k)}$ by

$$(l \| u^{(k)} \| l')=\delta(l,l'), \tag{29}$$

and take (20) into account, we may also write

$$(-1)^{l_1+l_2-L}(l_1 l_2 L \| [T^{(k_1)}\times U^{(k_2)}]^{(K)} \| l_2 l_1 L')$$
$$=(l_1 \| T^{(k_1)} \| l_2)(l_1 \| U^{(k_2)} \| l_2)\sum_{r,s}(-1)^{k_2-s}(2r+1)(2s+1)$$
$$\cdot U\begin{pmatrix} l_1 & l_1 & r \\ l_2 & l_2 & s \\ k_1 & k_2 & K \end{pmatrix}\cdot(l_1 l_2 L \| [u_1^{(r)}\times u_2^{(s)}]^{(K)} \| l_1 l_2 L'). \tag{30}$$

Therefore, the calculation of the coefficients of the exchange integrals can be carried out in the same way as for obtaining those of direct integrals with respect to the operator

$$(l_1 \| T^{(k_1)} \| l_2)(l_1 \| U^{(k_2)} \| l_2) \sum_{r,s} (-1)^{k_2-s}(2r+1)(2s+1)$$
$$\cdot U\begin{pmatrix} l_1 & l_1 & r \\ l_2 & l_2 & s \\ k_1 & k_2 & K \end{pmatrix} [\boldsymbol{u}_1^{(r)} \times \boldsymbol{u}_2^{(s)}]^{(K)}, \tag{31}$$

in place of $[\boldsymbol{T}^{(k_1)} \times \boldsymbol{U}^{(k_2)}]$. In a similar way, another irreducible tensor of degree K can be obtained as the spin part, which becomes usually much simpler. Therefore, the complete operator necessary for the calculation of coefficients of exchange integrals is given by contructing scalar products of these two irreducible tensors of degree K with respect to the ordinary and spin spaces and reversing the total sign due to the antisymmetry of the wave functions.

As a trivial example of this procedure, the formula RII (59) of the coefficient $g_k(l_1 l_2 L)$ for the electrostatic interaction is derived by putting $K=0$ and $T^{(k_1)}=U^{(k_2)}=C^{(k)}$ in (31) where $C_q^{(k)}=[4\pi/(2k+1)]^{1/2}\Theta(kq)\Phi(q)$. Another trivial example is given by the construction of Dirac's exchange operator with (30). Letting $l_1=l_2=1/2$, $\boldsymbol{T}^{(0)}=\boldsymbol{U}^{(0)}=\mathbf{1}$, and noting that $\mathbf{1}=(2)^{1/2}\boldsymbol{u}^{(0)}$ and $\boldsymbol{s}=(3/2)^{1/2}\boldsymbol{u}^{(1)}$, we immediately have Dirac's exchange operator

$$(-1)^{1-S}=1/2\cdot[1+2(\boldsymbol{s}_2\cdot\boldsymbol{s}_1)]. \tag{32}$$

The simplest example of the spin-dependent interactions is given by the spin-spin interaction between electrons.[7),8)] In this case, we need not change the form of the spin part since it is symmetrical with respect to the spin variables of two electrons. This holds also for the tensor interaction with arbitrary radial dependence. The coefficients of exchange integrals of the spin-spin interaction between electrons

$$\int\int R_{l_1}(r_1) R_{l_2}(r_2) \frac{r_1^{\,k}}{r_1^{\,k+3}} R_{l_1}(r_2) R_{l_2}(r_1)\, dr_1 dr_2$$

are given by the matrix elements of the operator

$$-2f_k(l_1 \| C^{(k)} \| l_2)(l_1 \| C^{k+2} \| l_2) \sum_{r,s}{}' (-1)^{k-s}(2r+1)(2s+1)$$
$$\cdot U\begin{pmatrix} l_1 & l_1 & r \\ l_2 & l_2 & s \\ k & k+2 & 2 \end{pmatrix} ([\boldsymbol{s}_1 \times \boldsymbol{s}_2]^{(2)} \cdot [\boldsymbol{u}_1^{(r)} \times \boldsymbol{u}_2^{(s)}]^{(2)}), \tag{33}$$

where the prime on the summation symbol denotes that the summation is extended only over those values of r and s which satisfy $r+s=$ even and the coefficients f_k is given by $(-1)^{k+1}4\,[(k+1)(k+2)(2k+1)(2k+3)(2k+5)/5]^{1/2}$.

§ 4. Transformation coefficients between *LS*- and *jj*-coupling schemes in l^n configuration

The transformation coefficient between *LS*- and *jj*-coupling schmes of two particle system can be obtained at once from the general expression for the transformation function between two different couplings in pairs of four angular momenta given in sec. 1 and 2. For two equivalent particles (identical particles which are contained in the same shell), however, the formula does not hold without modification on account of the Pauli exclusion principle as will be seen in the following. Here we consider the case in which there are n equivalent particles in the same shell with azimuthal quantum number l. The states of l^n configuration are characterized by $\alpha SLJM$ in *LS*-oupling scheme, where α is the quantum number other than S, L, J and M. On the other hand, the states with the same J and M are characterized by $j_1^{n_1}(\beta_1 J_1) j_2^{n_2}(\beta_2 J_2) JM$ in *jj*-coupling scheme, where $j_1 = l + 1/2$, $j_2 = l - 1/2$ and $n_1 + n_2 = n$, β's being the quantum number other than J and M. The quantum number of the isotopic spin employed sometimes in nuclear shell model can be included in α and in β. Taking into account the antisymmetry property of wave functions, the transformation function between *LS*- and *jj*-coupling schemes for n equivalent particles can be obtained in terms of those for $(n-1)$ equivalent particles, the coefficients of fractional parentages, and the U and W coefficients:

$$
\begin{aligned}
&(l^n \alpha SLJM | j_1^{n_1}(\beta_1 J_1) j_2^{n_2}(\beta_2 J_2) JM) \\
&= (-1)^{n_2} (n_1/n)^{1/2} \sum (l^n \alpha SL \{| l^{n-1}(\alpha' S' L') lsL) \left(S' \tfrac{1}{2}(S) L' l(L) J \,|\, S' L'(J') \tfrac{1}{2} l(j_1) J \right) \\
&\cdot (l^{n-1} \alpha' S' L' J' | j_1^{n_1-1}(\beta_1' J_1') j_2^{n_2}(\beta_2 J_2) J') (J_1' J_2 (J') j_1 J \,|\, J_1' j_1 (J_1) J_2 J) \\
&\cdot (j_1^{n_1-1}(\beta_1' J_1') j_1 J_1 |\} j_1^{n_1} \beta_1 J_1) + (n_2/n)^{1/2} \sum (l^n \alpha SL \{| l^{n-1}(\alpha' S' L') lSL) \\
&\cdot \left(S' \tfrac{1}{2}(S) L' l(L) J \,|\, S' L'(J') \tfrac{1}{2} l(j_2) J \right) (l^{n-1} \alpha' S' L' J' | j_1^{n_1}(\beta_1 J_1) j_2^{n_2-1}(\beta_2' J_2') J') \\
&\cdot (J_1 J_2' (J') j_2 J \,|\, J_1, J_2' j_2 (J_2) J) (j_2^{n_2-1}(\beta_2' J_2') j_2 J |\} j_2^{n_2} \beta_2 J_2). \qquad (34)
\end{aligned}
$$

The tables of the coefficients of fractional parentages were given by RIII for the atomic p^n and d^n configurations, by Jahn and van Wieringen[2] for the nuclear p^n, by Jahn[2] for d^3 and d^4 configurations in *LS*-coupling, and Edmond and Flowers[3] for $(3/2)^3$, $(3/2)^4$, $(5/2)^5$ $(7/2)^3$ and $(7/2)^4$ configurations. The tables of W coefficients were given by Biedenharn[12] and Obi *et al.*[13]

For two equivalent particles which have only states with $S + L =$ even in *LS*-coupling scheme, eq. (34) reduces to

$$
(l^2 SLJM | j^2 JM) = \left(\tfrac{1}{2} \tfrac{1}{2} (S) ll(L) J \,\middle|\, \tfrac{1}{2} l(j) \tfrac{1}{2} l(j) J \right), \qquad j = j_1 \text{ or } j_2,
$$

$$
(l^2 SLJM | j_1 j_2 JM) = (2)^{1/2} \left(\tfrac{1}{2} \tfrac{1}{2} (S) ll(L) J \,\middle|\, \tfrac{1}{2} l(j_1) \tfrac{1}{2} l(j_2) J \right).
$$

Acknowledgement: The authors wish to express their sincere thanks to Professor T. Yamanouchi and Dr. T. Ishidzu for their kind interest and valuable discussions. This work is partly supported by the Grant of Aid for the Fundamental Scientific Research. One of the authors (A. A.) is indebted to the Yomiuri Fellowship for the financial aid.

Appendix. Identities and recurrence formulae for the U coefficients

The values of the U coefficients can be obtained by inserting the values of the W coefficients in the formula (6). However, there are also some recurrence formulae between the U coefficients which may be available for the evaluation of the coefficients.

In order to obtain an identity from which recurrence formulae can be derived, we consider the following two different coupling schemes of five angular momenta and the transformation function between them. It is evident that

$$
\begin{aligned}
&(j_1 j_2(J_{12}) j_3 j_4(J_{34})(J_{1234}) j_5 J \mid j_1 j_3(J_{13}), j_2 j_4(J_{24}) j_5(J_{245}) J) \\
&= (j_1 j_2(J_{12}) j_3 j_4(J_{34}) J_{1234} \mid j_1 j_3(J_{13}) j_2 j_4(J_{24}) J_{1234}) \\
&\cdot (J_{13} J_{24}(J_{1234}) j_5 J \mid J_{13}, J_{24} j_5(J_{245}) J). \qquad \text{(A. 1)}
\end{aligned}
$$

But this transformation function can be expressed in another way by employing two intermediate state as

$$
\begin{aligned}
&\sum_{J_{345} J_{45}} (J_{12} J_{34}(J_{1234}) j_5 J \mid J_{12}, J_{34} j_5(J_{345}) J)(j_3 j_4(J_{34}) j_5 J_{345} \mid j_3, j_4 j_5(J_{45}) J_{345}) \\
&\cdot (j_1 j_2(J_{12}) j_3 J_{45}(J_{345}) J \mid j_1 j_3(J_{13}) j_2 J_{45}(J_{245}) J) \\
&\cdot (j_2, j_4 j_5(J_{45}) J_{245} \mid j_2 j_4(J_{24}) j_5 J_{245}). \qquad \text{(A. 2)}
\end{aligned}
$$

Equating this with (A. 1) and expressing the result by the U and W coefficients, we obtain the relation between them as follows;

$$
\begin{aligned}
U\begin{pmatrix} a & b & e \\ c & d & e' \\ f & f' & g \end{pmatrix} W(ff'\bar{g}h;\, g\bar{f}') = \sum_{\lambda\mu} (2\lambda+1)(2\mu+1)\, W(ee'\bar{g}h;\, g\lambda) \\
\cdot W(cd\lambda h;\, e'\mu)\, W(bd\bar{f}'h;\, f'\mu)\, U\begin{pmatrix} a & b & e \\ c & \mu & \lambda \\ f & \bar{f}' & \bar{g} \end{pmatrix}. \qquad \text{(A. 3)}
\end{aligned}
$$

Applications of this identity to give recurrence formulae are immediate. Take, for example, $h=1/2$. Then in the summation on the right hand side of (A. 3) λ and μ take only two values $\lambda=e'\pm 1/2$ and $\mu=d\pm 1/2$ respectively. Then, we can choose the values of $\bar{f}'$ and $\bar{g}$ as $\bar{f}'=f'\pm 1/2$ and $\bar{g}=g\pm 1/2$. Therefore, the values of $U\begin{pmatrix} a & b & e \\ c & d-1/2 & e'-1/2 \\ f & f'-1/2 & g-1/2 \end{pmatrix}$, for example, can be evaluated in terms of four U coefficients

$$U\begin{pmatrix} a & b & e \\ c & d & e' \\ f & f' & g \end{pmatrix},\ U\begin{pmatrix} a & b & e \\ c & d & e'-1 \\ f & f' & g \end{pmatrix},\ U\begin{pmatrix} a & b & e \\ c & d-1 & e' \\ f & f' & g \end{pmatrix} \text{ and } U\begin{pmatrix} a & b & e \\ c & d-1 & e'-1 \\ f & f' & g \end{pmatrix}.$$

Coefficients of the relation which come from the W coefficients with one variable equal to 1/2, are simple algebraic functions. In a similar way, we can have a relation betwen the following nine U coefficients :

$$U\begin{pmatrix} a & b & e \\ c & d & e' \\ f & f' & g \end{pmatrix},\ U\begin{pmatrix} a & b & e \\ c & d & e'\pm 1 \\ f' & f' & g \end{pmatrix},\ U\begin{pmatrix} a & b & e \\ c & d\pm 1 & e' \\ f & f' & g \end{pmatrix} \text{ and } U\begin{pmatrix} a & b & e \\ c & d\pm 1 & e'\pm 1 \\ f & f' & g \end{pmatrix}.$$

Furthermore, it is easily shown that

$$(ab(e)c(e')dg \mid ad(f)c(f')eg)$$

$$=(-1)^{e-f-e'+f'}[(2e+1)(2e'+1)(2f+1)(2f'+1)]^{1/2}U\begin{pmatrix} a & c & b \\ f & c & f' \\ d & e' & g \end{pmatrix}. \quad \text{(A.4)}$$

Using the definition (10) and (A, 4), we can find

$$U\begin{pmatrix} a & b & e \\ c & d & e' \\ f & f & g \end{pmatrix}=(2g+1)\sum_{\lambda,\mu,\nu}(2\lambda+1)(2\mu+1)(2\nu+1)$$

$$\cdot U\begin{pmatrix} a & b & e \\ c & \bar{d} & \lambda \\ f & \mu & \nu \end{pmatrix} U\begin{pmatrix} \lambda & \nu & e \\ c & d & e' \\ \bar{d} & \bar{h} & g \end{pmatrix} U\begin{pmatrix} \mu & b & \bar{d} \\ \nu & d & \bar{h} \\ f & f' & g \end{pmatrix}, \quad \text{(A. 5)}$$

and

$$U\begin{pmatrix} a & b & e \\ c & d & e' \\ f & f' & g \end{pmatrix} U\begin{pmatrix} \bar{a} & \bar{b} & \bar{c} \\ \bar{c} & \bar{d} & \bar{e}' \\ f & f' & g \end{pmatrix}=\sum_{\gamma_1,\gamma_2,\delta,\varepsilon}(-1)^{e'+\bar{a}-f'-\delta}(2\gamma_1+1)(2\gamma_2+1)(2\delta+1)(2\varepsilon+1)$$

$$\cdot U\begin{pmatrix} a & b & e \\ c & \varepsilon & \delta \\ f & \bar{a} & \bar{c} \end{pmatrix} U\begin{pmatrix} \bar{a} & \bar{b} & \bar{e} \\ \varepsilon & \bar{d} & \gamma_2 \\ b & f' & d \end{pmatrix} U\begin{pmatrix} c & d & e' \\ \varepsilon & \gamma_2 & \bar{d} \\ \delta & \bar{e} & \gamma_1 \end{pmatrix} U\begin{pmatrix} \bar{c} & \bar{d} & \bar{e}' \\ e & e' & g \\ \delta & \gamma_1 & \bar{c} \end{pmatrix}. \quad \text{(A6)}$$

References

1) G. Racah, Phys. Rev. **62** (1942), 438 and **63** (1943), 367. These are referred to as RII and RIII, respectively. Also "Lectures on Group Theory and Spectroscopy" Inst. Adv. Study, Princeton, 1951.
2) H. A. Jahn, Proc. Roy. Soc. **A201** (1950), 516; **205** (1951), 192 and H. A. Jahn and H. van Wieringen, ibid **A209** (1951), 502.
3) B. H. Flowers, Proc. Roy. Soc. **A210** (2951), 498; **212** (1952) 248; **215** (1951), 398 and A. R. Edmond and B. H. Flowers, ibid **A214** (1952), 515; **215** (1952), 398.
4) J. M. Blatt and L. C. Biedenharn, Revs. Mod. Phys. **24** (1952), 258.
5) L. C. Biedenharn and M. E. Roes, Revs. Mod. Phys. **25** (1953), 729.
6) Biedenharn, Blatt and Rose, Revs. Mod. Phys. **24** (1952), 249, referred to as BBR.
7) I. Talmi, Phys. Rev. **89** (1953), 1065.
8) H. Horie, Prog. Theor. Phys. **10** (1953), 296.
9) F. Innes, Phys. Rev. **91** (1953), 31.
10) A. Moszkowski, Phys. Rev. **89** (1953), 474.
11) A. Simon and T. A. Welton, Phys. Rev. **90** (1953), 1036.
12) L. C. Biedenharn, ORNL, No. 9098 (1952).
13) Obi, Ishidzu, Horie, Yanagawa, Tanabe and Sato, Ann. Tokyo Astr. Obs. Second Series, III, No. 3 (1953), 89.

IL NUOVO CIMENTO Vol. X, N. 3 1° Novembre 1958

Symmetry Properties of Clebsch-Gordan's Coefficients.

T. REGGE
Istituto Nazionale di Fisica Nucleare - Sezione di Torino

(ricevuto il 23 Settembre 1958)

It has been known since long that the coefficients for the composition of angular momenta possess a symmetry group of 12 elements. This group consists of the 6 permutations of 3 angular momenta and of the space reflection.

The physical significance of these symmetries is evident and it led very early to their discovery. However a more elaborate analysis shows that they are part of a much larger group of 72 elements. Unnecessary details of the proof will be omitted here. For a better understanding of the symmetry properties we shall employ a different notation for the coefficients as follows:

$$(1)\qquad \begin{pmatrix} a & b & c \\ \alpha & \beta & \gamma \end{pmatrix} = \boxed{\begin{matrix} b+c-a & c+a-b & a+b-c \\ a-\alpha & b-\beta & c-\gamma \\ a+\alpha & b+\beta & c+\gamma \end{matrix}}\,.$$

The 9 elements of the square symbol at the right are bound by 4 equations. In fact all sums by rows and columns are equal to $a+b+c=J$. Moreover all elements are non-negative integers. We shall define also a square symbol in the case when it does not satisfy the above restrictions by letting it vanish. Clebsch-Gordan coefficients are defined by the expansion (1):

$$(2)\qquad (v_1u_2-u_1v_2)^{a+b-c}(w_1v_2-w_2v_1)^{b+c-a}(u_1w_2-w_1u_2)^{c+a-b} =$$

$$= \sqrt{(J+1)!\,(a+b-c)!\,(b+c-a)!\,(c+a-b)!}\sum_{\alpha\beta\gamma}\begin{pmatrix} a & b & c \\ \alpha & \beta & \gamma \end{pmatrix}\cdot$$

$$\cdot\frac{u_1^{a-\alpha}u_2^{a+\alpha}}{\sqrt{(a+\alpha)!\,(a-\alpha)!}}\,\frac{v_1^{b-\beta}v_2^{b+\beta}}{\sqrt{(b-\beta)!\,(b+\beta)!}}\,\frac{w_1^{c-\gamma}w_2^{c+\gamma}}{\sqrt{(c-\gamma)!\,(c+\gamma)!}}\,.$$

By multiplying both sides of this equation by:

$$u_3^{b+c-a}\,v_3^{c+a-b}\,w_3^{a+b-c}\,\frac{1}{(a+b-c)!\,(b+c-a)!\,(c+a-b)!}\,,$$

where u_3, v_3, w_3 are parameters and summing upon all values of a, b, c compatible with the triangular inequality, the left term, if J is kept fixed, becomes the J-th power of a determinant. The right term assumes a most symmetrical form upon introduction of the square symbol:

$$\begin{vmatrix} u_1 & v_1 & w_1 \\ u_2 & v_2 & w_2 \\ u_3 & v_3 & w_3 \end{vmatrix}^J = \tag{3}$$

$$= \left(\frac{2J+1}{J!}\right)^{\frac{1}{2}} \cdot \sum_{\sum_{is} A_{is} = 3J} \boxed{\begin{matrix} A_{11} & A_{21} & A_{31} \\ A_{12} & A_{22} & A_{32} \\ A_{13} & A_{23} & A_{33} \end{matrix}} \frac{u_1^{A_{11}} u_2^{A_{12}} u_3^{A_{13}} v_1^{A_{21}} v_2^{A_{22}} v_3^{A_{23}} w_1^{A_{31}} w_2^{A_{32}} w_3^{A_{33}}}{(A_{11}!\, A_{12}!\, A_{13}!\, A_{21}!\, A_{22}!\, A_{23}!\, A_{31}!\, A_{32}!\, A_{33}!)^{\frac{1}{2}}} .$$

Under this form the properties of $C \cdot G$ symbols appear in full light:

a) The square is multiplied by $(P)^J$ $(P = \pm 1)$ by permutations of columns of parity P. These are known symmetries.

b) The same happens for rows. The exchange of the lowest rows is space reflection. The others are entirely new.

c) Rows can be exchanged with columns. Also this symmetry is new.

Wigner's $9-j$ symbol shows similar properties.

Thus far we cannot justify these symmetries using simple physical arguments. Work is in progress in order to check whether these additional properties entail some consequences for Racah's W coefficients and Wigner's $9-j$ symbols.

* * *

I wish to thank particularly Prof. G. RACAH for many fruitful discussions and kind encouragement.

IL NUOVO CIMENTO   VOL. XI, N. 1   1° Gennaio 1959

Symmetry Properties of Racah's Coefficients.

T. REGGE

Istituto Nazionale di Fisica Nucleare, Sezione di Torino - Torino

(ricevuto il 9 Ottobre 1958)

We have shown in a previous letter (1) that the true symmetry of Clebsch-Gordan coefficients is much higher that is was before believed. A similar result has been now obtained for Racah's coefficients. Although no direct connection has been established between these wider symmetries it seems very probable that it will be found in the future. We shall merely state here the results which can be checked very easily with the help of the well known Racah's formula:

$$\begin{Bmatrix} a & b & c \\ d & e & f \end{Bmatrix} = \left[\frac{(a+b-c)!\,(a+c-b)!\,(b+c-a)!\,(d+b-f)!\,(d+f-b)!\,(f+b-d)!\cdot \cdot(d+e-c)!\,(e+c-d)!\,(c+d-e)!\,(a+e-f)!\,(a+f-e)!\,(f+e-a)!}{(a+b+c+1)!\,(a+e+f+1)!\,(d+e+c+1)!\,(b+d+f+1)!}\right]^{\frac{1}{2}} \tag{1}$$

$$\sum_z (-1)^z \frac{(z+1)!}{(a+b+d+e-z)!\,(b+c+e+f-z)!\,(a+c+d+f-z)!\cdot \cdot(z-a-b-c)!\,(z-a-e-f)!\,(z-b-d-f)!\,(z-d-e-c)}\,.$$

From the usual tethrahedral symmetry group of $\begin{Bmatrix} a & b & c \\ d & e & f \end{Bmatrix}$ we know already that:

$$\begin{Bmatrix} a & b & c \\ d & e & f \end{Bmatrix} = \begin{Bmatrix} b & a & c \\ e & d & f \end{Bmatrix} = \begin{Bmatrix} a & e & f \\ d & b & c \end{Bmatrix} = \begin{Bmatrix} c & e & d \\ f & b & a \end{Bmatrix} = \text{etc.}. \tag{2}$$

---

(1) T. REGGE: *Nuovo Cimento*, **10**, 544 (1958).

Our results can be put into the following form:

$$(3)\qquad \begin{Bmatrix} a & b & c \\ d & e & f \end{Bmatrix} = \left\{ \begin{array}{ccc} a & \dfrac{b+e+c-f}{2} & \dfrac{b+c+f-e}{2} \\ d & \dfrac{b+e+f-c}{2} & \dfrac{c+e+f-b}{2} \end{array} \right\} =$$

$$= \left\{ \begin{array}{ccc} \dfrac{a+c+f-d}{2} & b & \dfrac{a+c+d-f}{2} \\ \dfrac{c+d+f-a}{2} & e & \dfrac{a+d+f-c}{2} \end{array} \right\} = \left\{ \begin{array}{ccc} \dfrac{a+b+e-d}{2} & \dfrac{a+b+d-e}{2} & c \\ \dfrac{b+d+e-a}{2} & \dfrac{a+d+e-b}{2} & f \end{array} \right\} =$$

$$= \left\{ \begin{array}{ccc} \dfrac{b+e+c-f}{2} & \dfrac{a+f+c-d}{2} & \dfrac{a+d+b-e}{2} \\ \dfrac{b+f+e-c}{2} & \dfrac{c+d+f-a}{2} & \dfrac{a+d+e-b}{2} \end{array} \right\} =$$

$$= \left\{ \begin{array}{ccc} \dfrac{b+c+f-e}{2} & \dfrac{a+c+d-f}{2} & \dfrac{a+b+e-d}{2} \\ \dfrac{c+e+f-b}{2} & \dfrac{a+d+f-c}{2} & \dfrac{d+e+b-a}{2} \end{array} \right\}.$$

Only the first of these symmetries is essentially new, the others can be obtained from it and (2). We see therefore that there are 144 identical Racah's coefficients. These new symmetries should reduce by a factor 6 the space required for the tabulation of *W*. It should be pointed out that this wider 144-group is isomorphic to the direct product of the permutation groups of 3 and 4 objects.

REVIEWS OF MODERN PHYSICS VOLUME 34, NUMBER 4 OCTOBER 1962

On the Representations of the Rotation Group

V. BARGMANN
Princeton University, Princeton, New Jersey

THE present paper contains hardly any new result and can claim only a methodological interest. In a recent article[1] I studied a family of Hilbert spaces $\mathfrak{F}_n$, whose elements are entire analytic functions of n complex variables. The methods developed there appear appropriate for a fairly effortless treatment of the representation theory of the rotation group, and this paper is offered in the hope that it may suggest further applications of these methods. (Here, and in the following, the term "rotation group" actually refers to the group $\mathfrak{U}$ of unitary unimodular transformations of a two-dimensional vector space, the spin space of quantum mechanics. It is this group that is basic for the quantum mechanical applications.)

The application of the function spaces $\mathfrak{F}_n$ to the study of the rotation group is related to the long known fact that its irreducible representations may be obtained by considering homogeneous polynomials in two complex variables. (All these polynomials are elements of $\mathfrak{F}_2$, and may thus be treated simultaneously.) This fact has been used, in one form or another, in almost every treatment of the representation theory of the rotation group. It has been most systematically exploited by Kramers and his school,[2] who have applied the concepts and the methods of the theory of binary invariants. Van der Waerden also used it very effectively in his book[3]—for example, in the derivation of the vector coupling coefficients.

It was shown by Wigner—in his profound investigation of simply reducible groups[4]—that remarkably many properties of the 3-j symbols, 6-j symbols, etc. and of their interrelations are shared by all simply reducible groups, and are not confined to the rotation group. By contrast, the present paper is restricted to the rotation group. Naturally, this restriction permits simplifications and short cuts. In addition, we know from Regge's intriguing discovery of unsuspected symmetries of the 3-j and the 6-j symbols[5] that there are important relations which do no longer hold for all simply reducible groups. While the following analysis does not lead to a deeper understanding of the Regge symmetries it yields, at least, a fairly transparent formulation and derivation of the symmetries.

Ten years ago Schwinger published a highly ingenious treatment of the rotation group based on a certain operator method.[6] In a strict mathematical sense, the Hilbert space method of the present paper is isomorphic to Schwinger's operator method. (For a detailed comparison see Sec. 2e below.) The generating functions for the 3-j and the 6-j symbols, in particular, are due to Schwinger.

There are, however, characteristic differences in our approach. (1) Schwinger introduces certain operators $a_\mathfrak{k}$ (and their adjoints) for which the commutation rules of the annihilation and creation operators of boson fields are postulated. All other objects to be studied are defined in terms of the $a_\mathfrak{k}$, including the orthonormal vector basis of the Hilbert space on which the operators $a_\mathfrak{k}$ act. In the present paper, however, the Hilbert space is *a priori* given as a function space, and the standard methods of analysis are available at each step. (2) Schwinger is primarily concerned with angular momenta—in group theoretical terms: with infinitesimal rotations—and he constructs the representations from their infinitesimal generators, while in the present paper the representations are directly defined on the function space $\mathfrak{F}$.

The present paper may be read without any knowledge of the content of the paper (H) of reference 1. To the extent that they are needed the results of (H) are reproduced in Sec. 1. Sections 2 through 4 deal with the rotation group. The representation theory of the rotation group is developed from its beginning—for the convenience of the reader, for the

[1] V. Bargmann, Comm. Pure Appl. Math. **14**, 187 (1961). Hereafter quoted as (H).
[2] For a survey of these methods see H. C. Brinkmann, *Applications of Spinor Invariants in Atomic Physics* (Interscience Publishers, Inc., New York, 1956).
[3] B. L. van der Waerden, *Die gruppentheoretische Methode in der Quantenmechanik* (Verlag Julius Springer, Berlin, Germany, 1932).
[4] An excellent exposition of this investigation is given by W. T. Sharp, "Racah Algebra and the Contraction of Groups." CRT—935 (AECL—1098) Atomic Energy of Canada Ltd., Chalk River, Ontario, 1960 (unpublished).
[5] T. Regge, Nuovo cimento **10**, 544 (1958); **11**, 116 (1959).
[6] J. Schwinger, "On Angular Momentum," U.S. Atomic Energy Commission, NYO—3071, 1952 (unpublished).

sake of logical coherence, and also in order to show that those definitions and constructions which appear natural in the framework of the function space $\mathfrak{F}$ are, at the same time, useful and relevant from a group theoretical point of view. The decomposition of the direct product and the 3-j symbols are treated in Sec. 3, the 6-j symbols in Sec. 4.—No loss in generality is caused by the fact that the representations are constructed on $\mathfrak{F}$, because the main results—for example, the properties of the 3-j and the 6-j symbols —depend only on the representations and not on the vector space on which the representations are realized.

Remarks on the notation. I adopt the definitions and the notation of Wigner's book,[7] with a few exceptions. (1) Complex conjugation is indicated by a bar ($\bar{\alpha}$ is the conjugate of α). (2) The (Hermitian) adjoint of an operator or a matrix A is denoted by A^*. (3) The transpose of a matrix A is denoted by ${}^tA$, and A's determinant by det A. (4) The product of a vector f by a scalar λ will be written either λf or $f\lambda$, whichever appears more convenient.

1. THE HILBERT SPACE $\mathfrak{F}_n$

a. Introductory remarks. The elements of $\mathfrak{F}_n$ are entire analytic functions $f(z)$, where $z = (z_1, z_2, \cdots, z_n)$ is a point of the n-dimensional complex Euclidean space C_n. Every entire $f(z)$ may be expanded in an everywhere converging power series

$$f(z) = \sum_{h_1, \cdots, h_n} \alpha_{h_1 h_2 \cdots h_n} z_1^{h_1} z_2^{h_2} \cdots z_n^{h_n} . \tag{1.1}$$

It will be convenient to use the following shorthand notation. We set

$$h = (h_1, \cdots, h_n)$$

for an ordered set of non-negative integers h_i, and $h = 0$ if all $h_i = 0$. We write α_h for the coefficient $\alpha_{h_1 \cdots h_n}$ and denote the power products in (1.1) by

$$z^{[h]} = z_1^{h_1} z_2^{h_2} \cdots z_n^{h_n} ,$$

so that the power series (1.1) takes the form

$$f(z) = \sum_h \alpha_h z^{[h]} . \tag{1.2}$$

We shall also use the abbreviations

$$|h| = h_1 + h_2 + \cdots + h_n , \quad [h!] = h_1! h_2! \cdots h_n! . \tag{1.3}$$

The elements of the n-dimensional space C_n will be called points or vectors (synonymously); $a \cdot b = \sum_{k=1}^n a_k b_k$ is the scalar product of a and b. In particular, $\bar{a} \cdot a = \sum_k |a_k|^2$.

[7] E. P. Wigner, *Group Theory* (Academic Press Inc., New York, 1959).

b. Definition of the Hilbert space $\mathfrak{F}_n$. The inner product of two elements f, f' of $\mathfrak{F}_n$ is

$$(f, f') = \int \overline{f(z)} f'(z) d\mu_n(z) , \tag{1.4}$$

where

$$d\mu_n(z) = \pi^{-n} \exp(-\bar{z} \cdot z) \prod_k dx_k dy_k , \quad (z_k = x_k + iy_k) . \tag{1.4a}$$

Here and in the following all integrals are extended over the whole space C_n.

The definition (1.4) is meant to imply that an entire function $f(z)$ belongs to $\mathfrak{F}_n$ if and only if

$$(f, f) = \int |f(z)|^2 d\mu_n(z) < \infty . \tag{1.4b}$$

[The norm of f is $\|f\| = (f, f)^{1/2}$.] Separating the Gaussian in (1.4a) we shall occasionally write

$$d\mu_n(z) = \rho_n(z) d^n z , \quad \rho_n(z) = \pi^{-n} \exp(-\bar{z} \cdot z) , \tag{1.5}$$

$$d^n z = \prod_{k=1}^n dx_k dy_k . \tag{1.5a}$$

In order to express the inner product of f and f' in the expansion coefficients of their power series, we first compute $(z^{[h]}, z^{[h']})$. Introducing polar coordinates, $z_k = r_k e^{i\phi_k}$, we have $(z^{[h]}, z^{[h']}) = \omega_1 \omega_2 \cdots \omega_n$,

$$\omega_k = \frac{1}{\pi} \int_0^{2\pi} \exp(i(h_k' - h_k)\phi_k) d\phi_k \times \int_0^\infty r_k^{h_k + h'_k + 1} e^{-r_k^2} dr_k .$$

It follows that $\omega_k = \delta_{h_k, h'_k} h_k!$ Hence

$$(z^{[h]}, z^{[h']}) = \begin{cases} 0 & , h \neq h' , \\ [h!] , & h = h' . \end{cases} \tag{1.6}$$

For two functions of $\mathfrak{F}_n$, $f(z) = \sum \alpha_h z^{[h]}$ and $f'(z) = \sum \alpha'_h z^{[h]}$, one now readily obtains

$$(f, f') = \sum_h [h!] \bar{\alpha}_h \alpha_h' . \tag{1.7}$$

In particular,

$$(f, f) = \sum_h [h!] |\alpha_h|^2 . \tag{1.8}$$

This last equation may be interpreted as follows. For an entire function $f(z)$ either both sides are infinite—in which case f does not belong to $\mathfrak{F}_n$—or both have the same finite value.

The orthonormal set u_h. According to (1.6), the simplest orthonormal set in $\mathfrak{F}_n$ is given by

$$u_h = z^{[h]} / [h!]^{1/2} , \tag{1.9}$$

and Eq. (1.8) expresses its completeness.

The subspaces $\mathfrak{P}_\nu$. Let $\mathfrak{P}_\nu$ be the set of all homogeneous polynomials in $\mathfrak{F}_n$ of order s. It is spanned by those u_h for which $|h| = h_1 + \cdots + h_n = s$. $\mathfrak{P}_s$ and $\mathfrak{P}_{s'}$ are clearly orthogonal if $s \neq s'$, and

$$\mathfrak{F}_n = \mathfrak{P}_0 + \mathfrak{P}_1 + \mathfrak{P}_2 + \cdots . \qquad (1.10)$$

is a decomposition into mutually orthogonal subspaces. It will be useful to introduce

$$\mathfrak{Q}_j = \mathfrak{P}_{2j}\,, \quad (j = 0,\tfrac{1}{2},1,\cdots)\,. \qquad (1.10a)$$

An element f of $\mathfrak{F}_n$ belongs to $\mathfrak{P}_s$ if and only if

$$f(\lambda z) = \lambda^s f(z) \qquad (1.10b)$$

for every constant λ, or alternatively if and only if Euler's equation

$$\textstyle\sum_k z_k(\partial f/\partial z_k) = s \cdot f \qquad (1.10c)$$

is satisfied.

c. The principal vectors $\mathbf{e}_a$. Define for every a in C_n the function $\mathbf{e}_a$ by

$$\mathbf{e}_a(z) = \exp\,(\bar{a} \cdot z)\,. \qquad (1.11)$$

It is clear that $\mathbf{e}_a$ belongs to $\mathfrak{F}_n$. Its power series is

$$\mathbf{e}_a(z) = \sum_h \frac{\bar{a}^{[h]} z^{[h]}}{[h!]}\,. \qquad (1.11a)$$

It follows therefore from (1.7) that for any f in $\mathfrak{F}_n$

$$(\mathbf{e}_a,f) = \textstyle\sum_h a^{[h]}\alpha_h = f(a)\,, \qquad (1.12)$$

or, in integral form

$$\int \exp\,(a \cdot \bar{z}) f(z) d\mu_n(z) = f(a)\,. \qquad (1.12a)$$

The existence of these "principal vectors" $\mathbf{e}_a$ is a characteristic feature of $\mathfrak{F}_n$. It is seen that they play here a role similar to that of the δ functions $\delta(q - a)$ in the standard Hilbert space of quantum mechanics, but unlike the δ functions they are elements of Hilbert space.

Applying (1.12) to $f = \mathbf{e}_b$ we have

$$(\mathbf{e}_a,\mathbf{e}_b) = \mathbf{e}_b(a) = \exp\,(\bar{b} \cdot a) \qquad (1.13)$$

and hence $(\mathbf{e}_a,\mathbf{e}_a) = \exp\,(\bar{a} \cdot a)$.

By Schwarz's inequality we conclude from (1.12) that

$$|f(z)| \leq ||f|| \cdot ||\mathbf{e}_z|| \leq ||f|| \exp\,(\tfrac{1}{2}\,\bar{z} \cdot z) \qquad (1.13a)$$

Conversely, if an entire function $f(z)$ satisfies the inequality

$$|f(z)| \leq c \exp\,(\tfrac{1}{2}\,\gamma \bar{z} \cdot z) \qquad (1.13b)$$

where c and γ are positive constants and $\gamma < 1$, then, by the integral definition (1.4b), f belongs to $\mathfrak{F}_n$. (The constant $\gamma < 1$ must not be omitted!)

d. Product decomposition of $\mathfrak{F}_n$. To every decomposition of n into the sum of two positive integers, $n = n' + n''$, corresponds a decomposition of $\mathfrak{F}_n$ into the direct product

$$\mathfrak{F}_n = \mathfrak{F}_{n'} \otimes \mathfrak{F}_{n''}\,. \qquad (1.14)$$

Set $z' = (z_1, \ldots, z_{n'})$ and $z'' = (z_{n'+1}, \ldots, z_n)$. If $f'(z')$ and $f''(z'')$ belong to $\mathfrak{F}_n'$ and $\mathfrak{F}_n''$, respectively, the product $f(z) = f'(z')\,f''(z'')$ belongs to $\mathfrak{F}_n$. Furthermore, $d\mu_n(z) = d\mu_{n'}(z')\,d\mu_{n''}(z'')$ by (1.4a), and for the inner product of f with $g(z) = g'(z')\,g''(z'')$ one obtains

$$(f,g) = (f',g')(f'',g'')\,,$$

the two factors (f',g') and (f'',g'') being taken on $\mathfrak{F}_{n'}$ and $\mathfrak{F}_{n''}$. The orthonormal functions u_h as well as the principal vectors $\mathbf{e}_a$ are decomposed accordingly.

Similarly one can form products of subspaces of $\mathfrak{F}_{n'}$ and $\mathfrak{F}_{n''}$, for example,

$$\mathfrak{P}_{s's''} = \mathfrak{P}'_{s'} \otimes \mathfrak{P}''_{s''}\,, \quad \mathfrak{Q}_{j'j''} = \mathfrak{Q}'_{j'} \otimes \mathfrak{Q}''_{j''} \qquad (1.14a)$$

[see (1.10) and (1.10a)], which contains all polynomials homogeneous in z' of order s' and in z'' of order s''. The functions f in $\mathfrak{P}_{s's''}$ are characterized by

$$f(\lambda' z',\lambda'' z'') = \lambda'^{s'} \lambda''^{s''} f(z',z'')$$

for any complex constants λ',λ''.

e. Operators on $\mathfrak{F}_n$. We turn now to a brief review of some operators which occur in the following.

(α) *The operators* z_k *and* d_k. Here d_k stands for the differential operator $\partial/\partial z_k$. Since the elements f of $\mathfrak{F}_n$ are analytic, $z_k f$ and $d_k f$ are always defined as analytic functions, but they do not necessarily belong to $\mathfrak{F}_n$. We shall apply, however, the operators z_k and d_k only to polynomials, so that no difficulties arise.

The d_k, z_l evidently satisfy the commutation rules

$$[z_k,z_l] = 0\,, \quad [d_k,d_l] = 0\,, \quad [d_k,z_l] = \delta_{kl}\,. \qquad (1.15)$$

Furthermore, z_k and d_k are *adjoint* [with respect to the inner product (1.4)],

$$z_k = d_k^*\,, \qquad (1.15a)$$

i.e., for any f,g in $\mathfrak{F}_n$,

$$(z_k f,g) = (f,d_k g) \qquad (1.16)$$

whenever $z_k f$ and $d_k g$ are in $\mathfrak{F}_n$. For simplicity, set $k = 1$. Write, for any $h = (h_1,h_2,\cdots,h_n)$, $h' = (1$

$+ h_1, h_2, \cdots, h_n)$. If $f = \sum \alpha_h z^{[h]}$ and $g = \sum \beta_h z^{[h]}$, we have

$$z_1 f = \sum \alpha_h z^{[h']}, \quad d_1 g = \sum (1 + h_1)\beta_{h'} z^{[h]},$$

$$(z_1 f, g) = \sum_h [h'!]\bar{\alpha}_h \beta_{h'},$$

$$(f, d_1 g) = \sum_h (1 + h_1)[h!]\bar{\alpha}_h \beta_{h'},$$

which proves (1.16) because $(1 + h_1)[h!] = [h'!]$.

It follows from (1.15) and (1.15a) that the operators d_k, z_k satisfy the defining relations for the annihilation and creation operators of boson fields.[8]

(β) *The unitary transformations* T_U. For every unitary transformation U on C_n we define an operator T_U on $\mathfrak{F}_n$ by[9]

$$(T_U f)(z) = f({}^tUz) \tag{1.17}$$

where ${}^tU$ is the transpose of the matrix U. T_U is clearly a linear operator (i.e., linear in f), and for two unitary transformations U, U'

$$T_U T_{U'} = T_{UU'}. \tag{1.17a}$$

If $U = 1$, then $T = 1$ (identity), so that $T_{U^{-1}} = T_U^{-1}$.

In addition T_U is *unitary*. Introducing the variables $z' = {}^tUz$ in the integral (1.4) one finds that

$$(T_U f, T_U g) = (f, g) \tag{1.18}$$

because the measure $d\mu_n(z)$ is invariant under unitary transformations of the z.

It follows that the T_u form a *unitary representation* of the n-dimensional unitary group, and also of any of its subgroups.

The representation is decomposed because any subspace $\mathfrak{P}_s$ is clearly carried into itself [apply, for example, the criterion (1.10b)]. In the case $n = 2$ this will provide the basis for our discussion of the rotation group.

(γ) *The conjugation* K. The last operator to be considered is the conjugation K, which is defined as follows. Let $g = Kf$, then

$$g(z) = \overline{f(\bar{z})}, \tag{1.19}$$

where the bar, as before, denotes complex conjugation. For $f = \sum \alpha_h z^{[h]}$ we find

$$g(z) = \sum \bar{\alpha}_h z^{[h]}, \tag{1.19a}$$

i.e., the power series with complex conjugate coefficients.

We note the following properties of K:

(1) K is *antilinear*, i.e.,

$$K(f_1 + f_2) = Kf_1 + Kf_2, \quad K(\lambda f) = \bar{\lambda} Kf$$

for any complex constant λ.

(2) $K^2 = 1$.

(3) $(Kf, Kf') = (f', f) = \overline{(f, f')}$,

i.e., K is *antiunitary*. [(3) follows from either definition of the inner product, (1.4) or (1.6).]

A function f may be called *real* if $Kf = f$ (so that its power series has real coefficients). Thus $z^{[h]}$ and u_h are real.

With the help of K we may also define the complex conjugate of a linear operator A on $\mathfrak{F}_n$ by setting

$$\bar{A} = KAK. \tag{1.20}$$

$\bar{A}$ itself is *linear* since K appears an even number of times in the definition (1.20). If $B = \bar{A}$, then $\bar{B} = A$. Let

$$Au_h = \sum_{h'} u_{h'} a_{h'h},$$

where $a_{h'h}$ are the matrix elements of A in the system u_h. Then, since $Ku_h = u_h$,

$$\bar{A}u_h = K(Au_h) = \sum_{h'} u_{h'} \overline{a_{h'h}}. \tag{1.21}$$

Thus, $\bar{A}$'s matrix elements are complex conjugate to those of A.

Application to T_U. If $\bar{U}$ is the matrix complex conjugate to U,

$$\overline{T_U} = T_{\bar{U}}. \tag{1.22}$$

Proof. Let $g = \overline{T_U} f$ and set, successively, $f_1 = Kf, f_2 = T_U f_1, g = Kf_2$. By definition, $g(z) = \overline{f_2(\bar{z})}$, $f_2(\bar{z}) = f_1({}^tU\bar{z}) = f_1(y)$, and, finally, $f_1(y) = \overline{f(\bar{y})} = \overline{f({}^t\bar{U}z)}$. Hence $h(z) = f({}^t\bar{U}z)$, Q.E.D.

2. THE REPRESENTATIONS $\mathfrak{D}_j$

a. The group $\mathfrak{U}$. We start with a brief review of the group $\mathfrak{U}$ of unimodular unitary transformations in two dimensions and its connection with the rotation group.

The vectors in C_2 will be denoted by ζ, with components ζ_1, ζ_2. (In dealing with several vectors ζ, we shall often denote their components by ξ, η instead of ζ_1, ζ_2 in order to avoid a profusion of indices.) The (Hermitian) inner product of two vectors ζ, ζ' is

$$\bar{\zeta} \cdot \zeta' = \bar{\zeta}_1 \zeta_1' + \bar{\zeta}_2 \zeta_2'.$$

Denoting the Hermitian Pauli spin matrices by

$$\sigma_1 = \begin{pmatrix} 0 & 1 \\ 1 & 0 \end{pmatrix} \quad \sigma_2 = \begin{pmatrix} 0 & -i \\ i & 0 \end{pmatrix} \quad \sigma_3 = \begin{pmatrix} 1 & 0 \\ 0 & -1 \end{pmatrix} \tag{2.1}$$

---

[8] I. E. Segal has used a generalization of $\mathfrak{F}_n$ to $\mathfrak{F}_\infty$ for a comprehensive study of the canonical operators of quantum field theory, where infinitely many d_k, z_k occur. (Lectures at the Summer Seminar on Applied Mathematics, 1960, Boulder, Colorado, unpublished.)

[9] This differs somewhat from the corresponding definition in (H) (reference 1), Eq. (3.4), p. 205.

we write

$$b_1\sigma_1 + b_2\sigma_2 + b_3\sigma_3 = \mathbf{b}\cdot\sigma \; ; \quad \mathbf{b} = (b_1,b_2,b_3)$$

for a three-vector **b** with real or complex components. Every 2×2 matrix B may be expressed in the form

$$B = b_0\cdot 1 + \mathbf{b}\cdot\sigma \tag{2.1a}$$

with uniquely determined $b_0,\mathbf{b}$.

The algebraic properties of the spin matrices are summarized in

$$(\mathbf{a}\cdot\sigma)(\mathbf{b}\cdot\sigma) + (\mathbf{b}\cdot\sigma)(\mathbf{a}\cdot\sigma) = 2(\mathbf{a}\cdot\mathbf{b})1$$
$$(\mathbf{a}\cdot\sigma)(\mathbf{b}\cdot\sigma) - (\mathbf{b}\cdot\sigma)(\mathbf{a}\cdot\sigma) = 2i(\mathbf{a}\times\mathbf{b})\cdot\sigma \tag{2.2}$$

for any two vectors $\mathbf{a},\mathbf{b}$, where $\mathbf{a}\times\mathbf{b}$ denotes the vector product.

In the following, the matrix

$$\Gamma = \begin{pmatrix} 0 & -1 \\ 1 & 0 \end{pmatrix} \tag{2.3}$$

will play an important role. (It is the basic matrix ϵ of the spinor calculus.) We note that

$${}^t\Gamma = -\Gamma\,, \quad \Gamma^2 = -1\,, \quad {}^t\Gamma\cdot\Gamma = 1\,, \quad \det\Gamma = 1\,, \tag{2.3a}$$

where "det" denotes the determinant.

For every 2×2 matrix B we define the *associate* matrix B_a by

$$B_a = \Gamma B \Gamma^{-1} \tag{2.4}$$

If $B = \binom{\alpha\ \beta}{\gamma\ \delta}$, then $B_a = \binom{\delta\ -\gamma}{-\beta\ \alpha}$. (The elements of B_a are the minors of B.) It follows that

$$({}^tB)_a = {}^t(B_a)\,, \quad (B^{-1})_a = (B_a)^{-1}\,, \quad (BC)_a = B_a\cdot C_a\,, \tag{2.4a}$$

$$B\cdot{}^tB_a = {}^tB\cdot B_a = (\det B)\cdot 1\,. \tag{2.4b}$$

Since for the spin matrices $\sigma_k^2 = 1$, $\det\sigma_k = -1$, we obtain from (2.4b)

$$({}^t\sigma_k)_a = -\sigma_k\,. \tag{2.4c}$$

Hence, for any B written in the form (2.1a),

$$B\cdot{}^tB_a = (b_0 + \mathbf{b}\cdot\sigma)(b_0 - \mathbf{b}\cdot\sigma) = (b_0^2 - b^2)\cdot 1$$
$$\det B = b_0^2 - b^2\,, \quad b^2 = \mathbf{b}\cdot\mathbf{b}\,. \tag{2.4d}$$

The group $\mathfrak{U}$. A matrix U belongs to $\mathfrak{U}$ if and only if ${}^tU\cdot\bar{U} = 1$, and $\det U = 1$. In view of (2.4b) these conditions may be replaced by

$$U_a = \Gamma U \Gamma^{-1} = \bar{U}\; ; \quad \det U = 1\,. \tag{2.5}$$

Let $U = b_0 + \mathbf{b}\cdot\sigma\epsilon\mathfrak{U}$. Then $U^* = \bar{b}_0 + \bar{\mathbf{b}}\cdot\sigma = U^{-1}$. By (2.4d), $U^{-1} = b_0 - \mathbf{b}\cdot\sigma$. Hence b_0 is real, and $\mathbf{b}$ imaginary. Setting $b_0 = a_0$, $\mathbf{b} = -i\mathbf{a}$, we find that U belongs to $\mathfrak{U}$ if and only if

$$U = a_0 - i\mathbf{a}\cdot\sigma\,, \quad \det U = a_0^2 + a^2 = 1\,, \tag{2.6}$$

a_0, **a** real. In matrix form

$$U = \begin{pmatrix}\alpha & \beta \\ \gamma & \delta\end{pmatrix} = \begin{pmatrix} a_0 - ia_3 & -ia_1 - a_2 \\ -ia_1 + a_2 & a_0 + ia_3\end{pmatrix}$$
$$\delta = \bar{\alpha}\,, \quad \gamma = -\bar{\beta}\,, \quad \alpha\bar{\alpha} + \beta\bar{\beta} = 1\,. \tag{2.6a}$$

Connection with the rotation group. Every U in $\mathfrak{U}$ defines a rotation $\mathbf{r}' = R_U\mathbf{r}$ by

$$\mathbf{r}'\cdot\sigma = U(\mathbf{r}\cdot\sigma)U^{-1}\,, \tag{2.7}$$

so that $R_{U_1}R_{U_2} = R_{U_1U_2}$, and $R_{-U} = R_U$. Using (2.2) one obtains by straightforward computation

$$\mathbf{r}' = R_U\mathbf{r} = (a_0^2 - a^2)\mathbf{r} + 2(\mathbf{a}\cdot\mathbf{r})\mathbf{a} + 2a_0(\mathbf{a}\times\mathbf{r})\,, \tag{2.7a}$$

the well known expression of a rotation in terms of Euler's *homogeneous parameters*. Specifically,

$$a_0 = \cos(\tfrac{1}{2}\phi)\,, \quad \mathbf{a} = \sin(\tfrac{1}{2}\phi)\mathbf{n}\,, \quad (\mathbf{n}\cdot\mathbf{n} = 1)\,, \tag{2.7b}$$

where $\mathbf{n}$ is the axis and ϕ the angle of the rotation R_U.

To the one-parametric subgroup of rotations about the axis $\mathbf{n}$ corresponds the subgroup

$$U(\phi) = \cos(\tfrac{1}{2}\phi) - i\sin(\tfrac{1}{2}\phi)\mathbf{n}\cdot\sigma$$
$$= \exp[-\tfrac{1}{2}i\phi(\mathbf{n}\cdot\sigma)] \tag{2.7c}$$

of $\mathfrak{U}$.

b. The representations $\mathfrak{D}^j$. It is now easy to obtain some of the basic results concerning the representations $\mathfrak{D}^j$ of $\mathfrak{U}$.

On the Hilbert space $\mathfrak{F}_2$ of analytic functions $f(\zeta)$ (we write now ζ instead of z) the operators T_U,

$$(T_Uf)(\zeta) = f({}^tU\zeta)\,, \tag{2.8}$$

provide a unitary representation of the group $\mathfrak{U}$, as was shown in Sec. 1e.

The subspace $\mathfrak{Q}_j = \mathfrak{P}_{2j}$ of homogeneous polynomials of order $2j$—where $2j = 0,1,2,\cdots$—is *invariant* under the transformations T_U, and $\mathfrak{D}^j$ is the representation of $\mathfrak{U}$ defined by the *restriction* of T_U to $\mathfrak{Q}_j$. Since different $\mathfrak{Q}_j$ have different dimensions, the various representations $\mathfrak{D}^j$ are clearly *inequivalent.*

According to the first section—see Eq. (1.9)—$\mathfrak{Q}_j$ is spanned by the $2j+1$ orthonormal functions

$$\zeta_1^\kappa\zeta_2^\lambda/(\kappa!\lambda!)^{1/2} = \xi^\kappa\eta^\lambda/(\kappa!\lambda!)^{1/2}\,, \quad (\kappa + \lambda = 2j) \tag{2.9}$$

or, with $m = j, j-1, \cdots, -j$,

$$v_m^j = \xi^{j+m}\eta^{j-m}/[(j+m)!(j-m)!]^{1/2}\,, \quad (\kappa - \lambda = 2m) \tag{2.9a}$$

If U is given by (2.6a), then

$$T_U v_m^j = \frac{(\alpha\xi + \gamma\eta)^{j+m}(\beta\xi + \delta\eta)^{j-m}}{[(j+m)!(j-m)!]^{1/2}} . \quad (2.10)$$

The matrix elements $\mathfrak{D}_m^{jm'}(U)$ are defined by

$$T_U v_m^j = \sum_{m'} v_{m'}^j \mathfrak{D}_m^{jm'}(U) , \quad \mathfrak{D}_m^{jm'}(U) = (v_{m'}^j, T_U v_m^j) , \quad (2.10a)$$

and their explicit form may be deduced from (2.10).

For rotations about the z-axis, $U(\phi) = \cos(\frac{1}{2}\phi) - i \sin(\frac{1}{2}\phi)\sigma_3$, so that, in (2.6a), $\alpha = e^{-i\phi/2}$, $\delta = e^{i\phi/2}$, $\beta = \gamma = 0$, and

$$T_U v_m^j = e^{-im\phi} v_m^j \quad (2.10b)$$

c. Infinitesimal transformations. Consider the one-parametric subgroup (2.7c), and the corresponding transformations $T_{U(\phi)}$. The infinitesimal generator of $T_{U(\phi)}$ may then be defined by

$$(\mathbf{n}\cdot\mathbf{M})f = i(d/d\phi)T_{U(\phi)}f|_{\phi=0} . \quad (2.11)$$

One obtains from (2.8) the expression

$$((\mathbf{n}\cdot\mathbf{M})f)(\zeta) = \tfrac{1}{2}\sum_{\alpha,\beta=1}^{2} \zeta_\alpha(\mathbf{n}\cdot\sigma)_{\alpha\beta}\frac{\partial f(\zeta)}{\partial\zeta_\beta} , \quad (2.11a)$$

where $(\mathbf{n}\cdot\sigma)_{\alpha\beta}$ are the matrix elements of $\mathbf{n}\cdot\sigma$. Hence

$$\mathbf{n}\cdot\mathbf{M} = n_1M_1 + n_2M_2 + n_3M_3$$
$$M_k = \tfrac{1}{2}\sum_{\alpha,\beta}\zeta_\alpha(\sigma_k)_{\alpha\beta}d_\beta , \quad d_\beta = \partial/\partial\zeta_\beta \quad (2.12)$$

The operators M_k transform each $\mathfrak{Q}_j$ into itself. [If f is a homogeneous polynomial of order $2j$, so is M_kf, by (2.11a).] Furthermore, they are self-adjoint. This may be inferred from the fact that $-i(\mathbf{n}\cdot\mathbf{M})$ is an infinitesimal unitary operator, or from the explicit expression (2.12) because σ_k is a Hermitian matrix, and $(\zeta_\alpha d_\beta)^* = \zeta_\beta d_\alpha$, by (1.15a).

For the *commutator* of $\mathbf{n}\cdot\mathbf{M}$ and $\mathbf{n}'\cdot\mathbf{M}$ one readily obtains

$$\begin{aligned}[\mathbf{n}\cdot\mathbf{M}, \mathbf{n}'\cdot\mathbf{M}] &= \tfrac{1}{4}\sum_{\alpha,\beta}\zeta_\alpha[\mathbf{n}\cdot\sigma, \mathbf{n}'\cdot\sigma]_{\alpha\beta}d_\beta \\ &= (i/2)\sum\zeta_\alpha((\mathbf{n}\times\mathbf{n}')\cdot\sigma)_{\alpha\beta}d_\beta \\ &= i(\mathbf{n}\times\mathbf{n}')\cdot\mathbf{M}\end{aligned}$$

where (2.2) has been used. Thus

$$[M_1,M_2] = iM_3 , \quad [M_2,M_3] = iM_1 ,$$
$$[M_3,M_1] = iM_2 . \quad (2.12a)$$

From (2.12),

$$M_1 + iM_2 = \zeta_1 d_2 , \quad M_1 - iM_2 = \zeta_2 d_1 ,$$
$$M_3 = \tfrac{1}{2}(\zeta_1 d_1 - \zeta_2 d_2) , \quad (2.13)$$

so that, for example

$$M_3 v_m^j = m v_m^j$$

in accordance with (2.10b).

Lastly,

$$\begin{aligned}M^2 = \sum_{k=1}^{3} M_k^2 &= M_3^2 + M_3 \\ &+ (M_1 - iM_2)(M_1 + iM_2) \\ &= \tfrac{1}{4}(\zeta_1d_1 + \zeta_2d_2)^2 + \tfrac{1}{2}(\zeta_1d_1 + \zeta_2d_2) \\ &= N(N+1) ,\end{aligned}$$

where

$$N = \tfrac{1}{2}(\zeta_1d_1 + \zeta_2d_2) .$$

On $\mathfrak{P}_{2j}$, $Nf = jf$ [see (1.10c)], hence $M^2f = j(j+1)f$.

Remark. Two questions have not yet been considered, (1) the *irreducibility*, (2) the *completeness* of the representations constructed so far. (1) To prove the irreducibility of $\mathfrak{D}^j$ it suffices to show that every linear operator A defined on $\mathfrak{Q}_j$ which commutes with all T_U is necessarily of the form $A = \alpha\cdot 1$. If A commutes with all T_U, it also commutes with all M_k [by (2.11)], and a standard computation, using (2.13), shows that this indeed implies $A = \alpha\cdot 1$. (2) The completeness is a much deeper problem, and it is doubtful whether the existing proofs by integral methods (Wigner, reference 7, p. 166) or by differential (Lie group) methods (Waerden, reference 3, Sec. 17) can be essentially simplified. In any event, the particular method of this paper does not seem to contribute anything to this problem.

d. Complex conjugation. At the end of the first section we saw that $T_{\bar{U}} = \overline{T_U}$. Since the transition to $\overline{T_U}$ implies also the transition to the *complex conjugate* matrix elements in the system v_m^j, we have

$$\mathfrak{D}^j(\bar{U}) = \overline{\mathfrak{D}^j(U)} . \quad (2.14)$$

It follows from the unitarity of the matrices $\mathfrak{D}^j$ that

$$\mathfrak{D}^j(U^*) = \mathfrak{D}^j(U^{-1}) = (\mathfrak{D}^j(U))^{-1} = (\mathfrak{D}^j(U))^*$$

and hence

$$\mathfrak{D}^j({}^tU) = \mathfrak{D}^j(\overline{U^*}) = \overline{(\mathfrak{D}^j(U))^*} = {}^t\mathfrak{D}^j(U) . \quad (2.14a)$$

The matrix Γ introduced in (2.3) belongs to the group $\mathfrak{U}$. Therefore the relation $\Gamma U\Gamma^{-1} = \bar{U}$ implies that $\overline{T_U} = T_\Gamma T_U T_\Gamma^{-1}$, in particular[10]

$$\overline{\mathfrak{D}^j(U)} = C^j\mathfrak{D}^j(U)(C^j)^{-1} ; \quad C^j = \mathfrak{D}^j(\Gamma) . \quad (2.15)$$

[10] On $\mathfrak{Q}_{1/2}$, $D^{1/2}(U) = U$, and $C^{1/2} = \Gamma$.

The relations ${}^t\Gamma = -\Gamma = \Gamma^{-1}$, $\Gamma^2 = -1$ imply

$$ {}^tC^j = (-1)^{2j}C^j = (C^j)^{-1}, \quad (C^j)^2 = (-1)^{2j}, \quad (2.15a)$$

because $\mathfrak{D}^j(-1) = (-1)^{2j}$.

Setting

$$w^j_m = T_\Gamma^{-1} v^j_m , \quad (2.16)$$

we obtain a new orthonormal system for which

$$T_U w^j_m = \sum_{m'} w^j_{m'} \overline{\mathfrak{D}^{jm'}_m(U)} . \quad (2.16a)$$

In fact,

$$T_U w^j_m = T_U T_\Gamma^{-1} v^j_m = T_\Gamma^{-1} \overline{T}_U v^j_m$$

$$= T_\Gamma^{-1}(\sum_{m'} v^j_{m'} \overline{\mathfrak{D}^{jm'}_m(U)}) = \sum_{m'} w^j_{m'} \overline{\mathfrak{D}^{jm'}_m(U)}$$

For any function $f(\zeta)$, set $T_{\Gamma^{-1}}f = g$. Then $g(\zeta) = f({}^t\Gamma^{-1}\zeta) = f(\Gamma\zeta)$, i.e.,

$$g(\zeta_1,\zeta_2) = f(-\zeta_2,\zeta_1) . \quad (2.16b)$$

Thus,

$$w^j_m = (-1)^{j+m} v^j_{-m} . \quad (2.16c)$$

Now $\quad w^j_m = \sum_{m'} v^j_{m'} (C^j)^{-1}_{m'm} = \sum_{m'} C^j_{mm'} v^j_{m'} ,$

$$v^j_m = \sum_{m'} w^j_{m'} C^j_{m'm} \quad (2.16d)$$

(where ${}^tC^j = (C^j)^{-1}$ has been used). Hence

$$C^j_{mm'} = (-1)^{j+m}\delta_{m,-m'} = (-1)^{j-m'}\delta_{m,-m'} . \quad (2.16e)$$

e. Comparison with Schwinger's method. Schwinger starts with the introduction of operators a which correspond to the d_α, ζ_α introduced above:

$$a_+ \to d_1 , \quad a_- \to d_2 , \quad a_+^+ \to \zeta_1 , \quad a_-^+ \to \zeta_2 . \quad (2.17)$$

For them he postulates the commutation rules (1.15) as well as the adjointness (1.15a). In terms of the operators a he next defines the operators J_k corresponding to the M_k of (2.12) above, as well as the orthonormal system of vectors which span the Hilbert space on which the a operate. The basic vector is ψ_0 which corresponds to $v^0_0 = 1$ used here, since $a_+\psi_0 = a_-\psi_0 = 0$ (or $\partial\psi_0/\partial\zeta_1 = \partial\psi_0/\partial\zeta_2 = 0$), and $\psi(jm)$ is defined by

$$\psi(jm) = \frac{(a_+^+)^{j+m}(a_-^+)^{j-m}}{[(j+m)!(j-m)!]^{1/2}}\psi_0$$

which, by (2.17), corresponds to

$$\frac{\zeta_1^{j+m}\zeta_2^{j-m}}{[(i+m)!(j-m)!]^{1/2}}\cdot 1 ,$$

i.e., to v^j_m. In addition, the action of the operators a on the ψ_{jm} is precisely the same as the action of the cooresponding d_α, ζ_α on the v^j_m, so that the *isomorphism* of the two methods is established. One may say that the function space $\mathfrak{F}$ with its operators d_α, ζ_α is a realization of Schwinger's more abstractly defined system.

3. THE DECOMPOSITION OF THE DIRECT PRODUCT AND THE 3-j SYMBOLS

In terms of the quantum-mechanical vector addition model the decomposition of the direct product $\mathfrak{D}^{j_1} \otimes \mathfrak{D}^{j_2}$ answers the question how two angular momenta $\mathbf{j}_1,\mathbf{j}_2$ combine to a third one, $\mathbf{j}' = \mathbf{j}_1 + \mathbf{j}_2$. The details of the answer are contained in the vector coupling coefficients. Setting $\mathbf{j}_3 = -\mathbf{j}'$ one may, alternatively, ask under what conditions $\mathbf{j}_1 + \mathbf{j}_2 + \mathbf{j}_3 = 0$. This latter problem leads to Wigner's 3-j symbols, and its greater symmetry (in $\mathbf{j}_1,\mathbf{j}_2,\mathbf{j}_3$) is the cause for the greater symmetry of the 3-j symbols.

a. Preliminary remarks on representation theory. We recall the following facts. Let V_α be a family of unitary operators defined on the unitary vector space $\mathfrak{V}$, and let $e_1,e_2,\cdots,e_m$ and $f_1,f_2,\cdots,f_n$ be two sets of vectors in $\mathfrak{V}$ which transform under V_α as follows:

$$V_\alpha e_i = \sum_{j=1}^{m} e_j\rho_{ji}(\alpha) \; ; \quad V_\alpha f_r = \sum_{s=1}^{n} f_s\sigma_{sr}(\alpha) . \quad (3.1)$$

(The case $m = n$, $f_i = e_i$ is not excluded!) The matrices $\rho_{ji}(\alpha)$ and $\sigma_{sr}(\alpha)$ are assumed *unitary* and *irreducible.*

Consider the inner products

$$\beta_{ir} = (e_i,f_r)$$

By the unitarity of V_α we obtain from (3.1)

$$\beta_{ir} = (V_\alpha e_i, V_\alpha f_r) = \sum_{j,s} \overline{\rho_{ji}(\alpha)}\beta_{js}\sigma_{sr}(\alpha)$$

In matrix form $\beta = \rho^*(\alpha)\beta\sigma(\alpha)$, and since ρ is unitary,

$$\rho(\alpha)\beta = \beta\sigma(\alpha) .$$

Schur's lemma now implies the following:

(1) If ρ and σ are *inequivalent*, then $\beta = 0$, i.e.,

$$(e_i,f_r) = 0 , \quad \text{for all } i,r . \quad (3.2)$$

(2) If $\rho = \sigma$ (hence $m = n$), $(e_i,f_r) = \beta_{ir} = \lambda\delta_{ir}$. This holds in particular for $f_i = e_i$, so that

$$(e_i,e_j) = \lambda\delta_{ij} \; ; \quad ||e_1||^2 = ||e_2||^2 = \cdots = ||e_n||^2 = \lambda . \quad (3.2a)$$

b. The product representation $\mathfrak{D}^{j_1} \otimes \mathfrak{D}^{j_2}$. Our treatment of the direct products $\mathfrak{D}^{j_1} \otimes \mathfrak{D}^{j_2}$ is based on the decomposition of $\mathfrak{F}_n$ discussed in Sec. 1d, specifically the decomposition of $\mathfrak{F}_4$. Set $\zeta' = (\xi_1,\eta_1)$, $\zeta'' = (\xi_2,\eta_2)$ and let $\mathfrak{F}_2'$ and $\mathfrak{F}_2''$ be the Hilbert spaces of analytic functions $f(\zeta')$ and $f(\zeta'')$ respectively. Then $\mathfrak{F}_4 = \mathfrak{F}_2' \otimes \mathfrak{F}_2''$ is a Hilbert space of analytic functions $f(\zeta',\zeta'')$ or $f(z)$ where $z = (z_1,z_2,z_3,z_4) = (\xi_1,\eta_1,\xi_2,\eta_2)$.

For any U in $\mathfrak{U}$ the operators T_U' and T_U'' are defined on $\mathfrak{F}_2'$ and $\mathfrak{F}_2''$, respectively, by Eq. (2.8). For a function $f(\zeta',\zeta'')$ in $\mathfrak{F}_4$ we set correspondingly

$$(T_U^{(2)}f)(\zeta',\zeta'') = f({}^tU\zeta', {}^tU\zeta'') . \tag{3.3}$$

As shown in Sec. 1e the operators $T_U^{(2)}$ form a unitary representation of $\mathfrak{U}$, and furthermore

$$T_U^{(2)} = T_U' \otimes T_U'' , \tag{3.3a}$$

for if $f(\zeta',\zeta'') = f_1(\zeta')f_2(\zeta'')$, then $T_U^{(2)}f = (T_U'f_1)(T_U''f_2)$.

It follows from (3.3) that the infinitesimal transformations corresponding to $T_U^{(2)}$ are

$$M_k^{(2)} = M_k' + M_k'' , \quad (k = 1,2,3) \tag{3.3b}$$

where M'_k and M''_k are formed according to (2.12) for ζ' and ζ'', respectively. All M'_k commute with all M''_l.

The subspace $\mathfrak{Q}_{j_1j_2} = \mathfrak{Q}_{j_1}' \otimes \mathfrak{Q}_{j_2}''$ of $\mathfrak{F}_4$ (see (1.14)) is spanned by the $(2_{j_1} + 1)(2_{j_2} + 1)$ orthonormal functions

$$v_{m_1}^{j_1}(\zeta')v_{m_2}^{j_2}(\zeta'') = \frac{\xi_1^{j_1+m_1}\eta_1^{j_1-m_1}\xi_2^{j_2+m_2}\eta_2^{j_2-m_2}}{[(j_1+m_1)!(j_1-m_1)!(j_2+m_2)!(j_2-m_2)!]^{1/2}} = \frac{\xi_1^{\kappa_1}\xi_2^{\kappa_2}\eta_1^{\lambda_1}\eta_2^{\lambda_2}}{[\kappa_1!\kappa_2!\lambda_1!\lambda_2!]^{1/2}} , \tag{3.4}$$

$$\kappa_\alpha + \lambda_\alpha = 2j_\alpha , \quad \kappa_\alpha - \lambda_\alpha = 2m_\alpha , \quad (\alpha = 1,2) . \tag{3.4a}$$

$\mathfrak{Q}_{j_1j_2}$ is invariant under $T_U^{(2)}$, and $T_U^{(2)}$ restricted to $\mathfrak{Q}_{j_1j_2}$ provides the product representation $\mathfrak{D}^{j_1} \otimes \mathfrak{D}^{j_2}$.

It is clear how this is generalized to the product of more than two spaces, for example $\mathfrak{F}_6 = \mathfrak{F}_2 \otimes \mathfrak{F}_2'' \otimes \mathfrak{F}_2'''$, the Hilbert space of analytic functions $f(\zeta',\zeta'',\zeta''')$. The subspace $\mathfrak{Q}_{j_1j_2j_3} = \mathfrak{Q}_{j_1}' \otimes \mathfrak{Q}_{j_2}'' \otimes \mathfrak{Q}_{j_3}''' = \mathfrak{Q}_{j_1j_2} \otimes \mathfrak{Q}_{j_3}'''$ is spanned by

$$v_{m_1}^{j_1}(\zeta')v_{m_2}^{j_2}(\zeta'')v_{m_3}^{j_3}(\zeta''') = \frac{\xi_1^{\kappa_1}\xi_2^{\kappa_2}\xi_3^{\kappa_3}\eta_1^{\lambda_1}\eta_2^{\lambda_2}\eta_3^{\lambda_3}}{[\kappa_1!\kappa_2!\kappa_3!\lambda_1!\lambda_2!\lambda_3!]^{1/2}} = \frac{\xi^{[\kappa]}\eta^{[\lambda]}}{([\kappa!][\lambda!])^{1/2}} , \tag{3.5}$$

where $\xi = (\xi_1,\xi_2,\xi_3)$, $\eta = (\eta_1,\eta_2,\eta_3)$, $\kappa = (\kappa_1,\kappa_2,\kappa_3)$, $\lambda = (\lambda_1,\lambda_2,\lambda_3)$, and

$$\kappa_\alpha + \lambda_\alpha = 2j_\alpha , \quad \kappa_\alpha - \lambda_\alpha = 2m_\alpha , \quad (\alpha = 1,2,3) . \tag{3.5a}$$

Defining

$$(T_U^{(3)}f)(\zeta',\zeta'',\zeta''') = f({}^tU\zeta', {}^tU\zeta'', {}^tU\zeta''') , \tag{3.6}$$

we have for the representation $T_U^{(3)}$ of $\mathfrak{U}$

$$T_U^{(3)} = T_U^{(2)} \otimes T_U''' = T_U' \otimes T_U'' \otimes T_U''' , \tag{3.6a}$$

the infinitesimal transformations are $M_k^{(3)} = M_k' + M_k'' + M_k'''$, and $T_U^{(3)}$ restricted to the invariant subspace $\mathfrak{Q}_{j_1j_2j_3}$ yields the representation $\mathfrak{D}^{j_1} \otimes \mathfrak{D}^{j_2} \otimes \mathfrak{D}^{j_3}$.

c. The decomposition of $\mathfrak{D}^{j_1} \otimes \mathfrak{D}^{j_2}$. Suppose the representation $\mathfrak{D}^{j_3}$ is contained in $\mathfrak{D}^{j_1} \otimes \mathfrak{D}^{j_2}$, i.e., there are $2j_3 + 1$ orthonormal functions $\psi_m^{j_3}$ in $\mathfrak{Q}_{j_1j_2}$ such that

$$T_U^{(2)}\psi_m^{j_3} = \sum_{\mu=-j_3}^{j_3} \psi_\mu^{j_3}\mathfrak{D}_m^{j_3\mu}(U) , \quad m = j_3, j_3 - 1, \cdots, -j_3 \tag{3.7}$$

Consider the function

$$a = \sum_m \psi_m^{j_3} w_m^{j_3}(\zeta''')$$

in $\mathfrak{Q}_{j_1j_2j_3}$, where $w_m^{j_3}(\zeta''') = \sum_{m'} C_{mm'}^{j_3} v_{m'}^{j_3}(\zeta''')$ (see (2.16d)). As a sum of orthonormal functions, $a \neq 0$. Since

$$T_U'''w_m^{j_3} = \sum_\nu w_\nu^{j_3}\overline{\mathfrak{D}_m^{j_3\nu}(U)} ,$$

$$T_U^{(3)}a = \sum_m (T_U^{(2)}\psi_m^{j_3})(T_U'''w_m^{j_3}) = \sum_{m,\mu,\nu} \psi_\mu^{j_3}w_\nu^{j_3}\mathfrak{D}_m^{j_3\mu}(U)\overline{\mathfrak{D}_m^{j_3\nu}(U)} = \sum_{\mu,\nu} \psi_\mu^{j_3}w_\nu^{j_3}\delta_{\mu\nu} = a . \tag{3.7a}$$

Thus, a is invariant under $T_U^{(3)}$, and $M_k^{(3)}a = (M_k' + M_k'' + M_k''')a = 0$ (which is equivalent to saying that $\mathfrak{D}^{j_1} \otimes \mathfrak{D}^{j_2} \otimes \mathfrak{D}^{j_3}$ contains the identical representation). This is the precise mathematical content of the remarks at the beginning of this section.

Conversely, let h be a function of unit norm in $\mathfrak{Q}_{j_1j_2j_3}$ such that

$$T_U^{(3)}h = h . \tag{3.8}$$

As the $w_m^{j_3}(\zeta''')$ span $\mathfrak{Q}_{j_3}'''$, h has an expansion

$$h = \sum_m \chi_m w_m^{j_3} \tag{3.8a}$$

with *uniquely determined* χ_m in $\mathfrak{Q}_{j_1j_2}$. Now, by (3.7a),

$$T_U^{(3)}h = \sum_m (T_U^{(2)}\chi_m)(T_U'''w_m^{j_3}) = \sum_m \{\sum_{m'} (T_U^{(2)}\chi_{m'})\overline{\mathfrak{D}_{m'}^{j_3m}(U)}\} w_m^{j_3} .$$

Since $T_U^{(3)}h = h$,

$$\sum_{m'} (T_U^{(2)}\chi_{m'})\overline{\mathfrak{D}_{m'}^{j_3m}(U)} = \chi_m ,$$

and hence

$$T_U^{(2)}\chi_m = \sum_\mu \chi_\mu\mathfrak{D}_m^{j_3\mu}(U) .$$

By (3.2a), $(\chi_m,\chi_{m'}) = \lambda\delta_{mm'}$. Thus, h is a sum of orthogonal functions, and since it was assumed normalized, $||h||^2 = \sum_m ||\chi_m||^2 = (2j_3+1)\lambda = 1$. Thus

$$\psi_m^{j_3} = (2j_3+1)^{1/2}\chi_m \tag{3.8b}$$

are orthonormal functions in $\mathfrak{Q}_{j_1j_2}$ which transform under $\mathfrak{D}^{j_3}$.

d. The functions F_k, H_k, and the 3-j symbols. The invariant functions h in $\mathfrak{Q}_{j_1j_2j_3}$ [see (3.8)] may be constructed as follows.[11] Since the U are unimodular, the three determinants

$$\delta_1 = \xi_2\eta_3 - \xi_3\eta_2\,,\quad \delta_2 = \xi_3\eta_1 - \xi_1\eta_3\,,\quad \delta_3 = \xi_1\eta_2 - \xi_2\eta_1 \tag{3.9}$$

are invariant under $T_U^{(3)}$, and so is every monomial in δ_α,

$$F_k = \frac{\delta_1^{k_1}\delta_2^{k_2}\delta_3^{k_3}}{k_1!k_2!k_3!} = \frac{\delta^{[k]}}{[k!]} \quad k = (k_1,k_2,k_3) \tag{3.9a}$$

where k_α are any non-negative integers and the factorials in the denominator are included for convenience. [Depending on the circumstances we shall indicate the variables on which F_k depends by writing either $F_k(\xi,\eta)$ or $F_k(\zeta',\zeta'',\zeta''')$.]

F_k belongs to $\mathfrak{Q}_{j_1j_2j_3}$, i.e., it is homogeneous in ζ', ζ'', ζ''' of the orders $2j_1$, $2j_2$, $2j_3$ if and only if

$$k_2+k_3 = 2j_1\,,\quad k_3+k_1 = 2j_2\,,\quad k_1+k_2 = 2j_3 \tag{3.10}$$

or equivalently

$$k_\alpha = J - 2j_\alpha(\alpha = 1,2,3)\ ;\ J = j_1+j_2+j_3 \tag{3.10a}$$

$$k_1 = j_2+j_3-j_1\,,\quad k_2 = j_3+j_1-j_2\,,$$
$$k_3 = j_1+j_2-j_3\,. \tag{3.10b}$$

Note that

$$k_1+k_2+k_3 = J\,. \tag{3.10c}$$

As will be shown below [see (3.24)], $||F_k||^2 = (J+1)!/[k!]$. The corresponding normalized h is therefore

$$H_k = \Delta(j_1,j_2,j_3)F_k\ ;\quad \Delta(j_1,j_2,j_3) = ([k!]/(J+1)!)^{1/2} \tag{3.11}$$

where Δ is the so-called "quantum mechanical triangle coefficient."

Corresponding to every H_k there are $2j_3+1$ orthonormal functions $\psi_m^{j_3}$ in $\mathfrak{Q}_{j_1j_2}$ [see (3.8b)] which transform under $\mathfrak{D}^{j_3}$ provided that $j_3 = j_1+j_2-k_3$ for an integral $k_3 \geq 0$, and $j_3 \geq |j_1-j_2|$, as follows from (3.10b). Since the ψ belonging to different j_3 are *orthogonal* to each other [by (3.2)] we thus obtain altogether $n = (2j_1+1)(2j_2+1)$ orthonormal functions in $\mathfrak{Q}_{j_1j_2}$. As n is the dimension of $\mathfrak{Q}_{j_1j_2}$, the decomposition of $\mathfrak{D}^{j_1}\otimes\mathfrak{D}^{j_2}$ is thus completed.

The 3-j symbols. H_k may be expanded in the products (3.5):

$$H_k = \sum_{m_1,m_2,m_3}\begin{pmatrix} m_1 & m_2 & m_3\\ j_1 & j_2 & j_3\end{pmatrix} v_{m_1}^{j_1}(\zeta')v_{m_2}^{j_2}(\zeta'')v_{m_3}^{j_3}(\zeta'''), \tag{3.12}$$

and the expansion coefficients are the 3-j symbols.[12]

The invariance relation $T_U^{(3)}H_k = H_k$ is equivalent to the equations

$$\sum_{\mu_1\mu_2\mu_3}\mathfrak{D}_{\mu_1}^{j_1m_1}(U)\mathfrak{D}_{\mu_2}^{j_2m_2}(U)\mathfrak{D}_{\mu_3}^{j_3m_3}(U)\begin{pmatrix}\mu_1 & \mu_2 & \mu_3\\ j_1 & j_2 & j_3\end{pmatrix} = \begin{pmatrix} m_1 & m_2 & m_3\\ j_1 & j_2 & j_3\end{pmatrix}.$$

Using the relations $v^{j_3}_{\mu} = \sum_{m_3} w^{j_3}_{m_3}C^{j_3}_{m_3\mu}$, we have

$$H_k = \sum_{m_1m_2m_3}\begin{pmatrix} m_1 & m_2 & j_3\\ j_1 & j_2 & m_3\end{pmatrix} v_{m_1}^{j_1}(\zeta')v_{m_2}^{j_2}(\zeta'')w_{m_3}^{j_3}(\zeta''') \tag{3.13}$$

$$\begin{pmatrix} m_1 & m_2 & j_3\\ j_1 & j_2 & m_3\end{pmatrix} = \sum_\mu C^{j_3}_{m_3\mu}\begin{pmatrix} m_1 & m_2 & \mu\\ j_1 & j_2 & j_3\end{pmatrix} = (-1)^{j_3+m_3}\begin{pmatrix} m_1 & m_2 & -m_3\\ j_1 & j_2 & j_3\end{pmatrix} \tag{3.13a}$$

Hence, by (3.8a) and (3.8b),

$$\psi_{m_3}^{j_3} = (2j_3+1)^{1/2}\sum_{m_1m_2}\begin{pmatrix} m_1 & m_2 & j_3\\ j_1 & j_2 & m_3\end{pmatrix} v_{m_1}^{j_1}(\zeta')v_{m_2}^{j_2}(\zeta'')\,. \tag{3.14}$$

This last equation relates the vector coupling (V-C) coefficients to the 3-j symbols. [In standard form the V-C coefficients differ from those of (3.14) by the factor $(-1)^{k_1}$, see Wigner,[12] Eq. (24.16), p. 294.]

For later use we add here a few remarks. (1) If in (3.12) or (3.13), F_k is substituted for H_k, the co-

[11] We follow B. L. van der Waerden's derivation (reference 3, p. 69).

[12] The position of the indices m in (3.12) corresponds to Wigner's general definition of co- and contravariant indices (reference 7, pp. 292–296). Since, however, the fully contravariant and the fully covariant 3-j symbols are numerically equal [reference 7, Eq. (24.18a), p. 295] the coefficients in (3.12) are the same as the more familiar ones with the position of j and m reversed. We have also written the matrix elements of D^j in accordance with Wigner's rules, but we follow Wigner in writing v^j_m, w^j_m, etc., irrespective of their transformation properties.

efficients will be divided by $\Delta(j_1,j_2,j_3)$, and we shall write

$$\begin{pmatrix} m_1 & m_2 & m_3 \\ j_1 & j_2 & j_3 \end{pmatrix}_F = \left(\frac{(J+1)!}{[k!]}\right)^{1/2} \begin{pmatrix} m_1 & m_2 & m_3 \\ j_1 & j_2 & j_3 \end{pmatrix}, \tag{3.15a}$$

and similarly for the 3-j symbol in (3.13a). (2) By (2.16) and (2.16b), $w^j_m(\zeta) = v^j_m(\Gamma\zeta)$. Consequently, if in (3.13), H_k is evaluated for ζ', ζ'', $\Gamma^{-1}\zeta'''$, there appears on the right-hand side $w^{j_3}_{m_3}(\Gamma^{-1}\zeta''') = v^{j_3}_{m_3}(\zeta''')$. If a similar transformation is carried out on ζ'', one obtains

$$F_k(\zeta',\Gamma^{-1}\zeta'',\Gamma^{-1}\zeta''') = \sum_{m_1,m_2,m_3} \begin{pmatrix} m_1 & j_2 & j_3 \\ j_1 & m_2 & m_3 \end{pmatrix}_F \times v^{j_1}_{m_1}(\zeta')v^{j_2}_{m_2}(\zeta'')v^{j_3}_{m_3}(\zeta''') \tag{3.15b}$$

$$\begin{pmatrix} m_1 & j_2 & j_3 \\ j_1 & m_2 & m_3 \end{pmatrix}_F = (-1)^{j_2+m_2+j_3+m_3} \begin{pmatrix} m_1 & -m_2 & -m_3 \\ j_1 & j_2 & j_3 \end{pmatrix}_F \tag{3.15c}$$

e. Computation of the 3-j symbols. We introduce two closely related sets of coefficients f, h by setting

$$F_k(\xi,\eta) = \sum_{\kappa,\lambda} f_{k\kappa\lambda}\xi^{[\kappa]}\eta^{[\lambda]} \tag{3.16a}$$

$$H_k(\xi,\eta) = \sum_{\kappa,\lambda} h_{k\kappa\lambda} \frac{\xi^{[\kappa]}\eta^{[\lambda]}}{([\kappa!][\lambda!])^{1/2}} \tag{3.16b}$$

$$h_{k\kappa\lambda} = \left(\frac{\prod_{\alpha=1}^{3} k_\alpha!\kappa_\alpha!\lambda_\alpha!}{(J+1)!}\right)^{1/2} f_{k\kappa\lambda}\,. \tag{3.16c}$$

In view of (3.5), comparison of (3.12) and (3.16b) shows that

$$\begin{pmatrix} m_1 & m_2 & m_3 \\ j_1 & j_2 & j_3 \end{pmatrix} = h_{k\kappa\lambda} \tag{3.17}$$

$$k_\alpha = J - 2j_\alpha\,, \quad \kappa_\alpha = j_\alpha + m_\alpha\,,$$
$$\lambda_\alpha = j_\alpha - m_\alpha \; (\alpha = 1,2,3)\,. \tag{3.17a}$$

Although the nine integers k, κ, λ may seem highly redundant they are better suited to expressing the full symmetry of the 3-j symbols than are the customary j and m. (A similar situation prevails in the case of the 6-j symbols as will be seen in the next section.)

Equations (3.15) define the coefficients f and h for all k,κ,λ, but since [by (3.9)] F_k is homogeneous of order $k_1 + k_2 + k_3$ in the ξ_α as well as the η_α, f and h vanish unless

$$\kappa_1 + \kappa_2 + \kappa_3 = \lambda_1 + \lambda_2 + \lambda_3 = k_1 + k_2 + k_3 = J\,. \tag{3.18}$$

This condition corresponds to $m_1 + m_2 + m_3 = 0$.

To compute $f_{k\kappa\lambda}$ we simply apply the binomial theorem to the powers of δ_α. Let

$$\frac{\delta_1^{k_1}}{k_1!} = \sum_{p_1+q_1=k_1} \frac{(\xi_2\eta_3)^{p_1}(-\xi_3\eta_2)^{q_1}}{p_1!q_1!}\,,$$

$$\frac{\delta_2^{k_2}}{k_2!} = \sum_{p_2+q_2=k_2} \frac{(\xi_3\eta_1)^{p_2}(-\xi_1\eta_3)^{q_2}}{p_2!q_2!}\,,$$

$$\frac{\delta_3^{k_3}}{k_3!} = \sum_{p_3+q_3=k_3} \frac{(\xi_1\eta_2)^{p_3}(-\xi_2\eta_1)^{q_3}}{p_3!q_3!}\,.$$

Then,

$$f_{k\kappa\lambda} = \sum \frac{(-1)^{q_1+q_2+q_3}}{p_1!p_2!p_3!q_1!q_2!q_3!}\,. \tag{3.19}$$

The summation extends over all non-negative integers p_α, q_α which satisfy the conditions summarized in the following matrix equation:

$$L \equiv \begin{pmatrix} k_1 & k_2 & k_3 \\ \kappa_1 & \kappa_2 & \kappa_3 \\ \lambda_1 & \lambda_2 & \lambda_3 \end{pmatrix} = \begin{pmatrix} q_1+p_1 & q_2+p_2 & q_3+p_3 \\ q_2+p_3 & q_3+p_1 & q_1+p_2 \\ q_3+p_2 & q_1+p_3 & q_2+q_1 \end{pmatrix} \equiv Q\,. \tag{3.19a}$$

Equation (3.19) reduces to a simple sum, because all p_α, q_α may be expressed by any one of them. Let $q_3 = z$. Then $p_1 = \kappa_2 - z$, $p_2 = \lambda_1 - z$, $p_3 = k_3 - z$; $q_1 = k_1 - \kappa_2 + z$, $q_2 = k_2 - \lambda_1 + z$, and the sum extends over those z for which all p and q are non-negative. (This is Racah's expression.[13]) If μ is the minimum of the entries of L the sum has $\mu + 1$ terms.

In L the elements of each row as well as the elements of each column add up to J, [see (3.17a) and (3.18)]. In Q all row sums and column sums are equal by definition, the common value being $\sum_{\alpha=1}^{3}(p_\alpha + q_\alpha)$.

Finally, one may write f_L instead of $f_{k\kappa\lambda}$, and similarly h_L. Denoting L's matrix elements by $l_{i\alpha}$ (where i denotes the row and α the column), (3.16) and (3.17) may be summarized by

$$\begin{pmatrix} m_1 & m_2 & m_3 \\ j_1 & j_2 & j_3 \end{pmatrix} = h_L = \left(\frac{\prod_{i,\alpha} l_{i\alpha}!}{(J+1)!}\right)^{1/2} f_L\,. \tag{3.19b}$$

f. The generating function Φ and the symmetries of the 3-j symbol. The generating function of the 3-j symbols is defined by[14]

$$\Phi(\tau,\xi,\eta) = \sum_k \tau^{[k]}F_k(\xi,\eta) = \sum_{k,\kappa,\lambda} f_{k\kappa\lambda}\tau^{[k]}\xi^{[\kappa]}\eta^{[\lambda]} = \sum_L f_L\tau^{[k]}\xi^{[\kappa]}\eta^{[\lambda]}\,, \tag{3.20a}$$

[13] G. Racah, Phys. Rev. 62, 438 (1942), Eq. (16). See also A. R. Edmonds, *Angular Momentum in Quantum Mechanics* (Princeton University Press, Princeton, New Jersey, 1957), Eq. (3.6.11).

[14] This corresponds to the function defined by Schwinger (reference 6) in Eq. (3.42).

where $\tau = (\tau_1, \tau_2, \tau_3)$ is a triple of complex variables. It will be useful to arrange the nine variables τ, ξ, η in a matrix

$$\Xi = \begin{pmatrix} \tau_1 & \tau_2 & \tau_3 \\ \xi_1 & \xi_2 & \xi_3 \\ \eta_1 & \eta_2 & \eta_3 \end{pmatrix}$$

in analogy to L. It follows at once from (3.9) that

$$\Phi(\tau,\xi,\eta) \equiv \Phi(\Xi) = \exp\left(\sum_{\alpha=1} \tau_\alpha \delta_\alpha\right) = \exp\left(D(\tau,\xi,\eta)\right) = \exp(\det \Xi) \qquad (3.21)$$

$$D(\tau,\xi,\eta) = \det \Xi = \begin{vmatrix} \tau_1 & \tau_2 & \tau_3 \\ \xi_1 & \xi_2 & \xi_3 \\ \eta_1 & \eta_2 & \eta_3 \end{vmatrix}. \qquad (3.21a)$$

The elementary symmetries of the determinant D now yield corresponding symmetries of the coefficients f and h.[15] We combine the following facts.

(α) For any 3×3 matrix A, let $P(A)$ be the matrix obtained by some fixed permutation of A's elements (such as the transposition of two rows or two columns, etc.). Then[16]

$$\Phi(P(\Xi)) = \sum f_{P(L)} \tau^{[k]} \xi^{[\kappa]} \eta^{[\lambda]}$$

(β) Set $\exp[-D(\tau,\xi,\eta)] = \Phi'(\tau,\xi,\eta) \equiv \Phi'(\Xi)$.

Evidently

$$\Phi'(\Xi) = \Phi(-\tau,\xi,\eta) = \sum (-1)^J f_L \tau^{[k]} \xi^{[\kappa]} \eta^{[\lambda]}$$

because $k_1 + k_2 + k_3 = J$.

By comparing coefficients we conclude therefore

(a) If $\det[P(\Xi)] = \det \Xi$, then $\Phi(P(\Xi)) = \Phi(\Xi)$, and hence $f_{P(L)} = f_L$.

(b) If $\det[P(\Xi)] = -\det \Xi$, then $\Phi(P(\Xi)) = \Phi'(\Xi)$, hence $f_{P(L)} = (-1)^J f_L$.

This leads to the final result:

First case: If P is (1) an even permutation of rows, (2) an even permutation of columns, (3) the interchange of rows and columns then

$$f_{P(L)} = f_L \quad \text{aud} \quad h_{P(L)} = h_L. \qquad (3.22)_{\text{I}}$$

Second case: If P is (1) an odd permutation of rows, (2) an odd permutation of columns then

$$f_{P(L)} = (-1)^J f_L \quad \text{and} \quad h_{P(L)} = (-1)^J h_L. \qquad (3.22)_{\text{II}}$$

[The equations for the coefficients f, which follow immediately from the above analysis, imply those for the coefficients h because neither the numerator nor the denominator of the normalization constant in (3.19b) is affected by the operations P in (3.22).]

The operations listed under I and II generate the symmetry group of 72 elements discovered by Regge. Previously, only the following more evident symmetry operations had been noticed: (1) Permutation of the *columns of L*, i.e., simultaneous permutation of j_α and m_α. (2) Transposition of the second and third row in L, i.e., changing the sign of all m_α.

g. The norm of F_k. As follows from (1.13b), for fixed τ the generating function $\Phi_\tau \equiv \Phi(\tau,\xi,\eta)$ is an element of $\mathfrak{F}_6$ as long as the τ_α are small enough. (The precise condition, which is $\sum_{\alpha=1}^{3} |\tau_\alpha|^2 < 1$, need not concern us.) The inner product of two such functions Φ_τ and Φ_τ' (taken on $\mathfrak{F}_6$) is then, by (3.20a)

$$(\Phi_\tau, \Phi_{\tau'}) = \sum_{k,k'} \bar{\tau}^{[k]} \tau'^{[k']} (F_k, F_{k'}) \qquad (3.23)$$

In the computation of the inner product according to (1.4) we may separate the ξ and the η integrations, so that

$$(\Phi_\tau, \Phi_{\tau'}) = \int \left[\int \overline{\exp D(\tau,\xi,\eta)} \exp D(\tau',\xi,\eta) d\mu_3(\eta) \right] \times d\mu_3(\xi). \qquad (3.23a)$$

In ordinary vector notation $D(\tau,\xi,\eta) = (\tau \times \xi)\cdot\eta$. Thus the inner integral is of the form (1.13) (if η is identified with z), with

$$a = \overline{\tau \times \xi}, \quad b = \overline{\tau' \times \xi},$$

and it has the value $\exp(\bar{b}\cdot a)$, where

$$\bar{b}\cdot a = (\tau'\cdot\bar{\tau})\bar{\xi}\cdot\xi) - (\tau'\cdot\bar{\xi})(\bar{\tau}\cdot\xi) = \bar{\xi}\cdot A\xi,$$

A denoting the matrix with elements

$$a_{\alpha\beta} = (\tau'\cdot\bar{\tau})\delta_{\alpha\beta} - \tau'_\alpha \bar{\tau}_\beta.$$

Hence, (Φ_τ, Φ_τ') is a Laplacian integral of the form

$$(\Phi_\tau, \Phi_{\tau'}) = \int \exp(\bar{\xi}\cdot A\xi) d\mu_3(\xi)$$

and, by Eq. (A5) in the Appendix,

$$(\Phi_\tau, \Phi_{\tau'}) = [\det(1 - A)]^{-1} = (1 - \bar{\tau}\cdot\tau')^{-2}. \qquad (3.23b)$$

Expanding in a power series one obtains

$$(\Phi_\tau, \Phi_{\tau'}) = \sum_{\mu=0}^{\infty} (\mu + 1)(\bar{\tau}\cdot\tau')^\mu = \sum_k \frac{(|k| + 1)!}{[k!]} \bar{\tau}^{[k]} \tau'^{[k]},$$

[15] This proof is essentially the same as Regge's [reference 5(a)].

[16] Consider a power series in n variables $x_r, G(x_1, x_2, \ldots, x_n) = \Sigma_{i_1, i_2, \ldots} \gamma_{i_1 i_2 \ldots i_n} x_1^{i_1} x_2^{i_2} \ldots x_n^{i_n}$, and let $G'(x_1, x_2, \ldots, x_n) \equiv G(x_{\pi_1}, x_{\pi_2}, \ldots, x_{\pi_n})$ for some permutation $(\pi_1, \pi_2, \ldots, \pi_n)$ of the integers $1, 2, \ldots, n$. Then $G'(x_1, x_2, \ldots, x_n) = \Sigma_{i_1, i_2 \ldots} \times \gamma_{i\pi_1 i\pi_2 \ldots i\pi_n} x_1^{i_1} x_2^{i_2} \ldots x_n^{i_n}$. In our case the variables τ, ξ, η correspond to the x_r, and f_L to the coefficients $\gamma_{i_1 i_2 \ldots i_n}$.

where $|k| = k_1 + k_2 + k_3$. Comparison with (3.23) yields

$$(F_k,F_{k'}) = \begin{cases} 0 & \text{if } k' \neq k \\ (J+1)!/[k!] & \text{if } k' = k \end{cases} \quad (J = |k|)\,, \tag{3.24}$$

as announced in Sec. 3d.

h. Recursion relations. For the derivatives of Φ one finds

$$\partial\Phi/\partial\tau_1 = (\xi_2\eta_3 - \xi_3\eta_2)\Phi\,, \quad \partial\Phi/\partial\xi_1 = (\eta_2\tau_3 - \eta_3\tau_2)\Phi\,,$$

$$\partial\Phi/\partial\eta_1 = (\tau_2\xi_3 - \tau_3\xi_2)\Phi$$

and six more equations obtained by cyclic permutations. If the expansion (3.20b) is inserted numerous relations between the coefficients f result, most of which are of course known. We mention two examples.

(1) $\partial\Phi/\partial\tau_1 = (\xi_2\eta_3 - \xi_3\eta_2)\Phi$ leads to

$$k_1 f_{k\kappa\lambda} = f_{k_1-1\cdots\kappa_2-1\cdots\lambda_3-1} - f_{k_1-1\cdots\kappa_3-1\cdots\lambda_2-1}\,.$$

(On the right-hand side only those indices are marked which differ from the corresponding ones of the left-hand terms.)

(2) $\tau_3\,\partial\Phi/\partial\tau_2 + \xi_3\,\partial\Phi/\partial\xi_2 + \eta_3\,\partial\Phi/\partial\eta_2 = 0$ leads to

$$-k_2 f_{k\kappa\lambda} = (1+\kappa_2) f_{\cdot k_2-1,k_3+1,\kappa_2+1,\kappa_3-1\cdots} + (1+\lambda_2) f_{\cdot k_2-1,k_3+1,\cdots\lambda_2+1,\lambda_3-1}$$

Upon insertion of the normalization constant and "translation" into the (j,m)-notation one obtains the following two formulas for 3-j symbols.

$$(1)\quad [(J+1)(J-2j_1)]^{1/2}\begin{pmatrix} j_1 & j_2 & j_3 \\ m_1 & m_2 & m_3 \end{pmatrix} = [(j_2+m_2)(j_3-m_3)]^{1/2}\begin{pmatrix} j_1 & j_2-\frac{1}{2} & j_3-\frac{1}{2} \\ m_1 & m_2-\frac{1}{2} & m_3+\frac{1}{2} \end{pmatrix} - [(j_2-m_2)(j_3+m_3)]^{1/2}\begin{pmatrix} j_1 & j_2-\frac{1}{2} & j_3-\frac{1}{2} \\ m_1 & m_2+\frac{1}{2} & m_3-\frac{1}{2} \end{pmatrix}$$

$$(2)\quad [(J-2j_2)(J+1-2j_3)]^{1/2}\begin{pmatrix} j_1 & j_2 & j_3 \\ m_1 & m_2 & m_3 \end{pmatrix} + ([j_2+m_2+1)(j_3+m_3)]^{1/2}\begin{pmatrix} j_1 & j_2-\frac{1}{2} & j_3+\frac{1}{2} \\ m_1 & m_2-\frac{1}{2} & m_3+\frac{1}{2} \end{pmatrix} + [(j_2-m_2+1)(j_3-m_3)]^{1/2}\begin{pmatrix} j_1 & j_2-\frac{1}{2} & j_3+\frac{1}{2} \\ m_1 & m_2+\frac{1}{2} & m_3-\frac{1}{2} \end{pmatrix} = 0\,.$$

4. RACAH COEFFICIENTS

This section deals with the Racah coefficients (in the form of Wigner's 6-j symbols). The main objective is the construction and analysis of a generating function, and its application to a discussion of the symmetries of the Racah coefficients.

a. Formal preliminaries. In terms of 3-j symbols the 6-j symbol is defined as follows.

$$\begin{Bmatrix} j_{23} & j_{31} & j_{12} \\ j_{01} & j_{02} & j_{03} \end{Bmatrix} = \sum_{m_\alpha, m'_\alpha} \begin{pmatrix} m'_1 & m'_2 & m'_3 \\ j_{23} & j_{31} & j_{12} \end{pmatrix}\begin{pmatrix} j_{23} & m_2 & j_{03} \\ m'_1 & j_{02} & m_3 \end{pmatrix} \times \begin{pmatrix} j_{01} & j_{31} & m_3 \\ m_1 & m'_2 & j_{03} \end{pmatrix}\begin{pmatrix} m_1 & j_{02} & j_{12} \\ j_{01} & m_2 & m'_3 \end{pmatrix} \tag{4.1}$$

(The summation of the m_α and m'_α extends over all values compatible with the associated j.)

The notation introduced here is meant to emphasize the tetrahedral symmetry of the 6-j symbol. It alludes to a tetrahedron with vertices V_α ($\alpha = 0,1,2,3$) and edges $j_{\alpha\beta}$. The four 3-j symbols in (4.1) correspond, respectively, to the triangles opposite V_0, V_1, V_2, V_3. We define

$$j_{\alpha\beta} \equiv j_{\beta\alpha} \quad (\alpha \neq \beta; \alpha,\beta = 0,1,2,3)\,. \tag{4.2}$$

In the sequel *different* subscripts in an equation containing $j_{\alpha\beta}$ or $k_{\alpha\beta}$ denote *different* integers taken from the sequence 0,1,2,3. Thus the perimeter of the αth triangle (with vertices V_β, V_γ, V_δ) is

$$J_\alpha = j_{\beta\gamma} + j_{\gamma\delta} + j_{\delta\beta}\,. \tag{4.3}$$

In accordance with (3.10a) and (3.10b), we set

$$k_{\alpha\beta} = J_\alpha - 2j_{\gamma\delta} = j_{\gamma\beta} + j_{\delta\beta} - j_{\gamma\delta} \quad (\alpha \neq \beta)\,. \tag{4.4}$$

The twelve $k_{\alpha\beta}$ depend on the *ordered* pairs (α,β) while in the definition of $j_{\alpha\beta}$ the order is irrelevant; α refers to the triangle, β to the vertex opposite $j_{\gamma\delta}$. The inverse relation is

$$j_{\gamma\delta} = \tfrac{1}{2}(k_{\alpha\gamma} + k_{\alpha\delta})\,. \tag{4.5}$$

Since $j_{\gamma\delta}$ belongs to the two triangles opposite V_α and V_β, we have also $j_{\gamma\delta} = \frac{1}{2}(k_{\beta\gamma} + k_{\beta\delta})$, so that the k satisfy the *compatibility* conditions

$$k_{\alpha\gamma} + k_{\alpha\delta} = k_{\beta\gamma} + k_{\beta\delta}\,. \tag{4.6}$$

Further useful relations are

$$k_{\alpha\beta} + k_{\alpha\gamma} + k_{\alpha\delta} = J_\alpha\,, \tag{4.7}$$

$$k_{\alpha\beta} - k_{\beta\alpha} = k_{\gamma\beta} - k_{\delta\alpha} = J_\alpha - J_\beta\,. \tag{4.7a}$$

The (triangle) conditions on the $j_{\alpha\beta}$ to lead to nonvanishing 3-j symbols in (4.1) are simply: *$k_{\alpha\beta}$ are non-negative integers.*[17]

Any set of twelve numbers $k_{\alpha\beta}$ ($\alpha \neq \beta$) which satisfy the compatibility conditions (4.6) will be called "*tetrahedral.*" Given such a tetrahedral set. If $j_{\gamma\delta}$ is defined by (4.5) then $j_{\gamma\delta} = j_{\delta\gamma}$, and the relations (4.4) hold.

The ordered pairs $\alpha\beta$ are conveniently arranged in four triads T_α defined as follows:

$$T_\alpha \text{ contains the three pairs with first element } \alpha\,. \quad (4.8)$$

We shall also need the four "transposed" triads T^*_α:

$$T^*_\alpha \text{ contains the three pairs with second element } \alpha\,. \quad (4.8a)$$

Lastly we introduce three tetrads W_i:

$$W_1:(01,10,23,32)\,, \quad W_2:(02,20,31,13)\,,$$
$$W_3:(03,30,12,21)\,. \quad (4.8b)$$

In the functions F_k connected with the four 3-j symbols in (4.1) the numbers $k_{\alpha\beta}$ appear in the order which corresponds to the order of the $j_{\gamma\delta}$ by Eq. (4.4). Thus we have in succession

$$(k_{01},k_{02},k_{03})(k_{10},k_{13},k_{12})(k_{23},k_{20},k_{21})(k_{32},k_{31},k_{30})\,. \quad (4.9)$$

It is seen that the four triples correspond to the four triads T_α. Moreover, the first, second, and third element in each triple corresponds, respectively, to W_1, W_2, W_3.

Let $\alpha \to \pi_\alpha$ be a permutation of the four integers 0,1,2,3 (which may be interpreted as a permutation of the four vertices V_α of the tetrahedron), and define

$$j'_{\alpha\beta} = j_{\pi_\alpha,\pi_\beta} \quad k'_{\alpha\beta} = k_{\pi_\alpha,\pi_\beta} \quad (4.10)$$

Then Eqs. (4.4) to (4.6) remain valid. The triads are permuted accordingly ($T \to T_{\pi_\alpha}$, $T^* \to T^*_{\pi_\alpha}$) while the tetrads are subject to a permutation π' of three integers, which depends on π ($W_i \to W_{\pi'_i}$).

b. The generating function $R(\tau)$. The 6-j symbol (4.1) is a function $r(k)$, where k represents the $k_{\alpha\beta}$. If we replace, on the right-hand side of (4.1), the 3-j symbols ( ) by the corresponding symbols $(\ )_F$ [see (3.15a)], i.e., if we divide by all four triangle coefficients, we obtain a function $s(k)$ such that

$$r(k) = \left(\frac{\prod_{\alpha,\beta} k_{\alpha\beta}!}{\prod_\alpha (J_\alpha + 1)!}\right)^{1/2} s(k)\,. \quad (4.11)$$

[17] These conditions are, however, insufficient to insure the existence of a tetrahedron with edges $j_{\alpha\beta}$. We deal here with the combinatorial rather than with the metric properties of a tetrahedron.

So far $r(k)$ and $s(k)$ are defined for "tetrahedral" sets $k_{\alpha\beta}$. In all other cases we set $r(k) = s(k) = 0$.

In terms of 12 complex variables $\tau_{\alpha\beta}(\alpha \neq \beta)$ we now define the generating function of the 6-j symbols by

$$R(\tau) = \sum\nolimits_{k_{\alpha\beta}} s(k) \prod\nolimits_{\alpha,\beta} \tau_{\alpha\beta}^{k_{\alpha\beta}}\,. \quad (4.12)$$

The function $R(\tau)$ may be expressed as an integral over the product of four generating functions Φ [see (3.20a)], which are related to the 3-j symbols in (4.1). To this end we introduce six pairs of complex variables $\zeta^\alpha = (\xi_\alpha,\eta_\alpha)$ and $\theta^\alpha = (\xi'_\alpha,\eta'_\alpha)$ ($\alpha = 1,2,3$) corresponding, respectively, to the summation indices m_α and m'_α in (4.1). Then

$$R(\tau) = \int \Phi_0\Phi_1\Phi_2\Phi_3 d\mu_6(\xi',\eta')d\mu_6(\xi,\eta)\,, \quad (4.13)$$

$$\Phi_0 = \Phi(\tau_{01},\tau_{02},\tau_{03};\theta^1,\theta^2,\theta^3)\,,$$
$$\Phi_1 = \Phi(\tau_{10},\tau_{13},\tau_{12};{}^t\Gamma\overline{\theta^1},\zeta^2,{}^t\Gamma\overline{\zeta^3})\,,$$
$$\Phi_2 = \Phi(\tau_{23},\tau_{20},\tau_{21};{}^t\Gamma\overline{\zeta^1},{}^t\Gamma\overline{\theta^2},\zeta^3)\,,$$
$$\Phi_3 = \Phi(\tau_{32},\tau_{31},\tau_{30};\zeta^1,{}^t\Gamma\overline{\zeta^2},{}^t\Gamma\overline{\theta^3})\,. \quad (4.13a)$$

[The variables $\tau_{\alpha\beta}$ in the four functions Φ correspond to the four triples (4.9), to an upper index m_α in (4.1) corresponds ζ^α, to a lower index m_α corresponds ${}^t\Gamma\ \overline{\zeta^\alpha}$ as an argument of Φ, similarly for the m'_α. Note that ${}^t\Gamma\ \overline{\zeta^\alpha} = (\bar{\eta}_\alpha, -\bar{\xi}_\alpha)$.]

The proof for the integral representation (4.13) is straightforward, but writing it out in full would lead to rather unmanageable equations. It will suffice to consider the contribution of the ζ^1 integration to (4.13). Φ_0 and Φ_1 are free of ζ^1, so that only Φ_2 and Φ_3 need be considered. Now

$$\Phi_2 = \sum_{k_{23},k_{20},k_{21}} \tau_{23}^{k_{23}}\tau_{20}^{k_{20}}\tau_{21}^{k_{21}} F_{k_{23}\ k_{20},k_{21}}({}^t\Gamma\overline{\zeta^1},{}^t\Gamma\overline{\theta^2},\zeta^3)\,,$$
$$\Phi_3 = \sum_{k_{32},k_{31},k_{30}} \tau_{32}^{k_{32}}\tau_{31}^{k_{31}}\tau_{30}^{k_{30}} F_{k_{32},k_{31},k_{30}}(\zeta^1,{}^t\Gamma\overline{\zeta^2},{}^t\Gamma\overline{\theta^3})\,.$$

The problem is further simplified by studying the contribution of just one F chosen from Φ_2 and one F from Φ_3. By (3.15b), since ${}^t\Gamma = \Gamma^{-1}$,

$$F_{k_{23},k_{20},k_{21}}({}^t\Gamma\overline{\zeta^1},{}^t\Gamma\overline{\theta^2},\zeta^3) = \sum_{\mu_1,\mu_2,\mu_3} \begin{pmatrix} j_1 & j_2 & \mu_3 \\ \mu_1 & \mu_2 & j_3 \end{pmatrix}_F$$
$$\times\ \overline{v^{j_1}_{\mu_1}(\zeta^1)}\,\overline{v^{j_2}_{\mu_2}(\theta^2)}\,v^{j_3}_{\mu_3}(\zeta^3)\,,$$

where $2j_1 = k_{20} + k_{21}$, $2j_2 = k_{23} + k_{21}$, $2j_3 = k_{23} + k_{20}$. Similarly,

$$F_{k_{32},k_{31},k_{30}}(\zeta^1,{}^t\Gamma\overline{\zeta^2},{}^t\Gamma\overline{\theta^3}) = \sum_{\nu_1\nu_2\nu_3} \begin{pmatrix} \nu_1 & l_2 & l_3 \\ l_1 & \nu_2 & \nu_3 \end{pmatrix}_F$$
$$\times\ v^{l_1}_{\nu_1}(\zeta^1)\,\overline{v^{l_2}_{\nu_2}(\zeta^2)}\,\overline{v^{l_3}_{\nu_3}(\theta^3)}\,,$$

with $2l_1 = k_{30} + k_{31}$, $2l_2 = k_{32} + k_{30}$, $2l_3 = k_{32} + k_{31}$.

If one multiplies the two functions F and integrates over ζ^1, one finds, due to the orthonormality of the v^j_μ: (1) The result is zero if $j_1 \neq l_1$. (2) If $j_1 = l_1$, i.e.,

$$k_{20} + k_{21} = k_{30} + k_{31} (= 2j_{01}) \qquad (4.14)$$

the result is

$$\sum_{\mu_2,\mu_3,\nu_2,\nu_3} \left\{ \sum_\mu \begin{pmatrix} j_{01} & j_2 & \mu_3 \\ \mu & \mu_2 & j_3 \end{pmatrix}_F \begin{pmatrix} \mu & l_2 & l_3 \\ j_{01} & \nu_2 & \nu_3 \end{pmatrix}_F \right\}$$
$$\times \overline{v^{j_2}_{\mu_2}(\theta^2)}\, v^{j_3}_{\mu_3}(\zeta^3)\, \overline{v^{l_2}_{\nu_2}(\zeta^2)}\, \overline{v^{l_3}_{\nu_3}(\theta^3)} \, .$$

Continuing, step by step, with the remaining variables, one obtains, in analogy to (4.14), the remaining five compatibility relations (4.6), which shows that only the tetrahedral sets $k_{\alpha\beta}$ give a nonvanishing contribution, and in addition it is seen that the contribution of a tetrahedral set is precisely the 6-j symbol divided by the four triangle coefficients, i.e., $s(k)$, as it should be.

The computation of $R(\tau)$ carried out below gives a simple result,[18] viz.,

$$R(\tau) = [G(\tau)]^{-2} \, , \quad G(\tau) = 1 + \sum_{\alpha=0}^{3} a_\alpha + \sum_{i=1}^{3} b_i \qquad (4.15)$$

$$a_0 = \tau_{10}\tau_{20}\tau_{30} \, , \quad a_1 = \tau_{01}\tau_{31}\tau_{21} \, , \quad a_2 = \tau_{32}\tau_{02}\tau_{12} \, ,$$
$$a_3 = \tau_{23}\tau_{13}\tau_{03} \qquad (4.15a)$$

$$b_1 = \tau_{01}\tau_{10}\tau_{23}\tau_{32} \, , \quad b_2 = \tau_{02}\tau_{20}\tau_{13}\tau_{31} \, ,$$
$$b_3 = \tau_{03}\tau_{30}\tau_{12}\tau_{21} \, . \qquad (4.15b)$$

If the τ are sufficiently small (for example, $|\tau_{\alpha\beta}| \leq \frac{1}{2}$ for all α,β), the integral (4.13) converges absolutely, the operations carried out below are legitimate, and the power series (4.12) may be obtained from a term-by-term integration.

c. Computation of $R(\tau)$. By (3.21), the integrand in (4.13) has the form $\exp(D_0 + D_1 + D_2 + D_3)$, where

$$D_0 = \begin{vmatrix} \tau_{01} & \tau_{02} & \tau_{03} \\ \xi'_1 & \xi'_2 & \xi'_3 \\ \eta'_1 & \eta'_2 & \eta'_3 \end{vmatrix} , \quad D_1 = \begin{vmatrix} \tau_{10} & \tau_{13} & \tau_{12} \\ \bar\eta'_1 & \xi_2 & \bar\eta_3 \\ -\bar\xi'_1 & \eta_2 & -\bar\xi_3 \end{vmatrix} ,$$

$$D_2 = \begin{vmatrix} \tau_{23} & \tau_{20} & \tau_{21} \\ \bar\eta_1 & \bar\eta'_2 & \xi_3 \\ -\bar\xi_1 & -\bar\xi'_2 & \eta_3 \end{vmatrix} , \quad D_3 = \begin{vmatrix} \tau_{32} & \tau_{31} & \tau_{30} \\ \xi_1 & \bar\eta_2 & \bar\eta'_3 \\ \eta_1 & -\bar\xi_2 & -\bar\xi'_3 \end{vmatrix} .$$

The cyclic symmetry of the exponent $\sum_{\alpha=0}^{3} D_\alpha$ in the indices 1,2,3 greatly reduces the work in this computation. In fact, only a few terms need actually be calculated. We have

$$D_1 + D_2 + D_3 = \sum_{\alpha=1}^{3} (c_\alpha \bar\xi'_\alpha + d_\alpha \bar\eta'_\alpha) - E$$
$$c_1 = \tau_{12}\xi_2 - \tau_{13}\bar\eta_3 \, , \quad c_2 = \tau_{23}\xi_3 - \tau_{21}\bar\eta_1 \, ,$$
$$c_3 = \tau_{31}\xi_1 - \tau_{32}\bar\eta_2$$
$$d_1 = \tau_{12}\eta_2 + \tau_{13}\bar\xi_3 \, , \quad d_2 = \tau_{23}\eta_3 + \tau_{21}\bar\xi_1 \, ,$$
$$d_3 = \tau_{31}\eta_1 + \tau_{32}\bar\xi_2$$
$$E = \tau_{10}(\bar\xi_3\xi_2 + \bar\eta_3\eta_2) + \tau_{20}(\bar\xi_1\xi_3 + \bar\eta_1\eta_3)$$
$$+ \tau_{30}(\bar\xi_2\xi_1 + \bar\eta_2\eta_1) \, .$$

First step: Integration over ξ',η'. By (1.12a),

$$\int \exp(c\cdot\bar\xi' + d\cdot\bar\eta') \exp(D_0 - E) d\mu_6(\xi',\eta') = \exp f \, ,$$

$$f = \begin{vmatrix} \tau_{01} & \tau_{02} & \tau_{03} \\ c_1 & c_2 & c_3 \\ d_1 & d_2 & d_3 \end{vmatrix} - E \, .$$

Inserting c_α and d_α one obtains

$$f = \sum_{\alpha=1}^{3} u_\alpha \bar\delta_\alpha + \sum_{\alpha=1}^{3} v_\alpha \delta_\alpha - \bar\xi\cdot H\xi - \bar\eta\cdot H\eta \, , \qquad (4.16)$$

where δ_α are the determinants in (3.9),

$$u_1 = \tau_{02}\tau_{13}\tau_{32} \, , \quad u_2 = \tau_{03}\tau_{21}\tau_{13} \, , \quad u_3 = \tau_{01}\tau_{32}\tau_{21}$$
$$v_1 = \tau_{03}\tau_{12}\tau_{23} \, , \quad v_2 = \tau_{01}\tau_{23}\tau_{31} \, , \quad v_3 = \tau_{02}\tau_{31}\tau_{12}$$

and H is the matrix

$$H = \begin{pmatrix} a_1 & -\tau_{03}\tau_{12}\tau_{21} & \tau_{20} \\ \tau_{30} & a_2 & -\tau_{01}\tau_{23}\tau_{32} \\ -\tau_{02}\tau_{13}\tau_{31} & \tau_{10} & a_3 \end{pmatrix}$$

the a_α being defined in (4.15a).

Second step: Integration over ξ,η. By (A8) of the Appendix

$$\int \exp f d\mu_6(\xi,\eta) = [\det(1 + H) - u\cdot v - u\cdot Hv]^{-2} \, , \qquad (4.16a)$$

and a straightforward computation gives the expression (4.15).

d. Symmetries of the 6-j symbols. It is useful to arrange $k_{\alpha\beta}$ and $\tau_{\alpha\beta}$ in matrix form

$$\mathcal{K} = \begin{pmatrix} k_{10} & k_{20} & k_{30} \\ k_{01} & k_{31} & k_{21} \\ k_{32} & k_{02} & k_{12} \\ k_{23} & k_{13} & k_{03} \end{pmatrix} \qquad \mathfrak{T} = \begin{pmatrix} \tau_{10} & \tau_{20} & \tau_{30} \\ \tau_{01} & \tau_{31} & \tau_{21} \\ \tau_{32} & \tau_{02} & \tau_{12} \\ \tau_{23} & \tau_{13} & \tau_{03} \end{pmatrix} \qquad (4.17)$$

such that the rows correspond to the four transposed triads T^*_α, and the columns to the three tetrads W_1, W_2, W_3. In the generating function (4.15), a_α is the product of the elements in the αth row, and b_i the product of the elements in the ith column of $\mathfrak{T}$.

---

[18] Apart from the notation this coincides with Schwinger's Eq. (4.18) in reference 6.

By (4.7), J_α is the sum of all k belonging to the triad T_α. Similarly, we may introduce J^*_α as the sum of all k belonging to the transposed triad T^*_α, and w_i as the sum of all k belonging to the tetrad W_i. (Equivalently, J^*_α is the sum of the elements in the αth row, and w_i the sum of the elements in the ith column of $\mathcal{K}$.) Clearly,

$$\sum_{\alpha=0}^{3} J_\alpha = \sum_{\alpha=0}^{3} J^*_\alpha = \sum_{i=1}^{3} w_i = |\mathcal{K}| \equiv \sum_{\alpha,\beta} k_{\alpha\beta}\,. \quad (4.17a)$$

For a tetrahedral $\mathcal{K}$, by (4.7a),

$$J^*_\alpha - J_\alpha = \textstyle\sum_\beta (k_{\beta\alpha} - k_{\alpha\beta}) = \sum_\beta (J_\beta - J_\alpha)\,,$$

or

$$J^*_\alpha = |\mathcal{K}| - 3J_\alpha\,. \quad (4.17b)$$

In analogy to (3.20) we write

$$R(\mathfrak{J}) = \textstyle\sum_K s(\mathcal{K}) \prod_{\alpha,\beta} \tau_{\alpha\beta}^{k_{\alpha\beta}}\,.$$

If we denote, as in 3f, by $P(\mathfrak{J})$ and $P(\mathcal{K})$ the matrices obtained from $\mathfrak{J}$ and $\mathcal{K}$, respectively, by a fixed permutation of their elements, then

$$R(P(\mathfrak{J})) = \textstyle\sum_K s(P(\mathcal{K})) \prod_{\alpha,\beta} \tau_{\alpha\beta}^{k_{\alpha\beta}}\,.$$

Consequently, if $R(P(\mathfrak{J})) = R(\mathfrak{J})$, then $s(P(\mathcal{K})) = s(\mathcal{K})$ for all $\mathcal{K}$.

This remark yields the *Regge group of symmetry operations*. In fact, R is invariant (1) under any permutation of the rows (this permutes the a_α and leaves the b_i invariant), (2) under any permutation of the columns (this permutes the b_i, but leaves the a_α invariant).

Note that the J_α are permuted by the first type and left unchanged by the second type of operations, as follows from (4.17b). Hence the normalization factor in (4.11) is unaffected by all these operations, and what we proved for $s(\mathcal{K})$ holds also for $r(\mathcal{K})$, i.e., for the 6-j symbols. [For any permutation of the k, the corresponding transformation of the j may be derived from (4.4) and (4.5).]

The symmetry operations listed above generate the group $S_4 \times S_3$ (the direct product of the symmetric groups S_4 and S_3) of order $24 \cdot 6 = 144$. Its elements are the products of permutations π of the rows and σ, say, of the columns of $\mathcal{K}$, which may be chosen independently of each other. The previously known symmetry operations are the transformations (4.10), induced by a permutation of the vertices of the tetrahedron, where σ is no longer independent of π, but equals π'.

e. Explicit expression for the 6-j symbol. Expanding $R = G^{-2}$ in a power series one obtains

$$R = \sum_{s=0}^{\infty} (-1)^s (z+1) \Big(\sum_{\alpha=0}^{3} a_\alpha + \sum_{i=1}^{3} b_i \Big)^s$$
$$= \sum_{\nu_\alpha, \omega_i} (-1)^z (z+1)! \Big(\prod_{\alpha=0}^{3} \frac{a_\alpha^{\nu_\alpha}}{\nu_\alpha!} \prod_{i=1}^{3} \frac{b_i^{\omega_i}}{\omega_i!} \Big)\,,$$

where ν_α, ω_i run independently over all non-negative integers, and $z = \sum_\alpha \nu_\alpha + \sum_i \omega_i$. This leads to

$$s(\mathcal{K}) = \sum \frac{(-1)^z (z+1)!}{\nu_0!\nu_1!\nu_2!\nu_3!\omega_1!\omega_2!\omega_3!}\,. \quad (4.18)$$

The summation extends over all non-negative integers which satisfy the matrix equation

$$\mathcal{K} \equiv \begin{pmatrix} k_{10} & k_{20} & k_{30} \\ k_{01} & k_{31} & k_{21} \\ k_{32} & k_{02} & k_{12} \\ k_{23} & k_{13} & k_{03} \end{pmatrix} = \begin{pmatrix} \nu_0+\omega_1 & \nu_0+\omega_2 & \nu_0+\omega_3 \\ \nu_1+\omega_1 & \nu_1+\omega_2 & \nu_1+\omega_3 \\ \nu_2+\omega_1 & \nu_2+\omega_2 & \nu_2+\omega_3 \\ \nu_3+\omega_1 & \nu_3+\omega_2 & \nu_3+\omega_3 \end{pmatrix} \equiv \mathfrak{N} \quad (4.19)$$

or

$$k_{\alpha\beta} = \nu_\beta + \omega_i \quad (\alpha,\beta) \in W_i\,. \quad (4.19a)$$

The 6-j symbol is then given by

$$\begin{Bmatrix} j_{23} & j_{31} & j_{12} \\ j_{01} & j_{02} & j_{03} \end{Bmatrix} = \Big(\frac{\prod_{\alpha,\beta} k_{\alpha\beta}!}{\prod_\alpha (J_\alpha + 1)!} \Big)^{1/2} s(\mathcal{K}) \quad (4.19b)$$

[see (4.11)], the j and k being related by Eqs. (4.4) and (4.5).

Apart from its immediate application to the expression (4.18) the equation $\mathcal{K} = \mathfrak{N}$ (where we assume k,ν,ω to be non-negative integers) is of interest as a parametrization of *tetrahedral* $\mathcal{K}$. It is not difficult to show that $\mathcal{K}$ is tetrahedral if and only if it satisfies (4.19) for some $\mathfrak{N}$. Furthermore, for a given $\mathcal{K}$ the equation $\mathcal{K} = \mathfrak{N}$ has $\mu + 1$ solutions, where μ is the value of $\mathcal{K}$'s smallest element, and hence (4.18) has $\mu + 1$ terms.

Action of the symmetry operations. Let P be an element of the Regge group characterized by the permutations π and σ. If $\mathcal{K} = \mathfrak{N}$, then $P(\mathcal{K}) = \mathfrak{N}'$, the parameters of $\mathfrak{N}'$ being given by $\nu'_\alpha = \nu_{\pi_\alpha}$, $\omega'_i = \omega_{\sigma_i}$.

Racah's formula.[19] To obtain Racah's famous expression for $s(\mathcal{K})$ we have merely to express ν_α and ω_i by z. From (4.19) and (4.7a), $k_{\alpha 0} - k_{0\alpha} = \nu_0 - \nu_\alpha = J_\alpha - J_0$. Furthermore,

$$\nu_0 + J_0 = \nu_0 + k_{01} + k_{02} + k_{03} = \sum_{\alpha=0}^{3} \nu_\alpha + \sum_{i=1}^{3} \omega_i = z$$

[19] Racah, reference 13, Eq. (36.) Edmonds, reference 13, Eq. (6.3.7).

[by (4.19)], hence,

$$\nu_\alpha = z - J_\alpha . \qquad (4.20)$$

The first row in (4.19) yields $\omega_i = k_{i0} - \nu_0$, i.e.,

$$\omega_i = t_i - z \,, \quad t_i = k_{i0} + J_0 \,. \qquad (4.20a)$$

In terms of the j,

$$t_1 = j_{02} + j_{03} + j_{12} + j_{13} \,,$$
$$t_2 = j_{03} + j_{01} + j_{23} + j_{21} \,,$$
$$t_3 = j_{01} + j_{02} + j_{31} + j_{32} \,. \qquad (4.20b)$$

Inserting ν_α and ω_i in (4.18) one obtains Racah's formula

$$s(\mathcal{K}) = \sum_z \frac{(-1)^z (z+1)!}{\prod_\alpha (z - J_\alpha)! \prod_i (t_i - z)!}$$

the summation to be extended over those z for which all ν_α and $\omega_i \geq 0$.

It is readily shown that $4t_i = w_i + |\mathcal{K}|$. Hence the Regge operations are also described by $J'_\alpha = J_{\tau_\alpha}$, $t'_i = t_{\sigma_i}$ [see (4.17a)].

Remark. Schwinger has also computed the generating function for the 9-j symbol [reference 6, Eq. (4.37)]. This does not reveal any new symmetries—at least none to be obtained by a permutation of the relevant quantities $k_{\alpha\beta}$.

f. Recursion relations. Let $\Omega_{\alpha\beta}$ be the differential operator $\tau_{\alpha\beta}\, \partial/\partial\tau_{\alpha\beta}$. Then $\Omega_{\alpha\beta} G = g_{\alpha\beta}$, where $g_{\alpha\beta} = a_\beta + b_i$ if $(\alpha,\beta) \in W_i$. Hence

$$\Omega_{\alpha\beta} R = -2 g_{\alpha\beta} G^{-3}$$

and $g_{\gamma\delta}\Omega_{\alpha\beta} R = g_{\alpha\beta}\Omega_{\gamma\delta} R$, which leads to recursion relations for the $s(k)$. As an example consider $g_{32}\Omega_{01} R = g_{01}\Omega_{32} R$. Now $g_{32} = a_2 + b_1$, $g_{01} = a_1 + b_1$, so that

$$a_2 \Omega_{01} R = a_1 \Omega_{32} R + b_1(\Omega_{32} - \Omega_{01}) R \,.$$

From the power series for R we obtain

$$\begin{aligned} k_{01} s(k) = {} & (k_{32}+1) s(\cdots k_{02}+1, k_{12}+1, k_{32} \\ & +1, \cdots k_{01}-1, k_{21}-1, k_{31}-1, \cdots) \\ & + (k_{32}+1-k_{01}) s(\cdots k_{02}+1, k_{12} \\ & +1, \cdots k_{01}-1, k_{10}-1, k_{23}-1, \cdots) \,, \end{aligned}$$

where again on the right-hand side only those k are marked which differ from the corresponding ones on the left-hand side. For the 6-j symbols one finds

$$\begin{aligned} & [(J_2+1) k_{01} (k_{02}+1)(k_{12}+1)]^{1/2} \begin{Bmatrix} j_{23} & j_{31} & j_{12} \\ j_{01} & j_{02} & j_{03} \end{Bmatrix} \\ & = [(J_1+2) k_{21} k_{31} (k_{32}+1)]^{1/2} \begin{Bmatrix} j_{23}+\frac{1}{2} & j_{31}-\frac{1}{2} & j_{12} \\ j_{01}-\frac{1}{2} & j_{02}+\frac{1}{2} & j_{03} \end{Bmatrix} \\ & \quad + (k_{32}+1-k_{01})[k_{01}k_{23}]^{1/2} \\ & \quad \times \begin{Bmatrix} j_{23}+\frac{1}{2} & j_{31}-\frac{1}{2} & j_{12} \\ j_{01} & j_{02} & j_{03}-\frac{1}{2} \end{Bmatrix} \end{aligned}$$

APPENDIX. EVALUATION OF SOME LAPLACIAN INTEGRALS

(a) Let

$$\Lambda(B) = \pi^{-n} \int \exp(-\bar{z}\cdot Bz) d^n z \qquad (A1)$$

B is an $n \times n$ (complex) matrix with elements b_{kl}, so that

$$\bar{z}\cdot Bz = \sum_{k,l=1}^{n} \bar{z}_k b_{kl} z_l$$

The integral extends over all of C_n, and $d^n z = \prod_{k=1}^n dx_k\, dy_k$ $(z_k = x_k + iy_k)$.

Every B has a unique decomposition $B = B' + iB''$, with Hermitian B' and B'', and we call B' the *Hermitian part* of B.

If B' is *positive definite* the integral in (A1) converges absolutely, and

$$\Lambda(B) = (\det B)^{-1} \,. \qquad (A2)$$

Proof. We proceed in three steps. (1) If $B = 1$, then $\Lambda = 1$. [This is (1.6) for $h = h' = 0$] ,(2) If $B'' = 0$ and B is *positive definite* there exists a nonsingular matrix S such that

$$B = S^* S \,. \qquad (A3)$$

Introducing new variables $z' = Sz$ (and $\bar{z}' = \bar{S}\bar{z}$) we obtain $\bar{z}\cdot Bz = \bar{z}'\cdot z'$, which proves the absolute convergence of the integral. Setting $z'_k = x'_k + iy'_k$ we find for the Jacobian of the transformation

$$\frac{\partial(x_1,\cdots,x_n,y_1,\cdots,y_n)}{\partial(x'_1,\cdots,x'_n,y'_1,\cdots,y'_n)} = (\det S\cdot\det\bar{S})^{-1} = (\det B)^{-1} . \qquad (A3a)$$

Hence, $\Lambda(B) = (\det B)^{-1}\, \Lambda(1)$, Q.E.D. (3) Consider now $B = B' + iB''$ with positive definite B' and arbitrary B''. The modulus of the integrand is $\exp(-\bar{z}\cdot B'z)$, which establishes absolute convergence.

Introduce a complex parameter $\theta = \theta_1 + i\theta_2$, and set $C(\theta) = B' + i\theta B''$, so that

$$C(0) = B' \,, \quad C(1) = B \,. \qquad (A4)$$

$C(\theta)$ has the decomposition $C'(\theta) + iC''(\theta)$ with

$$C'(\theta) = B' - \theta_2 B'' \,, \quad C''(\theta) = \theta_1 B'' \qquad (A4a)$$

For small θ_2, $C'(\theta)$ is close to B' so that, for a suitable constant κ,

$$\bar{z}\cdot C'(\theta) z > \tfrac{1}{2}\, \bar{z}\cdot B'z \quad \text{if} \quad |\theta_2| < \kappa \,.$$

We now restrict θ to the strip $|\theta_2| < \kappa$, and show that $\Lambda(C(\theta))$ is *analytic* in θ. For this it is sufficient to observe that the integrand $\exp(-\bar{z}\cdot C(\theta)z)$ is obviously analytic in θ and that its modulus exp $(-\bar{z}\cdot C'(\theta)z)$ is uniformly bounded by the integrable function exp $(-\frac{1}{2}\bar{z}\cdot B'z)$. For imaginary θ, C is Hermitian and positive definite [see (A4a)], and in this case the equation $\Lambda(C(\theta)) = [\det C(\theta)]^{-1}$ has already been established. By analyticity it remains valid throughout the strip $|\theta_2| < \kappa$, in particular for $C(1) = B$.

Corollary.

$$I(A) = \int \exp(\bar{z}\cdot Az)d\mu_n(z) = [\det(1 - A)]^{-1} \quad \text{(A5)}$$

if $1 - A$ has a positive definite Hermitian part, in particular if A has sufficiently small matrix elements. In fact, by the definition of $d\mu_n(z)$ [see (1.5)], $I(A) = \Lambda(1 - A)$.

(b) Let

$$M(B,a,b) = \pi^{-6}\int \exp g(B,a,b;\xi,\eta)d^3\xi d^3\eta \quad \text{(A6)}$$

$$g = -\bar{\xi}\cdot B\xi - \bar{\eta}\cdot B\eta + D(\bar{a},\bar{\xi},\bar{\eta}) + D(b,\xi,\eta)\,. \quad \text{(A6a)}$$

Here, ξ and η are points inC_3, B is a 3×3 matrix, a,b are constant vectors in C_3, and D is a determinant as in Sec. 3f. As before, we proceed in three steps. (1) If $B = 1$, this is the integral in (3.23a), and for sufficiently small a,b, $M(1,a,b) = (1 - \bar{a}\cdot b)^{-2}$, by (3.23b). (2) If B is positive definite Hermitian, M is absolutely convergent for sufficiently small a,b (for example, $\bar{a}\cdot Ba < \det B$, and $\bar{b}\cdot Bb < \det B$). As before, set $B = S^*S$, let $\sigma = \det S$, and introduce new variables $\xi' = S\xi$, $\eta' = S\eta$. Set also $a' = Sa$ and $b' = Sb$. Then

$$\bar{\xi}\cdot B\xi = \bar{\xi}'\cdot\xi'\,, \quad \bar{\eta}\cdot B\eta = \bar{\eta}'\cdot\eta'$$

$$D(\bar{a},\bar{\xi},\bar{\eta}) = \bar{\sigma}^{-1}D(\bar{a}',\bar{\xi}',\bar{\eta}') = D(\bar{a}'',\bar{\xi}',\bar{\eta}')$$

$$D(b,\xi,\eta) = \sigma^{-1}D(b',\xi',\eta') = D(b'',\xi',\eta')\,,$$

where $a'' = \sigma^{-1}a'$, $b'' = \sigma^{-1}b'$. Thus,

$$g(B,a,b;\xi,\eta) = g(1,a'',b'';\xi',\eta')\,.$$

The Jacobian corresponding to (A3a) is now $(\sigma\bar{\sigma})^{-2}$. Hence $M(B,a,b) = (\sigma\bar{\sigma})^{-2}\, M(1,a'',b'') = [\sigma\bar{\sigma}(1 - \bar{a}''\cdot b'')^{-2}] = (\sigma\bar{\sigma} - \bar{a}'\cdot b')^{-2}$. Now $\sigma\bar{\sigma} = \det B$, and $\bar{a}'\cdot b' = \bar{a}\cdot Bb$. Therefore

$$M(B,a,b) = (\det B - \bar{a}\cdot Bb)^{-2}\,. \quad \text{(A7)}$$

(3) If B is no longer Hermitian, but has a positive definite Hermitian part, we may again show by analytic continuation that (A7) remains valid.

The integral to be evaluated in 4c is

$$N(H,\bar{u},v) = \int \exp g(H,\bar{u},v;\xi,\eta)d\mu_3(\xi)d\mu_3(\eta)\,.$$

Since $d\mu_3(\xi)\; d\mu_3(\eta)$ introduces the factor exp $(-\bar{\xi}\cdot\xi - \bar{\eta}\cdot\eta)$ it follows that $N(H,\bar{u},v) = M(1 + H,\bar{u},v)$, and hence

$$N(H,\bar{u},v) = [\det(1 + H) - u\cdot v - u\cdot Hv]^{-2}\,. \quad \text{(A8)}$$

ALGEBRAIC TABLES OF $J=\frac{1}{2}$, 1, $\frac{3}{2}$, AND 2 WIGNER COEFFICIENTS

$(j_1\ \tfrac{1}{2}\ m_1\ m_2|j_1\ \tfrac{1}{2}\ j\ m)$

$j=$	$m_2=\frac{1}{2}$	$m_2=-\frac{1}{2}$
$j_1+\frac{1}{2}$	$\sqrt{\frac{j_1+m+\frac{1}{2}}{2j_1+1}}$	$\sqrt{\frac{j_1-m+\frac{1}{2}}{2j_1+1}}$
$j_1-\frac{1}{2}$	$-\sqrt{\frac{j_1-m+\frac{1}{2}}{2j_1+1}}$	$\sqrt{\frac{j_1+m+\frac{1}{2}}{2j_1+1}}$

$(j_1\ 1\ m_1\ m_2|j_1\ 1\ j\ m)$

$j=$	$m_2=1$	$m_2=0$	$m_2=-1$
j_1+1	$\sqrt{\frac{(j_1+m)(j_1+m+1)}{(2j_1+1)(2j_1+2)}}$	$\sqrt{\frac{(j_1-m+1)(j_1+m+1)}{(2j_1+1)(j_1+1)}}$	$\sqrt{\frac{(j_1-m)(j_1-m+1)}{(2j_1+1)(2j_1+2)}}$
j_1	$-\sqrt{\frac{(j_1+m)(j_1-m+1)}{2j_1(j_1+1)}}$	$\frac{m}{\sqrt{j_1(j_1+1)}}$	$\sqrt{\frac{(j_1-m)(j_1+m+1)}{2j_1(j_1+1)}}$
j_1-1	$\sqrt{\frac{(j_1-m)(j_1-m+1)}{2j_1(2j_1+1)}}$	$-\sqrt{\frac{(j_1-m)(j_1+m)}{j_1(2j_1+1)}}$	$\sqrt{\frac{(j_1+m+1)(j_1+m)}{2j_1(2j_1+1)}}$

$(j_1\ \tfrac{3}{2}\ m_1\ m_2|j_1\ \tfrac{3}{2}\ j\ m)$

$j=$	$m_2=\frac{3}{2}$	$m_2=\frac{1}{2}$
$j_1+\frac{3}{2}$	$\sqrt{\frac{(j_1+m-\frac{1}{2})(j_1+m+\frac{1}{2})(j_1+m+\frac{3}{2})}{(2j_1+1)(2j_1+2)(2j_1+3)}}$	$\sqrt{\frac{3(j_1+m+\frac{1}{2})(j_1+m+\frac{3}{2})(j_1-m+\frac{3}{2})}{(2j_1+1)(2j_1+2)(2j_1+3)}}$
$j_1+\frac{1}{2}$	$-\sqrt{\frac{3(j_1+m-\frac{1}{2})(j_1+m+\frac{1}{2})(j_1-m+\frac{3}{2})}{2j_1(2j_1+1)(2j_1+3)}}$	$-(j_1-3m+\frac{3}{2})\sqrt{\frac{j_1+m+\frac{1}{2}}{2j_1(2j_1+1)(2j_1+3)}}$
$j_1-\frac{1}{2}$	$\sqrt{\frac{3(j_1+m-\frac{1}{2})(j_1-m+\frac{1}{2})(j_1-m+\frac{3}{2})}{(2j_1-1)(2j_1+1)(2j_1+2)}}$	$-(j_1+3m-\frac{1}{2})\sqrt{\frac{j_1-m+\frac{1}{2}}{(2j_1-1)(2j_1+1)(2j_1+2)}}$
$j_1-\frac{3}{2}$	$-\sqrt{\frac{(j_1-m-\frac{1}{2})(j_1-m+\frac{1}{2})(j_1-m+\frac{3}{2})}{2j_1(2j_1-1)(2j_1+1)}}$	$\sqrt{\frac{3(j_1+m-\frac{1}{2})(j_1-m-\frac{1}{2})(j_1-m+\frac{1}{2})}{2j_1(2j_1-1)(2j_1+1)}}$

$j=$	$m_2=-\frac{1}{2}$	$m_2=-\frac{3}{2}$
$j_1+\frac{3}{2}$	$\sqrt{\frac{3(j_1+m+\frac{3}{2})(j_1-m+\frac{1}{2})(j_1-m+\frac{3}{2})}{(2j_1+1)(2j_1+2)(2j_1+3)}}$	$\sqrt{\frac{(j_1-m-\frac{1}{2})(j_1-m+\frac{1}{2})(j_1-m+\frac{3}{2})}{(2j_1+1)(2j_1+2)(2j_1+3)}}$
$j_1+\frac{1}{2}$	$(j_1+3m+\frac{3}{2})\sqrt{\frac{j_1-m+\frac{1}{2}}{2j_1(2j_1+1)(2j_1+3)}}$	$\sqrt{\frac{3(j_1+m+\frac{3}{2})(j_1-m-\frac{1}{2})(j_1-m+\frac{1}{2})}{2j_1(2j_1+1)(2j_1+3)}}$
$j_1-\frac{1}{2}$	$-(j_1-3m-\frac{1}{2})\sqrt{\frac{j_1+m+\frac{1}{2}}{(2j_1-1)(2j_1+1)(2j_1+2)}}$	$\sqrt{\frac{3(j_1+m+\frac{1}{2})(j_1+m+\frac{3}{2})(j_1-m-\frac{1}{2})}{(2j_1-1)(2j_1+1)(2j_1+2)}}$
$j_1-\frac{3}{2}$	$-\sqrt{\frac{3(j_1+m-\frac{1}{2})(j_1+m+\frac{1}{2})(j_1-m-\frac{1}{2})}{2j_1(2j_1-1)(2j_1+1)}}$	$\sqrt{\frac{(j_1+m-\frac{1}{2})(j_1+m+\frac{1}{2})(j_1+m+\frac{3}{2})}{2j_1(2j_1-1)(2j_1+1)}}$

$(j_1\, 2\, m_1\, m_2 | j_1\, 2\, j\, m)$

$j=$	$m_2=2$	$m_2=1$	$m_2=0$
j_1+2	$\sqrt{\frac{(j_1+m-1)(j_1+m)(j_1+m+1)(j_1+m+2)}{(2j_1+1)(2j_1+2)(2j_1+3)(2j_1+4)}}$	$\sqrt{\frac{(j_1-m+2)(j_1+m+2)(j_1+m+1)(j_1+m)}{(2j_1+1)(j_1+1)(2j_1+3)(j_1+2)}}$	$\sqrt{\frac{3(j_1-m+2)(j_1-m+1)(j_1+m+2)(j_1+m+1)}{(2j_1+1)(2j_1+2)(2j_1+3)(j_1+2)}}$
j_1+1	$-\sqrt{\frac{(j_1+m-1)(j_1+m)(j_1+m+1)(j_1-m+2)}{2j_1(j_1+1)(j_1+2)(2j_1+1)}}$	$-(j_1-2m+2)\sqrt{\frac{(j_1+m+1)(j_1+m)}{2j_1(2j_1+1)(j_1+1)(j_1+2)}}$	$m\sqrt{\frac{3(j_1-m+1)(j_1+m+1)}{j_1(2j_1+1)(j_1+1)(j_1+2)}}$
j_1	$\sqrt{\frac{3(j_1+m-1)(j_1+m)(j_1-m+1)(j_1-m+2)}{(2j_1-1)2j_1(j_1+1)(2j_1+3)}}$	$(1-2m)\sqrt{\frac{3(j_1-m+1)(j_1+m)}{(2j_1-1)j_1(2j_1+2)(2j_1+3)}}$	$\frac{3m^2-j_1(j_1+1)}{\sqrt{(2j_1-1)j_1(j_1+1)(2j_1+3)}}$
j_1-1	$-\sqrt{\frac{(j_1+m-1)(j_1-m)(j_1-m+1)(j_1-m+2)}{2(j_1-1)j(j_1+1)(2j_1+1)}}$	$(j_1+2m-1)\sqrt{\frac{(j_1-m+1)(j_1-m)}{(j_1-1)j_1(2j_1+1)(2j_1+2)}}$	$-m\sqrt{\frac{3(j_1-m)(j_1+m)}{(j_1-1)j_1(2j_1+1)(j_1+1)}}$
j_1-2	$\sqrt{\frac{(j_1-m-1)(j_1-m)(j_1-m+1)(j_1-m+2)}{(2j_1-2)(2j_1-1)2j_1(2j_1+1)}}$	$\sqrt{\frac{(j_1-m+1)(j_1-m)(j_1-m-1)(j_1+m-1)}{(j_1-1)(2j_1-1)j_1(2j_1+1)}}$	$\sqrt{\frac{3(j_1-m)(j_1-m-1)(j_1+m)(j_1+m-1)}{(2j_1-2)(2j_1-1)j_1(2j_1+1)}}$

$j=$	$m_2=-1$	$m_2=-2$
j_1+2	$\sqrt{\frac{(j_1-m+2)(j_1-m+1)(j_1-m)(j_1+m+2)}{(2j_1+1)(j_1+1)(2j_1+3)(j_1+2)}}$	$\sqrt{\frac{(j_1-m-1)(j_1-m)(j_1-m+1)(j_1-m+2)}{(2j_1+1)(2j_1+2)(2j_1+3)(2j_1+4)}}$
j_1+1	$(j_1+2m+2)\sqrt{\frac{(j_1-m+1)(j_1-m)}{j_1(2j_1+1)(2j_1+2)(j_1+2)}}$	$\sqrt{\frac{(j_1-m-1)(j_1-m)(j_1-m+1)(j_1+m+2)}{j_1(2j_1+1)(j_1+1)(2j_1+4)}}$
j_1	$(2m+1)\sqrt{\frac{3(j_1-m)(j_1+m+1)}{(2j_1-1)j_1(2j_1+2)(2j_1+3)}}$	$\sqrt{\frac{3(j_1-m-1)(j_1-m)(j_1+m+1)(j_1+m+2)}{(2j_1-1)j_1(2j_1+2)(2j_1+3)}}$
j_1-1	$-(j_1-2m-1)\sqrt{\frac{(j_1+m+1)(j_1+m)}{(j_1-1)j_1(2j_1+1)(2j_1+2)}}$	$\sqrt{\frac{(j_1-m-1)(j_1+m)(j_1+m+1)(j_1+m+2)}{(j_1-1)j_1(2j_1+1)(2j_1+2)}}$
j_1-2	$-\sqrt{\frac{(j_1-m-1)(j_1+m+1)(j_1+m)(j_1+m-1)}{(j_1-1)(2j_1-1)j_1(2j_1+1)}}$	$\sqrt{\frac{(j_1+m-1)(j_1+m)(j_1+m+1)(j_1+m+2)}{(2j_1-2)(2j_1-1)2j_1(2j_1+1)}}$

BIBLIOGRAPHY

The purpose of the bibliography that follows is to provide a useful, and reasonably complete, compilation of journal and monograph references which can aid in tracing the ideas and the development of the quantum theory of angular momentum, as well as providing a guide to more detailed information on topics represented inadequately in our selection.

It is difficult to separate fully the applications of the theory from its development, but we have attempted to do so by categorizing the various references as noted below. Where the title of the reference is insufficiently descriptive, or the reason for inclusion too obscure, an explanatory comment has been added.

A. Books and Monographs

Primarily Concerned with the Quantum Theory of Angular Momentum

1. Brink, D. M., and Satchler, G. R. *Angular Momentum.* Oxford Univ. Press, London and New York (1962).
2. Born, M., and Jordan, P. *Elementare Quantenmechanik.* Springer, Berlin (1930). [This text is included because Chapter IV, "Die Sätze über den Drehimpuls" is the first—and classic—systematic treatment of the Quantum Theory of Angular Momentum.]
3. Edmonds, A. R. *Angular Momentum in Quantum Mechanics.* Princeton Univ. Press, Princeton, New Jersey (1957).
4. Fano, U., and Racah, G. *Irreducible Tensorial Sets.* Academic Press, New York (1959).
5. Feenberg, E., and Pake, G. E. *Notes on the Quantum Theory of Angular Momentum.* Addison-Wesley, Reading, Massachusetts (1953); reissued by Stanford Univ. Press, Stanford, California (1959).
6. Rose, M. E. *Elementary Theory of Angular Momentum.* Wiley, New York (1957).
7. Sharp, W. T. "The Quantum Theory of Angular Momentum" (unpublished). CRL-43 (AECL No. 465), Chalk River, Ontario (July, 1957).
8. Yutsis, A. P., Levinson, I. B., and Vanagas, V. V. *The Theory of Angular Momentum.* (Mathematicheskii apparat teorii momenta kolichestva dvizheniya.) Vilnius, U.S.S.R. (1960). Translated from Russian by A. Sen and A. R. Sen, Jerusalem, Israel (1962). Available from Office of Technical Services, U. S. Department of Commerce, Washington, D. C. [Of particular interest are the graphical methods and the explicit consideration of $(3n\text{-}j)$ symbols up to $n = 6$.]

Group Theoretical with Emphasis on Physics

Primarily mathematical texts, aside from the rotation group, have not been included.

9. Bauer, E. *Introduction a la théorie des groupes et ses applications a la physique quantique.* Presses Universitaires de France, Paris (1933).
10. Bayman, B. F. "Groups and their Applications to Spectroscopy." Lecture notes, Nordita, Copenhagen (1957); reissued (1960).
11. Brinkman, H. C. *Applications of Spinor Invariants in Atomic Physics.* North-Holland Publ. Amsterdam (1956).
12. Gelfand, I. M., Minlos, R. A., and Shapiro, Z. Ya. *Representations of the Rotation and Lorentz Groups.* Translated from the Russian edition [Predstavleniya gruppy

vrashchenii i gruppy Lorentsa. Fizmatgiz, Moscow (1958)] by G. Cummins and T. Boddington. Macmillan, New York (1963).

13. Heine, V. *Group Theory and Quantum Mechanics; an Introduction to its Present Usage*. Pergamon Press, New York (1960).
14. Hamermesh, M. *Group Theory and Its Application to Physical Problems*. Addison-Wesley, Reading, Massachusetts (1962).
15. Lyubarskii, G. Ya. *The Application of Group Theory in Physics*. Translated by S. Dedijer. Pergamon Press, Oxford (1960).
16. Meijer, H. E., and Bauer, E. *Group Theory, the Application to Quantum Mechanics*. Wiley, New York (1962).
17. Murnaghan, F. D. *The Unitary and Rotation Groups*, Vol. 3 of "Lectures on Applied Mathematics." Spartan Books, Washington (1962).
18. Racah, G. "Group theory and spectroscopy." *Engeb. Exakt. Naturw.* **37**, 28–84 (1965).
19. van der Waerden, B. L. *Die Gruppentheoretische Methode in der Quantenmechanik*. Springer, Berlin (1932).
20. Venkatarayudu, T. "Applications of Group Theory to Physical Problems" (unpublished). New York University (1953).
21. Weyl, H. *Gruppentheorie und Quantenmechanik*. Hirzel, Leipzig: 1st ed. (1928); 2nd ed. (1931) translated as *The Theory of Groups and Quantum Mechanics* by H. P. Robertson. Methuen, London (1931); reissued Dover, New York (1949).
22. Wigner, E. P. *Gruppentheorie und ihre Anwendung auf die Quantenmechanik der Atomspektren*. Vieweg, Braunschweig (1931); reprinted by Edwards Brothers, Ann Arbor, Michigan (1944). Translated by J. J. Griffin: *Group Theory and its Application to the Quantum Mechanics of Atomic Spectra*. Academic Press, New York (1959). [The translated version has been expanded by three additional chapters: "Racah Coefficients" (Chapter 24); "Time Inversion" (Chapter 26); and "Physical Interpretation and Classical Limits of Representation Coefficients, Three and Six-j Symbols" (Chapter 27)].

B. Angular Correlations

This has been a major field of application of the techniques of the Quantum Theory of Angular Momentum. Bibliographic information can be found from many reviews, for example, the following:

Devons, S., and Goldfarb, L. J. B. *Encylopedia of Physics* Vol. 42, pp. 362–554. Springer, Berlin (1957).

Ajzenberg-Selove, F. *Nuclear Spectroscopy*, Part B. Academic Press, New York (1960).

C. Applications to Atomic and Molecular Spectroscopy

The monograph of Condon and Shortley has been the standard reference in this field since 1935. The tremendous task of an up-to-date account has recently been undertaken by Slater. The inclusion of very extensive, topicalized, bibliographies is especially noteworthy in these volumes by Slater.

1. Condon, E. U., and Shortley, G. H. *Theory of Atomic Spectra*. Cambridge Univ. Press, London and New York (1935).
2. Judd, R. *Operator Techniques in Atomic Spectroscopy*. McGraw-Hill, New York (1963). [This monograph is especially concerned with Racah's methods.]
3. Slater, J. C. *Quantum Theory of Atomic Structure*, Volumes I and II. McGraw-Hill, New York (1960).

4. Slater, J. C. *Quantum Theory of Molecules and Solids*. McGraw-Hill, New York (1963).

Applications of the Quantum Theory of Angular Momentum in quantum chemistry are almost innumerable. References later than Slater's review, particularly to projection operator methods, density matrices (pre-eminent in the work of Löwdin and collaborators), may be found in the *Reviews of Modern Physics* **35** (1963).

D. Applications to Nuclear Spectroscopy Including the Coefficients of Fractional Parentage

Monographs

1. de-Shalit, A., and Talmi, I. *Nuclear Shell Theory*, (Pure and Applied Physics Series, Vol. 14). Adademic Press, New York (1963). [An extensive account with particular attention to the algebraic techniques of Racah. Unfortunately, there is no attempt at including a bibliography.]
2. Elliott, J. P., and Lane, A. M. "The Nuclear Shell Model" *Encyclopedia of Physics*, Vol. 39, pp. 241–410, Springer, Berlin (1957).
3. Feenberg, E. *Shell Theory of the Nucleus*. Princeton Univ. Press, Princeton, New Jersey (1955).
4. Mayer, M. G., and Jensen, H. D. *Elementary Theory of Nuclear Shell Structure*. Wiley, New York (1955).

In addition see also Bayman (A-10) and Racah (A-18).

Journal Articles

5. Bargmann, V., and Moshinsky, M. "Group theory of harmonic oscillators. I. The collective modes." *Nuclear Phys.*, **18**, 697–712 (1960); "Group theory of harmonic oscillators, II. The integrals of motion for the quadrupole–quadrupole interaction." *ibid.* **23**, 177–199 (1961).
6. Bayman, B. F., and Bohr, A. "On the connection between the cluster model and the SU_3 coupling scheme for particles in a harmonic oscillator potential." *Nuclear Phys.* **9**, 596–599 (1958/59).
7. Brody, T. A. "Brackets de transformacion para funciones de oscilador armonico." *Rev. Mex. Fis.* **8**, 139–227 (1959).
8. Brody, T. A., Jacob, G., and Moshinsky, M. "Matrix elements in nuclear shell theory." *Nuclear Phys.*, **17**, 16–29 (1960).
9. de-Shalit, A. "The energy levels of odd-odd nuclei." *Phys. Rev.* **91**, 1479-1486 (1963). [Application of the algebra of tensor operators to the two-particle contact interaction in the *j-j* shell model to obtain Nordheim's coupling rules.]
10. de-Shalit, A. "Energy levels of odd-mass and even nuclei." *Nuclear Phys.* **7**, 225–232 (1958).
11. Elliott, J. P., and Jahn, H. A. "Two body nuclear interaction, consistent within the limits of a single configuration with the spin and magnetic moment of the ground states of lithium-6, boron-10 and lithium-7." Letter in *Nature* **167**, 32–33 (Jan. 6, 1951). [Results corrected in Elliott, J. P., *Proc. Roy. Soc.* **A218**, 345–370 (1953)].
12. Elliott, J. P. "Collective motion in the nuclear shell model, I. Classification schemes for states of mixed configurations," and "—. II. The introduction of intrinsic wave-functions." *Proc. Roy. Soc.* **A245**, 128–145; 562–581 (1958).

13. Elliott, J. P. "The shell model" in *Nuclear Reactions*, Vol. I, edited by P. M. Endt and M. Demeur. North-Holland Publ., Amsterdam (1959).

14. Elliott, J. P. "The nuclear shell model and its relation to other models," in *Selected Topics in Nuclear Theory*, pp. 157–208. Intern. Atomic Energy Agency, Vienna (1963).

15. Feenberg, E., and Wigner, E. P. "On the structure of the nuclei between helium and oxygen." *Phys. Rev.* **51**, 95 (1937).

16. Flowers, B. H. "*jj* coupling in nuclei." *Phys. Rev.* **86**, 254 (L) (1952).

17. Flowers, B. H. "Excited states of nuclei in *jj*-coupling." *Physica* **18**, 1101–1104 (1952).

18. Flowers, B. H. "Studies in *jj*-coupling. I. Classification of nuclear and atomic states." *Proc. Roy. Soc.* **A212**, 248 (1952). Flowers, B. H., and Edmonds, A. R. "Studies in *jj*-coupling. II. Fractional parentage coefficients and the central force energy matrix for equivalent particles." *ibid.* **A214**, 515–530 (1952); "Studies in *jj*-coupling. III. Nuclear energy levels. *ibid.* **A215**, 120–132 (1952).

19. Ford, K. W., and Konopinski, E. J. "Evaluation of Slater integrals with harmonic oscillator wave functions." *Nuclear Phys.* **9**, 218–224 (1958/59).

20. French, J. B., and Fujii, A. "Angular momentum coupling in the nuclear p shell." *Phys. Rev.* **105**, 652–657 (1957).

21. French, J. B. "The nuclear shell model-I," University of Pittsburgh Report No. 9, Part 1, 105 pp. (1958) mimeographed.

22. French, J. B. "Symplectic symmetry in the nuclear shell model." *Nuclear Phys.* **15**, 393–410 (1960).

23. Gamba, A., Malvano, R., and Radicati, L. A. "Selection rules in nuclear reactions." *Phys. Rev.* **87**, 440 (1952). [Group theoretical discussion of selection rules from symmetric groups (Young patterns).]

24. Hassitt, A. "Fractional parentage coefficients and their explicit evaluation." *Proc. Roy. Soc.* **A229**, 110–119 (1955).

25. Helmers, K. "On symplectic invariance of the energy matrix in a nuclear j^n configuration." *Nuclear Phys.*, **12**, 647–656 (1959).

26. Helmers, K. "Symplectic invariants and Flowers' classification of shell model states." *Nuclear Phys.* **23**, 594–611 (1961).

27. Hope, J. "The tensor force interaction between a shell closed except for a single vacancy and an external nucleon." *Phys. Rev.* **89**, 884 (L) (1953).

28. Hund, F. "Symmetrieeigenschaften der Kräfte in Atomkernen und Folgen für deren Zustände, insbesondere der Kerne bis zu sechzehn Teilchen." *Z. Physik* **105**, 202–225 (1937).

29. Ishidzu, T., and Obi, S. "Structure of the f^n electron configurations. I. Coefficients of fractional parentage of f^3. II. Spin-orbit interaction of the configuration f^3." *J. Phys. Soc.* (*Japan*) **5**, 142–145 (1950).

30. Jahn, H. A. "Theoretical studies in nuclear structure. I. Enumeration and classification of the states arising from the filling of the nuclear d-shell." *Proc. Roy. Soc.* **A201**, 517–544 (1950). "—. II. Nuclear d^2, d^3, and d^4 configurations. Fractional parentage coefficients and central force matrix elements. *ibid.* A205, 192–237 (1951). "—. III. Radial integrals for the s, p and d shells." (By Swiatecki, W. J. *alone*) *ibid.* **A205**, 238–246 (1951). "—. IV. Wave functions for the nuclear p-shell. A. $\langle p^n|p^{n-1}p\rangle$ fractional parentage coefficients." (With Wieringen, H.) *ibid.* **A209**, 502–524 (1951). "—. B. $\langle p^n|p^{n-2}p^2\rangle$ fractional parentage coefficients." (With Elliott, J. P., and Hope, J.) *Phil. Trans. Roy. Soc.* **A246**, 241–279 (1953). "—. V. The matrix elements of non-central forces with an application to the 2 p-shell." (By Elliott, J. P. *alone*). *Proc. Roy. Soc.* A218, 345–370 (1953).

31. Jahn, H. A. "Recent developments in the theory of complex spectra with applications to nuclear structure." Three lectures, Institut Henri Poincaré, Paris (1952).
32. Jahn, H. A. "Direct evaluation of fractional parentage coefficients using Young operators, general theory and ⟨4/2, 2⟩ coefficients." *Phys. Rev.* **96**, 989 (1954).
33. Kaplan, I. G. "Coordinate fractional parentage coefficients for multishell configurations." *Soviet Phys.* JETP (*Engl. Transl.*) **14**, 568–573 (1962).
34. Kaplan, I. G. "The transformation matrix for the permutation group and the construction of coordinate wave functions for a multishell configuration." *Soviet Phys.* JETP (*Engl. Transl.*) **14**, 401–407 (1962).
35. Kelly, P. S., and Armstrong, B. H. "Fractional parentage coefficients for mixed configurations in LS coupling." *Astrophys. J.* **129**, 786–793 (1959).
36. Kretzchmar, M. "Zur Theorie der Wignerschen Supermultipletts." *Z. Physik* **157**, 558–567 (1960).
37. Kretzschmar, M. "Gruppentheoretische Untersuchungen zum Schalenmodell. I. Die Mathematische Theorie des Hamilton-Operators." *Z. Physik* **157**, 433–456 (1960). "Gruppentheoretische Untersuchungen zum Schalenmodell. II. Zum Problem der Translationsinvarianz." *ibid.* **158**, 284–303 (1960).
38. Lane, A. M., and Wilkinson, D. H. "Concept of parentage of nuclear states and its importance in nuclear reaction phenomena." *Phys. Rev.* **97**, 1199–1204 (1955).
39. Lawson, R. D., and Goeppert-Mayer, M. "Harmonic oscillator wave functions in nuclear spectroscopy." *Phys. Rev.* **117**, 174–184 (1960).
40. Lawson, R. D., and Zeidman, B. "Seniority mixing in $1f_{7/2}$ nuclei." *Phys. Rev.* **128**, 821–825 (1962).
41. Lipkin, H. J. Method for simplifying calculations involving products of operators in many-particle systems." *Phys. Rev.* **108**, 191 (1957).
42. Lipkin, H. J. "On the connection between degeneracy and rotational states in the nuclear shell model." *Nuclear Phys.* **8**, 421–427 (1958).
43. McIntosh, H. V. "Symmetry-adapted functions belonging to the symmetric groups." *J. Math. Phys.* **6**, 453-460 (1960).
44. Moshinsky, M. "Transformation brackets for harmonic oscillator functions." *Nuclear Phys.* **13**, 104–116 (1959).
45. Moshinsky, M., and Nagel, J. G. "Complete classification of states of supermultiplet theory." *Phys. Letters* **5**, 173 (1963).
46. Nesbet, R. K. "Construction of symmetry-adapted functions in the many-particle problem." *J. Math. Phys.* **2**, 701–709 (1961).
47. Neudachin, V. G. "On the structure of nuclear shell $1f_{7/2}$." *Vestnik Moskov. Univ.* **12**, No. 2, 111–116 (1957); *Nuclear Sci. Abstr.* 16645 (1958).
48. Neudachin, V. G., and Smirnov, Yu. F. "Parentage coefficients in the generalized nuclear model." *Zh. Eksptl. Teoret. Fiz.* **36**, 186–192 (1959).
49. Racah, G. "Nuclear coupling and shell model." *Phys. Rev.* **78**, 622 (L) (1950).
50. Racah, G. "Nuclear levels and Casimir operators," in *The L. Farkas Memorial Volume* (A. Farkas and E. P. Wigner, eds.), p. 294. Research Council of Israel Spec. Publ. No. 1, Jerusalem (1952).
51. Racah, G., and Talmi, I. "The pairing property of nuclear interactions." *Physica* **18**, 1097 (1952).
52. Racah, G. "A search for new quantum numbers in nuclear configurations." *Proc. of the 1954 Glasgow Conf. on Nuc. & Meson Phys.*, pp. 126–128. Pergamon Press, London and New York (1955).
53. Racah, G. "The seniority quantum number and its applications to nuclear spectroscopy." *Proc. Rehovoth Conf. on Nucl. Structure* (H. J. Lipkin, ed.), pp. 155–160. North Holland Publ., Amsterdam (1958).

54. Racah, G. "Mathematical techniques" in *Nuclear Spectroscopy*, Proc. Intern. School of Physics, Vol. 15. Academic Press, New York (1962).
55. Redmond, P. J. "An explicit formula for the calculation of fractional parentage coefficients." *Proc. Roy. Soc.* **A222**, 84–93 (1954).
56. Schwarz, C. and de-Shalit, A. "Many-particle configurations in a central field." *Phys. Rev.* **94**, 1257–1266 (1954).
57. Shapiro, I. S., "Symmetry properties in the theory of elementary particles and nuclear processes." *Uspekhi Fiz. Nauk* **57**, 7 (1954).
58. Swiaticki, W. J. "Radical integrals for the s, p and d shells." *Proc. Roy. Soc.* **A205**, 238–246 (1951).
59. Talmi, I. "Nuclear spectroscopy with harmonic oscillator wave-functions." *Helv. Phys. Acta* **25**, 185–234 (1952).
60. Talmi, I. "Symplectic invariance and the pairing property of nuclear reactions." *Nuclear Phys.* **16**, 153–157 (1960).
61. Talmi, I. "Seniority of the $f_{7/2}$ 4 levels in Cr^{52}." *Phys. Rev.* **126**, 1096–1098 (1962).
62. Tauber, G. E., and Wu, T.-Y "The J values of states in configurations j^n." *Nuovo cimento* [9] **10**, 677 (1953).
63. Trainor, L. E. H. "The formation of anti-symmetric wave functions." *Can. J. Phys.* **35**, 555 (1957).
64. Wigner, E. P. "On the consequences of the symmetry of the nuclear hamiltonian on the spectroscopy of nuclei." *Phys. Rev.* **51**, 106 (1937).
65. Wigner, E. P. "On the structure of nuclei beyond oxygen." *Phys. Rev.* **51**, 947 (1937).
66. Wigner, E. P., and Feenberg, E. "Symmetry properties of nuclear levels." *Repts. Progr. Phys.* **8**, 274 (1942).
67. Wigner, E. P. "On the shell model for nuclei" in *The L. Farkas Memorial Volume* (A. Farkas and E. P. Wigner, eds.), pp. 45–61. Research Council of Israel Special Publication No. 1, Jerusalem (1952).
68. Wigner, E. P. "Isotopic spin: A quantum number for nuclei" in *Proc. Robert A. Welch Foundation Conf. on Chem. Res.*, Vol. 1: *The Structure of the Nucleus.* Houston, Texas (1958).
69. Zel'tser, G. I. "Fractional parentage coefficients for the wave function of four particles." *Zh. Eksptl. i Teoret. Fiz.* **34**, 694–699 (1958); *Soviet Phys. JETP* (*Engl. Transl.*) **7**, 477–481 (1958).

The techniques based originally on the Quantum Theory of Angular Momentum have been absorbed into the larger field of group theory, and the applications of group theoretical methods to nuclear spectroscopy have become numerous. Besides the references cited above (of which those due to Wigner and Hund initiated the field), let us cite two recent reviews for further referencing:

Biedenharn, L. C. "Group theoretical approaches to nuclear spectroscopy," *Lectures in Theoretical Physics*, Vol. V. (Lectures delivered at the Summer Institute for Theoretical Physics, University of Colorado, Boulder, 1962). Edited by W. E. Brittin, B. W. Downs, and Joanne Downs. Wiley (Interscience), New York (1963).

Moshinsky, M. "Group theory and the many body problems," in *Physics of Many Particle Systems.* Edited by E. Meeron. Gordon and Breach, New York (1964).

E. Algebraic and Numerical Tabulations of the Wigner, Racah and Associated Coefficients

Books

1. Rotenberg, M., Metropolis, N., Bivins, R., and Wooten, J. K., Jr. *The* 3-*j* and 6-*j* *symbols*. M.I.T. Press, Cambridge, Massachusetts (1959).
2. Ishidzu, T., Horie, H., Obi, S., Sata, M., Tanabe, Y., and Yanagawa, S. *Tables of the Racah Coefficients*. Pan-Pacific Press, Tokyo, Japan (1960).
3. Wapstra, A. H., Nijgh, C. J., and van Lieshout, R. *Nuclear Spectroscopy Tables*. North-Holland Publ., Amsterdam (1959).

3a. Niktoforov, A. F., Uvarov, V. B., and Levitan, Yu. L. *Tables of Racah Coefficients*. Translated by P. Basu. Macmillan, New York (1965).

Articles

4. Alder, K. "Beiträge zur Theorie der Richtungskorrelation." *Helv. Phys. Acta* **25**, 235–258 (1952). [Algebraic table of Racah coefficients and some Wigner coefficients.]
5. Arima, A., Horie, H., and Tanabe, Y. "Generalized Racah coefficient and its applications." *Progr. Theoret. Phys.* (*Kyoto*) **11**, 143 (1954). [Definitions and properties of the X coefficient.]
6. Biedenharn, L. C., Blatt, J. M., and Rose, M. E. "Some properties of the Racah and associated coefficients." *Revs. Mod. Phys.* **24**, 249–257 (1952).
7. Biedenharn, L. C. "Tables of the Racah coefficients." Oak Ridge National Laboratory, Physics Division, ORNL-1098 (April, 1952).
8. Boys, S. F. "Electronic wave functions. VI. Some theorems facilitating the evaluation of Schrödinger integrals of vector-coupled functions." *Phil. Trans. Roy. Soc.* **A245**, 95 (1952). [Properties of W (*abcd*; *ef*) function (in different notation).]
9. Boys, S. F. and Sahni, R. C. "Electronic wave functions. XII. The evaluation of the general vector-coupling coefficients by automatic computation." *Phil. Trans. Roy. Soc.* **A246**, 463–479 (1954). [This paper contains further references to the series of papers by Boys *et al.*]
10. Bryant, P. E. "Tables of Wigner 3-*j* symbols." [With a note on new parameters for the Wigner 3-*j* symbols by Jahn, H. A., and Bryant, P. E.] University of Southampton Research Report 60–1, Southampton, England (1960).
11. Cohen, E. R. "Tables of Clebsch-Gordan coefficients," 37 pp. AEC Report NAA-SR-2123 (1958).
12. Falkoff, D. L., Colladay, G. S., and Sells, R. E. "Transformation amplitudes for vector addition of angular momentum $(j\, 3mm'|j\, 3\, JM)$." *Can. J. Phys.* **30**, 253–256 (1952).
13. Goldstein, M., and Kazek, C., Jr. "Table of the Racah and Z coefficients." Los Alamos Scientific Laboratory Report LAMS-1739 (1954).
14. Howell, K. M. "Revised tables of 6*j*-symbols." [With an introduction, "New parameters and symmetry relations for the Wigner 6*j*-symbol," by Jahn, H. A., and Howell, K. M.] University of Southampton Research Report 59–1, Southampton, England (1959).
15. Howell, K. M. "Tables of 9-*j* symbols." University of Southampton Research Report 59–2, Southampton, England (1959).
16. Jahn, H. A. "Theoretical studies in nuclear structure. II. Nuclear d^2, d^3, and d^4 configurations. Fractional parentage coefficients and central force matrix elements." *Proc. Roy. Soc.* **A205**, 192–237 (1951). [Algebraic tables for a (renormalized) Racah coefficient contained in appendix.]

17. Kennedy, J. M., and Cliff, M. J. "Transformation coefficients between *LS* and *jj* coupling." Atomic Energy of Canada, Ltd., Report AECL-224 (1955).
18. Kennedy, J. M., Sears, B. J., and Sharp, W. T. "Tables of the X-coefficients." Atomic Energy of Canada, Ltd., Reports AECL-106, CRT-569 (1954).
19. Kumar, K. "Tables of certain Clebsch-Gordan coefficients and of matrix elements of P_2, P_2^2, and P_2^3 between single-particle states." *Can. J. Phys.* **35**, 341–345 (1957); Errata, *ibid.* **35**, 1401 (1957).
20. Majumdar, S. D. "The Clebsch-Gordan coefficients." *Progr. Theoret. Phys.* (*Kyoto*) **20**, 798–803 (1958).
21. Matsunobu, H. and Takebe, H. "Tables of U coefficients." *Progr. Theoret. Phys.* (*Kyoto*) **14**, 589–605 (1955).
22. Melvin, M. A., and Swamy, N. V. V. J. "Algebraic table of vector-addition coefficients for $j_2 = 5/2$." *Phys. Rev.* **107**, 186 (1957).
23. Saito, R., and Morita, M. "Clebsch-Gordan coefficients for $j_2 = 5/2$." *Progr. Theoret. Phys.* (*Kyoto*) **13**, 540 (1955).
24. Sato, K. "General formula of the Racah coefficients." *Progr. Theoret. Phys.* (*Kyoto*) **13**, 405 (1955).
25. Sears, B. J., and Radtke, G. "Algebraic tables of Clebsch-Gordan coefficients." Atomic Energy of Canada, Ltd., TPI-75 (1954).
26. Sharp, W. T., Kennedy, J. M., Sears, B. J., and Hoyle, M. G. "Tables of coefficients for angular distribution analysis." Atomic Energy of Canada, Ltd., Reports CRT-556, AECL-97 (1953). [Besides an excellent summary of the physical applications, this compact report tabulates the Wigner, Racah, Z, and X coefficients as square roots of rational fractions in a concise notation giving the prime factors explicitly.]
27. Shimpuku, T. "General theory and numerical tables of Clebsch-Gordan coefficients." *Progr. Theoret. Phys.* (*Kyoto*) *Suppl.* **13**, 1–135 (1960).
28. Simon, A., Vander Sluis, J. H., and Biedenharn, L. C. "Tables of the Racah coefficients." Oak Ridge National Laboratory Report ORNL-1679 (1952).
29. Simon, A. "Numerical table of the Clebsch-Gordan coefficients." Oak Ridge National Laboratory, Physics Division, ORNL-1718 (1954). [Ten place decimal fractions for parameters 0 ($\frac{1}{2}$) 9/2.]
30. Smith, K., and Stephenson, J. W. "A table of Wigner 9*j* coefficients for integral and half-integral values of the parameters." Argonne National Laboratory Report ANL-5776 (1957).
31. Stephenson, J. W., and Smith, K. "Tabulation of the Wigner 9*j* coefficients." *Proc. Phys. Soc.* (*London*) **70A**, 571–572 (1957).
32. Yamada, M., and Morita, M. "On the β-ray angular correlations." *Progr. Theoret. Phys.* (*Kyoto*) **8**, 431 (1952). [Clebsch-Gordan coefficients for $j_2 = 3$ in appendix.]

F. General References Relevant to the Quantum Theory of Angular Momentum

1. Altmann, S. L., and Bradley, C. J. "A note on the calculations of the matrix elements of the rotation group." *Phil. Trans. Roy. Soc.* **A255**, 193–198 (1962).
2. Ambler, E., Eisenstein, J. C., and Schooley, J. F. "Traces of products of angular momentum matrices." J. Math. Phys. **3**, 118–130 (1962). "Traces of products of angular momentum matrices. II. Spherical Basis." *ibid.* **3**, 760–771 (1962).
3. Bacher, R. F., and Goudsmit, S. "Atomic energy relations. I." *Phys. Rev.* **46**, 948 (1934). [Original reference on the genealogical term classification.]
4. Bade, W. L., and Jehle, H. "An introduction to spinors." *Revs. Mod. Phys.* **25**, 714 (1953).

5. Bargmann, V. "Zur Theorie des Wasserstoffatoms. Bemerkungen zur gleichnämigen Arbeit von V. Fock." *Z. Physik* **99**, 576 (1936). [Symmetry properties of the non-relativistic hydrogen atom.]
6. Bargmann, V. "On the representations of the rotation group." *Revs. Mod. Phys.* **34**, 829–845 (1962).
7. Biedenharn, L. C. "An identity satisfied by Racah coefficients." *J. Math. and Phys.* **31**, 287 (1953).
8. Biedenharn, L. C. "A note on statistical tensors in quantum mechanics." *Ann. Phys.* (N. Y.) **4**, 104–113 (1958). Addendum: *ibid.* **6**, 399–400 (1959).
9. Biedenharn, L. C. and Brussaard, P. J. "A note on proper angular momentum variables and the implication for angular correlation theory." *Ann. Phys.* (N. Y.) **16**, 1 (1961).
10. Biedenharn, L. C. "Wigner coefficients for the R_4 group and some applications." *J. Math. Phys.* **2**, 433–441 (1961).
11. Bopp, F., and Haag, R. "Über die Möglichkeit von Spinmodellen." *Z. Naturforsch.*, **5a**, 644–653 (1950).
12. Born, M., Heisenberg, W., and Jordan, P. "Zur Quantenmechanik. II." *Z. Phys.* **35**, 557 (1926). [Commutation rules for angular momentum.]
13. Brauer, R., and Weyl, H. "Spinors in n Dimensions." *Am. J. Math.* **57**, 425–449 (1935). [Very clear and elegant presentation. Reprinted in the "Selecta Hermann Weyl," Birkhäuser, Basel (1956).]
14. Breit, G., and Darling, B. T. "Note on the calculation of angular distributions in resonance reactions." *Phys. Rev.* **71**, 402–405 (1947). [Points out certain symmetry relations for Wigner coefficients.]
15. Brussard, P. J., and Tolhoek, H. A. "Classical limits of Clebsch-Gordan coefficients, Racah coefficients and $D^l_{mn}(\phi, \theta, \phi)$-functions." *Physica* **23**, 955–971 (1957).
16. Calais, J.-L. "Derivation of the Clebsch-Gordan coefficients by means of projection operators." Tech. note 25. [AF-61(514)-1200] unpublished (June, 1959).
17. Casimir, H. B. G. "Rotation of a rigid body in quantum mechanics." Thesis, University of Leyden. Wolters Publ. Co., Groningen, The Netherlands (1931). *Koninkl. Ned. Akad. Wetenschap., Proc.* **34**, 844 (1931).
18. Casimir, H. B. G., and van der Waerden, B. L. "Algebraischer Beweis der vollständigen Reduzibilität der Darstellungen, halbeinfacher Liescher Gruppen. *Math. Ann.* **111**, 1–12 (1935) (See criticism, Pauli, F-71, p. 91.)
19. Casimir, H. B. G. "Interaction between atomic nuclei and electrons." *Arch. musée Teyler* **8**, 201–287 (1936).
20. Chiu, Y.-N. "Irreducible tensor expansion of solid spherical harmonic-type operators in quantum mechanics." *J. Math. Phys.* **5**, 283–288 (1964).
21. Corben, H. C., and Schwinger, J. S. "The electromagnetic properties of mesotrons." *Phys. Rev.* **58**, 953 (1940). [Appendix: vector spherical harmonics by operator methods.]
22. Darwin, C. G. "The Zeeman effect and spherical harmonics." *Proc. Roy. Soc.* **A115**, 1 (1927).
23. de-Shalit, A. "Angular momentum in non-spherical fields." *Bull. Research Council Israel* **3**, 359 (1954).
24. Dirac, P. A. M. "The elimination of the nodes in quantum mechanics." *Proc. Roy. Soc.* **A111**, 281–305 (1926). [Commutation relations for angular momentum.]
25. Dolginov, A. Z. "Relativistic spherical functions." *Soviet Phys. JETP* (*Engl. Trans.*) **3**, 589–596 (1956).
26. Dolginov, A. Z., and Toptygin, I. N. "Clebsch-Gordan expansion for infinite-dimensional representations of the Lorentz group." *Soviet Phys. JETP* (*Engl. Trans.*) **8**, 550–551 (1959).

27. Eckart, C. "The application of group theory to the quantum dynamics of monatomic systems." *Revs. Mod. Phys.* **2**, 305–380 (1930). [Paragraph 24 contains beginning of the "Wigner-Eckart" Theorem.]
28. Eriksson, K. B. S. "Coupling of electrons with high orbital angular momentum, illustrated by $2p\ nf$ and $2p\ ng$ in N II." *Phys. Rev.* **102**, 102 (1956).
29. Falkoff, D. L. and Uhlenbeck, G. E. "On the directional correlation of successive nuclear radiations." *Phys. Rev.* **79**, 323 (1950). [Introduce Weyl's techniques (polarization process for tensors, etc.) in angular correlation calculations.]
30. Fano, U. "Statistical matrix techniques and their application to the directional correlations of radiations." U. S. National Bureau of Standards Report 1214 (1951) (unpublished). [Defines X coefficient and emphasizes importance.]
31. Fano, U. "Description of states in quantum mechanics by density matrix and operator techniques." *Revs. Mod. Phys.* **29**, 74–93 (1957).
32. Fano, U. "Real representations of coordinate rotations." *J. Math. Phys.*, **1**, 417–423 (1960).
33. Fock, V. "Zur Theorie des Wasserstoffatoms." *Z. Physik* **98**, 145 (1936). [R_4 symmetry of hydrogen atom.]
34. Gabriel, J. R. "On the construction of irreducible representations of the symmetric group." *Proc. Cambridge Phil. Soc.* **57**, 330–340 (1961). [Tailored to machine computation; useful critique of earlier work.]
35. Gabriel, J. R. "New methods for reduction of group representations using an extension of Schur's lemma." *J. Math. Phys.* **5**, 494–504 (1964).
36. Gaunt, J. A. "IV. The triplets of helium." *Phil. Trans. Roy. Soc.* **A228**, 151–196 (1929). [The appendix, beginning at page 192, is the widely quoted work on "The integral of a product of 3 Tesseral harmonics."]
37. Granzow, K. D. "N-Dimensional total orbital angular-momentum operator." *J. Math. Phys.* **7**, 897–900 (1963).
38. Gray, N. M., and Wills, L. A. "Note on the calculation of zero order eigenfunctions." *Phys. Rev.* **38**, 248 (1931). [The use of spin projection operators.]
39. Güttinger, P., and Pauli, W. "Zur Hyperfeinstruktur von Li II." *Z. Physik* **67**, 743–765 (1931). [Mathematical Appendix (754–765) is the beginning of the algebraic technique for tensor operators.]
40. Güttinger, P. "Das Verhalten von Atomen im magnetischen Drehfeld." *Z. Physik* **73**, 169–184 (1932). [Appendix gives general rotation matrices.]
41. Heisenberg, W., and Jordan, P. "Anwendung der Quantenmechanik auf das Problem der anomalen Zeemaneffekte." *Z. Physik* **37**, 263 (1926).
42. Hill, E. L. "The theory of vector spherical harmonics." *Am. J. Phys.* **22**, 211–214 (1954).
43. Hulthén, L. "Uber die quantenmechanische Herleitung der Balmerterme." *Z. Physik* **86**, 21 (1933).
44. Hund, F. "Symmetriecharaktere von Termen bei Systemen mit gleichen Partikeln in die Quantenmechanik." *Z. Physik* **43**, 788 (1927).
45. Infeld, L., and Hull, T. E. "The factorization method." *Revs. Mod. Phys.* **23**, 21 (1951). [A general discussion of the raising-lowering operator technique.]
46. Jackson, T. A. S. "Accidental degeneracy of hydrogen." *Proc. Phys. Soc.* (*London*) **A66**, 958 (1953).
47. Jahn, H. A., and Hope, J. "Symmetry properties of the Wigner 9-j symbol." *Phys. Rev.* **93**, 318–321 (1954).
48. Jauch, J. M., and Hill, E. L. "'On the problem of degeneracy in quantum mechanics." *Phys. Rev.* **57**, 641 (1940).
49. Johnson, M. H., Jr. "Spectra of two-electron systems." *Phys. Rev.* **38**, 1628–1641

(1931). [Uses the results of Güttinger and Pauli to explicitly determine the complete matrix elements of **L**·**S** operator (i.e., "Racah coefficients").]

50. Jordan, P. "Der Zusammenhang der symmetrischen und linearen Gruppen und das Mehrkorperproblem." *Z. Physik* **94**, 531–535 (1935). [First introduction of general unitary group by explicit fermion and boson operator structures.]
51. Kramers, H. A. "Zur Ableitung der quantenmechanischen Intensitätsformeln." *Koninkl. Ned. Akad. Wetenschap, Proc.* **33**, 953 (1930).
52. Lomont, J. S., and Moses, H. E. "An angular momentum Helmholtz theorem." *Communs. Pure and Appl. Math.* **14**, 69–76 (1960). [See also the simpler proof by Keller, J. B. *ibid.* pp. 77–80].
53. Lehrer-Ilamed, Y. "On the direct calculations of the representations of the three dimensional pure rotation group." *Proc. Cambridge Phil. Soc.* **60**, 61–66 (1964).
54. Lipkin, H. J., and Goldstein, S. "Rotational properties of the two-dimensional anisotropic harmonic oscillator. I." *Nucl. Phys.* **5**, 202–210 (1958). "—. II. The collective-intrinsic description." *ibid.* **7**, 184–194 (1958).
55. Louck, J. D. "New recursion relation for Clebsch-Gordan coefficients." *Phys. Rev.* **110**, 815 (1958).
56. Louck, J. D. "Theory of angular momentum in N-dimensional space." Los Alamos Scientific Laboratory Report LA 2451 (1960) 298 pp. (unpublished). [Very thorough treatment.]
57. Louck, J. D. "Generalized orbital angular momentum and the n-fold degenerate quantum-mechanical oscillator. I. The twofold degenerate oscillator" (with W. H. Shaffer) *J. Mol. Spectry.* **4**, 285–298 (1960). "—. II. The n-fold degenerate oscillator." *ibid.* **4**, 298–333 (1960). "—. III. Radial integrals." *ibid.* **4**, 334–341 (1960).
58. McIntosh, H. V. "On accidental degeneracy in classical and quantum mechanics." *Am. J. Phys.* **27**, 620–625 (1959).
59. Meckler, A. "Majorana formula." *Phys. Rev.* **111**, 1447–1449 (1958). [The appendix demonstrates an interesting relation between Tchebichef polynomials and certain Wigner coefficients.]
60. Melvin, M. A. "Simplification in finding symmetry-adapted eigenfunctions." *Revs. Mod. Phys.* **28**, 18–44 (1956).
61. Meshkov, S., and Ufford, C. W. "A complete Bacher and Goudsmit method." *Phys. Rev.* **94**, 75 (1954).
62. Meyer, B., "On the symmetries of spherical harmonics." *Can. J. Math.* **6**, 125 (1954).
63. Moshinsky, M. "On Wigner coefficients for the SU_3 group and some applications." *Revs. Mod. Phys.* **34**, 813–828 (1962).
64. Murraghan, F. D. "On the decomposition of tensors by contraction." *Proc. Natl. Acad. Sci. U. S.* **38**, 973 (1952).
65. Neumann, J. v., and Wigner, E. "Zur Erklärung einiger Eigenschaften der Spektren aus der Quantenmechanik des Drehelektrons. I." *Z. Physik* **47**, 203 (1928). "—. II." *ibid.* **49**, 73 (1928). "—. III." *ibid.* **51**, 844 (1928).
66. Ord-Smith, R. J. "The symmetry relations of the 12-j symbol." *Phys. Rev.* **94**, 1227–1228 (1954).
67. Pais, A. "On spinors in n dimensions." *J. Math. Phys.* **6**, 1135–1139 (1962).
68. Parks, D. "Relation between the parabolic and spherical eigenfunctions of hydrogen." *Z. Physik* **159**, 155–157 (1960).
69. Pauli, W. "Uber das Wasserstoffspektrum vom Standpunkt der neuen Quantenmechanik." *Z. Physik* **36**, 336 (1926). [Treatment of non-relativistic hydrogen atom by vector operator techniques.]

70. Pauli, W. "Zur Quantenmechanik des magnetischen Elektrons." *Z. Physik* **43**, 601–623 (1927).
71. Pauli, W. "Continuous groups in quantum mechanics." CERN-56-31 (1956); *Ergeb. Exakt. Naturw.* **37**, 85–104 (1965).
72. Percus, J. K., and Rotenberg, A. "Exact eigenfunctions of angular momentum by rotational projection." *J. Math. Phys.* **3**, 928–932 (1962).
73. Peter, L., and Weyl, H. "Die Vollständigkeit der primitiven Darstellungen einer geschlossenen kontinuierlichen Gruppe." *Math. Ann.* **97**, 737–755 (1927).
74. Racah, G. "Theory of complex spectra. I." *Phys. Rev.* **61**, 186 (1942). "—. II." *ibid.* **62**, 438 (1942). "—. III." *ibid.* **63**, 367 (1943). "—. IV." *ibid.* **76**, 1352 (1949).
75. Racah, G. "On the configurations ll's." *Phys. Rev.* **62**, 523 (1942).
76. Racah, G. "On a new type of vector coupling in complex spectra." *Phys. Rev.* **61**, 537 (1942).
77. Racah, G. "On the decomposition of tensors by contraction." *Revs. Mod. Phys.* **21**, 494 (1949).
78. Racah, G. "Sulla caratterizzazione della reppresentazioni irreducibili dei gruppi semisemplici di Lie." *Atti Acad. Naz. Lincei, Rend., Classe Sci. Fiz. Mat. Nat.* **8**, 108 (1950).
79. Racah, G. "Theory of Lie groups." *Nuovo cimento Suppl.* **14**, 67–74 (1959).
80. Regge, T. "Symmetry properties of Clebsch-Gordan coefficients." *Nuovo cimento* [10] **10**, 544–545 (1958).
81. Regge, T. "Symmetry properties of Racah's coefficients." *Nuovo cimento* [10] **11**, 116–117 (1959).
82. Rose, M. E. "Spherical tensors in physics," *Proc. Phys. Soc. (London)* **67A**, 239–247 (1954).
82. Rose, M. E. "Statistical tensors for oriented nuclei." *Phys. Rev.* **108**, 362–365 (1957).
83. Rose, M. E. "Properties of the irreducible angular momentum tensors." *J. Math. Phys.* **3**, 409–413 (1962).
84. Rose, M. E., and Morita, M. "Formal consequences of conservation of angular momentum projections." *Progr. Theoret. Phys. (Kyoto)* **31**, 103–106 (1964).
85. Rose, M. E., and Yang, C. N. "Eigenvalues and eigenvectors of a symmetric matrix of $6j$ symbols." *J. Math Phys.* **3**, 106 (1962).
86. Sack, R. A. "Three-dimensional addition theorem for arbitrary functions involving expansions in spherical harmonics." *J. Math. Phys.* **5**, 252–259 (1964).
87. Salam, A. "The formalism of Lie groups," Lectures presented at the 1962 Trieste Seminar on Theoretical Physics. Intern. Atomic Energy Agency, Vienna (1963).
88. Saletan, E. J. "Contraction of Lie groups." *J. Math. Phys.* **2**, 1–21 (1961).
89. Schwinger, J. "On angular momentum." U. S. Atomic Energy Commission, NYO-3071 (1952).
90. Serber, R. "Extension of the Dirac vector model to include several configurations." *Phys. Rev.* **45**, 461 (1934).
91. Sharp, W. T. "Some formal properties of the 12-j symbol." A.E.C.L., Chalk River, Ontario Report TPI-81 (1955).
92. Sharp, W. T. "Racah algebra and the contraction of groups." Thesis, Princeton University (1960) issued as A.E.C.L. 1098, Chalk River, Ontario. [Very clear and careful treatment.]
93. Shortley, G. H. "The theory of complex spectra." *Phys. Rev.* **40**, 185 (1932).
94. Shortley, G. H. "Transformations in the theory of complex spectra." *Phys. Rev.* **43**, 451 (1933).
95. Shortley, G. H., and Kimball, G. E. "Analysis of non-commuting vectors with

application to quantum mechanics and vector calculus." *Proc. Natl. Acad. Sci. U. S.* **20**, 82 (1934).

96. Shortley, G. H., and Fried, B. "Extension of the theory of complex spectra." *Phys. Rev.* **54**, 739 (1938).
97. Slater, J. C. "The theory of complex spectra." *Phys. Rev.* **34**, 1293 (1929).
98. Smith, F. T. "Generalized angular momentum in many-body collisions." *Phys. Rev.* **120**, 1058–1069 (1960).
99. Stone, A. P. "Some properties of Wigner coefficients and hyperspherical harmonics." *Proc. Cambridge Phil. Soc.* **52**, 424 (1956).
100. Stone, A. P. "Tensor operators under semi-simple groups." *Proc. Cambridge Phil. Soc.* **57**, 460–468 (1961).
101. Stone, A. P. "Representations of the n-dimensional rotation group." *Proc. Cambridge Phil. Soc.* **57**, 469–475 (1961).
102. Stone, A. P. "Expressions for certain Wigner coefficients." *Proc. Phys. Soc. (London)* **70A** 908–909 (1957).
103. Trees, R. E. "Hyperfine structure formulas for *LS* Coupling." *Phys. Rev.* **92**, 308 (1953). [The appendix to this paper derives independently the X coefficients (or $9j$-symbol) for *LS* coupling and discusses some of their properties.]
104. van der Waerden, B. L. "Spinoren Analyse." *Nachr. Akad. Wiss. Göttingen, Math.-Phys. Kl.* p. 100 (1929).
105. van Vleck, J. H. "The Dirac vector model in complex spectra." *Phys. Rev.* **45**, 405 (1934).
106. van Vleck, J. H. "The coupling of angular momentum vectors in molecules." *Revs. Mod. Phys.* **23**, 213 (1951).
107. Weyl, H. "Zur Charakterisierung der Drehungsgruppe." *Math. Z.* **17**, 293–320 (1923).
108. Wick, G. C., Wightman, A. S., and Wigner, E. P. "The intrinsic parity of elementary particles." *Phys. Rev.* **88**, 101–105 (1952). [Super selection rule on integral vs. half-integral angular momentum.]
109. Wigner, E. "Uber nichtkombinierende Terme in der neueren Quantentheorie." *Z. Physik.* **40**, 492–500 (1926–27). "—. II." *ibid.* **40**, 883 (1927).
110. Wigner, E. "Einige Folgerungen aus der Schrödingerschen theorie für die Termstrukturen." *Z. Physik* **43**, 624 (1927).
111. Wigner, E. "Berichtigung zu der Arbeit: Einige Folgerungen aus der Schrödingerschen theorie für die Termstrukturen." *Z. Physik* **45**, 601 (1927).
112. Wigner, E. P. "Über die Operation der Zeitumkehr in der Quantenmechanik." *Nachr. Akad. Wiss. Göttingen, Math.-Phys. Kl.* p. 546 (1932).
113. Wigner, E. P. "On the representations of certain finite groups." *Am. J. Math.* **63**, 57–63 (1941).
114. Wigner, E. P. "On matrices which reduce Kronecker products of representations of S. R. groups." (1951) unpublished.
115. Wigner, E. P., and Inönu, E. "On the contraction of groups and their representations." *Proc. Natl. Acad. Sci. U. S.* **39**, 510–524 (1953).
116. Wigner, E. P. "The interpretation of Racah's coefficients," in *Symp. on New Res. Techniques in Phys., Rio de Janeiro* p. 297 (July 1954).
117. Wigner, E. P. "The application of group theory to the special functions of mathematical physics." Princeton University Lectures, Spring term (1955).
118. Wigner, E. P. "Normal form of antiunitary operators." *J. Math. Phys.* **1**, 409–413 (1960).
119. Wigner, E. P. "Phenomenological distinction between unitary and antiunitary symmetry operators." *J. Math. Phys.* **1**, 414–416 (1960).

120. van Winter, Clasine. "The asymmetric rotator in quantum mechanics." *Physica* **20**, 274–292 (1954). [Very fine review; corrects several errors in the literature.]

121. Yamanouchi, T. "Calculations of atomic energy levels." *Proc. Phys. Math. Soc. Japan* **17**, 274 (1935); **18**, 10 (1936); **18**, 623 (1936); **19**, 436 (1937).

122. Yamanouchi, T. "Tables useful for construction of irreducible representation matrices of symmetric group." *J. Phys. Soc. Japan* **3**, 245 (1948).

123. Young, R. C. "Conversion electron angular correlations: General K-shell formulation and threshold limit." *Phys. Rev.* **115**, 577 (1959). [The appendix to this paper discusses several identities obeyed by the Racah and Wigner Coefficients].